Innovation Issues in Water, Agriculture and Food

Innovation Issues in Water, Agriculture and Food

Special Issue Editors

Maria do Rosario Cameira
Luis Santos Pereira

MDPI • Basel • Beijing • Wuhan • Barcelona • Belgrade

MDPI

Special Issue Editors
Maria do Rosario Cameira
Universidade de Lisboa
Portugal

Luis Santos Pereira
Universidade de Lisboa
Portugal

Editorial Office
MDPI
St. Alban-Anlage 66
4052 Basel, Switzerland

This is a reprint of articles from the Special Issue published online in the open access journal *Water* (ISSN 2073-4441) from 2018 to 2019 (available at: https://www.mdpi.com/journal/water/special_issues/Water_Agriculture_Food)

For citation purposes, cite each article independently as indicated on the article page online and as indicated below:

LastName, A.A.; LastName, B.B.; LastName, C.C. Article Title. *Journal Name* **Year**, *Article Number, Page Range.*

ISBN 978-3-03921-165-4 (Pbk)
ISBN 978-3-03921-166-1 (PDF)

Contents

About the Special Issue Editors

Maria do Rosario Cameira (Professor) is an Associate Professor at Instituto Superior de Agronomia at the University of Lisbon, Portugal, where she teaches Irrigation and drainage, water resources, water engineering in rural areas, and fluid mechanics. Her research focuses on (1) physics of porous media applied to water flow and solute transport in the soil vadose zone; (2) integrated system modeling of water and nitrogen transfers in the continuum soil/plant/atmosphere at farm and watershed scales in the perspective of understanding and quantifying the impact of natural and human drivers on the agriculture ecosystems both temporally and spatially.

Luis Santos Pereira is Professor Emeritus of the University of Lisbon and research leader of the LEAF Research Centre of Instituto Superior de Agronomia. He is the honorary President of the International Commission of Agriculture and Biosystems Engineering, honorary Vice-President of the International Commission of Irrigation and Drainage, and member of various engineering and agricultural Academies. He has taught and directed research at the University of Lisbon, Porto and Azores in Portugal, the Agronomic Mediterranean Institute of Bari, Italy, the Institute of Water Resources and Hydrology Research, Beijing, the China Agricultural University, Beijing, the Inner Mongolia Agricultural University, Hohhot, and the Federal University of Santa Maria, Brazil. His main areas of research refer to crop evapotranspiration, crops water use and saving, irrigation design and modeling, coping with water scarcity, and droughts identification, characterization, and mitigation. He is the author or editor of numerous books, e.g., Crop Evapotranspiration, Guidelines for Computing Crop Water Requirements, FAO, Rome, and Coping with Water Scarcity. Springer. In addition, he is the author of hundreds of papers in journals and books.

Preface to "Innovation Issues in Water, Agriculture and Food"

There is currently an increased interest in the water–agricultural–food nexus, since there is the need for a close link between problems and issues relative to those three domains in terms of water use for agriculture and food production. Those domains are essential when aiming at the welfare of populations, food security, and environmental friendliness. Research slowly moves into that combination of challenges: providing for the welfare of the population, namely those living in the rural world, assuring food security and responding to the millennial compromises for future, and caring about our common home as opportunely challenged in the encyclical letter Laudato Si. Several researchers responded to the call and produced a good diversity of article contributions to that theme. Many anonymous colleagues assisted in the review of the papers and, thus, contributed to the quality of the published material. In addition, several members of editorial staff at MDPI played an essential role in the complex process of publishing. Our sincere thanks go to all, staff, reviewers, and authors, with special consideration of the latter: this book consists of their contributions! We hope this may be useful to researchers, professionals, and students, thus contributing to new innovation issues.

Maria do Rosario Cameira, Luis Santos Pereira
Special Issue Editors

water MDPI

Editorial

Innovation Issues in Water, Agriculture and Food

Maria do Rosário Cameira * and Luís Santos Pereira *

Centro de Investigação em Agronomia, Alimentos, Ambiente e Paisagem (LEAF),
Instituto Superior de Agronomia, Universidade de Lisboa, Tapada da Ajuda, 1349-017 Lisboa, Portugal
* Correspondence: roscameira@isa.ulisboa.pt (M.d.R.C.); lspereira@isa.ulisboa.pt (L.S.P.);
  Tel.: +351-2-1365-3478 (M.d.R.C.); +351-2-1365-3480 (L.S.P.)

Received: 27 May 2019; Accepted: 9 June 2019; Published: 12 June 2019

Abstract: The main challenge faced by agriculture is to produce enough food for a continued increase in population, however in the context of ever-growing competition for water and land, climate change, droughts and anthropic water scarcity, and less-participatory water governance. Such a context implies innovative issues in agricultural water management and practices, at both the field and the system or the basin scales, mainly in irrigation to cope with water scarcity, environmental friendliness, and rural society welfare. Therefore, this special issue was set to present and discuss recent achievements in water, agriculture, and food nexus at different scales, thus to promote sustainable development of irrigated agriculture and to develop integrated approaches to water and food. Papers cover various domains including: (a) evapotranspiration and crop water use; (b) improving water management in irrigated agriculture, particularly irrigation scheduling; (c) adaptation of agricultural systems to enhance water use and water productivity to face water scarcity and climate change; (d) improving irrigation systems design and management adopting multi-criteria and risk approaches; (e) ensuring sustainable management for anthropic ecosystems favoring safe and high-quality food production, as well as the conservation of natural ecosystems; (f) assessing the impact of water scarcity and, mainly, droughts; (g) conservation of water quality resources, namely by preventing contamination with nitrates; (h) use of modern mapping technologies and remote sensing information; and (i) fostering a participative and inclusive governance of water for food security and population welfare.

Keywords: water-agriculture-food nexus; crop water use and evapotranspiration; irrigation scheduling; design of irrigation systems; simulation models; droughts; irrigation water governance; economic and environmental issues

---

1. Introduction

Agriculture's first challenge is to produce enough food for a continued increase in population in a context where the increased demand for food is associated with an ever-growing competition for water and land, climate change and uncertainty, anthropic and droughts water scarcity, poor supply reliability, decline in critical ecosystem services, less-participatory water resources governance, and changing regulatory environments. Facing such a challenging context implies innovative issues in agricultural water management, particularly in the various facets of irrigation to cope with water scarcity, environmental friendliness, and rural society welfare.

Innovation issues in the water-agriculture-food nexus aim at various essential problems and objectives: (a) developing integrated approaches to water and food policies and practices; (b) improving water management in agriculture; (c) adaptation of agricultural systems to enhance water use and water productivity to face water scarcity and climate change; (d) ensuring sustainable management and conservation of natural and anthropic ecosystems favoring high-quality food production; and (e) fostering a participative and inclusive governance of water for food security and population welfare. These large themes cover a variety of challenges faced by the irrigated agriculture, which represents 16% of the world cropped area, but is expected to produce 44% of world food by 2050 [1,2].

Progresses observed in irrigation support a positive view on future responses to challenges faced by the water-agriculture-food complex as analyzed by Pereira [3]. Crop evapotranspiration processes are progressive and better known with consequent advances in understanding the complexity of climate and in estimating crop water and irrigation requirements. Irrigation scheduling has greatly advanced since knowledge on evapotranspiration and crops water use is improved, computer modeling eases supporting the development of well-focused programs, and modern technologies provide for timely advice to farmers. Remote sensing is progressively revealing its great potential to identify the vegetation water status, assessing crop water requirements and supporting irrigation management decisions. Irrigation systems design at both the farm and system levels is progressing with using various approaches for decision making, namely with combining water and economic performance issues. Progresses in irrigation methods and in canal and pipe conveyance and distribution systems contribute to improved irrigation and delivery schedules with impacts on water saving, yield improvements, and economic and social issues. Participatory governance keeps developing, however slowly, favoring sustainable water use, water productivity, and social welfare. Developments in better and joint approaches to water and fertilizer management are also providing for controlling environmental contamination and reducing greenhouse gas emissions.

This Special Issue is an opportunity to gather different achievements in the domains referred above, thus aimed at the sustainable use of water and other natural resources at different scales, so contributing to promote better development of irrigated agriculture when adopting a variety of well-performing technologies with good consideration of related social and economic environments. The studies presented herein focus on evapotranspiration, crop water requirements and modeling, remote sensing for assessing evapotranspiration, irrigation management for improved productivity and water saving, irrigation methods and systems design, control of soil salinity in irrigation, soil amendments for improved water use, surface water conveyance and distribution systems, participatory water governance, droughts and climate change issues, and environmental impacts of irrigated agriculture with focus on nitrates. Studies largely used innovative information and communication technologies, namely relative to data acquisition, modeling, decision support systems, remote sensing, and mapping.

This book presents the twenty-one selected contributions, consisting of 19 research papers and two revision papers, which were organized in accordance with the spatial scale of the study presented. Three levels were considered: the farm scale, the irrigation system scale, and the basin and regional scale.

2. Innovation Issues at Local Scale

Most research work relative to the local, farm scale is related to innovation on evapotranspiration and crop water requirements. The study by de Bruin and Trigo [4] refers to a new method to estimate reference crop evapotranspiration (ET_o) using products of the European geostationary satellite Meteosat Second Generation (MSG), thus analyzing benefits from computing ET_o with solar radiation data of the Satellite Application Facility on Land Surface Analysis (LSA-SAF). Not only the method allows computing ET_o when ground data on solar radiation are not available but also avoids advection effects. The authors used observations from two quite different sites, one in the Netherlands and the other in Spain. Results are comparable with those computed using the ET_o definition and are therefore appropriate to be used when radiation data are missing, or are of poor quality, or are affected by advection.

Paredes et al. [5] present an innovative approach for the determination of the bermudagrass water requirements in southern Brazil, relating the frequency of cuttings with a density coefficient (K_d) that depends upon the fraction of ground cover and crop height. The basal crop coefficient (K_{cb}) used to estimate crop evapotranspiration (ET_c) with the dual K_c approach is then computed from K_d using the SIMDualKc model. Two seasons of experimentation were used with four different cutting treatments, which provided field data for model calibration and validation. K_{cb} values relative to the various crop growth stages were therefore reported. In addition, results indicated a very high beneficial consumptive water use fraction, mainly when cuttings are not very frequent. The SIMDualKc model proved to be reliable in the evaluation of water saving practices after proper calibration and validation. This was also demonstrated by Zhang et al. [6] when comparing basin irrigation and drip irrigation of a tomato crop in the upper reaches of the Yellow River, China. They concluded that, besides saving irrigation water, drip could speed up the ripening of fruits and shorten the period of crop growth. Considering both water saving and crop yield, drip irrigation using plastic-film mulch was recommended. Moreover, appropriate K_{cb} values relative to the defined crop growth stages are proposed for computing ET_c in the region and to transfer elsewhere after appropriate climate adjustment. Paço et al. [7] confirm the appropriateness of using the SIMDualKc model for ET_c computations and the great importance of irrigation for super intensive olive orchards in an application performed in southern Portugal. Appropriate values for K_{cb} and K_c were determined from the calibration and validation of the model. The authors also concluded that the level of water stress allowed to the crop should be accurately controlled to maximize related benefits, i.e., adopting the so-called eustress. They recommended the use of the SIMDualKc or a similar model to properly schedule irrigation. The papers referred above [5–7] relative to a grass crop, a horticultural crop, and a tree crop all proved the appropriateness of simply computing K_{cb} from the fraction of ground cover and height.

Fuentes-Peñailillo et al. [8] assessed the potential for the use of the two-source Shuttleworth and Wallace (SW) model to compute the intra-orchard spatial variability of the actual ET of olive trees using satellite images and ground-based climate data. The performance of the SW model approach was tested using eddy covariance heat and vapor fluxes measurements performed in Chile. The model separately characterizes evaporation fluxes from the soil and transpiration from the trees and its practical application requires adequate parametrization of both sources of fluxes, the soil and the trees, thus the stomatal resistance and leaf area index of olive orchards under different management practices. Ramos et al. [9] describe the impact of assimilation of leaf area index (LAI) data derived from Landsat 8 imagery on MOHID-Land's simulations of the soil water balance and maize state variables using data collected in southern Portugal. The main conclusion is that the implementation of the MOHID-Land model at the regional scale cannot depend solely on inputs from the LAI data assimilation because estimates may diverge substantially from the reality, thus confirming the need to use a proper data set for calibration.

The study presented by Abrisqueta and Ayars [10] describes how traditionally grapes are fully irrigated in California, while alternative irrigation strategies to reduce applied irrigation water may be necessary in the future. The paper assesses water use and productivity of alternative deficit irrigation schedules with identification of advantages of using sustained deficit irrigation, corresponding to 80% of present growers water use, which provides for water saving and improved yield. Nevertheless, there is little acceptance and implementation of scientific irrigation management by the growers' community. This work demonstrates that simple approaches can be used to facilitate improving irrigation management. The application of technological advances should be investigated particularly in regions where water resources are scarce and agriculture is primarily based on irrigation. In their work, Abi-Saab et al. [11] assessed for two crop seasons the suitability of cloud-based irrigation technologies for durum wheat, a strategic Mediterranean crop of Lebanon. Their study focused on the on-field assessment of a smartphone irrigation scheduling tool—Bluleaf® (Bari, Italy)—with respect to traditional water application practices. A water saving of more than $1000 \text{ m}^{-3} \text{ ha}^{-1}$ (25.7%) was observed for farmers using Bluleaf® with respect to traditional irrigation scheduling.

The study by Miao et al. [12], applied to the Hetao Irrigation System, upper Yellow River basin, China, focuses on upgrading wheat basin irrigation through improved design using multi-criteria analysis to support selection of technical solutions focusing on both water saving and economic returns of the achievable yields. The design considers land parcels characteristics, soil infiltration, hydraulic simulation, precise land leveling, and crop irrigation scheduling, and evaluating solutions through assessing environmental and economic impacts. Alternative solutions refer to basin lengths and inflow rates for both level- and graded-basins under full and deficit irrigation. Better solutions aimed at the upper Yellow River were generally assigned to level-basins.

The application of organic mulches and amendments to the soil is a well-recognized water saving practice. Zhang et al. [13] provided evidence that straw amendments applied to a sloping field of northeastern China reduced water losses by evaporation and increased soil organic matter while decreasing carbon dioxide emissions. In addition, it positively influenced the dynamics of available nitrogen and phosphorous in soil. Therefore, straw amendments not only contributed to improved crop conditions and increased yields but also to control greenhouse gas (GHG) emissions. Furthermore, Zhao et al. [14] show that the applications of organic amendments like farmyard manure and bioorganic fertilizer to a maize crop in the Guanzhong Plain of northwest China improve soil properties and crop roots distribution, controlled soil water depletion, and favored the soil hydrothermal conditions and crop yields.

3. Innovation Issues at Project and Watershed Scale

Agricultural water use at project and watershed scale also drew interest in terms of utilization and conservation of both surface water and groundwater. Quite different approaches were presented. The study by Derardja et al. [15] presents the development of indicators that can help managers and designers to better understand the behavior of pressurized conveyance and distribution networks with respect to the perturbation due to sudden changes in pipe flow rates, namely due to hydrants closure. The indicators refer to pressure variation and associated risk. The study was applied to an on-demand pressurized irrigation project of Capitanata, Southern Italy, and provides a modeling approach for the identification of risky hydrants and potentially affected pipes. Generally, the model contributes to safe water use in collective pressurized systems operating on demand.

An organizational solution for improving the governance of water and land use and, consequently, improving the supply–demand water balance in the Mendoza River basin in Argentina, is described by Solomon-Sirolesi and Farinós-Dasí [16]. A strategic analysis of water organization was performed that produced a strategic map and provided for using the Water Evaluation and Planning (WEAP) model. The application of the organizational and governance model to various scenarios provided for reordering allocations to irrigation water users, so improving farm and irrigation system, which is expected to make it possible to accommodate water demand in 2030 better than at present. Moreover, users' participation is enhanced. Differently, but along the same line of searching for improved irrigation water governance, Playán et al. [17] reviewed the evolution of water governance and societal perception in large irrigation systems in developing countries since the 1980s which included participatory irrigation management, irrigation management transfer, and public-private partnerships or market instruments and, more important, that led to a generalized implementation of water users associations (WUAs). The paper, therefore, reviews recurrent problems and solutions in the governance of irrigated projects in various regions of the world. The authors used a semiquantitative approach to relate solutions to problems in WUAs. The solution vector indicates the adequacy of each solution to a case study WUA. The application of this approach to various case study WUAs demonstrated its potential.

A modified version of the distributed hydrological model Soil and Water Assessment Tool (SWAT) was used by Wei et al. [18] for accurately simulating daily evapotranspiration and crop growth at the hydrological response unit (HRU) scale in an irrigation district of the Hei River basin, China. The parameters of the modified SWAT2009 model were calibrated and validated using data on maize from various HRUs with satisfactory results. The influences of various optimal management practices on the hydrology of agricultural watersheds could, therefore, be effectively assessed using the modified model. Contrasting, but also considering a large irrigation system, Wang et al. [19] reviewed the performance of dry drainage systems (DDS) as an alternative technique for controlling soil salinization. A five-year field observation from 2007 to 2011 was developed in the Hetao Irrigation District, Inner Mongolia. Results showed that the groundwater table depth in the areas kept fallow quickly responded to the lateral recharge from the surrounding croplands during irrigation events. The groundwater salinity increased in the fallow areas whereas the groundwater electrical conductivity (GEC) just fluctuated in croplands. The water and salt balance showed that excess water moved to the fallow area, roughly four times the one that moved by surface drainage, with near 8 times the corresponding salt. However, the salt transfer slows down after three years and salt starts accumulating in irrigated croplands. Using halophytes in the fallow area may be a solution. DDS may be effective and sustainable in situations where the fallow areas can sustain an upward capillary flux.

4. Innovation Issues at Regional and Country Scale

The Pampa biome characteristic of southern Brazil is a complex ecosystem where the processes of surface–atmosphere interaction are dependent on climate, soil, and vegetation. In this way, the preservation of this biome is essential for the maintenance of biodiversity, animal, and plant species, as well as land use. In their work, Rubert et al. [20] observed the actual ET with eddy covariance measurements over three years and two sites of the Pampa biome, and results showed a strong seasonality, with approximately 65% of the net radiation used for ET. The water availability in the Pampa biome is not a limiting factor for ET, which resulted in a small difference between the reference ET and the actual ET.

Framing integrated policies that improve the efficient use of resources requires understanding the interdependencies in the water–energy–food nexus (WEF) as discussed by Martinez et al. [21]. These authors developed and applied a participatory modeling approach to identify the main interlinkages within the WEF nexus in the region of Andalusia, Spain. Stakeholder involvement in this process is crucial to represent multiple perspectives, ensure political legitimacy, and promote dialogue. Results show that climate change and water availability are key drivers in the WEF nexus in Andalusia. The scenario analysis reveals the interdependencies among nexus sectors that need to be considered to design integrated policies.

At the regional scale, now focusing on water quality problems associated with the agricultural activity, Cordovil et al. [22] present a simplified nitrogen (N) assessment review along the Tagus River basin, which covers a large area of Spain and Portugal in the Iberian Peninsula, mainly relative to agriculture, livestock, and urban activities. Nitrate vulnerable zones comprise approximately one-third of territories in both countries. Differently, sensitive zones are more extended in Spain, attaining 60% of the basin area, against only 30% in Portugal. The authors concluded that the Tagus River basin sustainability can only be guaranteed through reducing load inputs and considering the effective transnational management processes of water flows.

Moreira et al. [23] modeled drought class transitions over a large region using a log-linear approach. Temporal scales of 6 and 12 months were adopted when studying rainfall data for 15 grid points selected over the Prut basin in Romania over a period of 112 years (1902–2014). The modeling also took into account the impact of the North Atlantic Oscillation (NAO), thus exploring the potential influence of this large-scale atmospheric driver on the climate of the Prut region. The authors conclude that the adopted method performs consistently better than the persistence forecast and that it can be part of an early warning system aiming at informing farmers to support them in making their water management decisions.

Finally, returning to the water quality and nitrates issue, Cruz et al. [24] examined the state of surface waters in terms of nitrogen concentrations, its impact on water quality status, and the policy responses across different climatic and management conditions. Portugal and Denmark were chosen as contrasting case studies for the Driver-Pressure-State-Impact-Response (DPSIR) analysis. Results showed a reduction in the levels of N in water bodies which were attributed to the previous, historical implementation of policies aimed at reducing the N losses from land to water in both countries. However, climatic factors such as precipitation play a major role in reducing the N concentration in waters through a dilution effect, namely after infiltrating the soil, so contributing for differences between both countries.

5. Conclusions

The general message emanating from this Special Issue is that consistent progresses are continuously developing that contributes to the sustainability of water use in agriculture and to make available a variety of innovative issues in the water-agriculture-food nexus. This collection of 21 papers emphasizes the importance of understanding the various interdependencies among innovative issues.

It is clear that sets of coherent strategies and solutions need to be applied to mitigate the complexities of overall challenges placed to the sustainable use of water in agriculture and food production. Strategies are different at both the small scale, the field or farm level, and the large operational scale relative to the irrigation district, the river basin, the region, or the country. At a small scale, strategies and solutions are mostly of technological nature, with great importance to be given to the information and communication technologies (ICT), including remote sensing. At the large scale, main strategies refer to governance, to user's participation in water management, and to the ICT required to provide for planning and management decisions. Furthermore, scaling-up innovation remains a challenge. Linkages between the different scales of application are quite important and need to be better investigated. Included are the impact of introduction of innovative management and practices at small scales, with the improved estimation of yields and water use at high spatial resolution, or the impact of the political measures at farm scale and relative to the maintenance of the quality of surface and ground waters. Related knowledge should facilitate the widespread application of technologies and adaptation measures (e.g., agronomic and irrigation practices) aimed at achieving better performance of water use, agricultural production, and relative to the sustainability of water use. Regardless of the spatial scale, modeling is a widespread tool which however requires calibration and validation. The use of models should be framed to adapt easily to obtain data and information, e.g., using the simple crop coefficient approach when modeling crop evapotranspiration, however adopting modern and accurate ET observation instrumentation both at the ground surface and by satellite earth observation.

We hope that this Special Issue will provide for coherent directions of research and support to the scientific community in defining priorities for in the water-agriculture-food nexus research.

Acknowledgments: The authors of this paper, who served as the guest-editors of this special issue, wish to thank the journal editors, all authors submitting papers to this special issue, and the many referees who contributed to paper revision and improvement of the 21 published papers.

Conflicts of Interest: The authors declare no conflict of interest.

References

1. FAO. *FAO World Agriculture towards 2030/2050: The 2012 Revision*; ESA Working Paper 12-03; FAO: Rome, Italy, 2012. Available online: http://www.fao.org/3/a-ap106e.pdf (accessed on 11 April 2019).
2. World Business Council for Sustainable Development. *WBCSD Water, Food and Energy Nexus Challenges*; World Business Council for Sustainable Development: Geneva, Switzerland, 2014.
3. Pereira, L.S. Water, agriculture and food: Challenges and issues. *Water Resour. Manag.* **2017**, *31*, 2985–2999. [CrossRef]
4. De Bruin, H.A.; Trigo, I.F. A new method to estimate reference crop evapotranspiration from geostationary satellite imagery: Practical considerations. *Water* **2019**, *11*, 382. [CrossRef]
5. Paredes, P.; Rodrigues, G.; Petry, M.; Severo, P.; Carlesso, R.; Pereira, L.S. Evapotranspiration partition and crop coefficients of Tifton 85 bermudagrass as affected by the frequency of cuttings. Application of the FAO56 dual Kc Model. *Water* **2018**, *10*, 558. [CrossRef]
6. Zhang, H.; Huang, G.; Xu, X.; Xiong, Y.; Huang, Q. Estimating evapotranspiration of processing tomato under plastic mulch using the SIMDualKc model. *Water* **2018**, *10*, 1088. [CrossRef]
7. Paço, T.A.; Paredes, P.; Pereira, L.S.; Silvestre, J.; Santos, F.L. Crop coefficients and transpiration of a super intensive Arbequina olive orchard using the dual K_c approach and the K_{cb} computation with the fraction of ground cover and height. *Water* **2019**, *11*, 383. [CrossRef]

8. Fuentes-Peñailillo, F.; Ortega-Farías, S.; Acevedo-Opazo, C.; Fonseca-Luengo, D. Implementation of a two-source model for estimating the spatial variability of olive evapotranspiration using satellite images and ground-based climate data. *Water* **2018**, *10*, 339. [CrossRef]
9. Ramos, T.; Simionesei, L.; Oliveira, A.; Darouich, H.; Neves, R. Assessing the impact of LAI data assimilation on simulations of the soil water balance and maize development using MOHID-Land. *Water* **2018**, *10*, 1367. [CrossRef]
10. Abrisqueta, I.; Ayars, J. Effect of alternative irrigation strategies on yield and quality of Fiesta raisin grapes grown in California. *Water* **2018**, *10*, 583. [CrossRef]
11. Abi-Saab, M.T.; Jomaa, I.; Skaf, S.; Fahed, S.; Todorovic, M. Assessment of a smartphone application for real-time irrigation scheduling in Mediterranean environments. *Water* **2019**, *11*, 252. [CrossRef]
12. Miao, Q.; Shi, H.; Gonçalves, J.; Pereira, L.S. Basin irrigation design with multi-criteria analysis focusing on water saving and economic returns: Application to wheat in Hetao, Yellow River basin. *Water* **2018**, *10*, 67. [CrossRef]
13. Zhang, S.; Wang, Y.; Shen, Q. Influence of straw amendment on soil physicochemical properties and crop yield on a consecutive Mollisol slope in Northeastern China. *Water* **2018**, *10*, 559. [CrossRef]
14. Zhao, L.L.; Li, L.S.; Cai, H.J.; Shi, X.H.; Zhang, C. Organic amendments influence soil water depletion, root distribution, and water productivity of summer maize in the Guanzhong Plain of Northwest China. *Water* **2018**, *10*, 1640. [CrossRef]
15. Derardja, B.; Lamaddalena, N.; Fratino, U. Perturbation indicators for on-demand pressurized irrigation systems. *Water* **2019**, *11*, 558. [CrossRef]
16. Salomón-Sirolesi, M.; Farinós-Dasí, J. A new water governance model aimed at supply-demand management for irrigation and land development in the Mendoza River Basin, Argentina. *Water* **2019**, *11*, 463. [CrossRef]
17. Playán, E.; Sagardoy, J.; Castillo, R. Irrigation governance in developing countries: Current problems and solutions. *Water* **2018**, *10*, 1118. [CrossRef]
18. Wei, Z.; Zhang, B.; Liu, Y.; Xu, D. The application of a modified version of the SWAT model at the daily temporal scale and the hydrological response unit spatial scale: A case study covering an irrigation district in the Hei River Basin. *Water* **2018**, *10*, 1064. [CrossRef]
19. Wang, C.; Wu, J.; Zeng, W.; Zhu, Y.; Huang, J. Five-year experimental study on effectiveness and sustainability of a dry drainage system for controlling soil salinity. *Water* **2019**, *11*, 111. [CrossRef]
20. Rubert, G.; Roberti, D.; Pereira, L.S.; Quadros, F.; Campos Velho, H.; Leal de Moraes, O. Evapotranspiration of the Brazilian Pampa Biome: Seasonality and influential factors. *Water* **2018**, *10*, 1864. [CrossRef]
21. Martinez, P.; Blanco, M.; Castro-Campos, B. The Water-Energy-Food Nexus: A fuzzy-cognitive mapping approach to support nexus-compliant policies in Andalusia (Spain). *Water* **2018**, *10*, 664. [CrossRef]
22. Cordovil, C.; Cruz, S.; Brito, A.; Cameira, M.R.; Poulsen, J.; Thodsen, H.; Kronvang, B. A simplified nitrogen assessment in Tagus River Basin: A management focused review. *Water* **2018**, *10*, 406. [CrossRef]
23. Moreira, E.; Russo, A.; Trigo, R. Monthly prediction of drought classes using log-linear models under the influence of NAO for early-warning of drought and water management. *Water* **2018**, *10*, 65. [CrossRef]
24. Cruz, S.; Cordovil, C.; Pinto, R.; Brito, A.; Cameira, M.R.; Gonçalves, G.; Poulsen, J.; Thodsen, H.; Kronvang, B.; May, L. Nitrogen in water. Portugal and Denmark: Two contrasting realities. *Water* **2019**, *11*, 1114. [CrossRef]

water

MDPI

Article

A New Method to Estimate Reference Crop Evapotranspiration from Geostationary Satellite Imagery: Practical Considerations

Henk A. R. de Bruin [1] and Isabel F. Trigo [2,3,*]

[1]   Meteorology and Air Quality, Wageningen University, 6708 PB Wageningen, The Netherlands (emeritus); hardb@xs4all.nl
[2]   Instituto Português do Mar e da Atmosfera, IPMA, 1749-077 Lisboa, Portugal
[3]   Instituto Dom Luiz, University of Lisbon, IDL, 1749-016 Lisboa, Portugal
*   Correspondence: isabel.trigo@ipma.pt; Tel.: +351-218-447-108

Received: 14 December 2018; Accepted: 19 February 2019; Published: 22 February 2019

Abstract: Reference crop evapotranspiration (ETo) plays a role in irrigation advisory being of crucial importance for water managers dealing with scarce water resources. Following the ETo definition, it can be shown that total solar radiation is the main driver, allowing ETo estimates from satellite observations. As such, the EUMETSAT LSA-SAF operationally provides ETo primarily derived from the European geostationary satellite MSG. ETo estimations following the original FAO report require several meteorological observations gathered over actual well-watered grass. Here we will consider the impact of two effects on ETo using the LSA-SAF and FAO methodologies: (i) local advection, related to the impact of advection of surrounding warm dry air onto the reference non-water stressed surface; and (ii) the so-called surface aridity error, which occurs when calculating ETo according to FAO, but with input data not collected over well-watered grass. The LSA-SAF ETo is not sensitive to any of these effects. However, it is shown that local advection may increase evapotranspiration over a limited field by up to 30%, while ignoring aridity effects leads to a great overestimation. The practical application of satellite estimates of ETo provided by the LSA-SAF are discussed here, and, furthermore, water managers are encouraged to consider its advantages and ways for improvement.

Keywords: reference evapotranspiration; local advection; aridity effects; satellite observations

1. Introduction

This research note is a contribution to the special issue of Water on "Innovation Issues in Water, Agriculture and Food", which, amongst other things, deals with the fact that the continued increase in population provides a challenge for agriculture to produce enough food. This is at a time where, in many semi-arid regions, fresh water resources are depleted. Therefore, the special issue also addresses the ever-growing competition for scarce water resources. The solution for such problems requires legislation, precise water management and science-based irrigation advisory. In this context, the United Nations Food and Agriculture Organization (FAO) developed a methodology based on standard meteorological data to estimate crop water requirements. The latter are then determined using a semi-empirical crop-factor approach, namely as $[k_c. ETo]$, where k_c is a crop factor and ETo the reference crop evapotranspiration, which is calculated from meteorological data collected over well-watered short grass [1]. The FAO method relies on a version of the Penman–Monteith equation (PM-ETo) and is wildly used in current irrigation advisory. However, most operational weather stations do not comply with the instrumental requirements prescribed in [1], which constitutes an important drawback to the use of the FAO methodology. Often, local observations rely on low cost sensors set over non well-watered grass. Moreover, in many cases, some of the required input data are missing.

The objective of this research note is to present an alternative method to estimate ETo. Instead of dealing with the hypothetic quantity ETo, we recall the basic idea behind ETo was that it should be an estimate of actual evapotranspiration (ET) of short well-watered grass. Therefore, the objective of this note is to present a methodology that relies essentially on remotely sensed estimates of global radiation complemented with air temperature forecasts provided by the European Centre for Medium-Range Weather Forecast (ECMWF), to estimate the actual ET of well-watered grass closely resembling the reference surface defined in [1].

Over small fields, under arid conditions, local advection effects (LAE) may occur when warm, dry air formed over an upwind adjacent field is advected horizontally over the well-watered grass, transporting additional energy for evaporation. By defining ETo for an extensive reference grass field [1], LAE was excluded, which was later explicitly confirmed by [2] by stating that: "No local advection occurs over the surface, thus the flux between the two levels is only vertical. Therefore, the conditions at the reference level z_M (solar radiation, wind speed, and vapor pressure deficit) can be considered to be the same as that over a large surrounding area". Nevertheless, validation studies of PM-ETo have mostly concerned comparisons with the actual ET measured over small well-watered fields surrounded by dry terrain, which are clearly affected by LAE [3]. Particularly in areas where water available for irrigation is limited, the LAE topic becomes crucially important. Recent analyses of actual ET data sets revealed that LAE can yield an increase in ETo of 30%. This occurs essentially in arid regions at high temperature [4,5], i.e., under the conditions where crops need irrigation. Due to the significance of LAE matter, in this paper we will consider two cases with and without LAE, respectively.

Our approach is physically based and follows from the application of thermodynamics to estimate the ET of a 'saturated' surface, combined with atmospheric boundary layer (ABL) physics; this methodology will hereafter be denoted as the T-ABL model. As shown already in 1915 [6], actual ET is close to the so-called equilibrium ET, later derived from the original Penman formula by [7]. Later it was found that during daytime when most evaporation takes place, entrainment of warm and dry air aloft the ABL supplies extra energy that can be used for evaporation [8–11]. These findings lead to the conclusion that actual ET of well-watered grass is mainly determined by the available energy and, secondly, by the air temperature. In the case where LAE can be ignored, the main external energy source is global solar radiation. When LAE plays a role, additional sensible heat due to horizontal advection of the warm and dry adjacent fields, should be accounted for. The T-ABL model recently presented by [4,5] is physically sound with some minor empirical aspects. Theoretically, it differs from the physical reasoning behind PM-ETo, which is based on a combination of thermodynamics, a simplified description of the vegetation layer model (Monteith's 'big-leaf' model) and Monin-Obukhov's similarity theory leading to empirical flux-profile relationships. If well calibrated, the PM-ETo approach is physically sound, as is T-ABL. However, internal correlations between actual ET over well-watered grass and many of the input variables used by PM-ETo explain why a model such as T-ABL, requiring much less input data, performs equally as well. Further details on the T-ABL method and its derivation are referred to in [4].

The objective of this research note is to deal with practical aspects and differences with PM-ETo in that context. We confine ourselves to time periods of one day or longer. Then PM-ETo requires as input global radiation (R_s), wind speed at 2 m (u_2), maximum and minimum temperatures at 2 m to estimate daily mean temperature (T_a) and saturation water vapor, and finally mean, maximum and minimum relative humidity at 2 m (RH, RH_x and RH_n). These data should be gathered over well-watered grass. These requirements set the standards on weather stations to be very high, e.g., accurate observations of R_s, RH_x and RH_n need good quality sensors and respective maintenance. Moreover, educated personal is needed for data quality control and labor as well-watered grass maintenance is expensive.

Often, weather station sensors are not installed over well-watered grass. Then, under arid conditions, the temperature and humidity data must be adjusted for surface aridity [12]. If one fails to apply these adjustments, ETo will be overestimated significantly [12]. The T-ABL approach requiring only R_s and T, hardly suffers from surface aridity [4,5].

Recently, [5] presented an application of the T-ABL model where R_s estimates from Meteosat Second Generation (MSG) observations [13,14] combined with T weather forecasts were used to derive ETo. The approach is used by Satellite Applications Facility on Land Surface Analysis (LSA SAF; http://lsa-saf.eumetsat.int) to operationally generate daily ETo [5].

This note can be seen as an extension of [5], dealing with the practical aspects of the T-ABL model and its possible advantages for certain applications, when compared with PM-ETo. Particular attention is paid to LAE and the surface aridity adjustments. In a time of water scarcity these practical aspects are very important.

2. Material and Methods

2.1. Used Datasets

Two sites were considered in this study, namely: a grass field located in a polder region at Cabauw (The Netherlands) surrounded by similar grass where water stress is rare and local advection can be neglected [4]; and an irrigated grass site near Cordoba of about 1 ha, where summers are predominantly dry and hot. Latent heat flux, and hence actual evapotranspiration, were obtained at Cabauw from flux measurements [4,5,15], using a residual approach to ensure energy balance closure. As the surface characteristics at the Cabauw site and surrounding area are very close to the reference surface, we considered actual ET to be representative of ETo [4,5]. The Cordoba site, in turn, is equipped with a lysimeter (hereafter Cordoba lysimeter site) and standard meteorological instruments. In this case, the surroundings are characterized by dry terrain and thus, the observations taken at the Cordoba lysimeter site were often affected by LAE, as described by [5,16]. For more information on the Cabauw and Cordoba sites, readers are referred to [4,5,16].

On top of the above, we also considered, as an example of an operational weather station installed in climate regions with dry, hot summers, data gathered at a site in Andalucía in Spain (RIA, close to Cordoba) described in [17]. At the RIA station, despite being about 1km from the Cordoba lysimeter site, the observations were often collected over bare soil. These data are considered here to illustrate the aridity effect on local observations and their impact on PM-ETo.

For a number of selected days, we will use ten-minute observations from Cabauw and half-hourly data from the lysimeter in Cordoba and from the RIA sites (see also [18]).

2.2. The T-ABL Model

The physical reasoning behind the T-ABL model is that for well-watered surfaces actual ET is mainly determined by the available energy, since water availability is not a limiting factor. Using thermodynamics, it then can be shown that ET should be close to the so-called equilibrium ET (e.g., [4–6]). Atmospheric boundary layer (ABL) studies reveal that entrainment processes of warm and dry air at the top of the ABL provide an extra energy source available for ET [4,5,8,10]. This leads to the simple estimate of ET of well-watered grass, earlier found by [19].

$$L.ET = L.ET_{equilibrium} + \beta = \frac{\Delta}{\Delta + \gamma} Q^* + \beta \quad (\text{Wm}^{-2}) \tag{1}$$

where L is the latent heat of vaporization of water, $\Delta = de/dT|_{T=T_a}$ is the derivative of the saturation water vapor pressure (e), at air temperature (T_a), γ the psychrometric constant and Q^* net radiation of well-watered grass (see also [1]); the parameter β was estimated using observations collected at Cabauw and was set to $\beta_{Cab} = 20$ Wm^{-2}. Equation (1) applies to daily time steps by which ground flux may be ignored. Moreover, it is confined to LAE-free cases.

A next step is to apply the so-called Slob-deBruin formula for Q^* explicitly introduced to estimate net radiation for well-watered grass by [20]. Recently, [5] provided a validation of Equation (1), together with the Slob-DeBruin approach to derive Q^* (Equation (2) below) for several stations, including Cabauw and Cordoba.

$$Q^* = (1 - 0.23)R_s - C_s \frac{R_s}{R_{ext}} \qquad (2)$$

where R_s is the global (=down-welling shortwave) radiation and R_{ext} is the incoming shortwave radiation at the top of the atmosphere. The parameter C_s was also estimated using observations collected at Cabauw ($C_s = C_{sCab} = 110$ Wm^{-2}), although Equation (2) was validated for other sites [3]. Combining (1) and (2) yields the T-ABL model for actual ET of well-watered grass without LAE, operationally used by the LSA SAF [5].

In case LAE play a role, an extra energy term Q_{adv} has to be included in Equation (1), accounting for the sensible heat advected from upwind dry and warm fields. As shown in [5], this can be written as:

$$L.ET = L.ET_{No\_LAE} + Q_{adv} = L.ET_{No\_LAE} + f(T_a) \qquad (3)$$

where ET_{No_LAE} corresponds to ET without LAE (as in (1)) and $f(T_a)$ is an empirical function of the mean daily air temperature measured over dry upwind terrain. Empirically, [5] found that, for the Cordoba site, $f(T_a)$ should be kept negligible for temperatures below 15 °C (i.e., $f(T_a) = 0$, for $T_a \le$ 15 °C), while for warmer cases it increases linearly with temperature (i.e., $f(T_a) = 4 \times (T_a - 15)$, for $T_a >$ 15 °C). This result is based on the fact that, for the Cordoba site, also considered in [5], LAE occur only at temperatures higher than 15 °C. At lower temperatures actual ET appears to be well described by (1). For other sites with similar characteristics, LAE have also been found to be well described by a linear function of air temperature, although, the regression coefficients are site dependent. This is not surprising, since LAE depend on local features, such as the spatial distribution of land cover/vegetation type of the surroundings, or orography, which, amongst others, will determine horizontal temperature and humidity gradients.

It is seen that the T-ABL model provides estimates of actual ET of well-watered grass for cases with and without LAE, requiring R_s and T_a as input only. The next step is to make use of the existing LSA SAF product for R_s derived from the geostationary satellite MSG [5,13,14], allowing an estimate of ETo over Europe, Africa, and parts of South America and the Middle East. Note that the same can be applied for other geostationary satellites, and therefore a nearly worldwide coverage can be obtained [21]. Daily mean air temperature can be extracted routinely from weather forecasts, such as ECMWF in the LSA-SAF product [5], or from reanalyses [22], the latter available on an unrestricted basis.

2.3. LSA SAF Reference Evapotranspiration Data

The LSA SAF routinely generates, archives and disseminates daily reference evapotranspiration, using the T-ABL Equation (1). Net radiation, Q^*, is derived from daily incoming solar radiation following Equation (2). The latter, referring to the radiative energy in the wavelength interval [0.3 μm, 4.0 μm] reaching the Earth's surface during a complete 24-h period, is also an independent LSA SAF product. It is based on measurements provided by the three short-wave channels (centered at 0.6 μm, 0.8 μm, and 1.6 μm) of the Spinning Enhanced Visible and Infrared Imager (SEVIRI) onboard MSG, considering that top-of-the atmosphere reflectances measured by these bands are anti-correlated with solar radiation at the surface [13]. The daily solar radiation is estimated, on a pixel-by-pixel basis, by temporal integration of the respective instantaneous 30-min estimates, then yielding ETo (LSA SAF METREF product), which is also available per pixel. Near surface air temperature, which is used to estimate the slope of the saturation water vapor pressure (Δ) and the latent heat of vaporization (L) in Equation (1), is obtained from ECMWF operational weather forecasts.

The data are available for the whole MSG disk, nominally centered at 0° E, from 2004 to present, since these data are continuously generated. The spatial resolution depends on pixel location, and varies from 3 km at the sub-satellite point to 4–6 km over Europe. Users may request (off-line or near real time) access to the LSA-SAF METREF [5], as with any other LSA-SAF products (including solar radiation) by registering at http://lsa-saf.eumetsat.int.

3. Results

Figure 1 presents the time-series of ETo (in mm/day) determined using the LSA SAF algorithm for the LAE-free case, together with in situ observations gathered in Cabauw. Considering that the surface surrounding the site is very close to the reference grass, and therefore assuming that local observations can be directly compared to ETo estimates, we obtain a bias (the mean difference between measured and estimated values) of 0.1 mm/day, while the standard deviation of the differences between the LSA-SAF T-ABL estimates and local observations is 0.3 mm/day.

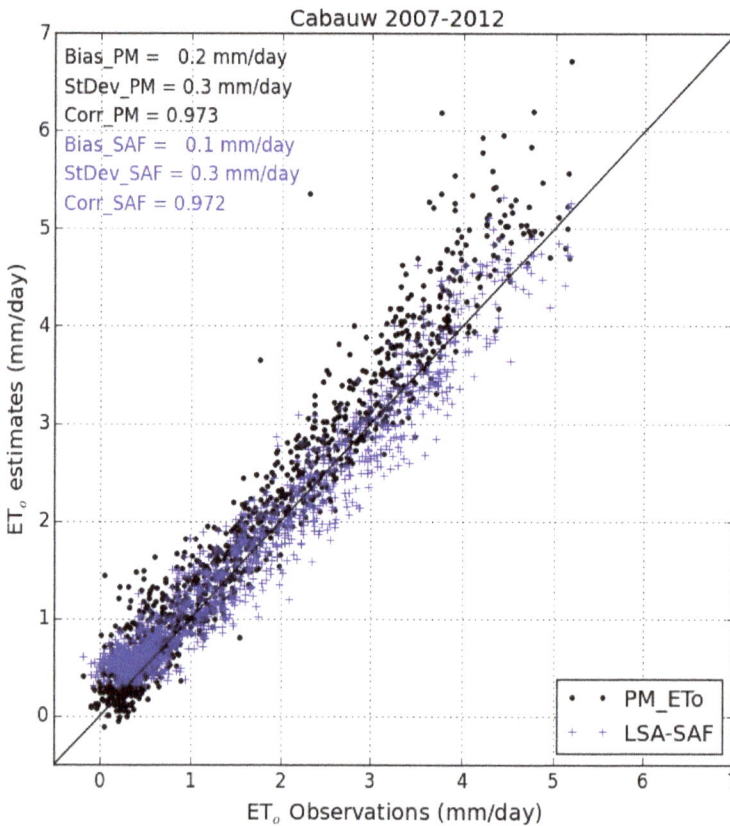

Figure 1. Daily estimates of reference evapotranspiration as obtained by the LSA-SAF (blue crosses) versus daily in situ observations obtained for Cabauw (from eddy-covariance flux measurements) for the 2007–2012 period. For reference, Penman-Monteith (PM_ETo) estimates obtained using in situ observations are also plotted (black dots). The mean difference (bias), standard deviation of differences and correlation between estimations and observed time-series are also shown.

A similar plot is shown for Cordoba in Figure 2, now showing the LSA SAF estimates without considering LAE (as in the case above), as well as an adjusted value to take into account LAE (as in Equation (3)). The mean difference between LSA-SAF estimates and in situ observations is reduced from −0.7 mm/day to 0.1 mm/day, when the correction for LAE is introduced, while the standard deviation of differences also decreases from 0.8 mm/day to 0.5 mm/day.

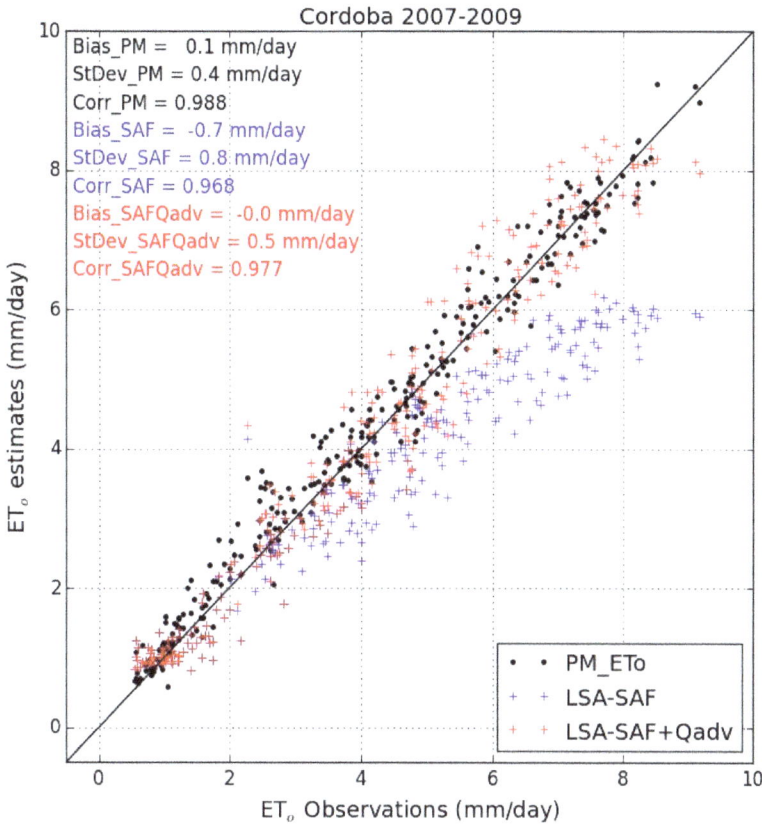

Figure 2. Daily estimates of reference evapotranspiration as obtained by the LSA-SAF without (blue crosses) and with an adjustment term to account for local advection effects (red crosses), versus daily in situ observations obtained for Cordoba Lysimeter site, for the 2007–2009 period. For reference, Penman-Monteith (PM_ETo) estimates obtained using in situ observations at the lysimeter site are also plotted (black dots). The mean difference (bias), standard deviation of differences and correlation between estimations and observed time-series are also shown.

These results were first published in [5]. A discussion of these results and implications for the practical use of the LSA SAF ETo product and Equation (3) are presented in the next section.

4. Discussion

The main purpose of this work is to draw the attention of water managers and irrigation advisors to the LSA SAF operational ETo product (being METREF its official acronym within the LSA-SAF) as an appropriate alternative for the FAO methodology. The physical basis of the two approaches are both sound, but they are derived along different routes. The LSA SAF product considers the T-ABL approach, based on thermodynamics of well-watered surfaces, whereas the FAO model is derived from a number of concepts postulated for the vegetation layer and the turbulent vertical transfer of water vapor in the atmosphere. T-ABL ETo implicitly includes entrainment processes at the top of the ABL, which are hidden in the empirical model constants in PM-ETo.

Because the thermodynamics-based T-ABL model does not account for local advection effects (LAE), it yields a correct estimate of ETo, which is consistent with this variable definition. As shown by [4,5] the T-ABL methodology can be easily extended to quantify LAE, by an additional energy term

13

in (1). It is found that when the temperature is higher than 15 °C, this extra energy term is proportional to $(T_a$-15) in the Cordoba site, and that at the peak of the dry season it accounts for an extra 30% of ETo (see largest differences between blue and red lines in Figure 2).

To illustrate the impact of LAE, the analyses by [5] show that at high temperatures the difference between Equations (1) and (3) can be as large as 30%. In order to discuss the LAE issue in more detail, Figure 3 shows the diurnal cycle of the main components of the surface energy balance, namely net radiation (Q^*), sensible (H) and latent heat flux ($L.ET$). All of these were measured in a typical clear sky day in Cabauw (no LAE) and in Cordoba (with LAE). It is seen that during daytime $L.ET$ is lower than Q^* in Cabauw, whereas, in the Cordoba case, $L.ET$ becomes greater than Q^* after about 12 UTC (close to local noon) and then H becomes negative during the local afternoon. Although not shown, the additional energy that enhances $L.ET$, via the process described by the term Q_{adv} in Equation (3), leads to evaporative cooling of the surface, which eventually drops below air temperature. A stable layer is then formed during daytime, consistent with the negative H values observed in Figure 3b.

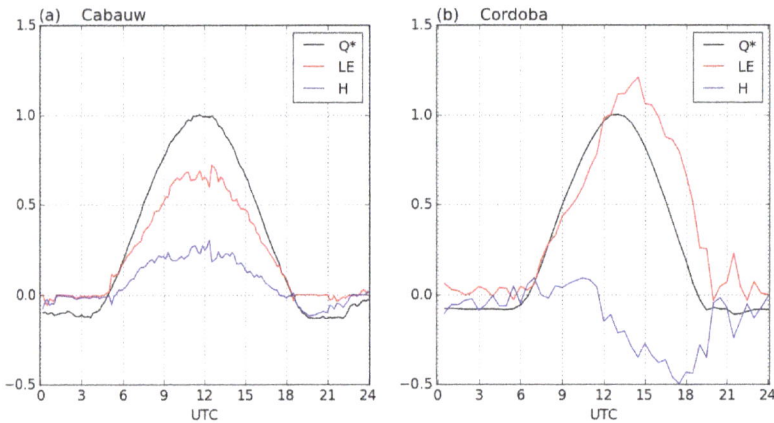

Figure 3. Diurnal variation of the terms of the energy budget equation—latent (red line) and sensible (blue line) heat fluxes and net radiation (black line)—on a sunny day for (**a**) Cabauw (28 May 2012, no local advection) and (**b**) the lysimeter Cordoba site (29 July 2009, with local advection). All values are based on observations gathered at both sites and are normalized by the local net radiation maximum.

Confining ourselves to daily values in the growing season, i.e., between about April and September, $0.8Q^*$ can be considered a suitable indicator for the occurrence of LAE, i.e., in cases where LAE is absent, daily $L.ET$ will not differ much from $0.8Q^*$, whereas for the LAE case, $L.ET$ will most likely exceed $0.8Q^*$. This value of $L.ET$ being roughly 80% of net radiation was found for the Cabauw data, during the growing season. Using Equation (2) to determine Q^*, a simple diagnostic tool is obtained to determine whether or not measured or calculated $L.ET$ are affected by LAE, as also shown by [16]. To illustrate this, Figure 4 presents the so-called advection index, defined as $I_a =$ $L.ET/Q^*$ (both $L.ET$ and Q^* correspond to daily averages), for the Cordoba site. I_a is plotted for a period encompassing the growing season, revealing that, for most of it, the measured $L.ET$ is greater than $0.8Q^*$ (for days-of-year between 147 and 300 shown in Figure 4, $L.ET$ is always above this threshold), with I_a reaching values up to 1.5. This supports the notion that the Cordoba site is clearly affected by LAE, and particularly, as we move ahead of day-of-year 150 (i.e., from June onwards), $L.ET$ becomes permanently higher than Q^*, indicating an intensification of LAE.

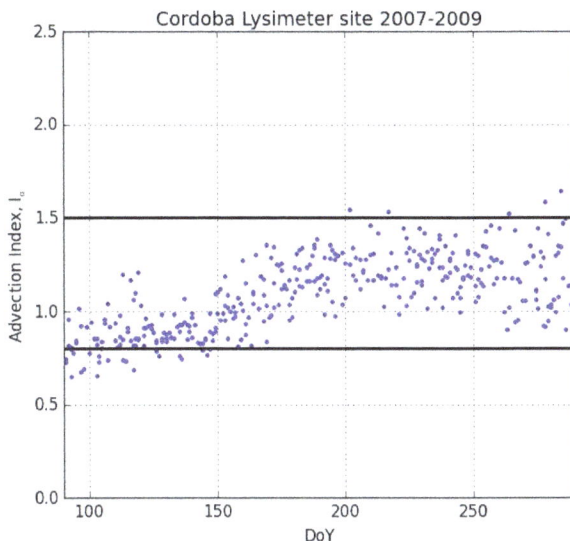

Figure 4. Advection-index being the ratio of daily means of actual *L.ET* and *Q**, as a function of day of year (DoY). I_a is estimated as the ratio between daily net radiation (obtained via Equation (2)) and daily latent heat flux obtained from observations taken over the well-watered Cordoba site (see also Figure 1 in [16]). The lower line represents the 0.8*Q** threshold beyond which LAE should be taken into account.

It is noted that for Cabauw, the revised Makkink formula, where *L.ET* over the reference surface is proportional to solar radiation ($L.ET = 0.65\Delta/(\Delta + \gamma) \times R_s$), yields estimates very close to local observations [4,23,24]. On the other hand, a closer inspection of the literature revealed that for the Cordoba site, as is the case with LAE, the Hargreaves formula ($L.ET = 0.0135\ (T_a + 17.8)R_s$) dealt with in [1,25] also provides similar results to Equation (3) with the LAE adjustment (shown in Figure 2 for Cordoba). Both the Makkink and Hargreaves formulations are easier to use and can be applied as alternatives to PM-ETo in the LAE-free case and in the case LAE is meant to be accounted for, respectively.

The advection index can also be used to reveal effects of surface aridity. These refer to the impact on the estimation of ETo with the FAO-method, but when the input data are not gathered over reference grass. Because the surface atmospheric layer is itself affected by the radiation budget and its portioning between sensible and latent heat fluxes, measurements of temperature and humidity variables will necessarily be affected by the underling surface. As such, the input data should be adjusted in those cases, a practice that is often omitted. To illustrate the impact of this, we analyzed ETo estimated using data from the RIA Cordoba station [17] that is located close to the Cordoba lysimeter site, for which the Slob-deBruin estimate for *Q** was tested by [5]. All input data for the FAO method were then obtained from in situ observations, but over non-irrigated ground.

Figure 5 shows the advection index I_a plotted for spring and early summer days, but now given by the calculated PM-ETo not adjusted for surface aridity, over the estimated net radiation using Equation (2). It is seen that in this case, I_a deviates even more generally from the 0.8 threshold than the values found for well-watered nearby grass (Figure 4). Adopting the rule of thumb that the LAE estimate of ETo is about 0.8*Q**, this example shows that a combination of LAE and surface aridity effects can lead to significantly higher values, up to 100% or more of *Q**. The LSA SAF method is not affected by surface aridity. From a practical point of view this is an enormous advantage, which is added to the fact that the LSA SAF approach does not need ground based-stations. The latter aspect is also relevant, when new underground water aquifers are explored, such as those present under the Sahara, or in locations where accurate weather data are simply absent.

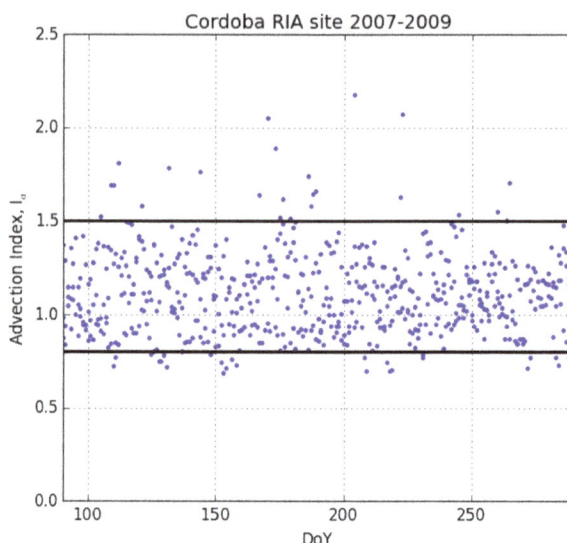

Figure 5. As in Figure 4, but I_a is estimated using observations gathered in the RIA station close to Cordoba (in situ observations gathered over non-irrigated surface).

It is clear from this example that PM-ETo values calculated with input data gathered over non-reference surfaces should be corrected for surface aridity. We note that in the literature there is no consensus about correction procedures. For this we refer to [12] and Annex 6 in [1] for example. It is outside the scope of this paper to discuss this topic. However, because errors in ETo due to surface aridity effects combined with LAE can be as high as 100% it is a very important aspect for users of PM-ETo. Note that our methodology does not suffer from these errors.

The LSA SAF ETo product does not account for LAE as required per definition [1,2], nor is it affected by aridity effects, and therefore can be used for irrigation advisory. Nevertheless, if one is interested in the actual ET of well-watered grass affected by LAE, then an adjustment is required for the LSA SAF values. As referred to above for the Cordoba lysimeter site and previously found by the authors for other similar sites, such correction may be parameterized as an additional term (Q_{adv}), which is a linear function of near surface air temperature, i.e., with a term $f(T_a)$, as shown in Equation (3), given by:

$$Q_{adv} = f(T_a) = \begin{cases} 0 & \text{if } T_a < T_o \\ m(T_a - T_o) & \text{if } T_a \geq T_o \end{cases} \tag{4}$$

where the parameter T_o corresponds to the temperature above which the site systematically experiences LAE. Both regression coefficients, m and T_o, may vary with location and must be locally tuned using either local observations as described above for the lysimeter Cordoba site, or considering PM-ETo and LSA-SAF ETo as approximations of ET with and without LAE, respectively. A threshold of 0.8 may also be considered for the advection index to determine conditions with LAE. The index may be estimated using lysimeter ET observations (or PM-ETo estimates) and Slob-deBruin Q*. The latter is also distributed by the LSA SAF, as an auxiliary dataset within its METREF product.

5. Concluding Remarks

This research note concerns some aspects that may concern applications of the recently published LSA SAF product for ETo (METREF in the LSA SAF catalogue). Using observations from two very different sites and for situations where local advection effects (LAE) can be ignored and where they play a significant role, we showed that the LSA SAF provides ETo estimates that are basically LAE-free, thus fully consistent with the definition of ETo given in [1] and fulfilling the requirements specified in [2], as referred to above. The FAO PM methodology requires input data collected over a reference-like surface, otherwise estimated values may far exceed measured ones due to the so-called aridity effect; these do not affect the LSA SAF estimates based on the T-ABL approach.

There are other issues that should be considered in the practical application of T-ABL or FAO methods. For example, we ignored elevation effects on air temperature data, particularly when these were obtained from a numerical weather prediction model. The LSA SAF makes use of ECMWF global fields, which have a spatial resolution ranging from about 9 km (current operational model) to 75 km (ERA-Interim reanalyses [22]). Near surface (2 m) air temperature is then interpolated to each MSG/SEVIRI pixel, taking into account height differences and performing an orography adjustment [23]. Nevertheless, air pressure may also affect both T-ABL-ETo and PM_ETo, as it is hidden in the psychrometric constant γ, as discussed in [1] and [23]. We invite water managers and researchers involved in irrigation advisory to apply the LSA SAF ETo (METREF) product, pinpointing caveats and a way forward. Through interaction with the LSA SAF consortium the operational applicability of LSA SAF products can be improved.

Author Contributions: The authors contributed to the article as follows: Conceptualization, H.A.R.d.B. and I.F.T.; Methodology, H.A.R.d.B. and I.F.T.; Software, I.F.T.; Writing, H.A.R.d.B. and I.F.T.; Visualization, I.F.T.

Funding: This research received no external funding.

Acknowledgments: We thank Fred Bosveld (KNMI) and Pedro Gavilàn (IFAPA) for providing the Cabauw and Cordoba data. Satellite reference evapotranspiration and solar radiation products are made available by the EUMETSAT Satellite Applications Facility on Land Surface Analysis (LSA SAF; http://lsa-saf.eumetsat.int).

Conflicts of Interest: The authors declare no conflict of interest.

Annex—Terminology

Name/Symbol	Meaning
ETo	Reference crop as defined in [1] and [2]; local advection effects are excluded.
ETc	Crop water requirement as defined in [1], and obtained via k_c.ETo, with k_c being a crop factor. Note that [1] introduced ETo and ETc to avoid the use of potential evapotranspriation, since the latter was generally used in as a maximum ET, without specifying the crop of surface.
PM-ETo	Version of the Penman-Monteith equation for estimating ETo, introduced by [1].
LSA SAF ETo	Estimated using Equation (1) derived from thermodynamics and PBL-physics in [4] and validated in [5].
METREF	The LSA SAF ETo estimates, as specified in the LSA SAF catalogue.
T-ABL	The model to estimate ETo derived from thermodynamics and atmospheric boundary layer (ABL) physics, leading to Equation (1).
Q*	Net radiation, i.e., the sum of the down-welling short and long wave radiation reaching the surface minus reflected (short and long wave) and the emitted (long wave) radiation.
R_s	Global radiation, or down-welling short wave radiation reaching the surface
R_{ext}	Incoming shortwave radiation at the top of the atmosphere, often denoted as the extra-terrestrial radiation.

References

1. Allen, R.G.; Pereira, L.A.; Raes, D.; Smith, M. *Crop Evapotranspiration—Guidelines for Computing Crop Water Requirements*; FAO Irrigation and Drainage Paper 56; FAO: Rome, Italy, 1998; p. 293.
2. Pereira, L.S.; Perrier, A.; Allen, R.G.; Alves, I. Evapotranspiration: Concepts and future trends. *J. Irrig. Drain.* **1999**, *125*, 45–51. [CrossRef]
3. Allen, R.G.; Pruitt, W.O.; Wright, J.L.; Howell, T.A.; Ventura, F.; Snyder, R.; Itenfisu, D.; Steduto, P.; Berengena, J.; Baselga, J.; et al. A recommendation on standardized surface resistance for hourly calculation of reference ETo by the FAO56 Penman-Monteith method. *Agric. Water Manag.* **2006**, *81*, 1–22. [CrossRef]

4.   De Bruin, H.A.R.; Trigo, I.F.; Bosveld, F.C.; Meirink, J.F. A thermodynamically based model for actual evapotranspiration of an extensive grass field close to FAO reference suitable for remote sensing application. *J. Hydrometeor.* **2016**, *17*, 1373–1382. [CrossRef]

5.   Trigo, I.F.; de Bruin, H.; Beyrich, F.; Bosveld, F.C.; Gavilán, P.; Groh, J.; López-Urrea, R. Validation of reference evapotranspiration from Meteosat Second Generation (MSG) observations. *Agric. For. Meteorol.* **2018**, *259*, 271–285. [CrossRef]

6.   Schmidt, W. Strahlung und Verdunstung an freienWasserflächen; ein Beitrag zum Wärmehaushalt des Weltmeers und zum Wasserhaushalt der Erde (Radiation and evaporation over open water surfaces; a contribution to the heat budget of the world ocean and to the water budget of the Earth). *Ann. Calender Hydrogr. Marit. Meteorol.* **1915**, *43*, 111–124.

7.   Slatyer, R.O.; McIlroy, I.C. Evaporation and the principles of its measurement. In *Practical Micrometeorology*; CSIRO (Australia); UNESCO: Paris, France, 1961.

8.   De Bruin, H.A.R. A model for the Priestley-Taylor parameter. *J. Appl. Meteorol.* **1983**, *22*, 572–578. [CrossRef]

9.   Monteith, J.L. Accommodation between transpiring vegetation and the convective boundary layer. *J. Hydrol.* **1995**, *166*, 251–263. [CrossRef]

10.   McNaughton, K.G.; Spriggs, T.W. A mixed-layer model for regional evaporation. *Bound.-Layer Meteorol.* **1986**, *34*, 243–262. [CrossRef]

11.   McNaughton, K.G.; Jarvis, P.G. Predicting effects of vegetation changes on transpiration and evaporation. In *Water Deficits and Plant Growth*; Kozlowski, T.T., Ed.; Academic Press: New York, NY, USA, 1983; Volume 7, pp. 1–47.

12.   Temesgen, B.; Allen, R.G.; Jensen, D.T. Adjusting temperature parameters to reflect well-watered conditions. *J. Irrig. Drain. Eng.* **1999**, *125*, 26–33. [CrossRef]

13.   Geiger, B.; Meurey, C.; Lajas, D.; Franchistéguy, L.; Carrer, D.; Roujean, J.L. Near real-time provision of downwelling shortwave radiation estimates derived from satellite observations. *Meteorol. Appl.* **2008**, *15*, 411–420. [CrossRef]

14.   Greuell, W.; Meirink, J.F.; Wang, P. Retrieval and validationof global, direct, and diffuse irradiance derived from SEVIRI satellite observations. *J. Geophys. Res. Atmos.* **2013**, *118*, 2340–2361. [CrossRef]

15.   Monna, W.; Bosveld, F. In Higher Spheres: 40 Years of Observations at the Cabauw Site. 2013, p. 56. Available online: http://publicaties.minienm.nl/documenten/in-higher-spheres-40-years-of-observations-at-the-cabauw-site (accessed on 21 February 2019).

16.   Berengena, J.; Gavilán, P. Reference Evapotranspiration Estimation in a Highly Advective Semiarid Environment. *J. Irrig. Drain. Eng.* **2005**, *131*, 147–163. [CrossRef]

17.   Cruz-Blanco, M.; Gavilán, P.; Santos, C.; Lorite, I.J. Assessment of reference evapotranspiration using remote sensing and forecasting tools under semi-arid conditions. *Int. J. Appl. Earth Obs. Geoinf.* **2014**, *33*, 280–289. [CrossRef]

18.   Castellvi, F.; Gavilan, P.; Gonzalez-Dugo, M.P. Combining the bulk transfer formulation and surface renewal analysis for estimating the sensible heat flux without involving the parameter kB^{-1}. *Water Resour. Res.* **2014**, *50*, 8179–8190. [CrossRef]

19.   De Bruin, H.A.R.; Holtslag, A.A.M. A simple parameterization of the surface fluxes of sensible and latent heat during daytime compared with the Penman-Monteith concept. *J. Appl. Meteorol.* **1982**, *21*, 1610–1621. [CrossRef]

20.   De Bruin, H.A.R. From Penman to Makkink. In *Proceedings and Information: TNO Committee on Hydrological Research N° 39, Den Haag, The Netherlands, 25 March 1987*; Hooghart, J.C., Ed.; The Netherlands Organization for Applied Scientific Research TNO: Den Haag, The Netherlands, 1987; pp. 5–31.

21.   Pinker, R.T.; Tarpley, J.D.; Laszlo, I.; Mitchell, K.E.; Houser, P.R.; Wood, E.F.; Schaake, J.C.; Robock, A.; Lohmann, D.; Cosgrove, B.A.; et al. Surface radiation budgets in support of the GEWEX Continental-Scale International Project (GCIP) and the GEWEX Americas Prediction Project (GAPP), including the North American Land Data Assimilation System (NLDAS) project. *J. Geophys. Res.* **2003**, *108*, 8844. [CrossRef]

22.   Dee, D.P.; Simmons, A.J.; Berrisford, P.; Poli, P.; Kobayashi, S.; Andrae, U.; Balmaseda, M.A.; Balsamo, G.; Bauer, P.; Bechtold, P.; et al. The ERA-Interim reanalysis: configuration and performance of the data assimilation system. *Q. J. R. Meteorol. Soc.* **2011**, *137*, 553–597. [CrossRef]

23. De Bruin, H.A.R.; Trigo, I.F.; Gavilán, P.; Martínez-Cob, A.; González-Dugo, M.P. Reference crop evapotranspiration estimated from geostationary satellite imagery. In Proceedings of the Remote Sensing and Hydrology, Jackson Hole, WY, USA, 27–30 September 2010; pp. 111–114.

24. Makkink, G.F. Testing the Penman formula by means of lysimeters. *J. Inst. Water Eng.* **1957**, *11*, 277–288.

25. Hargreaves, G.H.; Samani, Z. A Reference crop evapotranspiration from temperature. *Appl. Eng. Agric.* **1985**, *1*, 96–99. [CrossRef]

water

MDPI

Article

Evapotranspiration Partition and Crop Coefficients of Tifton 85 Bermudagrass as Affected by the Frequency of Cuttings. Application of the FAO56 Dual K_c Model

Paula Paredes [1], Geraldo J. Rodrigues [2], Mirta T. Petry [2,*], Paula O. Severo [2], Reimar Carlesso [2] and Luis Santos Pereira [1]

[1] Centro de Investigação em Agronomia, Alimentos, Ambiente e Paisagem (LEAF),
Instituto Superior de Agronomia, Universidade de Lisboa, Tapada da Ajuda, 1349-017 Lisboa, Portugal;
pparedes@isa.ulisboa.pt (P.P.); luis.santospereira@gmail.com (L.S.P.)

[2] Centro de Ciências Rurais, Universidade Federal de Santa Maria, Cidade Universitária, Bairro Camobi,
97105-900 Santa Maria, RS, Brazil; zootecniagjr@hotmail.com (G.J.R.); paulasevero.zoot@gmail.com (P.O.S.);
reimar.carlesso@gmail.com (R.C.)

* Correspondence: mirta.petry@gmail.com; Tel.: +55-553-220-8399

Received: 21 March 2018; Accepted: 23 April 2018; Published: 26 April 2018

Abstract: This study aims to model the impacts of the frequency of cuttings of Tifton 85 bermudagrass on the dynamics of evapotranspiration (ET_c) and to derive crop coefficients appropriate for grass water management. Two seasons of experimentation were used with four different cutting treatments which provided field data for calibration and validation of the soil water balance model SIMDualKc for all treatments. Cuttings were performed after the cumulative growth degree days (CGDD) attained 124 °C, 248 °C and 372 °C, thus from short to very long intervals between cuttings. SIMDualKc adopts the Food and Agriculture Organization (FAO) dual K_c approach for partitioning ET into crop transpiration and soil evaporation, thus providing for an assessment of their dynamics. All treatments were irrigated to avoid water stress. Grass ET_c was modelled adopting a K_{cb} curve to describe the ET variation for each cutting cycle, that is, using the FAO K_c curve that consists of a series of K_{cb} curves relative to each cutting cycle. Each individual K_{cb} curve consisted of three segments constructed when knowing the K_{cb} values at the initial, at the end of rapid growth, and at cutting, respectively $K_{cb\ ini}$, $K_{cb\ gro}$ and $K_{cb\ cut}$. These K_{cb} values were first estimated using the equation relating K_{cb} to the density coefficient (K_d), which is computed from the fraction of ground cover (f_c) and canopy height (h) at the same dates. The goodness of fit indicators relative to the calibration and validation of the SIMDualKc model were rather good, with the normalized root mean square error (RMSE) ranging from 4.0% to 6.7% of the mean available soil water. As an example, the standard K_{cb} values obtained after model calibration relative to the cuttings treatment with CGDD of 248 °C are: $K_{cb\ ini}$ = 0.86, $K_{cb\ gro}$ = 0.91 and $K_{cb\ cut}$ = 0.96. K_{cb} values were smaller when the frequency of cuts was larger because h and f_c were smaller, and were larger for reduced cuttings frequency since h and f_c were then larger. Because the soil was wet most of the time, the soil evaporation K_e varied little but its value was small due to the combined effects of the fraction of crop cover and plant litter covering the soil. The values of $K_c = K_{cb} + K_e$ also varied little due to the influence of K_e and the K_c curve obtained a form different from the K_{cb} curves, and a single K_c value was adopted for each cutting frequency, e.g., K_c = 0.99 for the treatment with CGDD of 248 °C. Results of the soil water balance have shown that, during the experimental periods, likely due to the effects of the El Niño Southern Oscillation (ENSO), runoff and deep percolation exceeded ET_c. Moreover, the soil evaporation ratio was small: 14% in case of frequent cuttings and less for more spaced cuttings, thus with a transpiration ratio close to 90%, which indicates a very high beneficial consumptive water use, mainly when cuttings are not very frequent.

Keywords: basal crop coefficients; crop coefficient curves; crop transpiration; K_{cb} from ground cover; SIMDualKc model; soil evaporation

1. Introduction

The landscape of southern Brazil is characterized by the Pampa biome, which occupies 63% of the State of Rio Grande do Sul. Grassland is the dominant vegetation and livestock production is a main economic activity in the area, but large areas have been converted into cropland, mainly for soybean production, thus suppressing the native grass vegetation [1]. Therefore, the sustainability of this biome for livestock production requires the planting of new and highly productive grasses such as the Tifton 85 bermudagrass [2,3], recently introduced in the region. Assessing grass evapotranspiration and water requirements as influenced by the frequency of cuttings is required to support an upgraded management of those grasslands. An innovative approach used is to relate the frequency of cuttings with the density coefficient (K_d) and then estimate the basal crop coefficient from K_d following the Allen and Pereira approach [4].

Studies on evapotranspiration (ET) of grasslands are numerous. Research has commonly been devoted to assessing the dynamics and abiotic driving factors of ET, mainly relative to climate influences on the processes of energy partition into latent and sensible heat. Such studies often use eddy covariance and/or Bowen ratio energy balance (BREB) observations, data which are commonly used in analysis performed with the Penman–Monteith (PM) combination equation [5] and/or the Priestley–Taylor (PT) equation [6], thus using the canopy resistance or the PT parameter α as behavioral indicators [7–9]. As an alternative to those measurement techniques, scintillometer measurements [10] and satellite images [11–13] were also used. Adopting similar research approaches, other studies compared the ET of grasslands with ET of forests or shrublands [14–16]. In addition to the available energy for evaporation, soil water availability and crop ground cover or the leaf area index (LAI) were often identified as main driving factors influencing grass ET [14–17].

Studies such as those referred to above are likely of great importance for understanding the variability of grass ET when focusing on the Pampa biome but different, operational research approaches are required when aiming at knowing grassland water requirements and/or grassland water management issues. Related operational studies are also numerous and refer to various climates, grass species and diverse herbage uses for hay or for grazing with different frequency of cuttings. However, such ET studies are lacking in southern Brazil and for Tifton 85 bermudagrass. ET research aiming at improved farm water management generally uses the grass reference ET (ET_o) proposed in the Food and Agriculture Organization guidelines for computing crop water requirements (FAO56) [17]; nevertheless, recent studies [3,18] relative to irrigated bermudagrass yields used the climatic potential ET equation of Thornthwaite [19], developed in 1948.

The reference ET_o was defined after parameterizing the PM combination equation [5] for a cool season grass, thus resulting that ET_o is defined as the rate of evapotranspiration from a hypothetical reference crop with an assumed crop height h = 0.12 m, a fixed daily canopy resistance r_s = 70 s m^{-1}, and an albedo of 0.23, closely resembling the evapotranspiration from an extensive surface of green grass of uniform height, actively growing, completely shading the ground and not short of water [17]. This definition is described by the daily PM-ET_o equation [17], which represents the climatic demand of the atmosphere. Thus, following FAO56 [17], the ET_o is to be used with a crop coefficient (K_c) when estimating or predicting the ET of a given surface, that is, the K_c-ET_o approach. K_c is the ratio between the crop ET and ET_o and varies with the crop surface characteristics and the crop growth stage, and is influenced by the climate and management. Single and dual K_c may be used [17]. The dual K_c consists of the sum $K_e + K_{cb}$ of the soil evaporation coefficient (K_e) and the basal crop coefficient (K_{cb}), and thus with consideration of both processes included in evapotranspiration. The K_c values are standard or potential when the crop is not stressed, while actual K_c values ($K_{c\,act}$ or $K_{cb\,act}$) are often smaller than the standard K_c or K_{cb} when water, salt, disease or management stresses affect crop transpiration. These effects may be considered using a stress coefficient applied to K_c or K_{cb}. Tabulated values of standard K_c and K_{cb} are provided by Allen et al. [17] for a variety of crops, including grasses and pastures. The standard K_c and K_{cb} values are transferable to other sites considering adjustments for climate described by Allen et al. [17]. The time variation of the K_c and K_{cb} values are described by K_c curves [17] that describe in a simplified way the dynamics of LAI and vegetation ET. The form of these curves varies from one crop to another; for grasses with cuttings, several successive K_c curves should be considered, each representing the dynamics of ET during each crop growth cycle between cuttings [17].

The operational use of the ET_o equation implies, therefore, the use of crop coefficients and the build-up of K_c curves to describe the respective time variation. It could be observed that several papers reporting on the use of the grass reference ET_o did not follow the concepts described above. A few authors directly compared ET obtained with eddy covariance, BREB, or a soil water balance with ET_o but not searching for a K_c value [20], or even assumed equality between grass ET and ET_o [21]. Other authors just computed daily $K_{c\,act}$ values (often using different designations for that parameter) but did not search for a K_c curve that would describe their seasonal variation [22,23], or just identified a mean seasonal $K_{c\,act}$ [24]. The lack of search for a K_c curve led some authors to consider the K_c-ET_o approach as non-useful [25]. By contrast, Pronger et al. [26] did not clearly assume the concepts behind actual vs. potential K_c and, in addition to K_s, adopted a correction factor to ET_o for highly stressed grass. This is theoretically not appropriate because it contradicts the concepts of reference ET, which depends solely upon the climate and not the crop under study, and of the $K_{c\,act}$ that has to be derived from the standard K_c when adapting to the management and environmental conditions [17]. The approaches referred to above, like other research quoted before, may support an improved understanding of the dynamic behaviour of grass ET but are likely not appropriate for operational use in irrigation water management.

The FAO56 K_c-ET_o method [17] was first applied by Cancela et al. [27] to grass using the single $K_{c\,act}$ with successive four-stage curves relative to four cuttings using the ISAREG soil water balance model [28]. Single $K_{c\,act}$ four stage curves were defined for remote-sensed grazed grasslands [29,30]. By contrast, other authors preferred replacing the typical four stages curve by average monthly $K_{c\,act}$ values [31]. The FAO dual K_c approach [32] was successfully used by Greenwood et al. [33], who reported on a large number of $K_{cb\,act}$ curves to represent numerous grass cuttings using the FAO56 spreadsheet [17]. The FAO dual K_c approach was adopted by Wu et al. [34] to represent a natural groundwater dependent grassland, then using a seasonal four-growth stages $K_{cb\,act}$ curve using the SIMDualKc model [35]. Krauß et al. [36] also used the dual K_c method to estimate the footprint of milk production but did not report about the K_c curves used.

The dual K_c approach has the advantage of partitioning ET into crop transpiration (T_c) and soil evaporation (E_s). Knowing T_c and E_s provides for a more detailed water balance and a better approach to understanding the functioning of the ecosystems. In addition, partitioning ET allows estimating T_c and therefore better calculating yields [37] since it directly relates to biomass production. Moreover, good results were obtained with the soil water balance SIMDualKc model [35] for the partition of ET using the FAO dual Kc approach, namely when applied to crops that nearly fully cover the ground, such as wheat, barley and soybean [38–40] whose E_s estimated values compared well with E_s observations using microlysimeters. T_c simulated also compared well with sap flow observations in tree crops [41,42], thus confirming the goodness of ET partitioning.

The application of various ET partition methods to grass is often reported, namely using the two source Shuttelworth and Wallace [43] model (SW). It is very precise when an appropriate parameterization is achieved, which is a quite demanding task that limits the operational use of SW in agricultural water management practice; various examples of the application of SW to grass are reported in the literature [44,45]. Other double source models were applied to grass, such as the one reported by Huang et al. [46], which is based upon the estimation of gross ecosystem productivity using

CO_2 fluxes observed with the eddy covariance method, and that proposed by Wang and Yamanaka [47], which consists of a modification of the SW model. Empirical ET partitioning approaches include the use of time series of soil surface temperature [15], and the adoption of a radiation extinction coefficient (k_{rad}) combined with a ground cover index [20] or with LAI when using the PT equation [48]. These approaches using k_{rad} are comparable with the FAO dual K_c approach [17,32]. Partitioning ET fluxes using stable isotopes is another proved alternative [49,50].

The dual K_c approach has been shown to be appropriate to partitioning grass ET, as reported above [33,34], and has shown to be less demanding in terms of parameterization and field and laboratory instrumentation than other ET partition methods. In addition, it is easily implemented using the referred SIMDualKc model, which has been extensively tested as reported above. Therefore, the objectives of this study consist of (a) assessing and partitioning evapotranspiration of Tyfton 85 bermudagrass in southern Brazil as influenced by the frequency of cuttings using the model SIMDualKc applied to two years of field data; and (b) deriving crop coefficient curves adapted to grass cuttings of various frequencies. Moreover, the study aims to contribute to the sustainability of grass uses in the Pampa biome, and to create the knowledge required to cope further with climate change in the region.

2. Material and Methods

2.1. Site Characteristics and Treatments

Field experiments aimed at assessing Tifton 85 bermudagrass ET and herbage production under different regimes of cutting were developed in the campus of the Federal University of Santa Maria, Rio Grande do Sul State, Southern Brazil (29°43′ S, 53°43′ W and altitude of 103 m); however, production is not analyzed in this article. The experiments were conducted during the growing seasons—spring, summer and autumn—of two cropping years, from 23 October 2015 to 11 May 2016 and from 27 October 2016 to 26 June 2017. The Tifton 85 bermudagrass was planted earlier, in 2011.

The climate of the region, according to the Köppen climatic classification, is a "Cfa", that is, humid subtropical without a defined dry season and with hot summers [51]. The meteorological conditions during the experimentation are given in Figure 1 and were observed in an automatic weather station located at 300 m from the experimental area which is in the charge of the National Institute of Meteorology. The reference evapotranspiration (ET_o) was computed with the PM-ET_o equation [17].

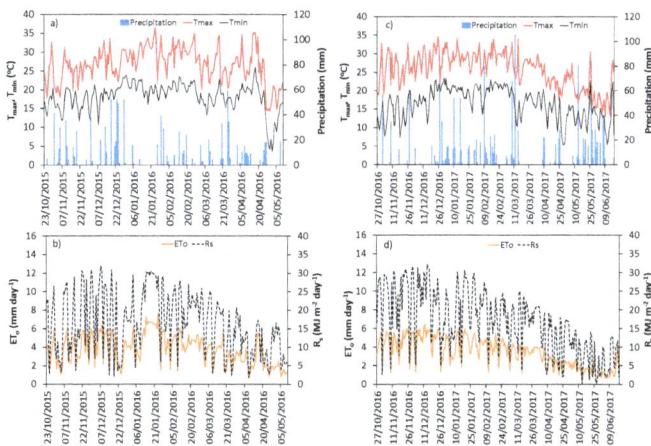

Figure 1. Daily weather data during the crop seasons of 2015/16 and 2016/17: (**a,c**) maximum (—) and minimum (—) temperatures, and rainfall (—); (**b,d**) solar radiation (..) and reference evapotranspiration (—).

The textural and hydraulic properties of the soil of the experimental site are given in Table 1. Disturbed and undisturbed soil samples were collected at the initiation of the experimentation. The particle size distribution was obtained using an ASTM 151H soil hydrometer (Chase Instruments Co., Swedesboro, NJ, USA). Soil water retention at matric potentials between -10 and -5000 cm were determined with a pressure plate apparatus (Soil Moisture Equipment Corp., S. Barbara, CA, USA) and at potentials of -5000 and $-15,000$ cm were determined with a WP4 dewpoint potentiometer (Decagon Devices, Pullman, WA, USA). The soil water content at field capacity (FC) and at wilting point (WP) were measured for the matric potentials at -10 kPa and $-15,000$ kPa, respectively.

Table 1. Soil physical characteristics of the experimental field.

Depth (cm)	Bulk Density (g cm^{-3})	Total Porosity (%)	Field Capacity (FC) (cm$^3 \cdot$cm^{-3})	Wilting Point (WP) (cm$^3 \cdot$cm^{-3})	Clay %	Silt %	Sand %
0–5	1.49	41.3	0.364	0.089	18	36	46
5–10	1.56	38.6	0.337	0.085	18	36	46
10–15	1.47	43.2	0.298	0.081	20	35	45
15–30	1.51	42.0	0.375	0.087	20	37	43
30–50	1.50	43.0	0.308	0.092	24	35	41

2.2. Grass and Field Observations

Three treatments of grass cuttings were used in the current study and three replications were adopted; the area of each experimental unit was 16 m^2. In agreement with information relative to cuttings of Tifton 85 [2,3], the stubble height (SH) of 0.15 m was adopted for all cuttings. The latter were executed with an electric lawnmower with adjusted cutting height, which provided for a precise SH. The cuttings were performed when the cumulative growing degree days (CGDD) attained 124, 248 and 372 °C after the start of each cutting cycle. The base and cut-off temperatures were 10 °C and 30 °C, respectively. The cuttings dates are given in Table 2. It can be observed that defining the intervals between cuttings for selected CGDD results in time intervals that are different within each treatment and among treatments, varying from a minimum near the summer solstice to a maximum by the winter solstice (Table 3).

Table 2. Cuttings dates of the various treatments having different cumulative growing degree days (CGDD) intervals between cuttings.

Events	Cutting Treatments			Dates	
	CGDD of 124 °C	CGDD of 248 °C	CGDD of 372 °C	Season of 2015/16	Season of 2016/17
Spring (t$_0$)				23 October 2015	27 October 2016
Scheduled cuttings	1st			4 November 2015	10 November 2016
	2nd	1st		15 November 2015	22 November 2016
	3rd		1st	27 November 2015	2 December 2016
	4th	2nd		8 December 2015	12 December 2016
	5th			17 December 2015	22 December 2016
	6th	3rd	2nd	28 December 2015	29 December 2016
Summer-Autumn (t$_0$)				9 January 2016	17 January 2017
Scheduled cuttings	1st			16 January 2016	27 January 2017
	2nd	1st		24 January 2016	5 February 2017
	3rd		1st	3 February 2016	15 February 2017
	4th	2nd		10 February 2016	23 February 2017
	5th			18 February 2016	6 March 2017
	6th	3rd	2nd	26 February 2016	19 March 2017
	7th			8 March 2016	2 April 2017
	8th	4th		18 March 2016	13 April 2017
	9th		3rd	30 March 2016	27 April 2017
	10th	5th		9 April 2016	15 May 2017
	11th			18 April 2016	5 May 2017
	12th	6th	4th	08 May 2016	26 June 2017

Table 3. Time intervals between cuttings for all treatments.

Cutting Treatments	Minimum Intervals (Days)		Maximum Intervals (Days)	
	2015–2016	2016–2017	2015–2016	2016–2017
CGDD of 124 °C	9	10	20	21
CGDD of 248 °C	20	17	29	42
CGDD of 372 °C	31	27	39	60

The K_c curves adopted for each cutting cycle considered only three growth stages: initial, from initiation after a cutting until rapid growth starts; rapid growth, from then on until growth slows down; and before cutting, from then on until a cutting is performed. Curves are described by three K_c or K_{cb} values corresponding to the three growth stages identified, respectively $K_{cb\ ini}$, $K_{cb\ gro}$ and $K_{cb\ cut}$. Cutting cycles were, therefore, described with the following observations:

(a) The time duration (days) of the three phases of the cutting cycles: initial, rapid growth and before cutting, whose time durations are respectively t_{ini}, t_{gro} and t_{cut}, with their sum equalling the time interval between cuttings, t_{int}. Their values varied among treatments and with air temperature, that is with climate.

(b) The grass height at the end of the periods referred to above, thus h_{ini}, h_{gro} and h_{cut} in the day of cutting, with h_{ini} = SH. Values varied among treatments but not within each treatment due to the great dependence of crop growth relative to temperature.

(c) The fraction of ground cover at the same days referred for h, thus $f_{c\ ini}$, $f_{c\ gro}$ and $f_{c\ cut}$. Alternatively, LAI could be observed by the same days. As for h, values of f_c varied among treatments but very little within each treatment.

The canopy height was measured with a millimeter graduated ruler. The fraction of ground cover was observed visually using frames with an area of 0.625 m². Photos taken vertically were used to count the percent of ground covered. Errors of observations of h and f_c did not exceed 10%. Average h and f_c values are presented in Table 4. The effective root depth (Z_r, m) was observed using nine soil samples taken at each 0.10 m layer, down to the depth of 0.6 m, which were washed, sieved and observed for the root material. Results have shown that 90% of the roots were above the 0.3 m depth and only a very small fraction was below the 0.5 m depth. Thus, in agreement with literature [9,13,14,46], Z_r = 0.5 m was adopted for the simulations.

Table 4. Average of observed canopy height (h, m) and fraction of ground cover (fc, dimensionless) relative to treatments with various intervals between cuttings defined by observed CGDD.

Variables and Treatments	Grass Development Stages		
	Initial	End of Rapid Growth	Before Cutting
Canopy height (h, m)	h_{ini}	h_{gro}	h_{cut}
CGDD of 124 °C	0.15	0.18	0.19
CGDD of 248 °C	0.15	0.22	0.23
CGDD of 372 °C	0.15	0.27	0.30
Fraction of ground cover (f_c, dimensionless)	$f_{c\ ini}$	$f_{c\ gro}$	$f_{c\ cut}$
CGDD of 124 °C	0.81	0.85	0.90
CGDD of 248 °C	0.81	0.88	0.92
CGDD of 372 °C	0.85	0.90	0.93

All treatments used sprinkler irrigation to supplement rainfall and assure that the crop was not water stressed, so allowing for potential ET and crop coefficients to be determined. Full circle sprinklers Pingo® (Fabrimar Ltda., Joinville, SC, Brazil), spaced 6 m and operating at the pressure of

180 kPa were used. The coefficient of uniformity averaged 82% and the applied net irrigation depths varied from 12–22 mm in 2015–2016 and from 7.5–12.5 mm in 2016–2017. Irrigations were performed whenever the soil moisture in the 0.50 m layer reached less than 85% of the FC. The soil water content (SWC, cm^3 cm^{-3}) was daily monitored with Frequency Domain Reflectometry (FDR) sensors installed in the center of each unit in the 0.00–0.20 m and 0.20–0.50 m depth layers. The CS616 sensors were connected to a CR10X data logger with the AM16/32 channel relay multiplexer (all from Campbell Scientific, Logan, UT, USA). The FDR system was calibrated for soil water contents ranging from near the wilting point up to saturation.

2.3. The SIMDualKc Model

The soil water balance SIMDualKc model [35] uses a daily time step to compute the grass crop evapotranspiration (ET$_c$) using the dual crop coefficient approach [17,32]. This model has previously been applied to a variety of crops as referred to in the "Introduction", in particular to a *Leymus chinensis* (Trin.) Tzvel. grassland [34], however not submitted to cuttings.

The SIMDualKc model computes the daily soil water balance in the crop root zone as:

$$D_{r,i} = D_{r,\,i-1} - (P - RO)_i - I_i - CR_i + ET_{c,i} + DP_i \tag{1}$$

where $D_{r,i}$ and $D_{r,i-1}$ are the root zone depletion [mm] at the end of day i and i − 1, respectively, P_i is precipitation, RO_i is runoff, I_i is net irrigation depth, CR_i is capillary rise from the groundwater table, $ET_{c,i}$ is crop evapotranspiration, and DP_i is deep percolation, all referring to day i and expressed in mm. In the present application, the water table is quire deep and CR is null. The soil water balance refers to a soil in which total available water (TAW, mm) is:

$$TAW = 1000\,Z_r(FC - WP) \tag{2}$$

In the current application, with Z_r = 0.50 m, FC = 0.336 m^3 m^{-3} and WP = 0.088 m^3 m^{-3}, this results in TAW = 124 mm. The readily available soil water (RAW, mm) is the fraction of TAW that may be depleted without causing any water stress, and thus RAW = (1 − p) TAW. A constant value p = 0.55 was used for all cycles, thus resulting in RAW = 56 mm.

ET$_{c\,act}$ is computed as a function of the available soil water in the root zone (ASW, mm). If the depletion exceeds the depletion fraction for no stress (p), i.e., ASW < RAW, then the stress coefficient becomes K$_s$ < 1.0, otherwise K$_s$ = 1.0 [17]. Thus, in general, we have:

$$ET_{c\,act} = (K_e + K_s\,K_{cb})\,ET_o \tag{3}$$

where K$_e$ is the soil evaporation coefficient and K$_{cb}$ is the basal crop coefficient. As referred to before, K$_{cb\,act}$ = K$_s$ K$_{cb}$. The actual daily grass transpiration is, therefore, T$_{act}$ = K$_{cb\,act}$ ET$_o$ and the soil evaporation is E$_s$ = K$_e$ ET$_o$.

K$_e$ are daily computed through a daily water balance of the soil evaporative layer, whose thickness is (m), when knowing the total and readily evaporable water, respectively TEW (mm) and REW (mm). Considering the two-stage evaporation process, the first is energy limiting and the corresponding evaporable amount is REW; the second stage is water limiting and evaporation is linearly decreasing until TEW is depleted [17,32,52]. The thickness Z$_e$ = 0.15 m was adopted for the evaporation layer as commonly occurs for medium to heavy textured soils. TEW and REW are optimized during the process of model calibration. The water balance of the evaporation layer, that considers the referred evaporative characteristics of the soil, takes into consideration the fraction f$_c$ of ground shaded by the crop, which determines the fraction of the soil that is both exposed to solar radiation and wetted by rain or irrigation, and from where most of the soil evaporation originates, as well as effects of mulching in reducing the energy available at the soil surface [35].

The deep percolation DP is computed by the model using the respective parametric function proposed by Liu et al. [53], which is a time decay function that relates the soil water storage near saturation after the occurrence of a heavy rain or irrigation with the draining time until FC is attained. The values of the parameters (a_D, b_D) are optimized during model calibration. Runoff was estimated using the curve number approach following the USDA-ARS Hydrology Handbook [54].

The initial $K_{cb\,ini}$, $K_{cb\,gro}$ and $K_{cb\,cut}$ were estimated from the f_c and h values observed using the equation proposed by Allen and Pereira [4]:

$$K_{cb} = K_{c\,min} + K_d(K_{cb\,full} - K_{c\,min}) \tag{4}$$

where K_d is the density coefficient, $K_{cb\,full}$ is the estimated K_{cb} during peak plant growth for conditions having nearly full ground cover (or LAI > 3), and $K_{c\,min}$ is the minimum basal K_c for bare soil, with $K_{cb\,min}$ = 0.15 under typical agricultural conditions and for native vegetation when rainfall frequency is high. K_d can be estimated as a function of measured or estimated leaf area index LAI:

$$K_d = \left(1 - e^{[-0.7LAI]}\right) \tag{5}$$

or as a function of the fraction of ground covered by vegetation, as in the present study,

$$K_d = \min\left(1,\ M_L f_{c\,eff},\ f_{c\,eff}^{\left(\frac{1}{1+h}\right)}\right) \tag{6}$$

where $f_{c\,eff}$ is the effective fraction of ground covered or shaded by vegetation [0.01–1] near solar noon, M_L is a multiplier on $f_{c\,eff}$ describing the effect of canopy density on shading and on maximum relative ET per fraction of ground shaded [1.5–2.0], and h is the mean height of the vegetation in m. For low and dense crops such as grass, it may be assumed that $f_{c\,eff} = f_c$. The M_L multiplier on $f_{c\,eff}$ in Equation (6) imposes an upper limit on the relative magnitude of transpiration per unit of ground area as represented by $f_{c\,eff}$ and is expected to range from 1.5 to 2.0, depending on the canopy density and thickness [4]. The value for M_L can be adjusted to fit the specific vegetation.

The input data required by the SIMDualKc model consist of: (i) daily meteorological data of rainfall (mm), ET_0 (mm), minimum relative humidity (RH_{min}, %) and wind speed at 2 m height (u_2, m s^{-1}); (ii) the soil water content at FC and WP for all root zone soil layers; (iii) the soil water evaporation parameters Z_e (m), TEW and REW (mm); (iv) the deep percolation parameters; (v) crop heights (h, m); (vi) the effective rooting depth Z_r; (vi) the fraction of ground cover $f_{c\,ini}$, $f_{c\,gro}$, and $f_{c\,cut}$; (vii) the water depletion fraction for no stress, p; (viii) the irrigation dates and net irrigation depths applied; (ix) the soil wetted fraction by irrigation (f_w); and (x) the runoff curve number (CN).

In this application, because the ground is covered by plant litter, the importance of which in Tifton 85 bermudagrass fields is well known [55], the effect of plant litter on E_s was considered in modelling [35]. Litter, like organic mulches, reduces the energy available at the soil surface and, consequently, soil evaporation. The respective model inputs consisted of: the fraction of mulched soil of 1, low thickness of the mulch, and 3% reduction in E_s for each 10% of soil surface covered. In former applications of SIMDualKc to soils with organic mulch or crop residuals [56,57] a larger reduction of E_s was considered.

The standard K_{cb} values should refer to the minimum relative humidity RH_{min} = 45% and the average wind speed at 2 m height u_2 = 2 m s^{-1} [17]. They were obtained from the calibrated ones by adjusting them for climate using the climate adjustment equation [17] inversely:

$$K_{cb} = K_{cb\,calib} - [0.04\,(u_2 - 2) - 0.004\,(RH_{min} - 45)]\left(\frac{h}{3}\right)^{0.3} \tag{7}$$

where the K_{cb} are the standard values, $K_{cb\,calib}$ are those values obtained from calibration, u_2 and RH_{min} are the average values observed during the calibration, and h are the observed crop heights. The current adjustment refers to $K_{cb\,ini}$, $K_{cb\,gro}$ and $K_{cb\,cut}$, as well as to K_c.

2.4. Model Calibration and Validation and Goodness of Fitting Indicators

Model calibration and validation for the Tifton 85 bermudagrass were performed using independent data sets of the referred two years of observations. The calibration of the model was performed with data collected in the summer and autumn seasons of 2016. The calibration of SIMDualKc aimed at optimizing the basal crop coefficients ($K_{cb\,ini}$, $K_{cb\,gro}$ and $K_{cb\,cut}$), which was performed independently for the three cutting treatments because respective growth characteristics are different, the soil evaporation parameters (TEW, REW), the deep percolation parameters (a_D and b_D) and the runoff CN value. An iterative trial-and-error procedure was applied in order to minimize the deviations between the available soil water data observed and simulated by the model. The procedure described by Pereira et al. [39] was adopted. The trial and error was first applied to the $K_{cb\,ini}$, $K_{cb\,gro}$ and $K_{cb\,cut}$ relative to the treatment of CGDD 248 °C, and then interactively applied to the K_{cb} values, the soil evaporation TEW and REW, the deep percolation parameters and CN. Using the DP, soil evaporation and runoff parameters already calibrated for the treatment of CGDD 248 °C, which are common to all treatments, the trial and error procedure was in the following applied independently to the other cutting treatments for calibration of the respective K_{cb} values. The model validation consisted in applying the calibrated $K_{cb\,ini}$, $K_{cb\,gro}$ and $K_{cb\,cut}$, TEW and REW, DP parameters and CN to the remaining observed data relative to the spring seasons of 2015 and 2016 and the summer and autumn seasons of 2017.

The initial parameter values were estimated as follows: (1) the K_{cb} values were computed for each treatment with Equation (4) assuming $K_{c\,min} = 0.15$, $K_{cb\,full} = 0.95$, and with the density coefficient K_d computed with Equation (6) using the observed data given in Table 4; (2) the depletion fraction for no stress p was estimated from the values tabulated in FAO56 [17]; (3) TEW was computed from the difference (FC-0.5 WP) relative to the top soil layer of depth Z_e (0.15 m), and REW was estimated from the textural characteristics (Table 1) of that same layer [17,32]; (4) the DP parameters a_D and b_D were estimated from those proposed by Liu et al. [53] for moderately permeable soils; and (5) CN was obtained from tabulated values for grasses in moderately permeable soils [54].

A set of goodness of fit indicators were used to assess model fitting during calibration and to evaluate the results of validation. As analyzed previously in various SIMDualKc applications, these indicators [39,58,59] are the following:

(i)   The regression coefficient (b_0) of the linear regression forced to the origin relating the observed and model predicted values, respectively O_i and P_i (i = 1, 2, ... , n), where b_0 close to 1.0 indicates that the predicted values are statistically close to the observed ones.
(ii) The determination coefficient (R^2) of the linear regression between observed and predicted values, where a R^2 close to 1.0 indicates that most of the variation of the observed values is explained by the model.
(iii) The root mean square error (RMSE), which measures the overall differences between observed and predicted values

$$RMSE = \left[\frac{\sum_{i=1}^{n}(P_i - O_i)^2}{n}\right]^{0.5} \qquad (8)$$

which should be as small as possible and has the same units of the variable under analysis.
(iv) the normalized RMSE (NRMSE, %), ratio of RMSE to the mean value of the variable observations, which expresses the relative size of the estimation errors and which target is a small value, at least smaller than 10%.

(v) The average relative error (ARE, %), which express the relative size of estimated errors in alternative to NRMSE:

$$ \text{ARE} = \frac{100}{n} \sum_{i=1}^{n} \left| \frac{O_i - P_i}{O_i} \right| \tag{9} $$

and which target is a value as small as possible, generally smaller than 10%.

(vi) The percent bias of estimation, PBIAS (%), is an indicator that measures the average tendency of the simulated data to be larger or smaller than the correspondent observations and is given by:

$$ \text{PBIAS} = 100 \frac{\sum_{i=1}^{n}(O_i - P_i)}{\sum_{i=1}^{n} O_i} \tag{10} $$

Its optimal value is 0.0; thus, values near 0.0 indicate that model simulation is accurate, while positive or negative values indicate under- or over-estimation bias.

(vii) The modelling efficiency (EF, dimensionless), that indicates the relative magnitude of the variance of residuals of estimation compared to the measured data variance:

$$ \text{EF} = 1.0 - \frac{\sum_{i=1}^{n}(O_i - P_i)^2}{\sum_{i=1}^{n}(O_i - \overline{O})^2} \tag{11} $$

the target value of which is 1.0 when the variance of residuals is negligible relative to the variance of observations; EF values close to 0 or negative indicate that the observations mean is as good or better predictor than the model. Therefore, achieving a positive EF is a must.

3. Results and Discussion

3.1. Model Calibration and Validation

The initial values of parameters used with the SIMDualKc model to simulate the three treatments of frequency of cuttings of Tifton 85 bermudagrass are presented in Table 5.

Table 5. Initial and calibrated parameters of SIMDualKc relative to the three treatments of Tifton 85 bermudagrass having different frequency of cuttings.

Parameters	Symbols	Initial Values			Calibrated Values		
		CGDD of 124 °C	CGDD of 248 °C	CGDD of 372 °C	CGDD of 124 °C	CGDD of 248 °C	CGDD of 372 °C
Basal crop coefficients	$K_{cb\ ini}$	0.77	0.82	0.84	0.80	0.83	0.84
	$K_{cb\ gro}$	0.80	0.87	0.89	0.82	0.88	0.90
	$K_{cb\ cut}$	0.84	0.90	0.91	0.84	0.93	0.94
Depletion fraction	p		0.55			0.55	
Soil evaporation	Z_e (m)		0.15			0.15	
	REW (mm)		10			10	
	TEW (mm)		37			44	
Deep percolation	a_D		335			325	
	b_D		−0.017			−0.005	
Runoff	CN		70			74	

Z_e = Depth of the soil evaporation layer; REW = Readily evaporable water; TEW =Total evaporable water; a_D and b_D = parameters of the deep percolation equation [53]; CN = Curve number

The calibration of the SIMDualKc model through minimizing the differences between simulated and observed available soil water (Figure 2) enabled the calibrated parameters also listed in Table 5 to be obtained for the three cutting treatments considered. These parameters were later used for validation of the model for other observation periods, whose results are shown in Figure 3.

The goodness of fit indicators relative to the calibration and the validation are presented in Table 6. It can be observed that the regression coefficient b_0 is close to 1.0 for all sets of data used both in the calibration and validation for all treatments, thus indicating a trend for equality of simulated and observed values, thus no trend to over- or underestimation of the ASW. Consequently, the PBIAS are quite small, thus confirming no trends for over- or underestimation. The determination coefficients are above 0.80 for the calibration, indicating that there is a dispersion of pairs P_i-O_i around the 1:1 line, i.e., a fraction of less than 20% of cases cannot be explained by the model. An explanatory hypothesis is that the FDR sensors used, which were previously tested for conditions where wettings consisted of controlled irrigation applications and not intense rainfall events [56], have not been shown to be adequate to record quick changes in ASW when heavy rains occur. This can be observed in Figures 2 and 3 in cases when peak increases of ASW occurred. However, R^2 values are generally high and, in combination with high b_0 values, confirm the adequacy of model simulations.

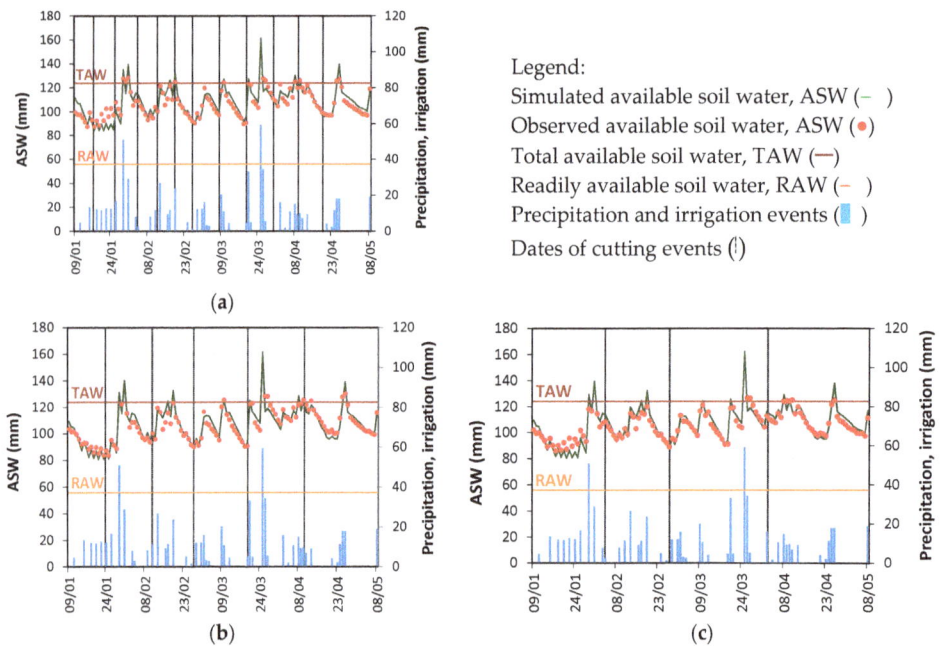

(a)

Legend:
Simulated available soil water, ASW (–)
Observed available soil water, ASW (•)
Total available soil water, TAW (—)
Readily available soil water, RAW (–)
Precipitation and irrigation events (▮)
Dates of cutting events (¦)

(b)   (c)

Figure 2. Simulated vs. observed available soil water during the Summer-Autumn of 2016 relative to the calibration of SIMDualKc model for treatments of CGDD of: (**a**) 124 °C, (**b**) 248 °C and (**c**) 372 °C.

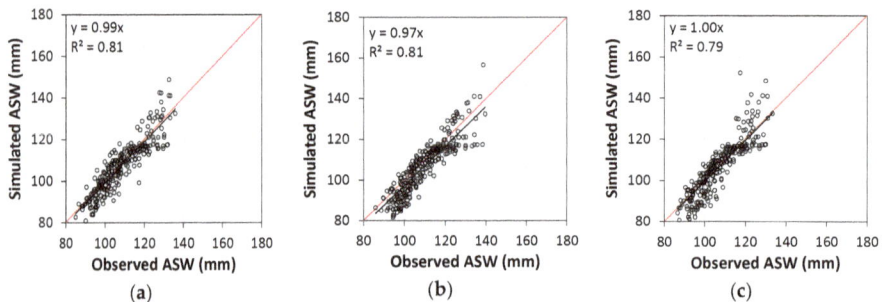

(a)   (b)   (c)

Figure 3. Simulated vs. observed available soil water (ASW) relative to the validation of SIMDualKc model for treatments of CGDD of: (**a**) 124 °C, (**b**) 248 °C and (**c**) 372 °C.

Errors of estimation are small (Table 6). On the one hand, the RMSE range from 4.2–5.2 mm at calibration and 5.0–7.2 mm at validation, corresponding to NRMSE in the range of 4.0–6.7% of observed ASW, as well as ARE also ranging from 3.2–5.2%, thus indicating good accuracy of model simulations. On the other hand, EF values were quite high for the calibration cases (0.76–0.86) and reasonably good for the validation simulations, which ranged from 0.42–0.79, therefore indicating that the variances of the residuals of estimation were much smaller than the variance of observations. Overall, the goodness of fit indicators point to the appropriateness of using SIMDualKc to simulate the soil water balance of Tifton 85 bermudagrass adopting the obtained calibrated parameters and, in particular, the adequacy of adopting a K_{cb} curve consisting of successive individual cutting K_{cb} lines designed with the respective $K_{cb\ ini}$, $K_{cb\ gro}$ and $K_{cb\ cut}$.

Table 6. Goodness-of-fit indicators relative to both the calibration and validation of all treatments.

Cuttings Interval	Period			Goodness-of-Fit Indicators				
		b_0	R^2	RMSE (mm)	NRMSE (%)	ARE (%)	PBIAS (%)	EF
Calibration								
CGDD of 124 °C	Summer–Autumn 2016	1.01	0.81	5.2	4.9	4.0	−0.9	0.76
CGDD of 248 °C	Summer–Autumn 2016	1.00	0.87	4.2	4.0	3.2	−0.3	0.86
CGDD of 372 °C	Summer–Autumn 2016	1.01	0.86	4.8	4.5	3.6	−0.7	0.77
Validation								
CGDD of 124 °C	Spring 2015	1.00	0.83	5.2	4.8	3.4	0.2	0.79
	Spring 2016	0.98	0.71	5.3	5.2	3.4	1.6	0.61
	Summer–Autumn 2017	0.99	0.80	6.0	5.4	4.5	0.8	0.75
CGDD of 248 °C	Spring 2015	0.97	0.76	6.7	5.9	5.2	3.0	0.47
	Spring 2016	0.97	0.85	5.0	4.9	4.0	2.7	0.71
	Summer–Autumn 2017	0.97	0.78	7.2	6.4	5.0	2.7	0.62
CGDD of 372 °C	Spring 2015	1.01	0.75	7.2	6.7	4.0	−1.3	0.52
	Spring 2016	0.98	0.67	6.5	6.3	4.7	2.1	0.42
	Summer–Autumn 2017	1.00	0.82	5.8	5.4	4.4	0.6	0.72

RMSE = Root mean square error; NRMSE = Normalized RMSE; ARE = Average relative error; PBIAS = Percent bias of estimation; EF = Modelling efficiency

The proximity of initial and calibrated K_{cb} values result from the goodness of Equation (4), which computes K_{cb} from the fraction of ground cover and crop height, as well as from $K_{cb\ full}$. In this application h and f_c were observed while values for $K_{cb\ full}$ were estimated from the K_{cb} values tabulated in FAO56 [17]. These results demonstrate that the Allen and Pereira equation 4 [4] is highly valuable to estimate K_{cb} from simple field observations.

An alternative K_{cb} curve with a non-variable K_{cb} was also assessed for the treatment with CGDD of 124 °C, i.e., with highly frequent cuttings. Results for crop height and fraction of cover of this treatment (Table 4) indicate small variation of both h and f_c for each cutting cycle, which do not imply quite distinctive K_{cb} values. Assuming this single $K_{cb\ sing} = K_{cb\ gro}$ (0.80), this results in goodness of fit indicators (Table 7) similar to those discussed above (Table 6). It may therefore be assumed that when h and f_c of bermudagrass vary little, a simple solution with a single K_{cb} value may be used. However, to better represent the dynamics of evapotranspiration and crop transpiration [17], the best solution is to use a 3-value K_{cb} curve for each cutting cycle.

Table 7. Goodness-of-fit indicators for the treatment relative to frequent cuttings with CGDD of 124 °C when adopting a single $K_{cb} = 0.80$.

	Period			Goodness-of-Fit Indicators				
		b_0	R^2	RMSE (mm)	NRMSE (%)	ARE (%)	PBIAS (%)	EF
Calibration	Summer–Autumn 2016	1.00	0.79	5.5	5.2	4.2	−0.4	0.73
Validation	Spring 2015	0.99	0.85	4.8	4.4	3.4	1.1	0.82
	Spring 2016	0.98	0.69	7.0	6.8	4.3	2.3	0.43
	Summer–Autumn 2017	0.99	0.80	5.9	5.4	4.4	1.0	0.75

3.2. Crop Coefficients

The K_{cb}, K_e and K_c (=$K_{cb} + K_e$) curves relative to all three treatments and the simulations of the Summer–Autumn periods of both years are shown in Figure 4. Also included is information on the frequency of cuttings. The K_{cb} curves show a regular variation for every cutting cycle, increasing from a minimum after each cutting up to a maximum just before it occurs. Apparently, a three-segments curve adapted well to each cutting cycle.

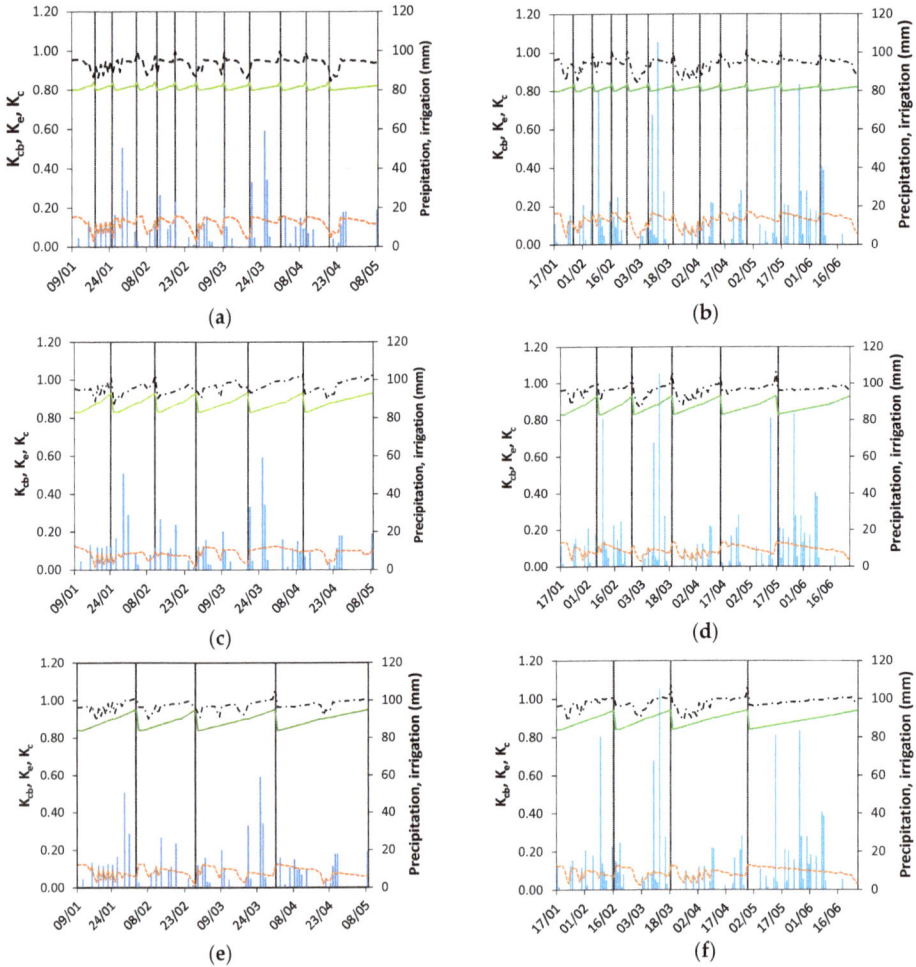

Figure 4. Basal crop coefficient (K_{cb}, —), soil evaporation coefficient (K_e, - - -) and average crop coefficient (K_c, — - —) curves of Tifton 85 bermudagrass and all cutting treatments with CGDD of 124 °C (**a,b**), 248 °C (**c,d**) and 372 °C (**e,f**) relative to the Summer–Autumn periods of 2016 (**a,c,e**) and 2017 (**b,d,f**).

The maximum K_{cb} values represented in Figure 4 are slightly smaller than those tabulated for bermudagrass in FAO56 [17]. This is likely due to the humid climate conditions prevailing at time of experiments, that were likely influenced by the El Niño Southern Oscillation (ENSO) in both years, when very abundant rainfall occurred, together with reduced air temperature and solar radiation, and high air humidity (Figure 1). These humid climatic conditions were less favorable to crop transpiration, thus lowering the K_{cb}. The standard K_{cb} summarized in Table 8 were computed by adjusting the calibrated values to climate with Equation (7). This adjustment was mainly due to the high RH_{min} observed during both years, consequently resulting in standard K_{cb} higher than the calibrated K_{cb} values of Figure 4. Therefore, it may be observed that the standard K_{cb} curves obtained in this study are comparable to those tabulated in FAO56 [17], which allows us to conclude that the standard K_{cb} values obtained in the present study (Table 8) may be transferable to other locations when adjusted to the local climate and considering the locally adopted frequency of cuttings. This assumption may be confirmed by observing that K_{cb} from this study are similar to those reported by Greenwood et al. [33] for a ryegrass/clover pasture and are larger than those computed by Wu et al. [34] for a groundwater-dependent grassland where *Leymus chinensis* (Trin.) Tzvel. is the dominant grass.

Table 8. Standard crop coefficients for Tifton 85 bermudagrass as dependent on the frequency of cuttings.

	Treatment		
	CGDD of 124 °C	CGDD of 248 °C	CGDD of 372 °C
$K_{cb\ ini}$	0.83	0.86	0.87
$K_{cb\ gro}$	0.85	0.91	0.93
$K_{cb\ cut}$	0.87	0.96	0.97
K_c	0.96	0.99	1.00

The K_e curves show a strong dependency upon the wetting events, well apparent in Figure 4. K_e are higher for the treatment with more frequent cuttings (CGDD of 124 °C) because the ground cover was smaller than for other treatments (Table 4), thus affecting less the energy available at the ground for soil evaporation. The K_c curves, representing the daily combination of K_{cb} and K_e, are more flat and irregular than the K_{cb} curves (Figure 4). This is likely due to the nearly constant values of K_e resulting from the abundant and frequent rains that kept the soil evaporation layer wet most of time for all three cutting treatments. The standard K_c values reported in Table 8 for this study with Tifton 85 bermudagrass are larger than those reported by Wherley et al. [23] for bermudagrass and by Graham et al. [48] for ryegrass, and slightly larger than those reported by Neal et al. [24] for ryegrass. These authors also adopted a single averaged K_c. Considering the behavior of K_c in this study (Figure 4), it may be appropriate to adopt a single K_c value for Tifton 85 bermudagrass when a dual K_c approach is not used.

3.3. Soil Water Balance and Transpiration and Soil Evaporation Ratios

The results of the soil water balance relative to all simulations performed during calibration and validation of the model are summarized in Table 9. Note first the unusual fact that the sum of runoff and deep percolation exceeds ET_c in the Spring of 2015 and during the Summer–Autumn of 2017 because rainfall was likely impacted by ENSO as previously stated. RO was particularly high in 2017 as well as DP. The latter was larger than irrigation in all the periods considered which leads us to realize that using irrigation was likely a wrong option, but the exceptional rainfall observed was not predicted at time of planning and starting the experiment. Differences in RO and DP among treatments are not notable. However, as expected from the differences in terms of f_c and K_{cb}, crop evapotranspiration increases when the frequency of cuttings is smaller, and the same happens with T_c. By contrast, soil evaporation is higher when the cutting frequency is also greater.

When analysing the evaporation and transpiration ratios (Table 10) it is evident that the E_s/ET_c ratio is much smaller than the T_c/ET_c ratio, particularly for the treatments with CGDD of 248 and 372 °C. Low values of the evaporation ratio relate to the high ground cover fraction f_c and to the effects of plant litter, which limit the energy available at the soil surface. This effect was reported by Wang and Yamanaka [47]. High f_c also indicates favorable conditions for plant transpiration. Low values for the E_s/ET_c ratio were reported by Greenwood et al. [33] and Wu et al. [34], while much higher values were reported for various meadows in northern China [28,44]. A small E_s/ET_c ratio of 13% was also referred to for native grasslands, however for arid conditions [49]. High transpiration ratios, but smaller than those observed in this study, were reported by various authors [15,46,50], namely influenced by soil texture [45]. Transpiration ratios for tropical and temperate grasslands of $62 \pm 19\%$ and $57 \pm 19\%$, respectively, were reported by Schlesinger and Jasechko [60], therefore values with upper limits that are smaller than those observed in this study. This behaviour may indicate that Tifton 85 grasslands are efficient in terms of beneficial water use [61] since results show a very high transpiration ratio of 90% when cuttings are not very frequent.

Table 9. Soil water balance terms (mm) for all cutting treatments and all observation periods.

Cuttings Treatments	Period	P	I	ΔSW	RO	DP	ET_c	E_s	T_c
CGDD of 124 °C	Spring, 2015	484	22	7	58	221	234	32	202
	Summer–Autumn, 2016	529	108	−3	17	206	411	53	358
	Spring, 2016	320	123	−20	16	134	273	37	236
	Summer–Autumn, 2017	1076	61	28	192	556	417	55	363
CGDD of 248 °C	Spring, 2015	484	22	21	58	229	240	23	217
	Summer–Autumn, 2016	529	108	−4	16	196	421	37	384
	Spring, 2016	320	123	−24	16	119	284	25	259
	Summer–Autumn, 2017	1076	61	26	192	547	428	39	389
CGDD of 372 °C	Spring, 2015	484	22	10	58	217	241	22	219
	Summer–Autumn, 2016	529	108	−5	15	192	425	36	389
	Spring, 2016	320	123	−20	16	125	282	25	257
	Summer–Autumn, 2017	1076	61	18	192	529	434	39	395

P = precipitation, I = irrigation, ΔSW = variation in stored soil water, DP = deep percolation, RO = runoff; E_s = soil evaporation, T_c = crop transpiration, ET_c = crop evapotranspiration.

Table 10. Crop evapotranspiration (ET_c, mm) and evaporation and transpiration ratios (E_s/ET_c and T_c/ET_c, %) for all cutting treatments and observation periods.

Cutting Treatments	Period	ET_c (mm)	E_s/ET_c (%)	T_c/ET_c (%)
CGDD of 124 °C	Spring, 2015	234	14	86
	Summer–Autumn, 2016	411	13	87
	Spring, 2016	273	14	86
	Summer–Autumn, 2017	417	13	87
CGDD of 248 °C	Spring, 2015	240	10	90
	Summer–Autumn, 2016	421	9	91
	Spring, 2016	284	9	91
	Summer–Autumn, 2017	428	9	91
CGDD of 372 °C	Spring, 2015	241	9	91
	Summer–Autumn, 2016	425	8	92
	Spring, 2016	282	9	91
	Summer–Autumn, 2017	434	9	91

4. Conclusions

The current study is a first application of the FAO dual crop coefficient approach to assess evapotranspiration and water use of a bermudagrass, more precisely the Tifton 85. It was performed using two years of field data relative to three cutting treatments where intervals between cuttings were defined by CGDD of 124 °C, 248 °C and 372 °C. These independent data sets were used to calibrate and validate the water balance model SIMDualKc, which allowed the K_{cb} and K_e curves for Tifton 85 to be obtained when managed with those three cutting intervals. Data of Summer and Autumn of 2016 were used for calibration and data for Spring 2015 and 2016, and Summer and Autumn of 2017 were used for validation. The procedure used led to quite small errors of estimation of the available soil water throughout both years, which allowed us to assume that the calibrated K_{cb} values were accurately estimated and may be considered as standard for the three cutting frequencies studied after adjustments with Equation (7). It is important to note that the estimation of the initial K_{cb} values from the crop cover and height has shown quite small differences to the calibrated values. Moreover, the Allen and Pereira [4] Equation (4) was revealed to be very accurate in estimating K_{cb} from h and f_c provided that $K_{cb\,full}$ is well estimated. The operational use of this Equation (4) is therefore recommended.

The K_{cb} curves consist of a series of K_{cb} curves relative to each cutting cycle, each one constructed with three linear segments. This approach follows the one proposed by the FAO56 guidelines [17] and differs from the commonly used single K_c curve. It was revealed to be accurate in describing crop transpiration and providing for the accuracy of soil water dynamics computed through the calibration and validation processes. However, for the case of very frequent cuttings (CGDD of 124 °C) there was no advantage over a single, averaged K_{cb}.

The K_e curve reflects the abundant and stormy rains that occurred in both years of field experiments, which made the soil evaporation layer wet most of the time. Thus, the K_e curve varied little throughout the periods under analysis. However, the soil water evaporation was mitigated due the organic mulch effect of the plant litter covering the ground, which reduced the energy available at the soil surface, thus reducing K_e and E_s. Comparing K_e for the three cutting treatments, this resulted in a larger value for that with very frequent cuttings (CGDD of 124 °C) because the ground cover was smaller than for cuttings with large intervals, thus giving more time for the crop to grow and crop density to increase during each cutting cycle. Due to the nearly flat form of the K_e curve, the K_c curve and the sum of the K_e and K_{cb} curves do not reflect the aggregation of individual K_c curves relative to each cycle of cutting. Therefore, by contrast with the K_{cb} curves, a single K_c value is appropriate, however specific for each cutting treatment.

Results for the soil water balance are marked by the enormous amount of rain observed during these two years, likely due to the impacts of ENSO. Thus, the amount of runoff and, mainly, deep percolation exceeded crop ET. Thus, the option of irrigating to avoid any water stress was shown to be inappropriate despite not being prejudicial to the experiments. However, the large number of rainy days and the large amount of rain were likely associated with reduced solar radiation and temperature, which could have contributed to reduce transpiration and the K_{cb} values. However, the latter, as well as the K_c values, are larger than most of K_{cb} and K_c values reported in literature, which support the assumption that K_{cb} values may be considered standard and transferable to other locations after appropriate adjustments.

The soil evaporation fraction (E_s/ET_c) for all cases was small, near 13% in the case of frequent cuttings and about 9% when cuttings were less frequent. These values could slightly decrease if soil wettings were less frequent. These results indicate that beneficial consumptive water use by the crop is high, with the transpiration ratio near 90%. These results agree well with those reported in literature relative to well-managed grasslands, particularly tropical ones. This may indicate that the Tifton 85 bermudagrass has the potential to contribute to the sustainability of the Pampa biome in southern Brazil. Adopting a median cuttings frequency (CGDD of 248 °C) is likely the most favorable. However, more studies are required, mainly relative to herbage production and water productivity, which are expected to be undertaken based upon the field data used in the current study.

Author Contributions: Reimar Carlesso and Geraldo J. Rodrigues conceived and designed the experiments. Geraldo J. Rodrigues and Paula Severo O. performed the field experiments under the supervision of Reimar Carlesso. Geraldo J. Rodrigues and Mirta T. Petry handled the data with advice of Reimar Carlesso. Paula Paredes performed modelling with the advice of Luis S. Pereira, and writing was undertaken by Luis S. Pereira with the contributions of Mirta T. Petry and Paula Paredes.

Funding: This research was funded by CAPES/CNPq Post-Graduation Cooperative Program in Agricultural Engineering, Brazil, grant number 88881.030480/2013-01; Fundação para a Ciência e a Tecnologia, Portugal through the research unit LEAF-Linking Landscape, Environment, Agriculture and Food (UID/AGR/04129/2013) and the first author post-doc fellowship SFRH/BPD/102478/2014.

Conflicts of Interest: The authors declare no conflict of interest.

References

1. Roesch, L.F.W.; Vieira, F.C.B.; Pereira, V.A.; Schünemann, A.L.; Teixeira, I.F.; Senna, A.J.T.; Stefenon, V.M. The Brazilian Pampa: A fragile biome. *Diversity* **2009**, *1*, 182–198. [CrossRef]
2. Liu, K.; Sollenberger, L.E.; Newman, Y.C.; Vendramini, J.M.B.; Interrante, S.M.; White-Leech, R. Grazing management effects on productivity, nutritive value, and persistence of 'Tifton 85' Bermudagrass. *Crop Sci.* **2011**, *51*, 353–360. [CrossRef]
3. Silva, V.J.; Pedreira, C.G.S.; Sollenberger, L.E.; Carvalho, M.S.S.; Tonato, F.; Basto, D.C. Seasonal herbage accumulation and nutritive value of irrigated 'Tifton 85', Jiggs, and Vaquero Bermudagrasses in response to harvest frequency. *Crop Sci.* **2015**, *55*, 1–9. [CrossRef]
4. Allen, R.G.; Pereira, L.S. Estimating crop coefficients from fraction of ground cover and height. *Irrig. Sci.* **2009**, *28*, 17–34. [CrossRef]
5. Monteith, J.L. Evaporation and environment. In *The State and Movement of Water in Living Organisms, Proceedings of the 19th Symposium of the Society for Experimental Biology, Swansea, 1964*; Cambridge University Press: Cambridge, UK, 1965; pp. 205–234.
6. Priestley, C.H.B.; Taylor, R.J. On the assessment of surface heat flux and evaporation using large-scale parameters. *Mon. Weather Rev.* **1972**, *100*, 81–92. [CrossRef]
7. Li, S.-G.; Lai, C.-T.; Lee, G.; Shimoda, S.; Yokoyama, T.; Higuchi, A.; Oikawa, T. Evapotranspiration from a wet temperate grassland and its sensitivity to microenvironmental variables. *Hydrol. Process.* **2005**, *19*, 517–532. [CrossRef]
8. Ryu, Y.; Baldocchi, D.D.; Ma, S.; Hehn, T. Interannual variability of evapotranspiration and energy exchange over an annual grassland in California. *J. Geophys. Res.* **2008**, *113*, D09104. [CrossRef]
9. Li, J.; Jiang, S.; Wang, B.; Jiang, W.-W.; Tang, Y.-H.; Du, M.-Y.; Gu, S. Evapotranspiration and its energy exchange in alpine meadow ecosystem on the Qinghai-Tibetan Plateau. *J. Integr. Agric.* **2013**, *12*, 1396–1401. [CrossRef]
10. Savage, M.J.; Odhiambo, G.O.; Mengistu, M.G.; Everson, C.S.; Jarmain, C. Measurement of grassland evaporation using a surface-layer scintillometer. *Water SA* **2010**, *36*, 1–8. [CrossRef]
11. Alfieri, J.G.; Xiao, X.; Niyogi, D.; Pielke, R.A., Sr.; Chen, F.; LeMone, M.A. Satellite-based modeling of transpiration from the grasslands in the Southern Great Plains, USA. *Glob. Planet. Chang.* **2009**, *67*, 78–86. [CrossRef]
12. Courault, D.; Hadria, R.; Ruget, F.; Olioso, A.; Duchemin, B.; Hagolle, O.; Dedieu, G. Combined use of FORMOSAT-2 images with a crop model for biomass and water monitoring of permanent grassland in Mediterranean region. *Hydrol. Earth Syst. Sci.* **2010**, *14*, 1731–1744. [CrossRef]
13. Krishnan, P.; Meyers, T.P.; Scott, R.L.; Kennedy, L.; Heuer, M. Energy exchange and evapotranspiration over two temperate semi-arid grasslands in North America. *Agric. Forest Meteorol.* **2012**, *153*, 31–44. [CrossRef]
14. Kurc, S.A.; Small, E.E. Dynamics of evapotranspiration in semiarid grassland and shrubland ecosystems during the summer monsoon season, central New Mexico. *Water Resour. Res.* **2004**, *40*, W09305. [CrossRef]
15. Moran, M.S.; Scott, R.L.; Keefer, T.O.; Emmerich, W.E.; Hernandez, M.; Nearing, G.S.; Paige, G.B.; Cosh, M.H.; O'Neill, P.E. Partitioning evapotranspiration in semiarid grassland and shrubland ecosystems using time series of soil surface temperature. *Agric. Forest Meteorol.* **2009**, *149*, 59–72. [CrossRef]
16. Zha, T.; Barr, A.G.; van der Kamp, G.; Black, T.A.; McCaughey, J.H.; Flanagan, L.B. Interannual variation of evapotranspiration from forest and grassland ecosystems in western Canada in relation to drought. *Agric. For. Meteorol.* **2010**, *150*, 1476–1484. [CrossRef]
17. Allen, R.G.; Pereira, L.S.; Raes, D.; Smith, M. *Crop Evapotranspiration—Guidelines for Computing Crop Water Requirements*; FAO Irrigation and Drainage Paper 56; FAO: Rome, Italy, 1998; p. 300.
18. Pequeno, D.N.L.; Pedreira, C.G.S.; Sollenberger, L.E.; de Faria, A.F.G.; Silva, L.S. Forage accumulation and nutritive value of brachiariagrasses and Tifton 85 Bermudagrass as affected by harvest frequency and irrigation. *Agron. J.* **2015**, *107*, 1741–1749. [CrossRef]
19. Thornthwaite, C.W. An approach toward a rational classification of climate. *Geogr. Rev.* **1948**, *38*, 55–94. [CrossRef]
20. Zhao, Y.; Peth, S.; Horn, R.; Krümmelbein, J.; Ketzer, B.; Gao, Y.; Doerner, J.; Bernhofer, C.; Peng, X. Modeling grazing effects on coupled water and heat fluxes in Inner Mongolia grassland. *Soil Till. Res.* **2010**, *109*, 75–86. [CrossRef]

21. Qassim, A.; Dunin, F.; Bethune, M. Water balance of centre pivot irrigated pasture in northern Victoria, Australia. *Agric. Water Manag.* **2008**, *95*, 566–574. [CrossRef]

22. Sumner, D.M.; Jacobs, J.M. Utility of Penman–Monteith, Priestley–Taylor, reference evapotranspiration, and pan evaporation methods to estimate pasture evapotranspiration. *J. Hydrol.* **2005**, *308*, 81–104. [CrossRef]

23. Wherley, B.; Dukes, M.D.; Cathey, S.; Miller, G.; Sinclair, T. Consumptive water use and crop coefficients for warm-season turfgrass species in the Southeastern United States. *Agric. Water Manag.* **2015**, *156*, 10–18. [CrossRef]

24. Neal, J.S.; Fulkerson, W.J.; Hacker, R.B. Differences in water use efficiency among annual forages used by the dairy industry under optimum and deficit irrigation. *Agric. Water Manag.* **2011**, *98*, 759–774. [CrossRef]

25. Zhang, F.; Zhou, G.; Wang, Y.; Yang, F.; Nilsson, C. Evapotranspiration and crop coefficient for a temperate desert steppe ecosystem using eddy covariance in Inner Mongolia, China. *Hydrol. Process.* **2012**, *26*, 379–386. [CrossRef]

26. Pronger, J.; Campbell, D.I.; Clearwater, M.J.; Rutledge, S.; Wall, A.M.; Schipper, L.A. Low spatial and inter-annual variability of evaporation from a year-round intensively grazed temperate pasture system. *Agric. Ecosyst. Environ.* **2016**, *232*, 46–58. [CrossRef]

27. Cancela, J.J.; Cuesta, T.S.; Neira, X.X.; Pereira, L.S. Modelling for improved irrigation water management in a temperate region of Northern Spain. *Biosyst. Eng.* **2006**, *94*, 151–163. [CrossRef]

28. Pereira, L.S.; Teodoro, P.R.; Rodrigues, P.N.; Teixeira, J.L. Irrigation scheduling simulation: The model ISAREG. In *Tools for Drought Mitigation in Mediterranean Regions*; Rossi, G., Cancelliere, A., Pereira, L.S., Oweis, T., Shatanawi, M., Zairi, A., Eds.; Kluwer Academic Press: Dordrecht, The Netherlands, 2003; pp. 161–180.

29. Pôças, I.; Cunha, M.; Pereira, L.S.; Allen, R.G. Using remote sensing energy balance and evapotranspiration to characterize montane landscape vegetation with focus on grass and pasture lands. *Int. J. Appl. Earth Observ.* **2013**, *21*, 159–172. [CrossRef]

30. Pakparvar, M.; Cornelis, W.; Pereira, L.S.; Gabriels, D.; Hafeez, M.; Hosseinimarandi, H.; Edraki, M.; Kowsar, S.A. Remote sensing estimation of actual evapotranspiration and crop coefficients for a multiple land use arid landscape of southern Iran with limited available data. *J. Hydroinform.* **2014**, *16*, 1441–1460. [CrossRef]

31. Jia, X.; Dukes, M.D.; Jacobs, J.M. Bahiagrass crop coefficients from eddy correlation measurements in central Florida. *Irrig. Sci.* **2009**, *28*, 5–15. [CrossRef]

32. Allen, R.G.; Pereira, L.S.; Smith, M.; Raes, D.; Wright, J.L. FAO-56 Dual crop coefficient method for estimating evaporation from soil and application extensions. *J. Irrig. Drain. Eng.* **2005**, *131*, 2–13. [CrossRef]

33. Greenwood, K.L.; Lawson, A.R.; Kelly, K.B. The water balance of irrigated forages in northern Victoria, Australia. *Agric. Water Manag.* **2009**, *96*, 847–858. [CrossRef]

34. Wu, Y.; Liu, T.; Paredes, P.; Duan, L.; Wang, H.; Wang, T.; Pereira, L.S. Ecohydrology of groundwater-dependent grasslands of the semi-arid Horqin sandy land of inner Mongolia focusing on evapotranspiration partition. *Ecohydrology* **2016**, *9*, 1052–1067. [CrossRef]

35. Rosa, R.D.; Paredes, P.; Rodrigues, G.C.; Alves, I.; Fernando, R.M.; Pereira, L.S.; Allen, R.G. Implementing the dual crop coefficient approach in interactive software. 1. Background and computational strategy. *Agric. Water Manag.* **2012**, *103*, 8–24. [CrossRef]

36. Krauß, M.; Kraatz, S.; Drastig, K.; Prochnow, A. The influence of dairy management strategies on water productivity of milk production. *Agric. Water Manag.* **2015**, *147*, 175–186. [CrossRef]

37. Paredes, P.; Rodrigues, G.C.; Alves, I.; Pereira, L.S. Partitioning evapotranspiration, yield prediction and economic returns of maize under various irrigation management strategies. *Agric. Water Manag.* **2014**, *135*, 27–39; Corrigendum in *Agric. Water Manag.* **2014**, *141*, 84. [CrossRef]

38. Zhao, N.; Liu, Y.; Cai, J.; Paredes, P.; Rosa, R.D.; Pereira, L.S. Dual crop coefficient modelling applied to the winter wheat–summer maize crop sequence in North China Plain: Basal crop coefficients and soil evaporation component. *Agric. Water Manag.* **2013**, *117*, 93–105. [CrossRef]

39. Pereira, L.S.; Paredes, P.; Rodrigues, G.C.; Neves, M. Modeling barley water use and evapotranspiration partitioning in two contrasting rainfall years. Assessing SIMDualKc and AquaCrop models. *Agric. Water Manag.* **2015**, *159*, 239–254. [CrossRef]

40. Wei, Z.; Paredes, P.; Liu, Y.; Chi, W.-W.; Pereira, L.S. Modelling transpiration, soil evaporation and yield prediction of soybean in North China Plain. *Agric. Water Manag.* **2015**, *147*, 43–53. [CrossRef]

41. Paço, T.A.; Ferreira, M.I.; Rosa, R.D.; Paredes, P.; Rodrigues, G.C.; Conceição, N.; Pacheco, C.A.; Pereira, L.S. The dual crop coefficient approach using a density factor to simulate the evapotranspiration of a peach orchard: SIMDualKc model vs. eddy covariance measurements. *Irrig. Sci.* **2012**, *30*, 115–126. [CrossRef]

42. Paço, T.A.; Pôças, I.; Cunha, M.; Silvestre, J.C.; Santos, F.L.; Paredes, P.; Pereira, L.S. Evapotranspiration and crop coefficients for a super intensive olive orchard. An application of SIMDualKc and METRIC models using ground and satellite observations. *J. Hydrol.* **2014**, *519*, 2067–2080. [CrossRef]

43. Shuttleworth, W.J.; Wallace, J.S. Evaporation from sparse crops-an energy combination theory. *Q. J. R. Meteorol. Soc.* **1985**, *111*, 839–855. [CrossRef]

44. Hu, Z.; Yu, G.; Zhou, Y.; Sun, X.; Li, Y.; Shi, P.; Wang, Y.; Song, X.; Zheng, Z.; Zhang, L.; Li, S. Partitioning of evapotranspiration and its controls in four grassland ecosystems: Application of a two-source model. *Agric. For. Meteorol.* **2009**, *149*, 1410–1420. [CrossRef]

45. Kochendorfer, J.P.; Ramírez, J.A. Modeling the monthly mean soil-water balance with a statistical-dynamical ecohydrology model as coupled to a two-component canopy model. *Hydrol. Earth Syst. Sci.* **2010**, *14*, 2099–2120. [CrossRef]

46. Huang, X.; Hao, Y.; Wang, Y.; Cui, X.; Mo, X.; Zhou, X. Partitioning of evapotranspiration and its relation to carbon dioxide fluxes in Inner Mongolia steppe. *J. Arid Environ.* **2010**, *74*, 1616–1623. [CrossRef]

47. Wang, P.; Yamanaka, T. Application of a two-source model for partitioning evapotranspiration and assessing its controls in temperate grasslands in central Japan. *Ecohydrology* **2014**, *7*, 345–353. [CrossRef]

48. Graham, S.L.; Kochendorfer, J.; McMillan, A.M.S.; Duncan, M.J.; Srinivasan, M.S.; Hertzog, G. Effects of agricultural management on measurements, prediction, and partitioning of evapotranspiration in irrigated grasslands. *Agric. Water Manag.* **2016**, *177*, 340–347. [CrossRef]

49. Ferretti, D.F.; Pendall, E.; Morgan, J.A.; Nelson, J.A.; LeCain, D.; Mosier, A.R. Partitioning evapotranspiration fluxes from a Colorado grassland using stable isotopes: Seasonal variations and ecosystem implications of elevated atmospheric CO_2. *Plant Soil* **2003**, *254*, 291–303. [CrossRef]

50. Hu, Z.; Wen, X.; Sun, X.; Li, L.; Yu, G.; Lee, X.; Li, S. Partitioning of evapotranspiration through oxygen isotopic measurements of water pools and fluxes in a temperate grassland. *J. Geophys. Res. Biogeosci.* **2014**, *119*, 358–371. [CrossRef]

51. Kottek, M.; Grieser, J.; Beck, C.; Rudolf, B.; Rubel, F. World Map of the Köppen-Geiger climate classification updated. *Meteorol. Z.* **2006**, *15*, 259–263. [CrossRef]

52. Ritchie, J.T. Model for predicting evaporation from a row crop with incomplete cover. *Water Resour. Res.* **1972**, *8*, 1204–1213. [CrossRef]

53. Liu, Y.; Pereira, L.S.; Fernando, R.M. Fluxes through the bottom boundary of the root zone in silty soils: Parametric approaches to estimate groundwater contribution and percolation. *Agric. Water Manag.* **2006**, *84*, 27–40. [CrossRef]

54. Hydrologic Soil-Cover Complexes. In *National Engineering Handbook Hydrology*; USDA-NRCS: Washington, DC, USA, 2004. Available online: https://www.wcc.nrcs.usda.gov/ftpref/wntsc/H&H/NEHhydrology/ch9.pdf (accessed on 14 March 2018).

55. Liu, K.; Sollenberger, L.E.; Silveira, M.L.; Newman, Y.C.; Vendramini, J.M.B. Grazing intensity and nitrogen fertilization affect litter responses in 'Tifton 85' Bermudagrass pastures: I. Mass, deposition rate, and chemical composition. *Agron. J.* **2011**, *103*, 156–162. [CrossRef]

56. Martins, J.D.; Rodrigues, G.C.; Paredes, P.; Carlesso, R.; Oliveira, Z.B.; Knies, A.E.; Petry, M.T.; Pereira, L.S. Dual crop coefficients for maize in southern Brazil: Model testing for sprinkler and drip irrigation and mulched soil. *Biosyst. Eng.* **2013**, *115*, 291–310. [CrossRef]

57. Fandiño, M.; Olmedo, J.L.; Martínez, E.M.; Valladares, J.; Paredes, P.; Rey, B.J.; Mota, M.; Cancela, J.J.; Pereira, L.S. Assessing and modelling water use and the partition of evapotranspiration of irrigated hop (*Humulus lupulus*), and relations of transpiration with hops yield and alpha-acids. *Ind. Crops Prod.* **2015**, *77*, 204–217. [CrossRef]

58. Moriasi, D.N.; Arnold, J.G.; Van Liew, M.W.; Bingner, R.L.; Harmel, R.D.; Veith, T.L. Model evaluation guidelines for systematic quantification of accuracy in watershed simulations. *Trans. ASABE* **2007**, *50*, 885–900. [CrossRef]

59. Nash, J.E.; Sutcliffe, J.V. River flow forecasting through conceptual models: Part 1. A discussion of principles. *J. Hydrol.* **1970**, *10*, 282–290. [CrossRef]

60. Schlesinger, W.H.; Jasechko, S. Transpiration in the global water cycle. *Agric. For. Meteorol.* **2014**, *189–190*, 115–117. [CrossRef]

61. Pereira, L.S.; Cordery, I.; Iacovides, I. Improved indicators of water use performance and productivity for sustainable water conservation and saving. *Agric. Water Manag.* **2012**, *108*, 39–51. [CrossRef]

water

MDPI

Article

Estimating Evapotranspiration of Processing Tomato under Plastic Mulch Using the SIMDualKc Model

Huimeng Zhang [1,2], Guanhua Huang [1,2,*], Xu Xu [1,2], Yunwu Xiong [1,2] and Quanzhong Huang [1,2]

[1] Center for Agricultural Water Research, China Agricultural University, Beijing 100083, China; zhanghmer@163.com (H.Z.); xushengwu@cau.edu.cn (X.X.); yxiong@cau.edu.cn (Y.X.); huangqzh@cau.edu.cn (Q.H.)

[2] Chinese-Israeli International Center for Research and Training in Agriculture, China Agricultural University, Beijing 100083, China

*   Correspondence: ghuang@cau.edu.cn; Tel.: +86-10-6273-7138

Received: 22 April 2018; Accepted: 10 August 2018; Published: 16 August 2018

Abstract: Accurate estimation of crop evapotranspiration (ET) is critical for agricultural water resource management and proper irrigation scheduling. The 2-year field experimental data of processing tomato under plastic-mulched drip and basin irrigation in the Hetao Irrigation District (Hetao), located in the upper reaches of the Yellow river, were used to calibrate and validate the SIMDualKc model. The model adopted the Food and Agriculture Organization (FAO) dual K_c method for partitioning ET into plant transpiration and soil evaporation. The results showed a good agreement between soil water observations and simulations throughout the growing seasons with a low error estimate and high model efficiency. The calibrated basal potential crop coefficients for the initial stage, mid-season stage, and late stage were 0.30, 0.92, and 0.60, respectively. ET during the two growing seasons was in the range of 284–331 mm for basin irrigation and 266–310 mm for drip irrigation. The average soil evaporation accounted for 5% of ET in 2015 and 14% of ET in 2016 for drip irrigation treatments, while it accounted for 4% and 13% of ET for basin irrigation treatments in the two experimental years, indicating that transpiration was the dominant component of ET of processing tomato under plastic mulch in Hetao. The highest water productivity was obtained from the drip irrigation treatment. The SIMDualKc model is an appropriate tool to estimate crop ET and may be further used to improve local irrigation scheduling for processing tomato in the upper reaches of the Yellow river.

Keywords: *Lycopersicon esculentum* Mill.; crop transpiration; soil evaporation; drip and basin irrigation; deficit irrigation

---

1. Introduction

The Hetao Irrigation District (Hetao), located in the upper reaches of the Yellow river, is one of the major grain and cash crops production regions in China. This region is characterized by a typical arid continental climate. The mean annual precipitation is approximately 150 mm, most of which occurs during monsoon season. The mean pan evaporation is in the range of 2200–2400 mm, which is 10 times as large as the precipitation [1,2]. Low precipitation and high potential evaporation make irrigation essential to agriculture development. Hetao covers an area of 1.12 million ha, and about 0.57 million ha of farmland are under irrigation. The amount of annual irrigation water derived from the Yellow river to Hetao is approximately 4.8 billion m³ [3]. Ninety percent (90%) of the total basin water resource is used by agriculture. Ninety-six percent (96%) of the agricultural water is used for irrigation while domestic supply accounts for only 4% [4]. However, the amount of water diverted from the Yellow river has decreased with rapid social economic development since 1999 [5]. As a result, the allocation of water for agriculture has also decreased. Thus, estimating crop evapotranspiration

(ET) accurately and improving crop water productivity (WP) are important to maintain sustainable agriculture development when the water resource is limited.

Processing tomato (*Lycopersicon esculentum* Mill.), mainly used for ketchup and tomato juice, is one of the most popular vegetable crops produced worldwide. As an important and long-term source of lycopene that is beneficial for human health, processing tomato accounts for approximately one third of the world's tomato production [6]. Due to the favorable weather conditions, lower cost of production, and better benefits for farmers, Hetao has become one of the most important processing-tomato-producing areas in China. The local government has reported that the cultivated area for processing tomato has expanded to 2.47×10^5 hm^2 in 2016 and the total yield has already reached 1.85×10^6 t. Processing tomato is a cash crop that consumes a large amount of water. A water shortage would affect the yield of processing tomato negatively, particularly during its sensitive growth stages [7,8]. Thus, irrigation is required throughout the growing season depending on the climatic conditions, especially in arid and semi-arid regions.

The traditional irrigation method used in Hetao is basin irrigation, which may result in low irrigation efficiencies and have negative impacts on crop growth and yield due to over-irrigation [1]. Basin irrigation is convenient for farmers to conduct; however, it also causes problems, such as a lower distribution uniformity and the alternation of soil wetting and drying between irrigation events with relatively long intervals, that may cause water stress to crops [9,10]. Optimizing irrigation strategies is thus important to reasonably utilize the limited water resources [11,12].

Efficient water saving irrigation technologies, such as drip irrigation, have played important roles when considering the improvement of WP. Drip irrigation can control the amount and position of soil moisture precisely to meet crop requirements and improve water management. Water and nutrients can be applied slowly and directly to the root zone or onto the soil surface [13]. Thus, crop yield can be maintained or improved and crop WP can be increased. Plants also suffer from less disease since water is not applied to the foliage directly. Drip irrigation is considered to be a water-saving technology, which can save 30–40% of water loss from deep percolation and water delivery compared with basin irrigation when appropriate irrigation management is applied. Soil evaporation is also decreased since the soil is only partially wetted by drip irrigation [14]. Presently, drip irrigation has been widely adopted in orchard, vegetable, and cereal crops around the world [15–18]. However, some studies have also demonstrated that the wide use of drip irrigation is still limited due to the high cost of installation and maintenance of drip irrigation systems, low economic returns, and poor technologies available to farmers [19–21]. Salt accumulation in the root zone is another concern for farmers using drip irrigation [22]. Plastic mulch is an agronomic technology widely used around the world [23–25]. Mulching has the advantages of increasing soil temperature, reducing weed pressure and certain insect pests, maintaining soil moisture, and improving nutrient use efficiency. Many studies in different soil types and climate conditions have indicated that mulching can effectively promote tomato yield and WP [26,27]. Plastic-mulched drip irrigation, an agronomic technology that combines drip irrigation and plastic mulch, is nowadays widely used and will be extended to 1.2 million ha around four provinces in northern China [28]. In recent years, local authorities in Hetao have encouraged farmers to convert from mulched basin irrigation to mulched drip irrigation through demonstration areas with financial support. Crop evapotranspiration depletes more than 90% of water in agricultural water consumption [29]. To better manage irrigation water inputs and improve crop WP, an accurate estimation of crop evapotranspiration is necessary for sustainable agricultural development and water management in arid and semi-arid areas.

Crop ET can be determined through direct measurement using experimental observations (e.g., a weighing lysimeter, the eddy covariance method, the Bowen ratio system) [30–32] or estimation with a model (e.g., Priestley-Taylor, Shuttleworth-Wallace, crop coefficient method) [33–36]. Due to the difficulties and cost of direct crop evapotranspiration measurement in field experiments, models provide an acceptable way of obtaining crop evapotranspiration values. The Food and Agriculture Organization 56 (FAO-56) crop coefficient method, which multiplies a reference crop evapotranspiration rate (ET_0) by a crop coefficient (K_c), is one of the most widely used methods for estimating crop evapotranspiration worldwide [37–39]. ET_0 is the evapotranspiration rate of the reference crop with an assumed crop height of 0.12 m, a fixed surface resistance of 70 s/m, and an albedo of 0.23, while K_c, the ratio of ET and ET_0, represents the effects of crop characteristics that distinguish specific field crops from the reference crop. K_c is deeply influenced by the soil and crop species and varieties. K_c is also adjusted to different meteorological elements (namely the minimum relative humidity (RH_{min}) and wind speed) as aerodynamic properties change due to various climatic conditions [40]. Different crops have different K_c. Even for the same crop, K_c varies throughout the growing season due to changes in growth stages and ground cover. Thus, local calibration of K_c is essential to facilitate sustainable water management, especially in arid and semi-arid environments with water-saving management practices. The FAO-56 crop coefficient method can be divided into the single K_c and dual K_c approaches [40]. In the dual K_c approach, K_c is segmented into two separate coefficients: the basal crop coefficient (K_{cb}) representing the crop transpiration, and the soil evaporation coefficient (K_e). Hence, the single coefficient K_c is replaced by the dual crop coefficients K_{cb} and K_e. Compared with the single K_c approach, the dual K_c approach makes it possible to better assess the impacts of soil wetting by rain or irrigation as well as the impacts of keeping part of the soil dry or using mulches for controlling soil evaporation [41], and has been widely applied for ET estimation with various technologies, such as drip irrigation, deficit irrigation, due to the simplicity and good performance of the approach [36,42–45].

The SIMDualKc model adopts the FAO dual K_c approach and computes ET with a daily time step. The model is easy to use and employs a helpful graphical user interface to assist users of various backgrounds [46]. Numerous studies have reported the successful application of the SIMDualKc model for various crops grown in different regions [12,40,47,48]. However, few studies have been conducted to assess the SIMDualKc model for estimating processing tomato ET, and the model should be properly calibrated and validated before use when water management options have not been previously tested. Not many investigations have been conducted on processing tomato ET partitioning under different plastic-mulched irrigation treatments in Hetao. Thus, the SIMDualKc model was selected for estimating processing tomato ET in Hetao. The purposes of this paper are: (1) to calibrate and validate the SIMDualKc model for processing tomato in Hetao; (2) to determine the K_{cb}, K_e, and variation trend of crop transpiration and soil evaporation during different crop growth stages for plastic-mulched basin and drip irrigation under various water stress conditions; and (3) to analyze the relationships between yield and crop ET (or T), the consumptive use water productivity (WP_{ET}), and the transpiration water productivity (WP_T) of processing tomato under different irrigation treatments.

2. Materials and Methods

2.1. Experimental Site

Field experiments were conducted at the Jiuzhuang agro-ecological station in Hetao (latitude 40°41′, longitude 107°18′, elevation 1041.2 m above sea level (a.s.l.)) during the crop-growing seasons (from 20 May to 31 August) in 2015 and 2016. The experimental site is characterized by a typical arid continental climate, and it belongs to dry desert climate (BWk) according to the Köppen climate classification. The soil particle size distribution at the experimental station was determined using a laser particle size analyzer and the soil is classified as a homogeneous silt loam soil. The dry bulk density, water content at field capacity, and wilting point were measured on undisturbed soil samples collected in the different horizons of each observation site. Soil samples were collected every 10 cm in the top 0–20 cm layer, and at 20 cm increments below the top layer to a depth of 60 cm. Additionally, the measurements were replicated three times within each sampling depth. The average bulk density of the soil is 1.50 g/cm^3. The average soil water content at field capacity and wilting point are 0.35 cm^3/cm^3 and 0.15 cm^3/cm^3 to a depth of 60 cm, respectively. Details of the properties of the soil at the experimental site are shown in Table 1.

Table 1. Soil physical properties of the soil profile at the experimental site.

| Soil Depths (cm) | Soil Particle Fraction (%) | | | Soil Texture | Bulk Density (g/cm^3) | Field Capacity (cm^3/cm^3) | Wilting Point (cm^3/cm^3) |
	Sand (>0.05 mm)	Silt (0.05–0.002 mm)	Clay (<0.002 mm)				
0–20	12.18	60.98	26.84	Silt loam	1.47	0.33	0.16
20–40	21.84	56.87	21.29	Silt loam	1.53	0.35	0.15
40–60	18.46	54.81	26.73	Silt loam	1.49	0.35	0.13

2.2. Weather Conditions

The meteorological variables, including maximum and minimum air temperature (°C), average relative humidity (%), and wind speed (m/s) and direction at 2 m above ground, were observed during the crop-growing seasons with an automatic weather station located close (1.0 km) by the experimental site. Data were sampled every 5 s, averaged over 30 min, and recorded by a data logger. Precipitation was measured by a tipping-bucket rain gauge. The meteorological conditions during the crop-growing seasons are shown in Figure 1. The average air temperature was 21.3, 24.1, and 22.8 °C in June, July, and August of the two years, respectively. The maximum air temperature during the growing season was 35.9 °C on 28 July 2015 (70 days after transplanting (DAT)) and 38.1 °C on 30 July 2016 (72 DAT) during the crop's rapid growth stages. The minimum average air temperature was 5.1 °C on 20 May 2015 and 6.7 °C on 25 May 2016 when the crop was transplanted to the field. The daily air temperature increased by a small amount during the crop's initial, development, and mid-season stages, but decreased gradually when the fruit was ripe. The average air relative humidity varied from 25% to 86% during the two years. The decline in relative humidity was usually accompanied by a rise in maximum temperature. Precipitation fluctuated significantly during the growing seasons. The total amount of precipitation during the growing season in 2016 was 107.9 mm. The maximum daily precipitation, with a value of 25.9 mm, occurred on 17 August when the fruit was ripe. However, the climate was very dry in 2015 with a precipitation of 8.4 mm.

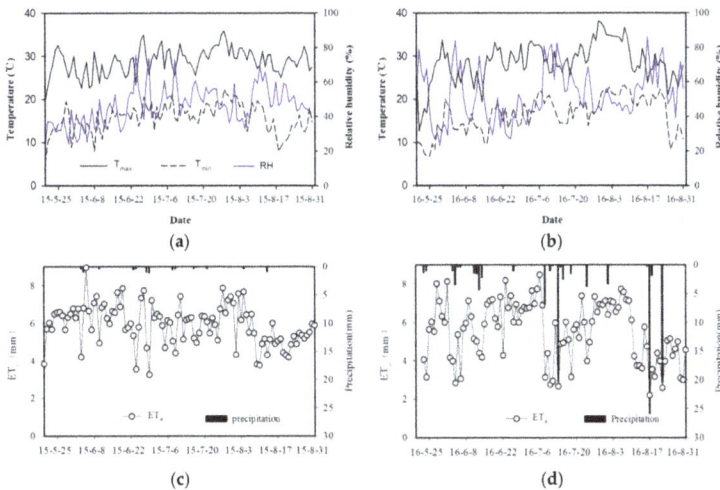

Figure 1. Daily meteorological data during the crop-growing seasons of maximum temperature (T_{max}), minimum temperature (T_{min}), average relative humidity (RH) (**a,b**), precipitation, and crop reference evapotranspiration (ET_0) (**c,d**) in 2015 and 2016.

The reference evapotranspiration (ET_0) was calculated using the FAO Penman-Monteith method from daily meteorological variables [40]. The variation of daily ET_0 during the crop-growing seasons is shown in Figure 1. The ET_o ranged between 3.25 and 8.94 mm/day in 2015 and between 2.19 and 8.48 mm/day in 2016, with an average value of 5.9 and 5.5 mm/day during the crop-growing seasons, respectively. In 2015, the maximum daily ET_o occurred at 17 DAT during the initial stage, whereas in 2016, the maximum daily ET_o occurred at 48 DAT during the mid-season stage. The cumulative ET_o during the growing season was slightly different for the two years, with a value of 614.2 mm in 2015 and 571.9 mm in 2016.

2.3. Experimental Design and Measurements

Field experiments were conducted in 2015 and 2016. Processing tomato (*Lycopersicon esculentum* Mill.) variety Tunhe No. 3, a local widely used variety, was transplanted on 20 May in both experimental years. Plants were arranged in a wide-narrow rows pattern, with a space of 90 cm for the wide row and 60 cm for the narrow row. Two rows of tomato were transplanted on the wide rows. The planting density was 40,000 plants per ha. The root depths at each growth stage were sampled with the root auger. According to our previous study in this area [9], the root zone soil depth was set as 30 cm in the development stage and was increased to 60 cm in the mid-season and late stages.

Two irrigation methods, drip irrigation and basin irrigation, were used in the experiment. For the drip irrigation treatments, drip lines were placed in the middle of each wide row to irrigate two rows of tomato. The emitter spacing was 0.3 m, and the discharge of each emitter was 2.7 L/h. For the basin irrigation treatments, polyethylene (PE) tubes were used as water pipelines supplying water to the plants. A flow meter, a pressure gauge, and a switch-valve were placed at the upstream end of each plot to control water applications.

A transparent plastic film was placed on the soil surface of wide rows to reduce soil evaporation and increase the soil temperature for both basin and drip irrigation. The plastic film was installed on the soil surface on 25 May (5 days after plant transplanting) for all of the treatments in both experimental years. The width of the plastic film was 0.9 m, and two processing tomato rows were covered. Seventy percent (70%) of the soil surface was covered with the plastic film applied. Then, the film was damaged due to agronomic practices (e.g., spraying pesticide, weeding) and the fraction of film covering soil decreased to 60% after 10 July (52 days after plant transplanting) in both experimental years. The diameter of the holes for transplanting was 0.05 m. Base fertilization was performed according to local custom and consisted of 150 kg/ha of urea ($CO(NH_2)_2$), 180 kg/ha of diammonium phosphate (($NH_4)_2HPO_4$), and 150 kg/ha of monopotassium phosphate (KH_2PO_4) before transplanting. Urea (40 kg/ha) was applied as a top dressing during the development and mid-season stages. Pesticide application and weed control were performed uniformly for all treatments.

Following the FAO-56 approach [40], the date of each crop growth stage corresponded to the time that 80% of the plants attained the stage. The crop growth stages were divided into four stages considering canopy coverage: (i) the initial stage, from plant transplanting to 10% canopy coverage; (ii) the development stage, from 10% to effective full canopy coverage; (iii) the mid-season stage, from full coverage to the start of fruit maturity; and (iv) the late stage, from the start of maturity to harvest. The crop growth stages of processing tomato observed in 2015 and 2016 are shown in Table 2. The fruits subject to basin irrigation treatments matured later than those subject to drip irrigation treatments. Thus, the late stage for drip irrigation was observed earlier than that for basin irrigation.

Table 2. The initial dates of different crop growth stages in 2015 and 2016.

Crop Growth Stages	Initial Date in 2015		Initial Date in 2016	
	Drip Irrigation	Basin Irrigation	Drip Irrigation	Basin Irrigation
Initial stage	20 May 2015	20 May 2015	20 May 2016	20 May 2016
Development stage	15 June 2015	15 June 2015	13 June 2016	13 June 2016
Midseason stage	1 July 2015	1 July 2015	1 July 2016	1 July 2016
Late stage	10 August 2015	16 August 2015	12 August 2016	16 August 2016
Harvest	31 August 2015	31 August 2015	31 August 2016	31 August 2016

The treatment plots were in complete random distribution with three replicates to minimize the effects of spatial heterogeneity. The size of each plot was 45 m^2 (4.5 m × 10 m). During the initial stage, each treatment was irrigated with the same amount water (55 mm in 2015 and 45 mm in 2016) to assure plant establishment.

Irrigation water was pumped from groundwater. Groundwater depth fluctuated, and was about 1.8–3.0 m below the ground surface at the experimental site during the two growing seasons. The ground surface was flat and irrigation water was evenly distributed. Both the basin and drip irrigation systems were observed to provide a relatively uniform water application throughout the growing seasons.

The irrigation treatments were set up as the following:

1. drip irrigation treatment (DI), where the irrigation schedule was based upon soil moisture measurements. The treatment plots were irrigated with the lower and upper limits of the soil water content at the root zone for irrigation, being 70% and 80–90% of field capacity, respectively. The irrigation treatments for 2015 and 2016 are shown in Table 3.
2. basin irrigation treatment (BI), where the treatment plots were irrigated during the development stage and mid-season stage with about 80 mm of water based on the experience of local farmers, whereas in 2016 only 50 mm of water were applied in the mid-season stage due to the relatively large rainfall that occurred during that stage (see Table 3).
3. reduced drip irrigation treatment (RDI) and a reduced basin irrigation treatment (RBI) were considered in 2016. The RDI and RBI treatments were irrigated with 60% of the amount of water of the DI and BI treatments at the same time, respectively. The schematic experimental setup of the 2016 growing season is displayed in Figure 2.

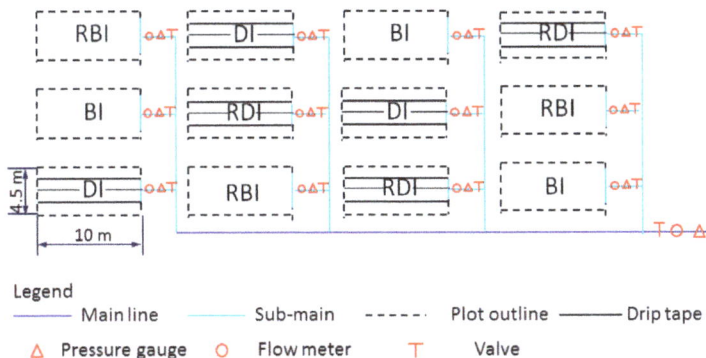

Figure 2. Schematic view of the experimental setup in 2016. Note: DI, Drip irrigation treatment; BI, Basin irrigation treatment; RDI, Reduced drip irrigation treatment; RBI, Reduced basin irrigation treatment.

Table 3. Irrigation depth (mm) during different growth stages for each irrigation treatment in 2015 and 2016.

Crop Growth Stages	Irrigation Depth in 2015 (mm)		Irrigation Depth in 2016 (mm)			
	DI	BI	DI	RDI	BI	RBI
Initial stage	55.0	55.0	45.0	45.0	45.0	45.0
Development stage	15.0	80.0	15.0	9.0	80.0	48.0
Midseason stage	108.2	80.0	90.4	54.2	50.0	30.0
Late stage	19.0	0.0	17.8	10.7	0.0	0.0
Harvest	197.2	215.0	168.2	118.9	175.0	123.0

Note: DI, Drip irrigation treatment; BI, Basin irrigation treatment; RDI, Reduced drip irrigation treatment; RBI, Reduced basin irrigation treatment.

2.4. Measurement

The soil water content in each treatment was measured every 7 days with the soil samples collected by a soil auger every 10 cm for the top 20 cm and every 20 cm from 20 to 60 cm during the crop-growing seasons. Three sampling points were selected in each treatment. Then, the soil water content at the root depth of 60 cm was averaged, and it was used to calibrate and validate the model.

The plant heights (h_c) of six randomly selected plants were measured every 7 days. The fraction of canopy cover (f_c) at each growth stage was estimated as the percentage of soil shaded by the crop with photographs taken above the plant canopy at near solar noon. Fruits were harvested by hand twice in both years (90 and 98 DAT in 2015 and 92 and 98 DAT in 2016). The yield of each plot was measured with the fruits of 30 randomly and consecutively selected plants. The mean value of three replicates was considered as the yield of each treatment.

Consumptive use water productivity (WP_{ET}) was expressed as the ratio between the fresh total yield and the actual crop evapotranspiration [49]:

$$WP_{ET} = \frac{Y \times 100}{ET} \tag{1}$$

where WP_{ET} is the consumptive use water productivity (kg/m³), Y is the fresh processing tomato yield measured in the field experiment (t/ha), and ET is the crop evapotranspiration simulated using the SIMDualKc model (mm).

Transpiration water productivity (WP_T) was calculated as the ratio of the fresh total yield and the crop transpiration:

$$WP_T = \frac{Y \times 100}{T} \tag{2}$$

where WP_T is the transpiration water productivity (kg/m³), and T is the crop transpiration simulated using the SIMDualKc model (mm).

2.5. SIMDualKc Model and Data Requirements

The SIMDualKc model was used to estimate the crop evapotranspiration in Hetao. The crop coefficient K_c consists of the crop basal coefficient K_{cb} and the soil evaporation coefficient K_e. Transpiration and evaporation are partitioned from crop evapotranspiration through multiplying K_{cb} and K_e by the reference evapotranspiration, respectively. When the model is applied to a new crop or to a new environment, the K_{cb} values have to be calibrated for various crop cultivars, and the K_e values need to be calculated through the daily soil water balance of the evaporative layer for different soils. The model computes the soil water balance in the root zone following [40]. Further descriptions of the model can be found in [46].

The actual crop ET was calculated using the model as follows:

$$ET = ET_o \times K_c = K_s K_{cb} ET_o + K_e ET_o \tag{3}$$

where $K_s K_{cb} ET_o$ represents the actual crop transpiration, while $K_e ET_o$ represents the soil evaporation. K_s is the water stress coefficient.

The soil water balance in the root zone was computed in terms of depletion at the end of every day [46]:

$$D_{r,i} = D_{r,i-1} - (P - RO)_I - I_I - CR_I + ET_i + DP_i \tag{4}$$

where $D_{r,i}$ and $D_{r,i-1}$ are the root zone depletion at the end of day i and the previous day, $i-1$ (mm), respectively, P_i is the precipitation (mm), RO_i is the runoff from the soil surface (mm), I_i is the net irrigation depth (mm), CR_i is the capillary rise from the groundwater (mm), ET_i is the actual crop evapotranspiration (mm), and DP_i is the deep percolation beyond the root zone (mm). In the present study, CR was not considered since the water table was relatively deep, about 1.8–3.0 m, while the maximum depth of the root zone was 0.6 m. No runoff was observed during the two experimental years; thus, RO was considered to be 0. DP was calculated with the default method described by [46].

The impact of plastic film on soil evaporation is considered in the SIMDualKc model. The model has successfully been applied to mulched conditions in various crops and regions [45,48]. Drip irrigation only wets part of the soil surface. Thus, the calculation of soil evaporation is computed in two fractions under drip irrigation: one is calculated for the fractions of soil wetted by precipitation only, and the other is calculated for the fraction of soil wetted by irrigation. Information about plastic mulch and drip irrigation can be found in [46].

The calibration of the SIMDualKc model included the process of adjusting significant model parameters to minimize differences between the observed and simulated soil water contents using data collected during the DI and BI treatments of 2015. The main adjusting parameters were: the crop parameters, e.g., the crop basal coefficient K_{cb} and the soil water depletion fraction for no stress (p), the soil evaporation parameters, e.g., the evaporative soil layer depth (Z_e), the total evaporable water (TEW), and the readily evaporable water (REW). Validation consisted of evaluating the accuracy of the model for soil water content using the calibrated parameter values with the independent data sets relative to the DI and BI treatments of 2016. Parameters were adjusted using a trial and error procedure.

2.6. Statistical Indicators for Model Performance

The determination coefficient (R^2), root mean square error ($RMSE$), mean relative error (MRE), and Nash–Sutcliffe coefficient (NS) were used to evaluate the performances of the model, which are expressed as follows:

$$R^2 = \left[\frac{\sum_{i=1}^{n} (O_i - \overline{O})(S_i - \overline{S})}{\sqrt{\sum_{i=1}^{n} (O_i - \overline{O})^2}\sqrt{\sum_{i=1}^{n}(S_i - \overline{S})^2}} \right]^2 \tag{5}$$

$$RMSE = \sqrt{\frac{1}{n}\sum_{i=1}^{n}(S_i - O_i)^2} \tag{6}$$

$$MRE = \frac{1}{n}\sum_{i=1}^{n}\frac{(P_i - O_i)}{O_i} \tag{7}$$

$$NS = 1 - \frac{\sum\limits_{i=1}^{n}(S_i - O_i)^2}{\sum\limits_{i=1}^{n}(O_i - \overline{O})^2} \tag{8}$$

where O_i (cm^3/cm^3) and S_i (cm^3/cm^3) are the observed and simulated values of soil water content at the i-th step, respectively; n is the number of the time steps; and $\overline{O}$ (cm^3/cm^3) and $\overline{S}$ (cm^3/cm^3) are the observed and simulated mean values, respectively. $RMSE$ (cm^3/cm^3) and MRE values closer to 0 indicate a more accurate model. The model's calibration is considered to be in a good situation when the simulated soil water content has an MRE value lower than 0.3 and an $RMSE$ value lower than

$0.03 \text{ cm}^3/\text{cm}^3$. R^2 values close to 1.0 represent that the variation of the observed values is well-captured by the model. The *NS* values range from $-\infty$ (poor model) to 1.0 (perfect model). A zero value of *NS* means the simulated value is as good as the observation mean.

3. Results and Discussion

3.1. Model Calibration and Validation

The measured and simulated soil water content in the root zone for the calibration and validation procedures is shown in Figure 3. It can be found that the simulated soil water content using SIMDualKc closely followed the observed soil water content. Table 4 presents the initial values and the calibrated results of the crop and soil evaporation parameters of the SIMDualKc model. For the drip irrigation treatments, the fraction of soil surface wetted by irrigation or precipitation (f_w) was set to 0.4, referring to [40], while for the basin irrigation treatments, f_w was set to 1.0. In the calibration process, the initial values adopted were recommended by FAO56. Greater calibrated values for *TEW* and Z_e may be due to the larger silt content of the soil at the experimental site as compared with the soil in FAO56, leading to greater field capacity. The calibrated p values were slightly smaller than those proposed in FAO56, which may be attributed to the local crop variety that is more sensitive to water stress. The experimental data of h_c and f_c were also used during the calibration process. The maximum h_c was 0.45 m for the DI treatment and 0.48 m for the BI treatment, whereas the maximum value of f_c was 0.7 for both treatments.

The model simulation had RMSE values of 0.01–$0.02 \text{ cm}^3/\text{cm}^3$, MRE values of -0.01–0.01, and *NS* values of 0.76–0.92 for the calibration and validation treatments in the two experimental years, respectively. Linear regression was performed between the simulated soil water content and the observed soil water content, and the determination coefficient R^2 values were in the range of 0.89–0.94. The results of both the model simulation and the linear regression supported that the variation of the observed soil water content throughout the crop-growing seasons could be well-explained by the SIMDualKc model. Several studies have also shown similar goodness-of-fit results on the SIMDualKc model's performance [12,50,51].

Table 4. Initial and calibrated values of crop parameters and soil evaporation parameters of the SIMDualKc model.

Types	Parameters (Units)	Initial Value	Calibrated Value
Crop	$K_{cb,ini}$ (/)	0.15	0.30
	$K_{cb,mid}$ (/)	1.10	0.92
	$K_{cb,end}$ (/)	0.60	0.60
	p_{ini} (/)	0.40	0.30
	p_{dev} (/)	0.40	0.30
	p_{mid} (/)	0.40	0.30
	p_{end} (/)	0.40	0.30
Soil evaporation	REW (mm)	8	10
	TEW (mm)	25	37
	Z_e (m)	0.10	0.15

Note: $K_{cb,ini}$, basal crop coefficient during the initial stage; $K_{cb,mid}$, basal crop coefficient during the mid-season stage; $K_{cb,end}$, basal crop coefficient during the late stage; p_{ini}, evapotranspiration depletion fraction during the initial stage; p_{dev}, evapotranspiration depletion fraction during the development stage; p_{mid}, evapotranspiration depletion fraction during the mid-season stage; p_{end}, evapotranspiration depletion fraction during the late stage; REW, readily evaporable water; TEW, total evaporable water; Z_e, depth of surface soil layer subjected to drying by evaporation.

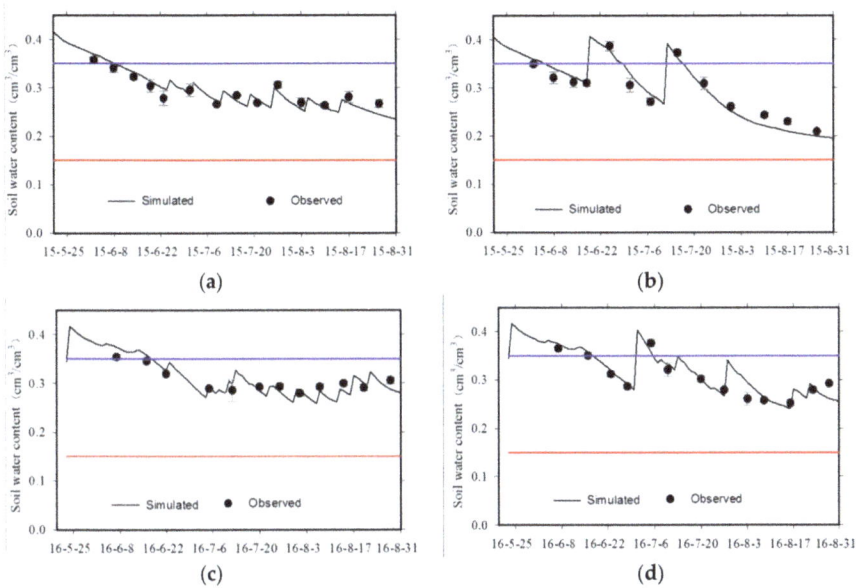

Figure 3. Comparison of the observed and simulated daily soil water content in the root zone: (**a**) DI in 2015 (calibration), (**b**) BI in 2015 (calibration), (**c**) DI in 2016 (validation), (**d**) BI in 2016 (validation). (The blue line (—) and red line (—) red the soil water content at field capacity and the wilting point. Note: DI, Drip irrigation treatment; BI, Basin irrigation treatment.

3.2. Basal Crop Coefficient and Soil Evaporation Coefficient Dynamics

The basal crop coefficient K_{cb} for processing tomato and the soil evaporation coefficient K_e over the crop-growing seasons are shown in Figure 4. The calibrated potential K_{cb} values were 0.30, 0.92, and 0.60, respectively (Table 4). The actual K_{cb} showed day-to-day fluctuation with values in the range of 0.27–0.30, 0.30–0.64, 0.29–0.81, and 0.19–0.68 (drip irrigation), 0.27–0.30, 0.30–0.89, 0.20–0.93, and 0.08–0.54 (basin irrigation) for the initial, development, mid-season, and late-season stages, respectively. The K_{cb} values in this study were lower than those proposed by FAO56 with one exception in the initial stage. This may be attributed to the shorter growing season (103 days in this study compared with 135–180 days in FAO56), the shallower root depth (0.6 m in this study compared with 0.7–1.5 m in FAO56), and the lower fraction of soil covered by the crop canopy (f_c) (0.7 in this study compared with 0.8–1.0 in FAO56).

The actual K_{cb} values of all the treatments were equal to the potential K_{cb} during the initial stage (Figure 4). This is attributed to the fact that the plots were irrigated with around 45–55 mm of water when the crop was transplanted. During the development, mid-season, and late stages, the actual K_{cb} values of the DI treatment were consistently smaller than the potential K_{cb}, indicating that the irrigation water could not meet the crop water requirement. The actual K_{cb} values of the BI treatment were smaller than those of the potential K_{cb} with the exception of the irrigation period during the development and mid-season stages, reflecting that basin irrigation could only satisfy the crop water requirement for a short time as a large amount of irrigation water was applied. For the reduced irrigation treatments, their actual K_{cb} values were smaller than those of the DI and BI treatments except for the initial stage. The reduced irrigation treatments also caused more serious water stress on crop growth due to the lower amount of water applied for irrigation. The peaks of K_e were related to irrigation and precipitation. The dynamics of $K_{cb,act}$ showed that the plants were under water-stressed conditions during most of the growing season. The above results indicated that the current drip irrigation regime for processing tomato in Hetao should be improved since the crop was under water-stressed conditions during most of the crop growth stages for the DI treatment.

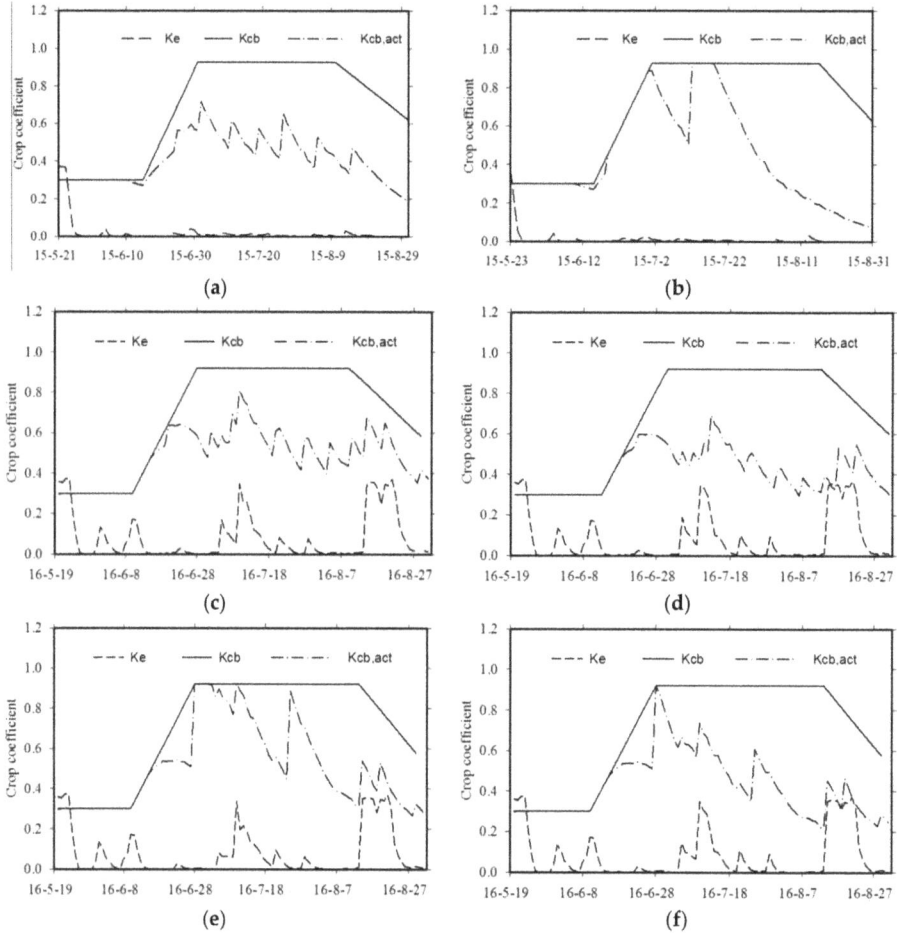

Figure 4. The daily variation of the potential and actual basal crop coefficients (K_{cb} and $K_{cb,act}$) and the evaporation coefficient (K_e) relative to (**a**) DI in 2015; (**b**) BI in 2015; (**c**) DI in 2016; (**d**) RDI in 2016; (**e**) BI in 2016; and (**f**) RBI in 2016. Note: DI, Drip irrigation treatment; BI, Basin irrigation treatment; RDI, Reduced drip irrigation treatment; RBI, Reduced basin irrigation treatment.

Figure 4 clearly shows that there was strong evidence of high water stress among all the treatments for almost the entire crop season in both experimental years, indicating that all of the irrigation schedules were inadequate to prevent water stress and should be improved to meet crop water requirements. The curve of $K_{cb,act}$ was underneath that of the potential K_{cb} during most of the growing season. Only one basin irrigation event was taken during the mid-season stage and no basin irrigation events took place during the late stage, resulting in severe water stress (Figure 4b). However, basin irrigation during the late stage may result in fruit rot and a decrease in the sugar content in the fruit. Thus, drip irrigation with partial soil wetting is an appropriate irrigation method for processing tomatoes in Hetao.

The soil evaporation coefficients were smaller than the basal crop coefficient (see Figure 4), implying that a very small proportion of water was lost by evaporation due to the plastic mulch. However, K_e increased significantly after large rainfalls occurred in July and August 2016. The damage of the plastic film by frequent agronomic practices resulted in a decrease in the amount of mulch covering the soil and an increase in soil evaporation. The highest K_e values were found during the initial stage of the two years due to the small fraction of crop cover. This also may be attributable to the fact that the soil surface was covered by plastic film 5 days after plant transplanting, resulting in a relatively larger amount of evaporation during this stage. No significant differences ($p < 0.05$) in K_e were observed among the treatments during the entire growing season. The variation in soil evaporation between the two years was mainly caused by the difference of rainfall.

Few publications are available about the crop coefficients of processing tomato under various water-stressed conditions. Based on the measurement of eddy covariance, Amayreh and Al-Abed found the $K_{c,act}$ values of field-grown tomato to be 0.65, 0.82, and 0.52, respectively, at the development stage, mid-season stage, and late stage in Jordan Valley [31]. Hanson and May reported that the average $K_{c,act}$ values of drip-irrigated processing tomato ranged from 0.19 at 10% canopy coverage to 1.08 for canopy coverage exceeding about 90% in San Joaquin Valley, California (USA) [52]. In this study, the average $K_{c,act}$ values at the initial, development, mid-season, and late stages were, respectively, 0.30, 0.49, 0.51, and 0.41 for drip irrigation and 0.30, 0.51, 0.59, and 0.29 for basin irrigation over the two years. The differences in the K_c values obtained by our study and those from the literature may be attributed to differences in (i) the methods used for ET and ET_o measurement and estimation; (ii) crop varieties, cultivars, and the length of their growing season and growth stages; (iii) soils, irrigation regimes, and cultivations; and (iv) weather conditions at specific locations (namely mean RH_{min} and wind speed).

3.3. Crop Evapotranspiration Partitioning

The simulated crop evapotranspiration (ET), transpiration (T), potential transpiration (T_p), and soil evaporation (E) of processing tomato are shown in Figure 5 and Table 5. The seasonal ET varied between 266 and 331 mm with the largest value for the BI treatment in 2016 and the smallest value for the DI treatment in 2015, and ET increased with an increase in irrigation amount. Compared with that in 2015, the seasonal ET in 2016 was higher due to the larger precipitation that occurred during the growing season. The daily ET ranged between 1.0 and 5.3 mm/day with a mean value of 2.8 mm/day for drip irrigation and 0.5 and 6.6 mm/day with a mean value of 2.9 mm/day for basin irrigation over the two years, respectively. In general, the daily ET was small during the initial stage, increased rapidly with its maximum values occurring during the mid-season stage, and then declined during the late stage.

Table 5. Crop yield (Y), evapotranspiration (ET), transpiration (T), water productivity (WP_{ET}), and transpiration water productivity (WP_T) over the two years.

Years	Treatments	Y (t/ha)	ET (mm)	T (mm)	WP_{ET} (kg/m^3)	WP_T (kg/m^3)	E_s/ET
2015	DI	99.4 ± 7.7 a	266.3	253.2	37.3 ± 2.9 a	39.2 ± 3.0 a	4.9%
	BI	91.6 ± 6.8 a	284.3	272.0	32.2 ± 2.4 b	33.7 ± 2.5 b	4.3%
2016	DI	107.9 ± 10.5 a	309.7	272.5	34.8 ± 3.4 a	39.6 ± 3.9 a	11.9%
	RDI	88.4 ± 7.5 b	270.9	233.3	32.6 ± 2.8 a	37.9 ± 3.2 a	13.8%
	BI	105.1 ± 8.3 a	331.2	295.0	31.7 ± 2.5 a	35.6 ± 2.8 a	10.9%
	RBI	94.6 ± 7.8 ab	283.9	246.9	33.3 ± 2.7 a	38.3 ± 3.2 a	12.9%

Note: DI, Drip irrigation treatment; BI, Basin irrigation treatment; RDI, Reduced drip irrigation treatment; RBI, Reduced basin irrigation treatment. The letters a, b, and ab are used to indicate the statistical significance according to the Duncan's multiple range tests at 0.05 P level. Values within columns followed by different letters are statistically significant at the 0.05 level.

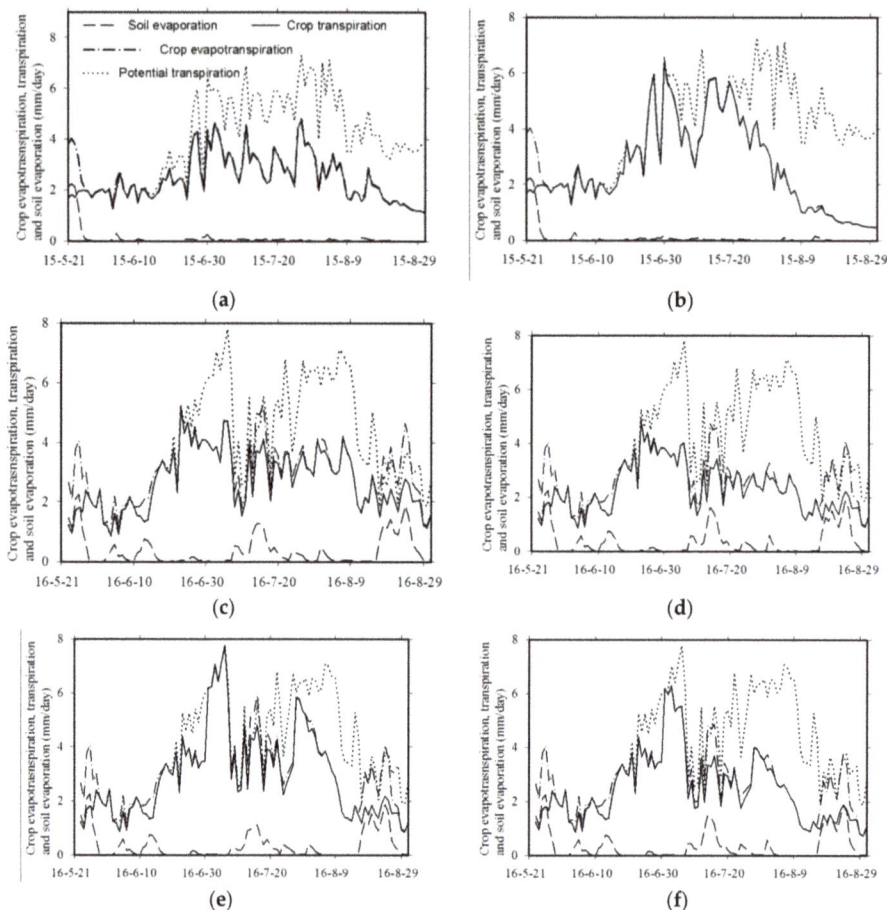

Figure 5. The daily variation of crop evapotranspiration, crop transpiration, potential transpiration, and soil evaporation relative to (**a**) DI in 2015; (**b**) BI in 2015; (**c**) DI in 2016; (**d**) RDI in 2016; (**e**) BI in 2016; and (**f**) RBI in 2016. Note: DI, Drip irrigation treatment; BI, Basin irrigation treatment; RDI, Reduced drip irrigation treatment; RBI, Reduced basin irrigation treatment; crop evapotranspiration (— · —); crop transpiration (—); potential transpiration (·····); soil evaporation (- - -).

The crop water requirement of processing tomato has been measured and estimated in many studies. With the measurement of the Bowen ratio, Hanson and May reported that the seasonal *ET* of tomato was 609 mm under subsurface drip irrigation and 678 mm under furrow irrigation in San Joaquin Valley, California, USA [53]. An *ET* value of 300 mm under non-irrigated conditions and *ET* values ranging from 556.5 to 621.2 mm under drip irrigation were reported for processing tomato in South Serbia [54]. Mukherjee et al. found *ET* values between 147 and 225 mm for polyethylene-mulched tomato under rainfed and furrow irrigation conditions in India [55]. Giuliani et al. reported an *ET* value of 558 mm for tomato with full-drip irrigation in southern Italy [56]. The lower *ET* of this study was mainly due to the various climatic conditions, the much shorter growing season, the smaller crop canopy fraction, and more severe water-stressed conditions as compared with those in the literature.

Seasonal *T* was in the range of 233–295 mm with the largest value for the BI treatment in 2016 and the smallest one for the RDI treatment in 2016. The proportion of *T* to *ET* in the two growing seasons was 86–96%, indicating that transpiration was the dominant component of crop evapotranspiration. Daily T ranged between 0.9 and 5.2 mm/day with a mean value of 2.5 mm/day for drip irrigation and 0.5 and 6.5 mm/day with a mean value of 2.6 mm/day for basin irrigation over the two years. The seasonal *T* values of the DI treatment were consistently smaller than those of the potential *T* except for the initial stage, while the seasonal *T* values of the BI treatment were smaller than those of the potential *T* with the exception of the irrigation period, indicating that the crop requirement was not satisfied by the irrigation water during the entire growing season. The total *E* was 13 and 37 mm during the growing seasons of 2015 and 2016, respectively, which accounted for only 5–14% of *ET* for the drip irrigation treatments and 4–13% of *ET* for the basin irrigation treatments. Soil evaporation did not show much difference, while the crop *ET* for drip irrigation was lower than that for basin irrigation. Thus, a higher ratio of E_s/ET was observed in the drip irrigation treatment. There was little soil evaporation after the application of the plastic film, indicating that plastic film could significantly reduce evaporative consumption.

3.4. Yield and Water Productivity

Crop yield and water productivity (WP) are listed in Table 5 for various irrigation treatments. The yield was in the range of 88.4–107.9 t/ha for the drip irrigation treatments and 91.6–105.1 t/ha for the basin irrigation treatments in 2015 and 2016, respectively. The highest yield in 2015 and 2016 was respectively 99.4 and 107.9 t/ha obtained by the DI treatment, which was slightly higher than those obtained by the BI treatment. In the study area, tomato fruits were hand harvested twice during the late stage. For the first harvest, the yield from plants subject to the drip irrigation treatments was 1.2–1.6 times as large as that from plants subject to the basin irrigation treatments, whereas the yield from plants subject to the drip irrigation treatments was 30% lower than that from plants subject to the basin irrigation treatments in the second harvest. This indicates that drip irrigation could speed up the ripening process of fruits due to the fact that soil temperatures in drip-irrigated plots are higher than those in the basin-irrigated plots. A similar result was also found by [28], who reported that crop growth could be significantly enhanced by drip irrigation. Various studies [19,57] have also demonstrated that high-frequency drip irrigation can promote crop yield for large varieties of crops and vegetables as compared with basin irrigation. As shown in Table 5, the yields of the reduced irrigation treatments were lower, indicating that yield increased as the irrigation amount increased.

The consumptive use water productivity (WP_{ET}) was in the range of 32.6–37.3 kg/m^3 for drip irrigation and 31.7–33.3 kg/m^3 for basin irrigation, whereas the transpiration water productivity (WP_T) was 37.9–39.6 kg/m^3 for drip irrigation and 33.7–38.3 kg/m^3 for basin irrigation. The highest WP_{ET} and WP_T were obtained by the DI treatment in both years, while the smallest WP_{ET} and WP_T were recorded by the BI treatment in both years. Larger WP_{ET} and WP_T values for drip irrigation treatments were associated with a reduction in crop *ET* and an increase in yield. Actually, more soil water was conserved and used for crop growth in drip-irrigated plots, which could thus benefit crop production. The slight differences of WP_{ET} and WP_T between the DI/BI and the reduced irrigation treatments may be attributed to the fact that both crop yield and *ET* were enhanced by increasing the irrigation amount; in contrast, reduced irrigation could inhibit crop yield and *ET*.

The relationship of relative yield to relative evapotranspiration (and/or relative transpiration) can be well-quantified using a quadratic function (see Figure 6) with a coefficient of determination R^2 of 0.90 for fitted curves. The relative values were obtained by dividing the actual values by the respective maximum value in each year. Thus, the values varied between 0 and 1. The relative yield generally first increased with relative *ET* and *T* in both years, and then decreased after reaching the highest value. The best result was achieved for the DI treatment when the relative *ET* was about 0.92. This result was different from the linear relationship between yield and *ET* obtained by [58] in an arid region of northwest China and the exponential relationship between yield and *ET* found by [59] under semi-arid Mediterranean climate conditions. In general, appropriate irrigation scheduling should be associated with high WP and high (or acceptable) yield. Thus, the irrigation schedule of the DI treatment could be proposed as a preferable irrigation regime for processing tomato due to its relatively high yield and WP in the study area under the condition of plastic mulch. However, such an irrigation regime for plastic-mulched processing tomato in the Hetao Irrigation District should be further formulated and validated with consideration for the water-stressed conditions of plants, the economic costs of drip irrigation systems, and the benefits of farmers.

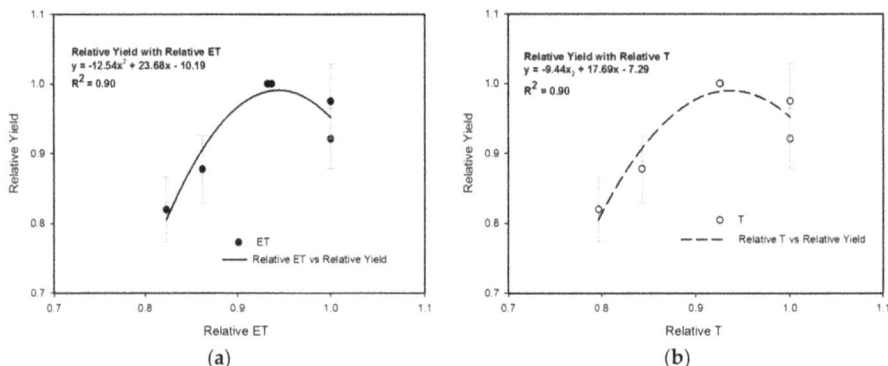

Figure 6. The relationships between the relative yield (Y) and the relative evapotranspiration (ET) (**a**) and the relative transpiration (T) (**b**).

4. Conclusions

The two-year field experiment on processing tomato under the conditions of plastic-mulched irrigation provided the basic data to validate and calibrate the high applicability of SimDualKc in the upper reaches of the Yellow river. The results of the crop coefficients adjusted to the actual conditions were successfully obtained from the model and were appropriate to be further used locally. The total ET increased with an increase in irrigation amount both in drip and basin irrigation. Transpiration accounted for over 86% of ET, whereas soil evaporation was less than 14% of ET during the two growing seasons.

The highest yield and WP in both years were achieved for the treatment with drip irrigation. A quadratic relationship was found between the relative yield and the relative ET (and/or the relative T). Compared with basin irrigation, drip irrigation could speed up the ripening of fruits and shorten the period of crop growth. Considering both WP and crop yield, plastic-film-mulched drip irrigation was recommended for processing tomato in the upper reaches of the Yellow River. However, the crop suffered from high water stress in all of the treatments during most of the growing season due to an inadequate irrigation schedule. Thus, a proper irrigation schedule, which can meet the crop water requirement and obtain high yield for processing tomato under plastic-film-mulched drip irrigation, should be established in the future research. The adoption of drip irrigation highly relies on farmers' decisions based on economic information on costs and benefits and the capability of using the new water-saving technologies. Further study is also required to obtain the optimal irrigation schedules for processing tomato using optimization approaches while considering farmers' willingness and social-economical situations.

Author Contributions: G.H. and H.Z. conceived and designed the experiments; H.Z. performed the experiments, analyzed the data, conducted the simulation, and wrote the paper. G.H. corrected and improved the paper. X.X., Y.X. and Q.H. made significant suggestions in data analysis and manuscript writing.

Funding: This research was jointly supported by the National Key R&D Program of China (Grant No. 2017YFC0403301) and the National Natural Science Foundation of China (Grant Nos. 51639009, 51621061, 51125036).

Acknowledgments: We are grateful to the editors and two anonymous reviewers for their constructive critiques on a previous version of this manuscript.

Conflicts of Interest: The authors declare no conflict of interest.

References

1. Feng, Z.Z.; Wang, X.K.; Feng, Z.W. Soil N and salinity leaching after the autumn irrigation and its impact on groundwater in Hetao Irrigation District, China. *Agric. Water Manag.* **2005**, *71*, 131–143. [CrossRef]
2. Lei, T.W.; Shainberg, I.; Yuan, P.J.; Huang, X.F.; Yang, P.L. Strategic considerations of efficient irrigation and salinity control on Hetao Plain in Inner Mongolia. *Transf. CSAE* **2001**, *17*, 48–52.
3. Xu, X.; Sun, C.; Qu, Z.Y.; Huang, Q.Z.; Ramos, T.B.; Huang, G.H. Groundwater recharge and capillary rise in irrigated areas of the upper Yellow river basin assessed by an agro-hydrological model. *Irrig. Drain.* **2016**, *64*, 587–599. [CrossRef]
4. Wu, Y.; Shi, X.H.; Li, C.Y.; Zhao, S.N.; Pen, F.; Green, T.R. Simulation of hydrology and nutrient transport in the Hetao Irrigation District, Inner Mongolia, China. *Water* **2017**, *9*, 169. [CrossRef]
5. Yang, Y.T.; Shang, S.H.; Jiang, L. Remote sensing temporal and spatial patterns of evapotranspiration and the responses to water management in a large irrigation district of North China. *Agric. For. Meteorol.* **2012**, *164*, 112–122. [CrossRef]
6. Szuvandzsiev, P.; Daood, H.G.; Posta, K.; Helyes, L.; Pék, Z. Application of VIS-NIR reflectance spectra for estimating soluble solid and lycopene content of open-field processing tomato fruit juice from irrigation and mycorrhiza treatments. *Acta Hortic.* **2017**, *1159*, 73–78. [CrossRef]
7. Patanè, C.; Cosentino, S.L. Effects of soil water deficit on yield and quality of processing tomato under a Mediterranean climate. *Agric. Water Manag.* **2010**, *97*, 131–138. [CrossRef]
8. Patane, C.; Saita, A. Biomass, fruit yield, water productivity and quality response of processing tomato to plant density and deficit irrigation under a semi-arid Mediterranean climate. *Crop Pasture Sci.* **2015**, *66*, 224–234. [CrossRef]
9. Zhang, H.M.; Xiong, Y.W.; Huang, G.H.; Xu, X.; Huang, Q.Z. Effects of water stress on processing tomatoes yield, quality and water use efficiency with plastic mulched drip irrigation in sandy soil of the Hetao Irrigation District. *Agric. Water Manag.* **2017**, *179*, 205–214. [CrossRef]
10. Xu, X.; Huang, G.H.; Qu, Z.Y.; Pereira, L.S. Assessing the groundwater dynamics and impacts of water saving in the Hetao Irrigation District, Yellow River basin. *Agric. Water Manag.* **2010**, *98*, 301–313. [CrossRef]
11. Hunsaker, D.J.; Barnes, E.M.; Clarke, T.R.; Fitzgerald, G.J.; Pinter, P.J., Jr. Cotton irrigation scheduling using remotely sensed and FAO-56 basal crop coefficients. *Trans. ASABE* **2005**, *48*, 1395–1407. [CrossRef]
12. Paredes, P.; D'Agostino, D.; Assif, M.; Todorovic, M.; Pereira, L.S. Assessing potato transpiration, yield and water productivity under various water regimes and planting dates using the FAO dual K_c approach. *Agric. Water Manag.* **2018**, *195*, 11–24. [CrossRef]
13. Sezen, S.M.; Yazar, A.; Eker, S. Effect of drip irrigation regimes on yield and quality of field grown bell pepper. *Agric. Water Manag.* **2006**, *81*, 115–131. [CrossRef]
14. Jensen, C.R.; Ørum, J.E.; Pedersen, S.M.; Andersen, M.N.; Plauborg, F.; Liu, F.; Jacobsen, S.E. A short overview of measures for securing water resources for irrigated crop production. *J. Agron. Crop Sci.* **2014**, *200*, 333–343. [CrossRef]
15. Gerçek, S.; Demirkaya, M.; Işik, D. Water pillow irrigation versus drip irrigation with regard to growth and yield of tomato grown under greenhouse conditions in a semi-arid region. *Agric. Water Manag.* **2017**, *180*, 172–177. [CrossRef]
16. Munoz-Carpena, R.; Bryan, H.; Klassen, W.; Dukes, M.D. Automatic soil moisture based drip irrigation for improving tomato production. *Proc. Fla. State Hortic. Soc.* **2003**, *116*, 80–85.
17. Ngouajio, M.; Wang, G.Y.; Goldy, R. Withholding of drip irrigation between transplanting and flowering increases the yield of field-grown tomato under plastic mulch. *Agric. Water Manag.* **2007**, *87*, 285–291. [CrossRef]
18. Sui, J.; Wang, J.D.; Gong, S.H.; Xu, D.; Zhang, Y.Q.; Qin, Q.M. Assessment of maize yield-increasing potential and optimum N level under mulched drip irrigation in the Northeast of China. *Field Crops Res.* **2018**, *215*, 132–139. [CrossRef]
19. Fang, Q.; Zhang, X.Y.; Shao, L.W.; Chen, S.Y.; Sun, H.Y. Assessing the performance of different irrigation systems on winter wheat under limited water supply. *Agric. Water Manag.* **2018**, *196*, 133–143. [CrossRef]

20. Hanson, B.R.; Schwankl, L.J.; Schulbach, K.F.; Pettygrove, G.S. A comparison of furrow, surface drip, and subsurface drip irrigation on lettuce yield and applied water. *Agric. Water Manag.* **1997**, *33*, 139–157. [CrossRef]

21. Namara, R.E.; Nagar, R.K.; Upadhyay, B. Economics, adoption determinants, and impacts of micro-irrigation technologies: Empirical results from India. *Irrig. Sci.* **2007**, *25*, 283–297. [CrossRef]

22. Burt, C.M.; Isbell, B. Leaching of accumulated soil salinity under drip irrigation. *Trans. ASABE* **2005**, *48*, 2115–2121. [CrossRef]

23. Zheng, J.H.; Huang, G.H.; Wang, J.; Huang, Q.Z.; Pereira, L.S.; Xu, X.; Liu, H.J. Effects of water deficits on growth, yield and water productivity of drip-irrigated onion (*Allium cepa* L.) in an arid region of Northwest China. *Irrig. Sci.* **2013**, *31*, 995–1008. [CrossRef]

24. Li, X.Y.; Shi, H.B.; Simunek, J.; Gong, X.W.; Peng, Z.Y. Modeling soil water dynamics in a drip-irrigated intercropping field under plastic mulch. *Irrig. Sci.* **2015**, *33*, 289–302. [CrossRef]

25. Vazquez, N.; Pardo, A.; Suso, M.L.; Quemada, M. Drainage and nitrate leaching under processing tomato growth with drip irrigation and plastic mulching. *Agric. Ecosyst. Environ.* **2006**, *112*, 313–323. [CrossRef]

26. Berihun, B. Effect of mulching and amount of water on the yield of tomato under drip irrigation. *J. Hortic. For.* **2011**, *3*, 200–206.

27. Biswas, S.K.; Akanda, A.R.; Rahman, M.S.; Hossain, M.A. Effect of drip irrigation and mulching on yield, water-use efficiency and economics of tomato. *Plant Soil Environ.* **2015**, *61*, 97–102.

28. Qin, S.J.; Li, S.E.; Kang, S.Z.; Du, T.S.; Tong, L.; Ding, R.S. Can the drip irrigation under film mulch reduce crop evapotranspiration and save water under the sufficient irrigation condition? *Agric. Water Manag.* **2016**, *177*, 128–137. [CrossRef]

29. Rana, G.; Katerji, N. Measurement and estimation of actual evapotranspiration in the field under Mediterranean climate: A review. *Eur. J. Agron.* **2000**, *13*, 125–153. [CrossRef]

30. Al-Omran, A.M.; Mohammad, F.S.; Al-Ghobari, H.M.; Alazba, A.A. Determination of evapotranspiration of tomato and squash using lysimeters in central Saudi Arabia. *Int. Agric. Eng. J.* **2004**, *13*, 27–36.

31. Amayreh, J.; Al-Abed, N. Developing crop coefficients for field-grown tomato (*Lycopersicon esculentum* Mill.) under drip irrigation with black plastic mulch. *Agric. Water Manag.* **2005**, *73*, 247–254. [CrossRef]

32. Hanson, B.R.; May, D.M. Crop evapotranspiration of processing tomato in the San Joaquin Valley of California, USA. *Irrig. Sci.* **2006**, *24*, 211–221. [CrossRef]

33. Colaizzi, P.D.; O'Shaughnessy, S.A.; Evett, S.R.; Mounce, R.B. Crop evapotranspiration calculation using infrared thermometers aboard center pivots. *Agric. Water Manag.* **2017**, *187*, 173–189. [CrossRef]

34. Howell, T.A.; Evett, S.R.; Tolk, J.A.; Schneider, A.D. Evapotranspiration of full-, deficit-irrigated and dryland cotton on the Northern Texas High Plains. *J. Irrig. Drain. Eng.* **2004**, *130*, 277–285. [CrossRef]

35. Valdés-Gómez, H.; Ortega-Farías, S.; Argote, M. Evaluation of water requirements for a greenhouse tomato crop using the Priestley-Taylor method. *Chil. J. Agric. Res.* **2009**, *69*, 3–11.

36. Zhao, P.; Li, S.E.; Li, F.S.; Du, T.S.; Tong, L.; Kang, S.Z. Comparison of dual crop coefficient method and Shuttleworth–Wallace model in evapotranspiration partitioning in a vineyard of Northwest China. *Agric. Water Manag.* **2015**, *160*, 41–56. [CrossRef]

37. Abrisqueta, I.; Abrisqueta, J.M.; Tapia, L.M.; Munguía, J.P.; Conejero, W.; Vera, J.; Ruiz-Sánchez, M.C. Basal crop coefficients for early-season peach trees. *Agric. Water Manag.* **2013**, *121*, 158–163. [CrossRef]

38. Allen, R.G.; Pereira, L.S.; Smith, M.; Raes, D.; Wright, J.L. FAO-56 dual crop coefficient method for estimating evaporation from soil and application extensions. *J. Irrig. Drain. Eng.* **2005**, *131*, 2–13. [CrossRef]

39. Sumner, D.M.; Jacobs, J.M. Utility of Penman–Monteith, Priestley–Taylor, reference evapotranspiration, and pan evaporation methods to estimate pasture evapotranspiration. *J. Hydrol.* **2005**, *308*, 81–104. [CrossRef]

40. Allen, R.G.; Pereira, L.S.; Raes, D.; Smith, M. *Crop Evapotranspiration: Guidelines for Computing Crop Water Requirements*; FAO: Rome, Italy, 1998; p. 56.

41. Zhang, B.Z.; Liu, Y.; Xu, D.; Zhao, N.N.; Lei, B.; Rosa, R.D.; Paredes, P.; Paço, T.A.; Pereira, L.S. The dual crop coefficient approach to estimate and partitioning evapotranspiration of the winter wheat–summer maize crop sequence in North China Plain. *Irrig. Sci.* **2013**, *31*, 1303–1316. [CrossRef]

42. Bodner, G.; Loiskandl, W.; Kaul, H.-P. Cover crop evapotranspiration under semi-arid conditions using FAO dual crop coefficient method with water stress compensation. *Agric. Water Manag.* **2007**, *93*, 85–98. [CrossRef]

43. Feng, Y.; Gong, D.Z.; Mei, X.R.; Cui, N.B. Estimation of maize evapotranspiration using extreme learning machine and generalized regression neural network on the China Loess Plateau. *Hydrol. Res.* **2017**, *48*, 1156–1168. [CrossRef]

44. Paredes, P.; Rodrigues, G.C.; Alves, I.; Pereira, L.S. Partitioning evapotranspiration, yield prediction and economic returns of maize under various irrigation management strategies. *Agric. Water Manag.* **2014**, *135*, 27–39. [CrossRef]

45. Zhao, N.N.; Liu, Y.; Cai, J.B.; Paredes, P.; Rosa, R.D.; Pereira, L.S. Dual crop coefficient modelling applied to the winter wheat-summer maize crop sequence in North China Plain: Basal crop coefficients and soil evaporation component. *Agric. Water Manag.* **2013**, *117*, 93–105. [CrossRef]

46. Rosa, R.D.; Paredes, P.; Rodrigues, G.C.; Alves, I.; Fernando, R.M.; Pereira, L.S.; Allen, R.G. Implementing the dual crop coefficient approach in interactive software. 1. Background and computational strategy. *Agric. Water Manag.* **2012**, *103*, 8–24. [CrossRef]

47. Paço, T.A.; Pôças, I.; Cunha, M.; Silvestre, J.C.; Santos, F.L.; Paredes, P.; Pereira, L.S. Evapotranspiration and crop coefficients for a super intensive olive orchard. An application of SIMDualKc and METRIC models using ground and satellite observations. *J. Hydrol.* **2014**, *519*, 2067–2080. [CrossRef]

48. Qiu, R.J.; Du, T.S.; Kang, S.Z.; Chen, R.Q.; Wu, L.S. Assessing the SIMDualKc model for estimating evapotranspiration of hot pepper grown in a solar greenhouse in Northwest China. *Agric. Syst.* **2015**, *138*, 1–9. [CrossRef]

49. Pereira, L.S.; Cordery, I.; Iacovides, I. Improved indicators of water use performance and productivity for sustainable water conservation and saving. *Agric. Water Manag.* **2012**, *108*, 39–51. [CrossRef]

50. Pereira, L.S.; Paredes, P.; Rodrigues, G.C.; Neves, M. Modeling malt barley water use and evapotranspiration partitioning in two contrasting rainfall years. Assessing AquaCrop and SIMDualKc models. *Agric. Water Manag.* **2015**, *159*, 239–254. [CrossRef]

51. Martins, J.D.; Rodrigues, G.C.; Paredes, P.; Carlesso, R.; Oliveira, Z.B.; Knies, A.E.; Petry, M.T.; Pereira, L.S. Dual crop coefficients for maize in southern Brazil: Model testing for sprinkler and drip irrigation and mulched soil. *Biosyst. Eng.* **2013**, *115*, 291–310. [CrossRef]

52. Hanson, B.R.; May, D.M. Crop coefficients for drip-irrigated processing tomato. *Agric. Water Manag.* **2006**, *81*, 381–399. [CrossRef]

53. Hanson, B.R.; May, D.M. Crop evapotranspiration of processing tomato under furrow and subsurface drip irrigation. *Acta Hortic.* **2004**, *664*, 303–307. [CrossRef]

54. Aksic, M.; Gudzic, S.; Deletic, N.; Gudzic, N.; Stojkovic, S. Tomato fruit yield and evapotranspiration in the conditions of South Serbia. *Bulg. J. Agric. Sci.* **2011**, *17*, 150–157.

55. Mukherjee, A.; Kundu, M.; Sarkar, S. Role of irrigation and mulch on yield, evapotranspiration rate and water use pattern of tomato (*Lycopersicon esculentum* L.). *Agric. Water Manag.* **2010**, *98*, 182–189. [CrossRef]

56. Giuliani, M.M.; Nardella, E.; Gagliardi, A.; Gatta, G. Deficit irrigation and partial root-zone drying techniques in processing tomato cultivated under Mediterranean climate conditions. *Sustainability* **2017**, *9*, 2197. [CrossRef]

57. Zaccaria, D.; Carrillo-Cobo, M.T.; Montazar, A.; Putnam, D.H.; Bali, K. Assessing the viability of sub-surface drip irrigation for resource-efficient alfalfa production in central and southern California. *Water* **2017**, *9*, 837. [CrossRef]

58. Zheng, J.H.; Huang, G.H.; Jia, D.D.; Wang, J.; Mota, M.; Pereira, L.S.; Huang, Q.Z.; Xu, X.; Liu, H.J. Responses of drip irrigated tomato (*Solanum lycopersicum* L.) yield, quality and water productivity to various soil matric potential thresholds in an arid region of Northwest China. *Agric. Water Manag.* **2013**, *129*, 181–193. [CrossRef]

59. Patanè, C.; Tringali, S.; Sortino, O. Effects of deficit irrigation on biomass, yield, water productivity and fruit quality of processing tomato under semi-arid Mediterranean climate conditions. *Sci. Hortic.* **2011**, *129*, 590–596. [CrossRef]

water

MDPI

Article

Crop Coefficients and Transpiration of a Super Intensive Arbequina Olive Orchard using the Dual K_c Approach and the K_{cb} Computation with the Fraction of Ground Cover and Height

Teresa A. Paço [1,*], Paula Paredes [1], Luis S. Pereira [1], José Silvestre [2] and Francisco L. Santos [3]

[1] Research Center on Linking Landscape, Environment, Agriculture and Food (LEAF), Instituto Superior de Agronomia, Universidade de Lisboa, Tapada da Ajuda, 1349-017 Lisboa, Portugal; pparedes@isa.ulisboa.pt (P.P.); lspereira@isa.ulisboa.pt (L.S.P.)

[2] Instituto Nacional de Investigação Agrária e Veterinária, I.P. (INIAV), 2565-191 Dois Portos, Portugal; jose.silvestre@iniav.pt

[3] Instituto de Ciências Agrárias e Ambientais Mediterrânicas (ICAAM), Universidade de Évora, Largo dos Colegiais, 7006-554 Évora, Portugal; f.lucio.santos@gmail.com

* Correspondence: tapaco@isa.ulisboa.pt; Tel.: +351-21-3653331

Received: 28 January 2019; Accepted: 19 February 2019; Published: 22 February 2019

Abstract: The SIMDualKc model was used to simulate crop water requirements for a super high density olive orchard in the region of Alentejo, Portugal. This model uses the dual crop coefficient approach to estimate and partitioning the actual crop evapotranspiration ($ET_{c\,act}$) and therefore to perform the soil water balance. The model was calibrated with 2011 tree transpiration using trunk sap flow measurements and was validated using similar data from 2012 and tested with 2013 data. Low root mean square errors (RMSE < 0.53 mm·d^{-1}) and acceptable modelling efficiency indicators (EF > 0.25) were obtained. Further validation was performed comparing modelled $ET_{c\,act}$ with eddy covariance measurements. These indicators support the appropriateness of using SIMDualKc to guide irrigation management. The basal crop coefficient (K_{cb}) curves obtained with SIMDualKc for those 3 years were compared with the K_{cb} values computed with the Allen and Pereira approach (A&P approach) where K_{cb} is estimated from the fraction of ground cover and plant height considering an adjustment factor for crop stomatal control (F_r). F_r values were obtained through a trial and error procedure through comparing the K_{cb} estimated with this approach and with SIMDualKc. The K_{cb} curves obtained by both methods resulted highly correlated, which indicates that the A&P approach may be used in the irrigation management practice to estimate crop water requirements. Results of performing the soil water balance with SIMDualKc have shown that soil evaporation is a large fraction of $ET_{c\,act}$, varying between 41% and 45% for the 3 years under study. Irrigation, applied with a drip system, represented 39 to 56% of $ET_{c\,act}$, which shows the great importance of irrigation to achieve the water requirements of super intensive olive orchards. Nevertheless, the analysis has shown that the irrigation management adopted at the orchard produces a water deficit larger than desirable, with a ratio of $ET_{c\,act}$ to non-stressed crop evapotranspiration (ET_c) varying from 70% to 94% during the mid-season, when that ratio for a eustress irrigation management could be around 90%.

Keywords: Evapotranspiration; Irrigation; Density coefficient; Dual crop coefficients; Row crops

---

1. Introduction

Olive orchards consist of the dominant permanent crops in Portugal, covering approximately 50% of the total area with tree crops, namely in Alentejo, southern Portugal, where olives are mainly cropped for oil production. Super high density olive orchards, with more than 1500 trees ha^{-1}, also known as hedgerow olive orchards, are growing fast due to their high yield and economic productivity [1–3]. However, they have higher water requirements than less intensive orchards [1]; it is therefore necessary to improve related knowledge about crop evapotranspiration to support appropriate irrigation management and scheduling [4,5]. Meanwhile, impacts of water deficit on olives growth and yield are relatively well known [6–11]. That knowledge about olives evapotranspiration and responses to water deficits is also essential to assess scenarios of climate change expected for the region, mainly referring to higher temperature during summer and less rainfall in winter and autumn [12,13].

Although information on crop evapotranspiration can be obtained through using various field measurements techniques [14], those involving direct measurements of ET are generally expensive, labour consuming, require appropriate skills of users and are more appropriate for research, that is, for example, measurements of soil water content [15,16], sap-flow [17–19] and eddy covariance [18–20], which are applied in this study. Differently, crop ET modelling, using commonly observed meteorological data [4], radiometric canopy temperature [21,22], web based sensors networks [23] and remote sensing information [24–28], including using unmanned aerial vehicles [29] might be useful irrigation management tools, namely for irrigation scheduling purposes and to generate mitigation scenarios to face drought and climate change. Example of models applications to olive orchards include WABOL [30], SIMDualKc [19] and HYDRUS 2D [31].

The most common approach to estimate potential crop evapotranspiration (ET$_c$) is the use of the K$_c$-ET$_0$ approach [32], where ET$_0$ is the grass reference evapotranspiration (PM-ET$_0$) and K$_c$ is a crop coefficient, which relates ET$_c$ with ET$_0$ relative to various crop characteristics [4]. K$_c$ changes throughout the crop season and a multi-stage linear approximation is commonly used to represent the K$_c$ curve through defining their values at the initial, mid-season and end-season stages, respectively K$_{c\,ini}$, K$_{c\,mid}$ and K$_{c\,end}$ [4]. Adopting the dual K$_c$ approach, both ET$_c$ components are considered, crop transpiration described by the basal crop coefficient K$_{cb}$ and the soil evaporation described by the coefficient K$_e$ [33]. When any crop stress occurs, K$_{cb}$ is corrected through a stress coefficient, K$_s$, resulting that the actual K$_{cb}$ (K$_{cb\,act}$ = K$_s$ K$_{cb}$) is smaller than the non-stress or standard one, that is, K$_{cb\,act}$ ≤ K$_{cb}$. It results that the actual crop ET is smaller than the potential, non-stressed ET, that is, ET$_{c\,act}$ ≤ ET$_c$. Therefore:

$$ET_{c\,act} = (K_s\,K_{cb} + K_e) \times ET_0 = (K_{cb\,act} + K_e) \times ET_0 = K_s \times K_c \times ET_0 = K_{c\,act} \times ET_0 \qquad (1)$$

Field research, as the one reported in this study, is required to determine K$_c$ and K$_{cb}$ in relation with the crop characteristics and crop management. The standard or non-stressed values of K$_{cb}$ and K$_c$ are transferable to other locations—where training of olives orchards are similar—after adjusting K$_c$ or K$_{cb}$ to the prevailing climate conditions as proposed by Allen et al. [4]. The use of a well calibrated model such as SIMDualKc [34], which adopts the dual K$_c$ approach to partitioning ET$_{c\,act}$ into actual plant transpiration and soil evaporation, using respectively K$_{cb\,act}$ and K$_e$ [4,33], is quite helpful to accurately computing ET$_{c\,act}$ and to derive the standard K$_{cb}$ and K$_c$ values as demonstrated in previous studies [35–37]. Moreover, the performance of that partition using SIMDualKc has been positively tested through comparing the model simulated transpiration against sap-flow measurements [19,38,39] or the simulated soil evaporation against micro-lysimeters' observations [40–42].

In alternative to modelling, Allen and Pereira [43] proposed predicting K$_{cb}$ or K$_c$ by adopting a density coefficient (K$_d$) computed from the fraction of ground cover (f$_c$) and the crop height (h), herein referred as A&P approach. It takes into consideration crop stomatal control through an adustment factor (F$_r$) that varies with crop characteristics and water management. The A&P approach has been used to ease the SIMDualKc modelling process applied to partial cover woody crops, such as vineyards [35], peach orchards [38] and olive orchards [19]. However, the A&P approach has not yet been used extensively for predicting K$_{cb}$ throughout the crop season.

Considering the analysis above, the objectives of this study consisted of (i) calibrating and validating the model SIMDualKc using both sap-flow estimates of transpiration and ET eddy covariance observations; (ii) determining K_{cb} and K_c from SIMDualKc calibration; (iii) testing the A&P approach for predicting K_{cb} from the fraction of ground cover and crop height; and (iv) determining the terms of the soil water balance, particularly relationships between soil evaporation, actual transpiration and $ET_{c\,act}$. The overall objective is to provide information for irrigation of olives' orchards with reduced irrigation, so accepting a yield-water beneficial stress, the "eustress."

2. Material and Methods

2.1. Experimental Site

The experimental site (38°24′N, 7°43′W, 143 m a.s.l.) is located in Alentejo, Southern Portugal, in a commercial super-high intensive hedgerow olive orchard farmed by "Olivais do Sul." The orchard has a total area of 78 ha, which land has a smooth undulation. The climate is dry sub-humid of Mediterranean type, with most of the rainfall in autumn and winter; according to the Köppen-Geiger classification [44], the climate is a Csa, characterized by mild rainy winters and dry hot summers. The annual rainfall ranged between 511 and 736 mm in the experimental years, while the average monthly temperature ranged from 9.6 °C in January to 23.3 °C in July and August. Prevailing winds are from the North-West direction. Main daily weather data characterizing the experimental years is presented in Figure 1.

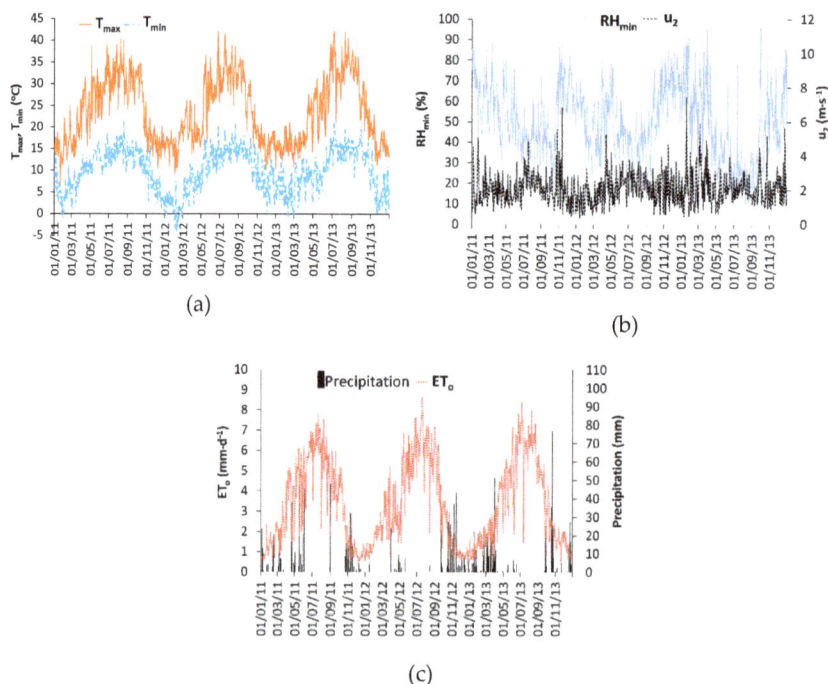

(a)

(b)

(c)

Figure 1. Daily weather data of Viana do Alentejo relative to (**a**) maximum (T_{max}) and minimum (T_{min}) temperatures, (**b**) minimum relative humidity (RH_{min}), wind speed at 2 m height (u_2) and (**c**) precipitation and reference evapotranspiration (ET_o) for the period 2011–2013.

The soil is a Cambissol [45] having a sandy loam texture, with moderate to low infiltration. Soil water at field capacity averages 0.24 cm^3·cm^{-3} through the soil profile down to 1.20 m, while the permanent wilting point to the same depth is 0.12 cm^3·cm^{-3}. The total available water to that depth is TAW = 160 mm. Soil water observations were performed weekly with a TDR probe (TRIME, IMKO, Ettlingen, Germany); data were used to perform the soil water balance considering a soil root zone depth Z_r = 1.10 m.

The olive trees (cv. Arbequina) were hedgerow planted in 2006 adopting a super-high density (1.35 m × 3.75 m, 1975 trees ha^{-1}). The observed crop growth stages dates along the crop seasons of 2011, 2012 and 2013 are presented in Table 1. The fraction of ground cover (f_c, dimensionless) was estimated from measurements of the crown diameter along the row direction and perpendicularly to it for 51 trees. In addition, f_c values estimated from remote sensing [28] were also considered. f_c ranged from 0.17 to 0.38. The tree height (h, m) ranged from 3.0 to 4.0 m. Both f_c and h varied through the crop season because trees were pruned during winter, by early January. However, in 2012, after a severe frost that occurred by late February, a heavy pruning was applied after, which caused larger changes in f_c and h in that year. Harvesting in the study plots were performed by 16 October in 2011 and 2012 and by 10 October in 2013 (Table 1). The yield average was 14 ton·ha^{-1} in 2011 and 18 ton·ha^{-1} in 2013. A low yield of 3 ton·ha^{-1} was observed in 2012 due to the referred frost and heavy pruning.

Table 1. Crop growth stages of the olive orchard under study.

Year	Crop Growth States					
	Non-Growing	Initiation	Crop Development	Mid-Season	Late-Season	Non-Growing
2011	1/1–28/2	1/3–15/3	16/3–30/4	1/5–13/9	14/9–11/11	12/11–31/12
2012	1/1–20/3	21/3–11/4	12/4–14/5	15/5–10/9	11/9–23/11	24/11–31/12
2013	1/1–2/3	3/3–17/3	18/3–20/4	21/4–19/9	20/9–15/11	16/11–31/12

Ground cover conditions affecting soil evaporation may be taken into consideration in the water balance simulations by SIMDualKc and were therefore observed. They refer to both crop residues and active ground cover [34,35,46]. In 2011, a reduced active ground cover was present from early January until May, covering 10% of the ground in the row and 5% in the inter-row. In 2012, following the frost by February, a heavy defoliation occurred and leaves formed an organic mulch from late February until early August; its ground cover fraction was estimated as 0.30, corresponding to a reduction of the soil evaporation of 30%. In 2013, an active ground cover of nearly 20% on both the row and the inter-row was observed from January until May, when it dried out and became a residues cover.

Irrigation took place nearly every day during spring and summer and was applied by the evening. A drip system was used, with emitters spaced of 0.75 m along the row. The emitters discharge was of 2.3 L h^{-1}. The wetted fraction (f_w = 0.23) was calculated from the area of a wetted ellipse circling each emitter. Irrigations were scheduled by the farmer, who adopted average daily irrigation depths close to 3 mm, however varying from 1 to 8 mm, approximately. Irrigation depths were measured with a tipping-bucket rain gauge (ARG100, Environmental Measurements Ltd., Sunderland, UK).

2.2. Eddy Covariance Measurements

The eddy covariance (EC) micrometeorological technique was used to measure evapotranspiration during short periods of the irrigation season (July–August in 2011 and June–August in 2012) as a strategy to calibrate the sap flow measurements and obtaining accurate transpiration estimates. This strategy has been successfully used in previous studies [38,47].

The EC system comprised a three-dimensional sonic anemometer and a krypton hygrometer (respectively Models CSAT3 and KH20, Campbell Scientific, Inc., Logan, UT, USA) mounted at a height of 4.8 m on a metallic tower, with a path separation of 0.1 m. The fetch were 470 m, 353 m, 455 m and 504 m for the north, west, south and east directions, respectively (Figure 2). EC raw data (H—sensible heat flux density and LE—latent heat flux density) were collected at a 10 Hz frequency to a datalogger (Model CR1000, Campbell Scientific, Inc., Logan, UT, USA) and afterwards analysed with the Software package TK3 [48] for correction and calculation of 30 min-averages. Data corrections were performed according to Foken et al. [49]. They included: despiking of the raw data [50]; coordinate rotation using the Double Rotation method [51] to account for the non-flat terrain conditions; corrections for oxygen cross-sensitivity of the krypton hygrometer [52] and spectral loss [53]; conversion of buoyancy flux to sensible heat flux [54]; and the WPL correction for density fluctuations [55]. To evaluate the surface energy balance, soil heat flux (G) and net radiation (R_n) were measured using eight soil heat flux plates (calibrated Peltier modules sealed 20 V, 4.4 A, 40 × 40 × 3.9 mm, RS Components, Madrid, Spain) and a net radiometer (Model NR-LITE, Campbell Scientific, Inc., Logan, Utah, USA), respectively. The soil heat flux plates were placed in the tree row and in the inter-rows at a depth of 2 cm. The daily energy balance equation error closure was determined by linear regression forced to origin.

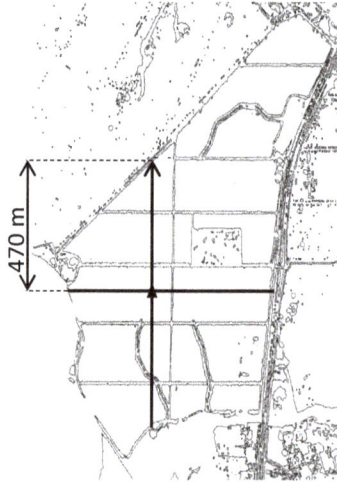

Figure 2. Olive orchard and location of the eddy covariance tower (black triangle, 38°24'46.3''N, 7°43'39.8''W).

The footprint analysis [56], performed to access the representativeness of the EC measurements, showed that over 90% of the fluxes sensed, determined as cumulative normalized flux (CNF), came from the region of interest, regarding the four main cardinal points. The fetch (distance to the edge of the plot) in the main directions, varying approximately between 350 and 500 m, was then considered adequate (Figure 3). Measurements were mainly affected by fluxes coming from an upwind area at a distance of 15 m from the tower (maximum of the one-dimensional footprint function, which provides the relative contribution to the vertical flux: $Q_f = (1/Q_o) \, dQ/dx$, for a given height z, being Q_o the latent heat flux density measured at point x = (0, z)).

Measurements in days with prevailing winds from the North-East direction were discarded given the vicinity of a building approximately at 200 m in that direction. Prevailing winds, calculated as average frequency for each of the EC measurement periods, were in agreement with historical climatic data, which indicates the North-West direction (Figure 4). Wind direction frequency was further analysed in detail for individual days and EC data screened accordingly for specific modelling purposes.

The energy balance equation closure error (Figure 5) was determined with a linear regression analysis forced to the origin using daily values of the measured fluxes (LE, H, R_n and G). The error, below 10% (H + LE = 0.91 (R_n − G), R^2 = 0.87), is similar to that found by other authors using this technique in orchards [57–59].

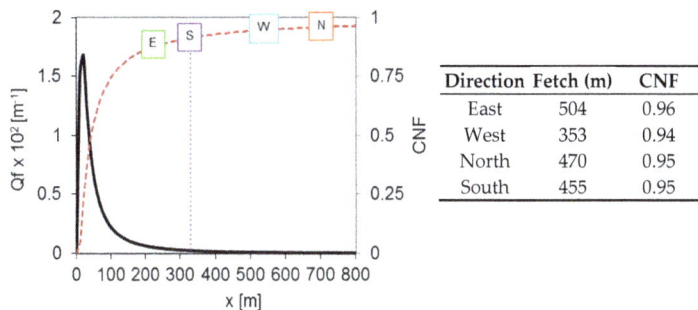

Figure 3. Cumulative normalized flux (CNF, dashed line) and relative contribution to the vertical flux for a given height z, (Q_f, continuous line) according to distance from the measurement point (EC tower) to the plot limit (x), obtained with footprint analysis; marked values for the fetch and CNF according to the cardinal directions; x is the distance between the observation point and the fluxes source region.

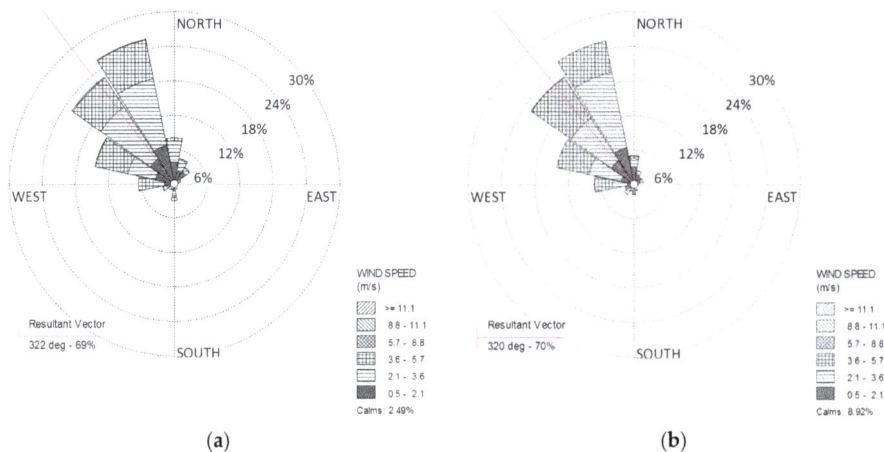

Figure 4. Wind prevailing directions during periods of eddy covariance measurements and resultant vector, (**a**) July–August 2011, (**b**) June–August 2012 (deg.: degrees).

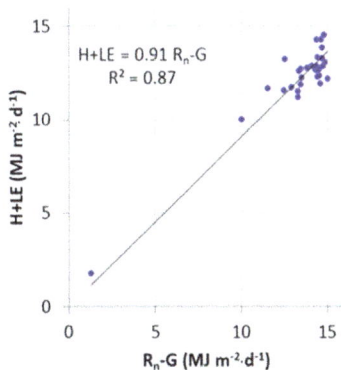

Figure 5. Surface energy balance equation for flux density daily values of LE (latent heat flux density), H (sensible heat flux density), G (soil heat flux density) and R_n (net radiation).

2.3. Transpiration from Sap Flow Measurements with the Granier Method

Plant transpiration was measured with thermal dissipation Granier probes [60] between DOY (day of year) 134 in 2011 and DOY 194 in 2013 (continuous half-hourly records). Six trees, chosen according to trunk diameter class frequency determined in a large sample of the orchard, were equipped with 1 cm length sensors (UP GmbH, Cottbus, Germany). Average data were stored each thirty minutes in a datalogger (Model CR1000, Campbell Scientific, Inc., Logan, UT, USA). Natural temperature gradients in the tree trunk were accounted for using data from non-heated sap flow sensors calculated during long periods. Daily transpiration was calculated from the half-hourly data. For flux calculation, the sapwood area in the cross section of the trunk was evaluated from core samples taken with an increment borer. Observation of the samples allowed considering the whole area conductive except the bark.

Sap flow methods are often referred as underestimating transpiration [19,61–65]. Although the original calibration of this method, which allows converting temperature differences between the two probes into crop transpiration, was initially considered universal, more recently, evidences show that it is advisable to verify its accuracy. Raw sap flow observations (SF, mm·d^{-1}) were therefore corrected to obtain calibrated sap-flow transpiration (T_{SF}, mm·d^{-1}), as referred in previous applications to tree crops [38,47,61]. This was done by mathematically relating sap-flow, SF, with transpiration computed from EC observations (T_{EC} = ET_{EC}−E_{sim}). T_{EC} represents the difference between ET obtained from EC measurements (ET_{EC}, mm·d^{-1}) and the soil evaporation (E_{sim}, mm·d^{-1}) simulated with the two phases Ritchie model [66], the latter computed following Allen et al. [4]. The computation of E_{sim} based upon the water balance of the soil evaporation layer using soil moisture observations. Parameters characterizing that soil layer, with a depth Z_e = 0.10 m, consisted of: total and readily evaporable water, respectively TEW = 18 mm and REW = 9 mm. Further information on E_{sim} data was provided by Paço et al. [19]. This procedure allows enlarging the available field information to the entire simulation period [67,68].

The best mathematical relationships between T_{EC} and SF (Figure 6) were of exponential form:

$$T_{EC} = K_{SF}\, e^{\alpha \cdot SF} \tag{2}$$

with K_{SF} = 0.34 and α = 1.40 (R^2 = 0.55; n = 13) for 2011 and K_{SF} = 0.34 and α = 1.72 (R^2 = 0.61; n = 28) for 2012, which was extended for 2013. The adjustment of SF was performed with Equation 2 assuming that the transpiration derived from sap-flow measurements (T_{SF}) equals that estimated from EC observations as referred above (T_{EC}). The approach used may be affected by uncertainties in computing E_{sim} as well as measurements errors affecting both T_{EC} and SF. Nevertheless, The T_{SF} values resulting from the application of Equation 2, which varied between 0.36 and 3.54 mm·d^{-1}, are comparable to transpiration values reported for other studies on olive orchards [18,59,69].

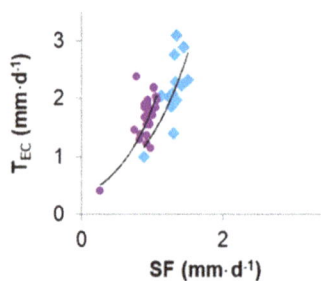

Figure 6. Relationships between the raw transpiration flux observed with the Granier sap-flow method (SF) and the transpiration rate computed from the eddy covariance observations (T_{EC}) relative to 2011 (squares) and 2012 (circles).

T_{SF} data were screened for outliers taking into consideration: (a) the range of values used to establish the sap flow calibration equation and the respective K_{cb} values ($K_{cb\,SF} = T_{SF}/ET_o$), which should be kept within an acceptable range. The common definition of outliers by Tukey [70] was used, thus considering as outliers the $K_{cb\,SF}$ values higher than 1.5 times the interquartile range, below the first quartile or above the third quartile. The application of these rules led to discarding a total of 177 T_{SF} values, with outliers representing 3.5% and 7% of observations in 2011 and 2012, respectively, contrasting with 56% of values computed for 2013. That high number of discarded values was most likely due to the fact that the SF equation used for 2013 was not purposefully calibrated. It was assumed that discarding the outliers decreased the uncertainties of observations and computations.

2.4. Basal Crop Coefficients Derived from Crop Height and Fraction of Ground Cover

To take into account fraction of ground cover by the crop (f_c) and the crop height (h), a density coefficient (K_d) is used with the A&P approach to estimate K_{cb} [43]. K_d (dimensionless) is defined as the minimum value among the effective ground cover corrected for the density of shading or corrected for the crop shading depending upon the crop height. Thus, K_d is given by:

$$K_d = \min(1, M_L\, f_{c\,eff}, f_{c\,eff}^{(1/(1+h))}) \qquad (3)$$

where $f_{c\,eff}$ is the effective fraction of ground covered or shaded by vegetation near solar noon (dimensionless), M_L is a multiplier on $f_{c\,eff}$ describing the effect of canopy density on shading the ground, thus on maximum relative ET per fraction of ground shaded and h (m) is the mean height of the crop vegetation.

K_{cb} is estimated from K_d as

$$K_{cb} = K_{c\,min} + K_d\,(K_{cb\,full} - K_{c\,min}) \qquad (4a)$$

when the inter-row is bare soil or, when an active ground cover is present, as

$$K_{cb} = K_{cb\,cover} + K_d\left(\max\left[K_{cb\,full} - K_{cb\,cover},\ \frac{K_{cb\,full} - K_{cb\,cover}}{2}\right]\right) \qquad (4b)$$

where $K_{c\,min}$ is the minimum K_c for bare soil (≈ 0.15 for typical agricultural conditions), $K_{cb\,cover}$ is the K_{cb} of the ground cover in the absence of tree foliage and reflects the density and vigour of the active ground cover crop and $K_{cb\,full}$ is the estimated basal K_{cb} during the peak plant growth for conditions having nearly full ground cover. The second term of the max function allows taking into consideration the impacts of shading of the canopy over the ground cover crop. Further information is provided by Allen and Pereira [43].

The $K_{cb\,full}$ represents a general upper limit on $K_{cb\,mid}$ for tall vegetation having full ground cover and LAI > 3 under full water supply. $K_{cb\,full}$ takes into consideration crop stomatal control and is approximated as a function of the mean plant height and adjusted for climate [43],

$$K_{cb\,full} = F_r\left(\min(1.0 + 0.1h,\ 1.20) + [0.04(u_2 - 2) - 0.004(RH_{min} - 45)]\left(\frac{h}{3}\right)^{0.3}\right) \qquad (5)$$

with F_r (0–1) is an adjustment factor relative to crop stomatal control, the min function assumes that 1.20 is the upper bound for $K_{cb\,full}$ prior to adjustment for climate. h (m) is the average maximum crop height, u_2 (m s^{-1}) is the average wind speed at 2 m height, RH_{min} (%) is the average minimum relative humidity during mid-season.

The parameter F_r applies a downward adjustment when the vegetation exhibits more stomatal control on transpiration than it is typical of most annual agricultural crops. $F_r < 1.0$ for various tree crops and natural vegetation. Standard values for F_r for various tree crops are provided by Allen and Pereira [43] but F_r values may be adjusted experimentally, namely through comparing K_{cb} obtained with the A&P approach with those derived through a different but accurate procedure, namely through the use of models. F_r may also be computed through comparing the leaf resistances for the crop (r_l, s m^{-1}) with that of grass (100 s m^{-1}) considering the climate conditions [43], thus:

$$F_r \approx \frac{\Delta + \gamma \, (1 + 0.34 \, u_2)}{\Delta + \gamma \, (1 + 0.34 \, u_2 \, \frac{r_l}{100})} \tag{6}$$

where r_l (s·m^{-1}) is mean leaf resistance for the vegetation in question (s·m^{-1}) Δ is the slope of the saturation vapour pressure curve (kPa·$^\circ$C^{-1}), γ is the psychometric constant (kPa·$^\circ$C^{-1}) and u_2 is the average wind speed during the relevant crop stage measured at 2 m height (m·s^{-1}). r_l values for various crops are tabulated by Allen e Pereira [43], who also referred that for most annual agricultural crops r_l is often not far from 100 s·m^{-1}.

2.5. The SIMDualKc Model

The SIMDualKc water balance model [34] uses a daily time step to perform the soil water balance at the field scale through computing crop ET with the dual K_c approach [4,33,43]. The SIMDualKc model adopts the approach described with Equation 1, thus assuming that $ET_{c\,act} = ET_c$ when the soil water depletion fraction does not exceeds the depletion fraction for no stress, p (dimensionless).

The model computes the soil water depletion at the end of every day:

$$D_{r,\,i} = D_{r,\,i-1} - (P_e - RO)_i - I_i - CR_i + ET_{cact,\,i} + DP_i \tag{7}$$

where $D_{r,i}$ and $D_{r,i-1}$ are the root zone depletion at the end of respectively day i and day i−1 (mm), P_e is precipitation (mm), RO is runoff (mm), I is the net irrigation depth that infiltrates the soil (mm), CR is capillary rise from the groundwater table (mm), $ET_{c\,act}$ is actual crop evapotranspiration (mm) and DP is deep percolation through the bottom of the root zone (mm), all referring to day i. CR and DP are calculated with parametric equations described by Liu et al. [71] and RO is estimated using the curve number approach [72]. Because the groundwater table is quite deep in the region, CR was assumed to be null.

The model computes the actual ET ($ET_{c\,act}$, mm) as a function of the available soil water in the root zone using a water stress coefficient (K_s, 0–1). K_s is computed daily as a linear function of the depletion D_r in the effective root zone [4,33]:

$$K_s = \frac{TAW - D_r}{TAW - RAW} = \frac{TAW - D_r}{(1 - p)\,TAW} \rightarrow \text{for } D_r > RAW \tag{8}$$

$$K_s = 1 \rightarrow \text{for } D_r \leqslant RAW \tag{9}$$

where TAW and RAW are, respectively, the total and readily available soil water (mm) relative to the root zone depth Z_r and p is the soil water depletion fraction for no stress, therefore with RAW = p TAW. Thus, $ET_{c\,act} = ET_c$ when $K_s = 1.0$; otherwise, when water and/or salinity stress occurs, $ET_{c\,act} < ET_c$ and $K_s < 1.0$. Then, because $K_{cb\,act} = K_s\,K_{cb}$, it results $K_{cb\,act} < K_{cb}$. Similarly, it results that actual transpiration will be smaller than its potential value, thus $T_{c\,act} < ET_c$ when $K_s < 1.0$.

The evaporation from the soil surface (E_s, mm·d^{-1}) is limited by the amount of energy available at that surface in conjunction with the energy consumed as crop transpiration [4,72]. The model computes the evaporation coefficient (K_e) performing a daily water balance of the evaporation soil layer, which is characterized by its depth (Z_e, m), the total evaporable water (TEW, mm) and the readily evaporable water (REW, mm). TEW is the maximum depth of water that can be evaporated from the evaporation soil layer when that layer has been fully wetted and REW is the depth of water that can be evaporated without water availability restrictions. E_s is maximum when the topsoil is fully wetted by rain or irrigation and the soil surface shadowed by the crop is minimum. Differently, E_s is minimum when the crop fully shadows the soil and energy available for evaporation is minimum [4,34,72].

The maximum value of K_e is attained when the soil is wet. That value is limited by the energy available at the soil surface, which corresponds to the difference between $K_{c\,max}$, representing maximum ET when the crop would fully cover the soil and K_{cb}, representing the effective transpiration of the crop. As the topsoil dries, less water is available for evaporation and E_s is reduced. This reduction in E_s is proportional to the amount of water remaining in the surface soil layer. K_e is then computed considering an evaporation reduction coefficient ($K_r \leqslant 1.0$) as

$$K_e = K_r (K_{cmax} - K_{cb}) \rightarrow \text{with } K_e \leqslant f_{ew} K_{cmax} \tag{50}$$

where $K_{c\,max}$ is the maximum value of K_c (i.e., $K_{cb} + K_e$) following a wetting event by rain or irrigation, generally 1.20 and f_{ew} is the fraction of the soil that is both exposed to radiation and wetted by rain or irrigation, that is, the fraction of soil surface from which most evaporation occurs. Thus, f_{ew} depends upon the fraction of ground covered by the crop (f_c) and of the fraction of soil wetted by irrigation (f_w), thus $f_{ew} = \min(1-f_c, f_w)$. K_r is calculated using the 2-stage drying cycle approach [4,66], where the first stage is the energy limiting stage and the second is the water limited stage or falling rate stage, where evaporation decreases as evaporable water decreases in the evaporation soil layer beyond the readily evaporable water (REW):

$$K_r = 1 \rightarrow \text{for } D_{e,i-1} \leqslant REW \tag{61}$$

$$K_r = \frac{TEW - D_{e,\,i-1}}{TEW - REW} \text{ for } D_{e,i-1} > REW \tag{72}$$

Further descriptions of the model and auxiliary equations are given by Rosa et al. [34]. The input data required by the model include:

i Daily climatic data: reference evapotranspiration (ET_o, mm), precipitation (P, mm), minimum relative humidity (RH_{min}, %) and wind speed at 2 m height (u_2, m·s^{-1}).

ii Soil data for a multi-layered soil: number of layers and related depths d (m); the respective soil water content at field capacity and at the wilting point (θ_{FC} and θ_{WP}, m^3·m^{-3}) or the total available water (TAW, mm); the characteristics of the soil evaporation layer (Z_e, REW and TEW); and the soil water content at planting in both the root zone and the evaporation layer expressed as a % of depletion of TAW and TEW, respectively.

iii Crop data: dates of the crop growth stages (non-growing, initial, crop development, mid-season and end season); the basal crop coefficients for the non-growing, initial, mid-season and end season ($K_{cb\,non-growing}$, $K_{cb\,ini}$, $K_{cb\,mid}$ and $K_{cb\,end}$); the soil water depletion fractions for no stress at the same stages ($p_{non-growing}$, p_{ini}, p_{mid} and p_{end}); root depths (Z_r, m), plant height (h, m) and the fraction of ground cover by the crop (f_c, %) throughout the crop season.

iv Irrigation scheduling data: dates and depths of observed irrigation events. In addition, the model requires data relative to the fraction of soil wetted by irrigation (f_w). In this application, $f_w = 0.23$ because the soil was wetted by micro-irrigation.

v Parameters required to compute capillary rise and deep percolation when using the parametric equations proposed by Liu et al [71]; in the present application only the parameters a_D and b_D characterizing DP were used.

vi Base data to compute surface runoff using the Curve Number method [72].

vii Information characterizing the active ground cover, soil residues and mulches and related effects on E_s [34].

viii soil and water salinity information, however not used in the present application.

2.6. Model Calibration and Validation and Goodness of Fit Indicators

So far, the calibration and validation of the SIMDualKc model have been performed either through comparing computed actual transpiration with sap-flow measurements [38,39], simulated $ET_{c\,act}$ with

EC observed evapotranspiration [73,74] or, more commonly, simulated with observed soil water content [35,37,41,75,76]. In this study, model calibration was considered as the process of adjusting influential model parameters and inputs within their reasonable ranges so that the simulated $T_{c\,act}$ values were in agreement with sap flow derived data (T_{SF}).

The calibration procedure consisted of adjusting the crop parameters—K_{cb} and p values relative to the non-growing, initial-, mid- and end-season growth stages-, the soil evaporation parameters—TEW, REW and Z_e-, the deep percolation parameters a_D and b_D and the CN of the runoff curve number algorithm, by minimizing the differences between T_{SF} and simulated $T_{c\,act}$. Daily data of 2011 was used. An initial set of parameters was selected: crop parameters (K_{cb} and p) from Allen et al. [4] and Paço et al [19]; soil evaporation parameters from Paço et al. [19], CN from Allen et al. [72] and DP parameters from Paço et al. [19]. The initial soil water conditions in 2011 were observed in the field and consisted of the initial depletion of the evaporation layer, that was 40% of TEW and the initial depletion in the entire root zone, that was 40% of TAW. A trial and error procedure was then developed for selecting first the K_{cb} values. In the following, the trial and error procedure was applied to the p values, then to the referred soil evaporation, deep percolation and, lastly, to the CN parameters.

The validation of SIMDualKc consisted of using the previously calibrated parameters ($K_{cb\,non\text{-}growing}$, $K_{cb\,ini}$, $K_{cb\,mid}$, $K_{cb\,end}$, $P_{non\text{-}growing}$, P_{ini}, P_{mid}, P_{end}, TEW, REW, Z_e, a_D, b_D and CN) with daily data of 2012. T_{SF} data relative to 2013 was used for model testing with the same calibrated parameters. The process of calibration and validation was considered satisfactory when the goodness-of-fit indicators relative to the validation were within 20% of variation relative to the calibration.

A set of goodness-of-fit indicators were used to assess the model accuracy, as in prior studies with SIMDualKc. These indicators are fully described by Pereira et al. [37]. They included a linear regression coefficient (b_0) of the regression forced through the origin between observed T_{SF} and predicted $T_{c\,act}$ and the determination coefficient R^2 of the ordinary least squares regression. A regression coefficient b_0 close to 1.0 indicates that the predicted values are statistically close to the observed ones and a determination coefficient R^2 near 1.0 indicates that most of the variance of the observed values is explained by the model estimates. Errors of estimation were assessed using the root mean square error (RMSE) and the average absolute error (AAE) as indicators. Targeted values for error indicators are 0.0 which corresponds to a perfect match between simulated and observed values. In addition, to assess any bias tendency of the estimations, the percent bias (PBIAS, %) was used. The PBIAS measures the average tendency of predictions to be larger or smaller than the corresponding observations, with positive values indicating an over-estimation bias and negative values indicating an under-estimation bias. To assess the modelling quality, the Nash and Sutcliff [77] modelling efficiency (EF, dimensionless) was used. The EF is a normalized statistic that determines the relative magnitude of the residual variance compared to the measured data variance or how well observations versus simulations fit the 1:1 line [78]. EF values close to 1.0 indicate that the variance of residuals is much smaller than the variance of observations, thus that model performance is excellent. Contrarily, when EF is negative this means that there is no gain in using the model, thus that the mean of observations is as good predictor as the model.

3. Results and Discussion

3.1. Calibration and Validation of SIMDualKc with Sap Flow Data

The initial and calibrated values of the parameters relative to the SIMDualKc model are presented in Table 2 and they refer to the conditions observed in field. The calibrated K_{cb} parameters are smaller than the initial ones, mainly those relative to the non-growing season and the initial periods ($K_{cb\,non\text{-}growing}$ and $K_{cb\,ini}$). Differences for $K_{cb\,mid}$ resulted relatively small. However, $K_{cb\,mid}$ adjusted to the climate are slightly smaller than reported in Table 2 (0.44 to 0.47) because local climate is dry and RH_{min} is small, while $K_{cb\,end}$ are slightly increased (0.44 to 0.48) since RH_{min} is also increased by the end-season. The calibrated p parameters closely follow those proposed by FAO56 [4]. Runoff CN changed little and deep percolation parameters did not change because these parameters were adjusted in a previous study [19].

Table 2. Initial and calibrated parameters used in SIMDualKc model.

Parameters		Initial	Calibrated
Crop	$K_{cb\ non-growing}$	0.50	0.30
	$K_{cb\ ini}$	0.50	0.30
	$K_{cb\ mid}$	0.55	0.48
	$K_{cb\ end}$	0.50	0.43
	$p_{non-growing}$	0.65	0.65
	p_{ini}	0.65	0.65
	p_{mid}	0.65	0.60
	p_{end}	0.65	0.65
Soil evaporation	TEW	18	18
	REW	9	9
	Z_e	0.10	0.10
Runoff	CN	72	75
Deep percolation	a_D	246	246
	b_D	−0.02	−0.02

$K_{cb\ non-growing}$—basal crop coefficient for the non-growing stage; $K_{cb\ ini}$—basal crop coefficient for the initial crop development stage, $K_{cb\ mid}$—basal crop coefficient for the mid stage, $K_{cb\ end}$—basal crop coefficient for the late-season stage, p—depletion fraction, M_L parameter, TEW—total evaporable water, REW—readily evaporable water, Z_e—thickness of the evaporation layer, CN—curve number, a_D and b_D—deep percolation parameters.

Most studies in literature report on actual K_{cb} values, that is, when the crop was affected by water stress since deficit irrigation is mostly used in olive orchards. In addition, most studies focus on the irrigation season only and data relative to other crop growth stages are quite scarce, namely for the non-growing period and initial crop stage. Moreover, K_{cb} values reported in literature are not adjusted to the climate. Therefore, it results difficult to compare our results with those relative to other studies. $K_{cb\ mid}$ values of this study are similar to those reported by Santos et al. [26] when using remote sensing observations of an olive orchard under water stress conditions. Slightly lower $K_{cb\ mid}$ values were reported by Conceição et al. [59]. Differently, López-Olivari et al [69] reported much lower $K_{cb\ mid}$ values, averaging 0.28, which is likely due to water stress as commonly imposed by farmers. Higher $K_{cb\ mid}$ values (0.58 ± 0.14) were reported by Cammalleri et al. [18] for an intensive olive orchard with f_c = 0.35. Er-Raki et al. [20] also reported a higher $K_{cb\ mid}$ of 0.54. Much higher K_{cb} values were proposed by Allen and Pereira [43]. The p values are similar to those reported by Er-Raki et al. [20] and by Rallo et al. [79].

Results in Figure 7, comparing model simulated $T_{c\ act}$ with transpiration derived from sap-flow measurements (T_{SF}), show a good agreement between $T_{c\ act}$ and T_{SF} along the three years of study. The dynamics of both $T_{c\ act}$ and T_{SF} are coherently described for all three years but fitting is less good for 2012 due to the heavy frost and subsequent pruning that occurred by that winter. Apparently, the trees responded well to pruning and that exceptional condition could have justified using specific K_{cb} and p values but that option was not adopted because validating the model requires using the calibrated parameters. However, despite the peculiar condition occurring in 2012, as well as the uncertainties associated with deriving T_{SF}, model predicted $T_{c\ act}$ described reasonably well the dynamics of transpiration of the olive orchard throughout the three experimental seasons.

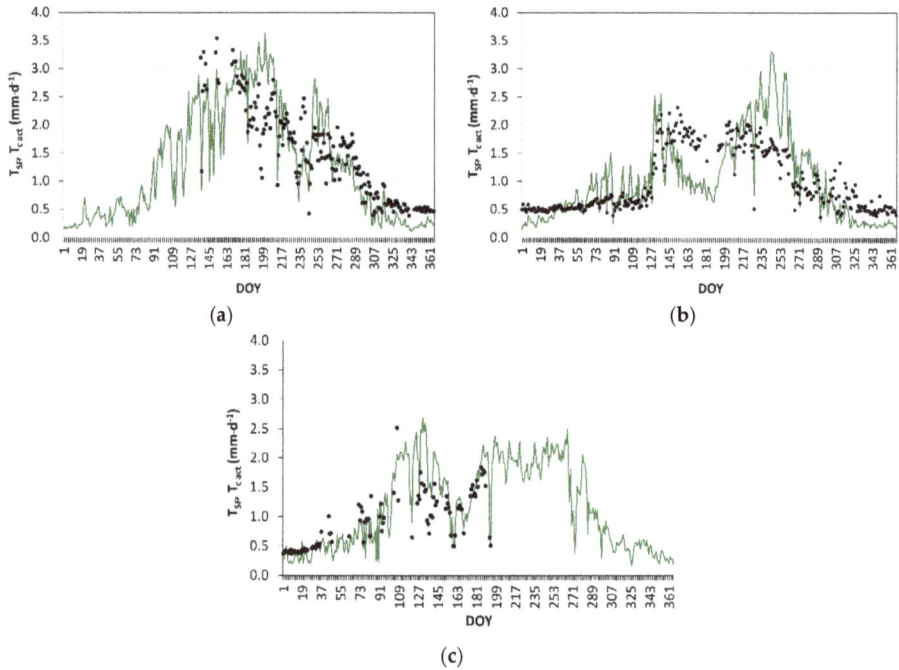

Figure 7. Simulated transpiration dynamics ($T_{c\,act}$, —) compared with sap flow adjusted transpiration (T_{SF}, •) in the (**a**) calibration (2011), (**b**) validation (2012) and (**c**) testing (2013).

The goodness-of-fit indicators (Table 3) show a slight tendency for the model to over-estimate T_{SF}, with $b_0 = 1.04$ and PBIAS = 3.9% in the calibration year; the same trend for over-estimation was observed in 2012 and 2013, with b_0 of 1.03 and 1.17 and PBIAS of 5.2% and 13.3%. The determination coefficients are high for the first and third years, with respectively $R^2 = 0.75$ and 0.78; due to the problems identified to 2012, it resulted $R^2 = 0.60$, however reasonably high. These R^2 values indicate, therefore, that a large fraction of the variance of T_{SF} is explained by the model. Estimation errors are reasonably low, with RMSE ranging from 0.37 to 0.53 mm·d^{-1} and AAE ranging from 0.25 to 0.42 mm·d^{-1}. The modelling efficiency is reasonably high for the calibration year, with EF = 0.57 and acceptable for the validation and testing years. Overall, results for the goodness-of-fit indicators show that the predicted $T_{c\,act}$ values were statistically close to the sap-flow observed ones and that most of the variation of the observed values is explained by the model. Higher errors of estimation were reported by Rallo et al. [79] when using a modification of the FAO56 approach for an intensive olive orchard, with RMSE ranging from 0.30 to 0.78 mm·d^{-1}. Er-Raki et al. [20], also reported higher $T_{c\,act}$ estimation errors, with RMSE = 0.59 mm·d^{-1}.

Table 3. Goodness-of-fit indicators relative to the SIMDualKc simulated transpiration ($T_{c\,act}$) when compared with transpiration obtained from sap flow (T_{SF}).

	n	b_0	R^2	PBIAS (%)	RMSE (mm·d^{-1})	AAE (mm·d^{-1})	EF
Calibration, 2011	201	1.04	0.75	2.0	0.53	0.42	0.57
Validation, 2012	338	1.04	0.60	3.0	0.47	0.36	0.25
Testing, 2013	111	1.17	0.78	13.3	0.37	0.25	0.35

n = number of observations, b_0 = regression coefficient, R^2 = determination coefficient, PBIAS = percent bias; RMSE = root mean square error, ARE = average relative error, AAE = average absolute error, EF = modelling efficiency.

The dynamics of the actual crop evapotranspiration ($ET_{c\,act}$) simulated along 2011 and 2012, which was computed with the calibrated model parameters of Table 2, is presented in Figure 8. $ET_{c\,act}$ are compared with evapotranspiration values observed with the eddy covariance system (ET_{EC}). Despite ET_{EC} observations were in reduced number, they could be used for further validate the model since results show that the model was able to adequately estimate the dynamics of $ET_{c\,act}$ during the mid-season stage when ET_{EC} was measured. The average ET_{EC} value during this stage was 2.5 mm·d^{-1} (± 0.41, n = 28). Goodness-of-fit indicators for 2012 show a tendency for the SIMDualKc model to under-estimate ET_{EC} (b_0 = 0.94 and PBIAS = −6.0%), thus contrary to the tendency observed when $T_{c\,act}$ was compared with T_{SF}. However, given the reduced number of EC observations, it is not possible to draw a conclusion about model trending. An acceptably high R^2 = 0.64 was obtained, which indicates that a large fraction of the variance of ET_{EC} observations is explained by the model. In addition, acceptable estimation errors of $ET_{c\,act}$ were obtained with RMSE = 0.42 mm·d^{-1}, AAE = 0.35 mm·d^{-1} and, consequently, a reasonably good EF value of 0.45 was obtained. Higher estimation errors for $ET_{c\,act}$, ranging from 0.54 to 0.71 mm·d^{-1}, were reported by Er-Raki et al. [20].

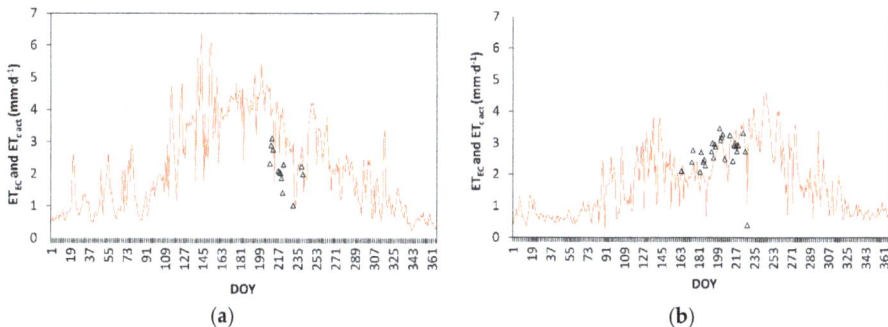

Figure 8. Actual crop evapotranspiration dynamics measured with Eddy covariance (ET_{EC}, Δ) and simulated with SIMDualKc ($ET_{c\,act}$, —) for (**a**) 2011 and (**b**) 2012.

3.2. Dual and Single Crop Coefficients

The standard and actual K_{cb} curves, the K_e curves and the standard and actual single K_c curves derived for the 3-year observed hedgerow olive orchard are shown in Figure 9. Precipitation and irrigation events are also depicted to easily interpreting the behaviour of the K_e, K_{cb} and K_c curves. The represented K_{cb} values were adjusted for climate as a function of wind speed, RH_{min} and crop height as proposed in FAO56 [4], so resulting slightly different from those given in Table 2 as referred before. The time-averaged standard K_c relative to the various crop stages was computed by summing the standard K_{cb} of the same crop stages with the daily K_e values computed for the same periods, that is, $K_c = K_{cb} + K_e$. Differently, $K_{c\,act} = K_{cb\,act} + K_e$, thus varying daily.

The K_{cb} and $K_{cb\,act}$ curves (Figure 9) are coincident during winter and most of autumn and spring periods when rainfall was enough to avoid water stress, that is, when $K_{cb\,act} = K_{cb}$ as discussed about Equation 1. Differently, high water stress occurred during most of the mid-season stage, particularly during 2012 and 2013, when those curves indicated $K_{cb\,act} < K_{cb}$. This water stress occurred because the orchard was under-irrigated following the common practice of deficit irrigation in olive orchards, in particular because irrigations were started too late in those years. To achieve a eustress irrigation management it would be desirable that irrigation would start earlier in both 2012 and 2013, eventually adopting larger irrigation depths applied every two or three days instead of 3 mm depths every day. However, further studies are required to better defining the best eustress management with consideration of impacts of transpiration deficits on yields.

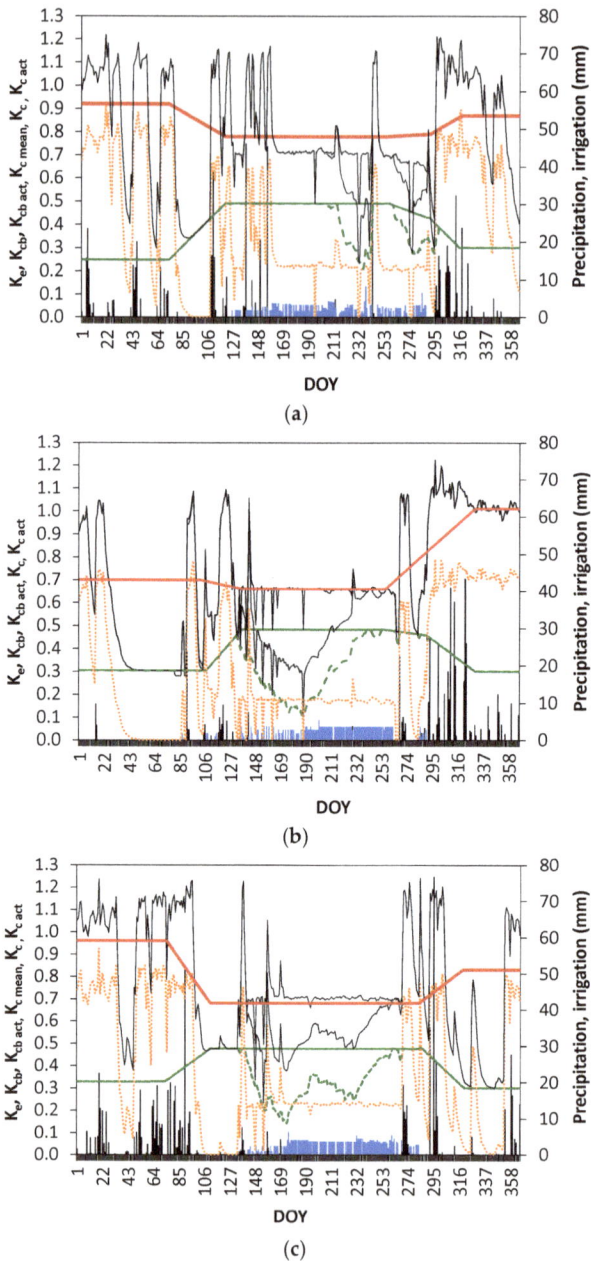

Figure 9. Standard and actual basal crop coefficients (K_{cb}, $K_{cb\ act}$), evaporation coefficient (K_e) and standard and actual single crop coefficients (K_c, $K_{c\ act}$) for a super-intensive olive orchard for: (**a**) 2011, (**b**) 2012 and (**c**) 2013, with depicting precipitation and irrigation events. K_{cb} (—) $K_{cb\ act}$ (- - -) K_e (⋯) $K_{c\ mean}$ (—) K_c (—) $K_{c\ act}$ (⋯) precipitation ( ), irrigation ( ).

Soil evaporation K_e curves show to react daily to the soil wetting events, mainly with various peaks occurring in response to rainfall events (Figure 9). During the rainfall season, from early autumn to early spring, K_e is often high because the inter-row is wide and solar energy is then available at the soil surface. During mid-season soil evaporation decreases and K_e drops because soil dries and residues of the ground cover vegetation act as an organic mulch to limit evaporation. K_e peaks are much small during most of the mid-season stage because irrigation was applied along the row of trees, so under crop shadow, thus resulting in limited energy available for evaporating the irrigation water.

Differently of the time-averaged standard K_c curves, the $K_{c\,act}$ curves show numerous peaks in response to the rainfall events (Figure 9) since these curves represent $K_{c\,act} = K_{cb\,act} + K_e$ as referred above. The K_c and $K_{c\,act}$ curves are coincident in the periods without water stress, thus during the rainy season, roughly from October to May. The standard and actual K_c values during the non-growing periods and the initial stage are highly dependent upon the precipitation amounts. Differently, the $K_{c\,act}$ curves lay bellow the K_c curve under water stress conditions during mid-season.

Figure 9 clearly shows the contrasting behaviour of standard K_c and K_{cb}. The above described dynamics of K_c curves, particularly the dependence of K_c values from soil evaporation, mainly due to soil wettings by precipitation, explains why the time-averaged standard $K_{c\,mid}$ is smaller than $K_{c\,non\text{-}growing}$ and $K_{c\,ini}$, that refer to the rainy season, while the higher standard K_{cb} values are for the mid-season, $K_{cb\,mid}$, thus when transpiration is higher under irrigation.

The time averaged K_c are presented in Table 4 for all crop stages of 2011, 2012 and 2013. $K_{c\,ini}$ values, ranging 0.70 to 0.96, are in the range of values reported for intensive olive orchards [43,80,81]. The $K_{c\,mid}$ values ranged 0.70 to 0.78, which are comparable with values reported by various authors [18,43,81,82]. K_c values for the late season (0.79 to 0.90) increased relative to the mid-season due to the occurrence of various precipitation events. High $K_{c\,end}$ values found in the present study are comparable with those reported by Testi et al. [80].

Table 4. Time-averaged crop coefficients (K_c) for the different crop growth stages.

Crop Growth Stages	2011	2012	2013
Non-growing *	0.91	0.77	0.75
Initial	0.96	0.70	0.96
Crop development	0.96–0.78	0.70–0.66	0.96–0.68
Mid-season	0.78	0.66	0.68
Late season	0.78–0.79	0.66–0.90	0.68–0.83
End season	0.79	0.90	0.83

* average value relative to both non-growing periods, by the end and the beginning of the year.

3.3. K_{cb} Predicted with the A&P Approach vs. K_{cb} Obtained with SIMDualKc

The A&P approach (Section 2.4, Allen & Pereira [43]) was applied with f_c and h data observed along the three seasons. The dynamics of f_c and h are presented in Figure 10a in conjunction with the corresponding computed values for K_{cb} ($K_{cb\,A\&P}$), which are compared with the standard K_{cb} obtained with SIMDualKc for the same days ($K_{cb\,SIMDualKc}$), resulting a close match (Figure 10b) except for the cases that follow the heavy frost with great defoliation and severe pruning in the winter of 2012.

When computing the density coefficient K_d (Equation (3)) the value $M_L = 1.5$ proposed by Allen and Pereira [43] was used. Nevertheless, because insufficient information on shadowing was available, it was assumed $f_c = f_{c\,eff} \cdot K_{cb\,full}$ (Equation (5)) was computed with observed weather data for the various crop stages and considering the observed crop height. The F_r values were obtained after applying a simple trial and error procedure against the values obtained by the model. That trial and error procedure was initiated with the value $F_r = 0.48$ proposed by Allen and Pereira [43] for the entire season and olive orchards without ground cover but just with residues of old leaves. The resulting value for the non-growing and the initial stages was $F_r = 0.49$, thus about the same as the proposed one, while values for both the mid- and late-season were respectively 0.55 and 0.53. These slightly higher values of F_r were likely due to the fact that less stomatal control was required by Arbequina olives under irrigation. Nevertheless, the estimated $F_r = 0.48$ proposed by Allen and Pereira [43] would also provide for quite good K_{cb} estimations.

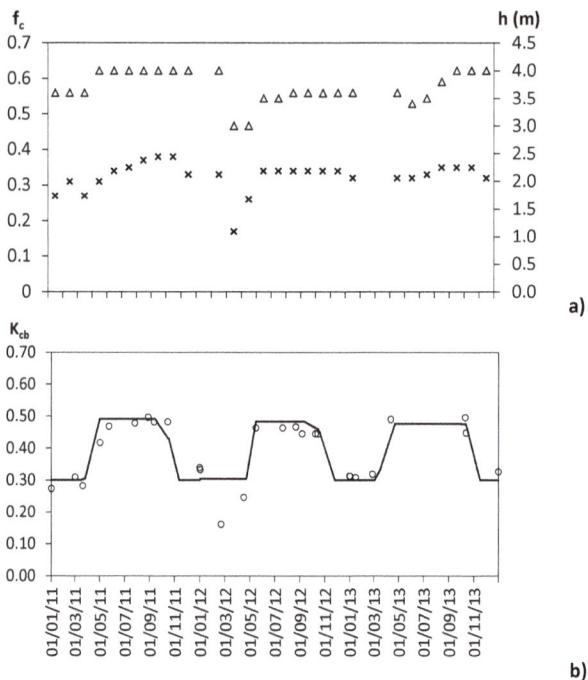

Figure 10. (**a**) Fraction of ground cover (f_c, x) measured and/or derived from SAVI and olives height (h, ∆) observed and (**b**) K_{cb} estimates using the A&P approach (○) compared with the K_{cb} curve computed with the SIMDualKc model (—) for the period 2011 to 2013.

For the period from November to April, K_{cb} were computed from K_d with Equation 4b relative to the case when active ground cover exists. A variable $K_{cb\ cover}$, from 0.004 to 0.01, was used. For May to October, the Equation 4a was used with $K_{c\ min} = 0.15$

Results in Figure 10b show that K_{cb} computed with the A&P approach match quite well the $K_{cb\ SIMDualKc}$ except for the cases that follow the heavy frost that occurred in the winter of 2012, which produced a great defoliation and required a severe pruning. Consequently, $K_{cb\ A\&P}$ were then smaller than the $K_{cb\ non-growing}$ obtained with SIMDualKc. For all other cases, $K_{cb\ A\&P}$ values were close to the $K_{cb\ SIMDualKc}$ estimates.

The regression coefficient relating both K_{cb} sets (Figure 11) results close to the unit ($b_0 = 0.97$), thus indicating a good overall match, while the determination coefficient is high ($R^2 = 0.77$). The estimation of $K_{cb\ A\&P}$ approach has a small RMSE of 0.05 relative to the modelled $K_{cb\ SIMDualKc}$ values. Comparing the mean square error of $K_{cb\ A\&P}$ with the variance of the $K_{cb\ SIMDualKc}$ values it resulted a high EF of 0.76. In conclusion, results of comparing both sets of K_{cb} values indicate that the A&P approach is adequate for estimating K_{cb} values to be used when assessing crop water requirements for olives and to support irrigation management decision making.

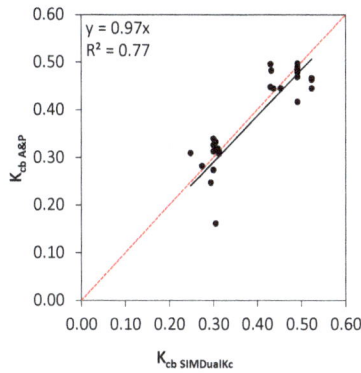

Figure 11. Linear regression forced to the origin of the K_{cb} values estimated with the A&P approach compared with the K_{cb} values obtained from the SIMDualKc model; also depicted the 1:1 line (- - -).

3.4. Water Balance and Respective Components

The terms of the soil water balance computed with the SIMDualKc model are presented in Table 5. There are appreciable differences among years which mainly steam from the differences in precipitation (P, mm). This is particularly the case of deep percolation and runoff (DP and RO, mm), which are much higher in the wet year of 2013, representing 47% of precipitation, which contrast with smaller DP and RO values in 2011 that represent 34% of the precipitation total. Those percent values indicate that most of precipitation, 53 to 66%, was used by the crop. As it could be expected, the larger amount of DP and RO occurred during the non-growing, initial, crop development and late season crop stages, that is, during the rainy season, which coincides with the period when crop water demand, evapotranspiration, is smaller as analysed relative to Figures 7 and 9. However, the distribution of DP and RO throughout the year varied enormously as a consequence of the inter- and intra-annual variability of the precipitation, as it may be noticed in Figure 1. DP and RO were very small during the mid-season and their occurrence refers only to a few rainfall events. The use of small drip irrigation applied depths makes that DP and RO are not originated by irrigation; moreover, the deficit irrigation management adopted also, likely, did not influence DP or RO occurrences.

The soil water storage (ΔSW, mm) at end of each crop stage and of the year greatly varied inter and intra-annually, mostly due to the referred variability of precipitation and likely less influenced by irrigation management as indicated by similar ΔSW values at end of the late season. This fact agrees with the conclusion that the adopted deficit irrigation management did not contribute to operational water losses.

Because rainfall was very small during the periods of large crop water demand, irrigation represented 39%, 56% and 50% of $ET_{c\,act}$ respectively in 2011, 2012 and 2013. Overall, irrigation occurred during the mid- and late season, mainly the former. These results show the great contribution of irrigation to achieve successful olives growth and yields, particularly in case of super-intensive olives orchards. Naturally, most of evapotranspiration occurred during the mid-season and, however less important, during the late-season. Soil evaporation exceeded transpiration during the non-growing and initial crop stages because E_s mostly originated from the inter-row when wetted by rainfall. The large values of E_s estimated for the mid-season were mostly due to rainfall events that occurred early in that season; similarly, most of E_s estimated for the late season also were due to rainfall events wetting the exposed inter-row. E_s originated from irrigation was reduced because energy available under the canopy shadow was little.

Table 5. Simulated soil water balance components (all variables in mm).

Crop Growth Stages	Year	P	I	ΔSW	RO	DP	E_s	$T_{c\,act}$	$ET_{c\,act}$
Non-growing	2011	189	0	−20	11	54	70	34	104
	2012	76	0	35	0	30	37	45	82
	2013	224	0	−20	12	85	63	43	106
Initial	2011	37	0	−16	0	3	11	7	18
	2012	36	5	8	0	0	20	30	50
	2013	106	0	−15	5	54	23	10	33
Development	2011	95	0	15	18	11	27	54	81
	2012	33	4	14	0	0	21	30	51
	2013	159	0	40	31	89	38	41	79
Mid-season	2011	203	260	63	1	28	186	311	497
	2012	13	299	−4	0	0	117	191	308
	2013	29	358	54	0	0	173	269	442
Late season	2011	141	59	−77	10	2	48	63	111
	2012	353	42	−89	70	102	60	74	134
	2013	218	34	−58	27	45	55	62	117
Full year	2011	665	319	−35	40	98	342	469	811
	2012	511	350	−36	70	132	255	370	625
	2013	736	392	1	75	273	352	425	777

P = precipitation, I = net irrigation depths, ΔSW = variation in stored soil water, RO = runoff, DP = deep percolation, E_s = soil evaporation, $T_{c\,act}$ = actual crop transpiration, $ET_{c\,act}$ = actual crop evapotranspiration.

The consumptive use of water is analysed in Table 6. The ratios of $E_s/ET_{c\,act}$ show that E_s represents 45 to 67% of $ET_{c\,act}$ during the non-growing stage and 40 to 70% during the initial crop stage. These periods are those when more precipitation occurs, so where the soil is frequently wetted by rain and soil cover is less, particularly in the inter-row. Thus, soil evaporation is the main consumptive water use during those stages. During the crop development, mid-season and late season stages, transpiration becomes the most important, corresponding to 52 to 67% of $ET_{c\,act}$. The variability of the ratio $E_s/ET_{c\,act}$ is then quite small, depending upon the occurrence of precipitation, which is the main source of soil wetting in the inter-row and of the subsequent evaporation. E_s could likely be reduce if less frequent but larger irrigation amounts would be used instead of daily irrigation events averaging 3 mm per event. $E_s/ET_{c\,act}$ ratios smaller than those computed in this study were reported by various researchers but referring to the irrigation season only [18,20,69]. Less frequent irrigations are also referred in these studies.

Table 6. Ratios of soil evaporation to actual evapotranspiration ($E_s/ET_{c\,act}$, %) and of actual to potential evapotranspiration ($ET_{c\,act}/ET_c$ %).

	Years	Non-Growing Periods	Initial	Development	Mid-Season	Late-Season	Whole Year
$E_s/ET_{c\,act}$ (%)	2011	67	61	33	37	43	42
	2012	45	40	41	38	45	41
	2013	59	70	48	39	47	45
$ET_{c\,act}/ET_c$ (%)	2011	100	100	100	94	88	97
	2012	100	100	98	70	99	82
	2013	100	100	100	78	100	86

Results relative to the ratio $ET_{c\,act}/ET_c$ (Table 6) show that $ET_{c\,act}$ was smaller by 3%, 18% and 14% relative to the potential ET_c in 2011, 2012 and 2013, respectively. During the mid-season, that ratio $ET_{c\,act}/ET_c$ was small for the years of 2012 and 2013, when $ET_{c\,act}$ was smaller by respectively 30% and 22% comparatively to the potential ET_c. Those decreases are likely excessive as analysed before (Figure 9) and may have affected yields, namely that of 2012. Adopting eustress irrigation, the ratio

$ET_{c\,act}/ET_c$ should likely not decrease below 90%. However, this threshold needs to be better assessed in terms of yield and economic impacts.

4. Conclusions

The calibration, validation and testing of the SIMDualKc model for a super high density Arbequina olive orchard was successfully performed using measurements of sap flow plant transpiration and eddy covariance evapotranspiration. Three years of field studies, from 2011 to 2013, were used, which made it possible to partitioning olive orchard ET into actual transpiration and soil evaporation. The results showed a good accuracy of the model to simulate transpiration during both the growing and non-growing seasons. Goodness of fit indicators were less good for 2012 when the orchard suffered an intense defoliation following a great frost and, later, a heavy pruning. However, the model could well represent the dynamics of evapotranspiration, transpiration and soil evaporation throughout the three years.

The calibration of the model and its validation and testing provided for updated basal and single crop coefficient curves for potential and actual crop water use, as well as soil evaporation coefficients. The standard K_{cb} were close to 0.30 during the non-growing and initial crop stages and 0.48 during the mid-season. Differently, the standard K_c varied from 0.70 to 0.96 during the non-growing and initial crop stages and decreased to 0.66 to 0.78 during the mid-season. The variability of standard K_c values results from the fact that, contrarily to K_{cb}, the K_c values depend upon soil evaporation, consequently also upon precipitation. It may be concluded that for perennial crops, as olive orchards and other woody crops, it is important to focus on using the dual instead of single K_c, thus in using basal K_{cb}.

The A&P approach for estimating K_{cb} values from the fraction of ground cover and crop height was tested against the K_{cb} values obtained with the model SIMDualKc after calibration. The K_{cb} obtained with A&P matched well the K_{cb} values obtained from SIMDualKc. Thus, results allow assuming that the A&P approach is useful to accurately estimating the K_{cb} values required to assess crop water requirements and deciding irrigation management practices of olives orchards. Likely, that conclusion may be extended to other crops under condition of properly parameterizing $K_{cb\,full}$ used for computing K_{cb} from f_c and h.

For the three years, deep percolation and runoff totalized 34% to 47% of the precipitation, thus with most of precipitation being used as crop evapotranspiration. Those percent values highly depend upon the annual distribution of precipitation and were not depending upon irrigation because frequent and small application depths were used. Soil water storage also revealed to vary along the year with precipitation. Soil evaporation ranged from 42% to 45% of annual olives ET, with $E_s > T_{c\,act}$ during the rainy season, the contrary occurring during the irrigation period. This fact also relates with the above referred irrigation management practiced. Irrigation amounts for 39% to 56% of the actual crop ET, which confirms the great importance of irrigation for super intensive olive orchards. However, those amounts correspond to deficit irrigation, with $ET_{c\,act}$ smaller than potential ET_c by 30% and 22% in 2012 and 2013, respectively. Those deficits are apparently excessive relative to a desirable eustress management that could lead to high yields and save water. Further research is required along these lines.

Author Contributions: T.A.P. and L.S.P. conceptualization, T.A.P., P.P. and L.S.P. methodology, P.P. calibration and validation of the model, T.A.P., P.P., J.S. and L.S.P. formal data analysis, J.S. and F.L.S. resources, T.A.P. and P.P. original draft preparation, L.S.P. review and editing, F.L.S. project supervision.

Funding: This study was supported by the projects PTDC/AGR-PRO/111717/2009 and EXPL/AGR-PRO/1559/2012 and the research unit LEAF (UID/AGR/04129/2013) funded by "Fundação para a Ciência e a Tecnologia," Portugal. The second author also acknowledges the same institution for a Post-Doc research grant (SFRH/BPD/79767/2011).

Conflicts of Interest: The authors declare no conflict of interest.

References

1. Connor, D.J.; Gómez-del-Campo, M.; Rousseaux, M.C.; Searles, P.S. Structure, management and productivity of hedgerow olive orchards: A review. *Sci. Hortic.* **2014**, *169*, 71–93. [CrossRef]
2. Trentacoste, E.R.; Puertas, C.M.; Sadras, V.O. Effect of irrigation and tree density on vegetative growth, oil yield and water use efficiency in young olive orchard under arid conditions in Mendoza, Argentina. *Irrig. Sci.* **2015**, *33*, 429–440. [CrossRef]
3. Ahumada-Orellana, L.E.; Ortega-Farías, S.; Searles, P.S. Olive oil quality response to irrigation cut-off strategies in a super-high density orchard. *Agric. Water Manag.* **2018**, *202*, 81–88. [CrossRef]
4. Allen, R.G.; Pereira, L.S.; Raes, D.; Smith, M. *Crop Evapotranspiration Guidelines for Computing Crop Water Requirements*; FAO: Rome, Italy, 1998; p. 300.
5. Pereira, L.S.; Oweis, T.; Zairi, A. Irrigation management under water scarcity. *Agric. Water Manag.* **2002**, *57*, 175–206. [CrossRef]
6. Palese, A.M.; Nuzzo, V.; Favati, F.; Pietrafesa, A.; Celano, G.; Xiloyannisa, C. Effects of water deficit on the vegetative response, yield and oil quality of olive trees (*Olea europaea* L., cv Coratina) grown under intensive cultivation. *Sci. Hortic.* **2010**, *125*, 222–229. [CrossRef]
7. Rallo, G.; Provenzano, G. Modelling eco-physiological response of table olive trees (*Olea europaea* L.) to soil water deficit conditions. *Agric. Water Manag.* **2013**, *120*, 79–88. [CrossRef]
8. Fernández, J.E.; Perez-Martin, A.; Torres-Ruiz, J.M.; Cuevas, M.V.; Rodriguez-Dominguez, C.M.; Elsayed-Farag, S.; Morales-Sillero, A.; García, J.M.; Hernandez-Santana, V.; Diaz-Espejo, A. A regulated deficit irrigation strategy for hedgerow olive orchards with high plant density. *Plant Soil* **2013**, *372*, 279–295. [CrossRef]
9. Gómez-del-Campo, M. Summer deficit-irrigation strategies in a hedgerow olive orchard cv. 'Arbequina': Effect on fruit characteristics and yield. *Irrig. Sci.* **2013**, *31*, 259–269. [CrossRef]
10. Padilla-Díaz, C.M.; Rodriguez-Dominguez, C.M.; Hernandez-Santana, V.; Perez-Martin, A.; Fernández, J.E. Scheduling regulated deficit irrigation in a hedgerow olive orchard from leaf turgor pressure related measurements. *Agric. Water Manag.* **2016**, *164*, 28–37. [CrossRef]
11. Padilla-Díaz, C.M.; Rodriguez-Dominguez, C.M.; Hernandez-Santana, V.; Perez-Martin, A.; Fernandes, R.D.M.; Montero, A.; García, J.M.; Fernández, J.E. Water status, gas exchange and crop performance in a super high density olive orchard under deficit irrigation scheduled from leaf turgor measurements. *Agric. Water Manag.* **2018**, *202*, 241–252. [CrossRef]
12. Tanasijevic, L.; Todorovic, M.; Pizzigalli, C.; Lionello, P.; Pereira, L.S. Impacts of climate change on olive crop evapotranspiration and irrigation requirements in the Mediterranean region. *Agric. Water Manag.* **2014**, *144*, 54–68. [CrossRef]
13. Haworth, M.; Marino, G.; Brunetti, C.; Killi, D.; De Carlo, A.; Centritto, M. The Impact of Heat Stress and Water Deficit on the Photosynthetic and Stomatal Physiology of Olive (*Olea europaea* L.)—A Case Study of the 2017 HeatWave. *Plants* **2018**, *7*, 76. [CrossRef] [PubMed]
14. Allen, R.G.; Pereira, L.S.; Howell, T.A.; Jensen, M.E. Evapotranspiration information reporting: I. Factors governing measurement accuracy. *Agric. Water Manag.* **2011**, *98*, 899–920. [CrossRef]
15. Egea, G.; Diaz-Espejo, A.; Fernández, J.E. Soil moisture dynamics in a hedgerow olive orchard under well-watered and deficit irrigation regimes: Assessment, prediction and scenario analysis. *Agric. Water Manag.* **2016**, *164*, 197–211. [CrossRef]
16. Rallo, G.; Provenzano, G.; Castellini, M.; Puig-Sirera, A. Application of EMI and FDR sensors to assess the fraction of transpirable soil water over an olive grove. *Water* **2018**, *10*, 168. [CrossRef]
17. Ramos, A.F.; Santos, F.L. Water use, transpiration and crop coefficients for olives (cv. Cordovil), grown in orchards in Southern Portugal. *Biosyst. Eng.* **2009**, *102*, 321–333. [CrossRef]
18. Cammalleri, C.; Rallo, G.; Agnese, C.; Ciraolo, G.; Minacapilli, M.; Provenzano, G. Combined use of eddy covariance and sap flow techniques for partition of ET fluxes and water stress assessment in an irrigated olive orchard. *Agric. Water Manag.* **2013**, *120*, 89–97. [CrossRef]
19. Paço, T.A.; Pocas, I.; Cunha, M.; Silvestre, J.C.; Santos, F.L.; Paredes, P.; Pereira, L.S. Evapotranspiration and crop coefficients for a super intensive olive orchard. An application of SIMDualKc and METRIC models using ground and satellite observations. *J. Hydrol.* **2014**, *519*, 2067–2080. [CrossRef]

20. Er-Raki, S.; Chehbouni, A.; Boulet, G.; Williams, D.G. Using the dual approach of FAO-56 for partitioning ET into soil and plant components for olive orchards in a semi-arid region. *Agric. Water Manag.* **2010**, *97*, 1769–1778. [CrossRef]

21. Sepulcre-Cantó, G.; Zarco-Tejada, P.J.; Jiménez-Muños, J.C.; Sobrino, J.A.; De Miguel, E.; Villalobos, F.J. Detection of water stress in olive orchard with thermal remote sensing imagery. *Agric. For. Meteorol.* **2006**, *136*, 31–44. [CrossRef]

22. Santos, F.L. Assessing olive evapotranspiration partitioning from soil water balance and radiometric soil and canopy temperatures. *Agronomy* **2018**, *8*, 43. [CrossRef]

23. Capraro, F.; Tosetti, S.; Rossomando, F.; Mut, V.; Serman, F.V. Web-Based System for the Remote Monitoring and Management of Precision Irrigation: A Case Study in an Arid Region of Argentina. *Sensors* **2018**, *18*, 3847. [CrossRef] [PubMed]

24. Hoedjes, J.C.B.; Chehbouni, A.; Jacob, F.; Ezzahar, J.; Boulet, G. Deriving daily evapotranspiration from remotely sensed instantaneous evaporative fraction over olive orchard in semi-arid Morocco. *J. Hydrol.* **2008**, *354*, 53–64. [CrossRef]

25. Minacapilli, M.; Agnese, C.; Blanda, F.; Cammalleri, C.; Ciraolo, G.; D'Urso, G.; Iovino, M.; Pumo, D.; Provenzano, G.; Rallo, G. Estimation of actual evapotranspiration of Mediterranean perennial crops by means of remote-sensing based surface energy balance models. *Hydrol. Earth Syst. Sci.* **2009**, *13*, 1061–1074. [CrossRef]

26. Santos, C.; Lorite, I.J.; Allen, R.G.; Tasumi, M. Aerodynamic parameterization of the satellite-based energy balance (METRIC) model for ET estimation in rainfed olive orchards of Andalusia, Spain. *Water Resour. Manag.* **2012**, *26*, 3267–3283. [CrossRef]

27. Pôças, I.; Paço, T.A.; Cunha, M.; Andrade, J.A.; Silvestre, J.; Sousa, A.; Santos, F.L.; Pereira, L.S.; Allen, R.G. Satellite based evapotranspiration of a super-intensive olive orchard: Application of METRIC algorithm. *Biosyst. Eng.* **2014**, *128*, 69–81. [CrossRef]

28. Pôças, I.; Paço, T.A.; Paredes, P.; Cunha, M.; Pereira, L.S. Estimation of actual crop coefficients using remotely sensed vegetation indices and soil water balance modelled data. *Remote Sens.* **2015**, *7*, 2373–2400. [CrossRef]

29. Ortega-Farías, S.; Ortega-Salazar, S.; Poblete, T.; Kilic, A.; Allen, R.G.; Poblete-Echeverría, C.; Ahumada-Orellana, L.; Zuñiga, M.; Sepúlveda, D. Estimation of energy balance components over a drip-irrigated olive orchard using thermal and multispectral cameras placed on a helicopter-based unmanned aerial vehicle (UAV). *Remote Sens.* **2016**, *8*, 638. [CrossRef]

30. Abazi, U.; Lorite, I.J.; Cárceles, B.; Martínez Raya, A.; Durán, V.H.; Francia, J.R.; Gómez, J.A. WABOL: A conceptual water balance model for analyzing rainfall water use in olive orchards under different soil and cover crop management strategies. *Comput. Electron. Agric.* **2013**, *91*, 35–48. [CrossRef]

31. Autovino, D.; Rallo, G.; Provenzano, G. Predicting soil and plant water status dynamic in olive orchards under different irrigation systems with Hydrus-2D: Model performance and scenario analysis. *Agric. Water Manag.* **2018**, *203*, 225–235. [CrossRef]

32. Pereira, L.S.; Allen, R.G.; Smith, M.; Raes, D. Crop evapotranspiration estimation with FAO56: Past and future. *Agric. Water Manag.* **2015**, *147*, 4–20. [CrossRef]

33. Allen, R.G.; Pereira, L.S.; Smith, M.; Raes, D.; Wright, J. FAO-56 Dual Crop Coefficient Method for Estimating Evaporation from Soil and Application Extensions. *J. Irrig. Drain. Eng.* **2005**, *131*, 2–13. [CrossRef]

34. Rosa, R.D.; Paredes, P.; Rodrigues, G.C.; Alves, I.; Fernando, R.M.; Pereira, L.S.; Allen, R.G. Implementing the dual crop coefficient approach in interactive software. 1. Background and computational strategy. *Agric. Water Manag.* **2012**, *103*, 8–24. [CrossRef]

35. Fandiño, M.; Cancela, J.J.; Rey, B.J.; Martínez, E.M.; Rosa, R.G.; Pereira, L.S. Using the dual-Kc approach to model evapotranspiration of albariño vineyards (*Vitis vinifera* L. cv. albariño) with consideration of active ground cover. *Agric. Water Manag.* **2012**, *112*, 75–87. [CrossRef]

36. Paredes, P.; Rodrigues, G.C.; Alves, I.; Pereira, L.S. Partitioning evapotranspiration, yield prediction and economic returns of maize under various irrigation management strategies. *Agric. Water Manag.* **2014**, *135*, 27–39. [CrossRef]

37. Pereira, L.S.; Paredes, P.; Rodrigues, G.C.; Neves, M. Modeling malt barley water use and evapotranspiration partitioning in two contrasting rainfall years. Assessing AquaCrop and SIMDualKc models. *Agric. Water Manag.* **2015**, *159*, 239–254, Corrigendum in **2015**, *163*, 408. [CrossRef]

38. Paço, T.A.; Ferreira, M.I.; Rosa, R.G.; Paredes, P.; Rodrigues, G.; Conceição, N.; Pacheco, C.; Pereira, L.S. The dual crop coefficient approach using a density factor to simulate the evapotranspiration of a peach orchard: SIMDualKc model versus eddy covariance measurements. *Irrig. Sci.* **2012**, *30*, 115–126. [CrossRef]

39. Ran, H.; Kang, S.; Li, F.; Tong, L.; Ding, R.; Du, T.; Li, S.; Zhang, X. Performance of AquaCrop and SIMDualKc models in evapotranspiration partitioning on full and deficit irrigated maize for seed production under plastic film-mulch in an arid region of China. *Agric. Syst.* **2017**, *151*, 20–32. [CrossRef]

40. Zhao, N.; Liu, Y.; Cai, J.; Paredes, P.; Rosa, R.D.; Pereira, L.S. Dual crop coefficient modelling applied to the winter wheat-summer maize crop sequence in North China Plain: Basal crop coefficients and soil evaporation component. *Agric. Water Manag.* **2013**, *117*, 93–105. [CrossRef]

41. Gao, Y.; Yang, L.; Shen, X.; Li, X.; Sun, J.; Duan, A.; Wu, L. Winter wheat with subsurface drip irrigation (SDI): Crop coefficients, water-use estimates and effects of SDI on grain yield and water use efficiency. *Agric. Water Manag.* **2014**, *146*, 1–10. [CrossRef]

42. Wei, Z.; Paredes, P.; Liu, Y.; Chi, W.-W.; Pereira, L.S. Modelling transpiration, soil evaporation and yield prediction of soybean in North China Plain. *Agric. Water Manag.* **2015**, *147*, 43–53. [CrossRef]

43. Allen, R.G.; Pereira, L.S. Estimating crop coefficients from fraction of ground cover and height. *Irrig. Sci.* **2009**, *28*, 17–34. [CrossRef]

44. Kottek, M.; Grieser, J.; Beck, C.; Rudolf, B.; Rubel, F. World Map of the Köppen-Geiger climate classification updated. *Meteorol. Z.* **2006**, *15*, 259–263. [CrossRef]

45. IUSS; Working Group WRB. World Reference Base for Soil Resources 2014, Update 2015. In *International Soil Classification System for Naming Soils and Creating Legends for Soil Maps*; World Soil Resources Reports No. 106; FAO: Rome, Italy, 2015; 192p.

46. Fandiño, M.; Olmedo, J.L.; Martínez, E.M.; Valladares, J.; Paredes, P.; Rey, B.J.; Mota, M.; Cancela, J.J.; Pereira, L.S. Assessing and modelling water use and the partition of evapotranspiration of irrigated hop (*Humulus lupulus*) and relations of transpiration with hops yield and alpha-acids. *Ind. Crop Prod.* **2015**, *77*, 204–217. [CrossRef]

47. Ferreira, M.I.; Paço, T.A.; Silvestre, J. Combining techniques to study evapotranspiration in woody crops: Application to small areas—Two case studies. *Acta Hortic. ISHS* **2004**, *664*, 225–232. [CrossRef]

48. Mauder, M.; Foken, T. Documentation and Instruction Manual of the Eddy-Covariance Software Package TK3 (Update). Arbeitsergebnisse Nr. 62; 2015 Univ Bayreuth, Abt Mikrometeorol. Available online: https://epub.uni-bayreuth.de/2130/1/ARBERG062.pdf (accessed on 5 July 2017).

49. Foken, T.; Leuning, R.; Oncley, S.P.; Mauder, M.; Aubinet, M. Corrections and data quality. In *Eddy Covariance: A Practical Guide to Measurement and Data Analysis*; Aubinet, M., Vesala, T., Papale, D., Eds.; Springer: Berlin/Heidelberg, Germany, 2011.

50. Vickers, D.; Mahrt, L. Quality control and flux sampling problems for tower and aircraft data. *J. Atmos. Ocean. Technol.* **1997**, *14*, 512–526. [CrossRef]

51. Kaimal, J.C.; Finnigan, J.J. *Atmospheric Boundary Layer Flows: Their Structure and Measurement*; Oxford University Press: New York, NY, USA, 1994.

52. Tanner, B.D.; Swiatek, E.; Greene, J.P. Density fluctuations and use of the krypton hygrometer in surface flux measurements. In *Management of Irrigation and Drainage Systems: Integrated Perspectives*; Allen, R.G., Ed.; American Society of Civil Engineers: New York, NY, USA, 1993; pp. 945–952.

53. Moore, C.J. Frequency-response corrections for eddy-correlation systems. *Bound.-Layer Meteorol.* **1986**, *37*, 17–35. [CrossRef]

54. Schotanus, P.; Nieuwstadt, F.T.M.; Debruin, H.A.R. Temperature-measurement with a sonic anemometer and its application to heat and moisture fluxes. *Bound.-Layer Meteorol.* **1983**, *26*, 81–93. [CrossRef]

55. Webb, E.K.; Pearman, G.I.; Leuning, R. Correction of flux measurements for density effects due to heat and water-vapor transfer. *Q. J. R. Meteorol. Soc.* **1980**, *106*, 85–100. [CrossRef]

56. Schuepp, P.H.; Leclerc, M.Y.; Macpherson, J.I.; Desjardins, R.L. Footprint prediction of scalar fluxes from analytical solutions of the diffusion equation. *Bound.-Layer Meteorol.* **1990**, *50*, 355–373. [CrossRef]

57. Testi, L.; Villalobos, F.J.; Orgaz, F. Evapotranspiration of a young irrigated olive orchard in southern Spain. *Agric. For. Meteorol.* **2004**, *121*, 1–18. [CrossRef]

58. Teixeira, A.H.D.C.; Bastiaanssen, W.G.M.; Moura, M.S.B.; Soares, J.M.; Ahmad, M.D.; Bos, M.G. Energy and water balance measurements for water productivity analysis in irrigated mango trees, Northeast Brazil. *Agric. For. Meteorol.* **2008**, *148*, 1524–1537. [CrossRef]

59. Conceição, N.; Tezza, L.; Häusler, M.; Lourenço, S.; Pacheco, C.A.; Ferreira, M.I. Three years of monitoring evapotranspiration components and crop and stress coefficients in a deficit irrigated intensive olive orchard. *Agric. Water Manag.* **2017**, *191*, 138–152. [CrossRef]

60. Granier, A. Une nouvelle méthode pour la mesure du flux de sève brute dans le tronc des arbres. *Ann. Sci. For.* **1985**, *42*, 193–200. [CrossRef]

61. Silva, R.M.; Paço, T.A.; Ferreira, M.I.; Oliveira, M. Transpiration of a kiwifruit orchard estimated using the Granier sap flow method calibrated under field conditions. *Acta Hortic. ISHS* **2008**, *792*, 593–600. [CrossRef]

62. Steppe, K.; De Pauw, D.J.W.; Doody, T.M.; Teskey, R.O. A comparison of sap flux density using thermal dissipation, heat pulse velocity and heat field deformation methods. *Agric. For. Meteorol.* **2010**, *150*, 1046–1056. [CrossRef]

63. Bush, S.E.; Hultine, K.R.; Sperry, J.S.; Ehleringer, J.R. Calibration of thermal dissipation sap flow probes for ring- and diffuse-porous trees. *Tree Physiol.* **2010**, *30*, 1545–1554. [CrossRef]

64. Hölttä, T.; Linkosalo, T.; Riikonen, A.; Sevanto, S.; Nikinmaa, E. An analysis of Granier sap flow method, its sensitivity to heat storage and a new approach to improve its time dynamics. *Agric. For. Meteorol.* **2015**, *211–212*, 2–12. [CrossRef]

65. Marañón-Jiménez, S.; Van den Bulcke, J.; Piayda, A.; Van Acker, J.; Cuntz, M.; Rebmann, C.; Steppe, K. X-ray computed microtomography characterizes the wound effect that causes sap flow underestimation by thermal dissipation sensors. *Tree Physiol.* **2017**, *38*, 288–302. [CrossRef]

66. Ritchie, J.T. Model for predicting evaporation from a row crop with incomplete cover. *Water Resour. Res.* **1972**, *8*, 1204–1213. [CrossRef]

67. Paço, T.A.; Conceição, N.; Ferreira, M.I. Measurements and estimates of peach orchard evapotranspiration in Mediterranean conditions. *Acta Hortic.* **2004**, *664*, 505–512. [CrossRef]

68. Paço, T.A.; Conceição, N.; Ferreira, M.I. Peach orchard evapotranspiration in a sandy soil: Comparison between eddy covariance measurements and estimates by the FAO 56 approach. *Agric. Water Manag.* **2006**, *85*, 305–313. [CrossRef]

69. López-Olivari, R.; Ortega-Farías, S.; Poblete-Echeverría, C. Partitioning of net radiation and evapotranspiration over a superintensive drip-irrigated olive orchard. *Irrig. Sci.* **2016**, *34*, 17–31. [CrossRef]

70. Tukey, J.W. *Exploratory Data Analysis*; Addison-Wesley: Reading, MA, USA, 1977.

71. Liu, Y.; Pereira, L.S.; Fernando, R.M. Fluxes through the bottom boundary of the root zone in silty soils: Parametric approaches to estimate groundwater contribution and percolation. *Agric. Water Manag.* **2006**, *84*, 27–40. [CrossRef]

72. Allen, R.G.; Wright, J.L.; Pruitt, W.O.; Pereira, L.S.; Jensen, M.E. Water requirements. In *Design and Operation of Farm Irrigation Systems*, 2nd ed.; Hoffman, G.J., Evans, R.G., Jensen, M.E., Martin, D.L., Elliot, R.L., Eds.; ASABE: St. Joseph, MI, USA, 2007; pp. 208–288.

73. Zhang, B.; Liu, Y.; Xu, D.; Zhao, N.; Lei, B.; Rosa, R.D.; Paredes, P.; Paco, T.A.; Pereira, L.S. The dual crop coefficient approach to estimate and partitioning evapotranspiration of the winter wheat-summer maize crop sequence in North China Plain. *Irrig. Sci.* **2013**, *31*, 1303–1316. [CrossRef]

74. Tian, F.; Yang, P.; Hu, H.; Dai, C. Partitioning of cotton field evapotranspiration under mulched drip irrigation based on a dual crop coefficient model. *Water* **2016**, *8*, 72. [CrossRef]

75. Rosa, R.D.; Paredes, P.; Rodrigues, G.C.; Fernando, R.M.; Alves, I.; Pereira, L.S.; Allen, R.G. Implementing the dual crop coefficient approach in interactive software: 2. Model testing. *Agric. Water Manag.* **2012**, *103*, 62–77. [CrossRef]

76. Zhang, H.; Huang, G.; Xu, X.; Xiong, Y.; Huang, Q. Estimating evapotranspiration of processing tomato under plastic mulch using the SIMDualKc model. *Water* **2018**, *10*, 1088. [CrossRef]

77. Nash, J.E.; Sutcliffe, J.V. River flow forecasting through conceptual models: Part 1. A discussion of principles. *J. Hydrol.* **1970**, *10*, 282–290. [CrossRef]

78. Moriasi, D.N.; Arnold, J.G.; Van Liew, M.W.; Bingner, R.L.; Harmel, R.D.; Veith, T.L. Model evaluation guidelines for systematic quantification of accuracy in watershed simulations. *Trans. ASABE* **2007**, *50*, 885–900. [CrossRef]

79. Rallo, G.; Baiamonte, G.; Manzano Juárez, J.; Provenzano, G. Improvement of FAO-56 model to estimate transpiration fluxes of drought tolerant crops under soil water deficit: Application for olive groves. *J. Irrig. Drain. Eng.* **2014**, *140*, A4014001. [CrossRef]

80. Testi, L.; Orgaz, F.; Villalobos, F.J. Variations in bulk canopy conductance of an irrigated olive (*Olea europaea* L.) orchard. *Environ. Exp. Bot.* **2006**, *55*, 15–28. [CrossRef]

81. Martínez-Cob, A.; Faci, J.M. Evapotranspiration of an hedge-pruned olive orchard in a semiarid area of NE Spain. *Agric. Water Manag.* **2010**, *97*, 410–418. [CrossRef]

82. Fernández, J.E.; Moreno, F. Water Use by the Olive Tree. *J. Crop Prod.* **2000**, *2*, 101–162. [CrossRef]

water

MDPI

Article

Implementation of a Two-Source Model for Estimating the Spatial Variability of Olive Evapotranspiration Using Satellite Images and Ground-Based Climate Data

Fernando Fuentes-Peñailillo [1], Samuel Ortega-Farías [1,*], César Acevedo-Opazo [1] and David Fonseca-Luengo [2]

[1]   Centro de Investigación y Transferencia en Riego y Agroclimatología (CITRA) and Research Program on Adaptation of Agriculture to Climate Change (A2C2), Universidad de Talca, Casilla 747, Talca 3460000, Chile; ffuentesp@utalca.cl (F.F.-P.); cacevedo@utalca.cl (C.A.-O.)
[2]   Faculty of Natural Resources, Universidad Católica de Temuco, Temuco 4780000, Chile; dfonseca@uct.cl
*   Correspondence: sortega@utalca.cl; Tel.: +56-71-220-0426

Received: 6 November 2017; Accepted: 16 March 2018; Published: 19 March 2018

Abstract: A study was carried out to evaluate the potential use of the two-source Shuttleworth and Wallace (SW) model to compute the intra-orchard spatial variability of actual evapotranspiration (ET) of olive trees using satellite images and ground-based climate data. The study was conducted in a drip-irrigated olive orchard using satellite images (Landsat 7 ETM+), which were acquired on clear sky days during the main phenological stages (2009/10 growing season). The performance of the SW model was evaluated using instantaneous latent heat flux (LE) measurements that were obtained from an eddy correlation system. At the time of satellite overpass, the estimated values of net radiation (Rn_i) and soil heat flux (G_i) were compared with ground measurements from a four-way net radiometer and soil heat flux plates, respectively. The results indicated that the SW model subestimated instantaneous LE (W m^{-2}) and daily ET (mm d^{-1}), with errors of 12% and 10% of observed values, respectively. The root mean square error (RMSE) and mean absolute error (MAE) values for instantaneous LE were 26 and 20 W m^{-2}, while those for daily values of ET were 0.31 and 0.28 mm d^{-1}, respectively. Finally, the submodels computed Rn_i and G_i with errors of between 4.0% and 8.0% of measured values and with RMSE and MAE between 25 and 39 W m^{-2}.

Keywords: crop water requirements; latent heat flux; remote sensing; olive orchard; spatial variability

---

1. Introduction

Better water management of irrigation is required to optimize the water productivity of olive oil production due to water scarcity. Under this scenario, sophisticated irrigation water management will be required to maintain sufficient levels of productivity and quality [1,2]. For these objectives, it is necessary to have an accurate estimation of daily actual evapotranspiration (ET) that is generally computed as a function of reference evapotranspiration (ET_o) and crop coefficient (Kc) [3]. Several researchers also suggested using the dual crop coefficient approach to describe the ratio of ET to ET_o by separating Kc into basal crop coefficient (Kcb) and soil evaporation coefficient (Ke) [3–5]. However, the values of Kc and Kcb reported in the literature for heteronomous canopies require local adjustment because they depend on canopy architecture and non-linear interaction of soil, cultivar, and climate [1,6,7]. For a hedge-pruned olive orchard, Martínez-Cob and Faci [8] indicated that Kc values depend on the geometric characteristics of the canopy (canopy shape, distance between trees, etc.) and fractional cover (fc). For a super intensive olive orchard, Paço et al. [7] suggested that the Kc value is affected by several factors, including the canopy architecture, fractional cover, crop management practices, and rainfall variability. For a drip-irrigated olive orchard, López-Olivari et al. [9] indicated that the ratios of transpiration and evaporation to ET_o could be significantly affected by the irrigation systems, which determine the percentage of wetted area (Aw) of the soil surface. Allen and Pereira [10] and Paço et al. [7] recommended adjusting the ET/ET_o ratios using fc and tree height, according to a density coefficient (Kd) that describes the increase in Kc with an increase in canopy size. Pôças et al. [11] for olive orchard improved the estimation of crop coefficients using remotely sensed vegetation indices and the SIMdualKc soil water balance model. Cammalleri et al. [6] also evaluated a remote sensing-based approach to estimate olive ET, combining a modified version of the standard FAO-56 dual crop coefficient procedure and Penman-Monteith (PM) equation with actual canopy characteristics (i.e., leaf area index, albedo, and canopy height) that were derived from optical remote sensing data. Finally, Er-Raki et al. [12] evaluated the potential of assimilating ET derived from satellite thermal infrared observations to improve the ET simulation performances from the FAO-56 single crop coefficient approach.

Several studies indicated that the two-layer model of Shuttleworth and Wallace (SW) could be used to estimate ET directly without using crop coefficients [13–17]. The SW model can estimate soil evaporation and transpiration separately and has been widely used to estimate ET for homogeneous (sorghum, corn, and wheat) and heterogeneous (vineyards and orchards) canopies [14,18–21]. For heterogeneous canopies, such as vineyards and orchards, the SW approach has been able to estimate ET with errors ranging between 6% and 25%. In a furrow-irrigated vineyard (fc = 0.35), Zhang et al. [20] indicated that the SW model overestimated latent heat flux (LE) with an index of agreement (I_a) = 0.75 and a mean absolute error (MAE) = 39 W m^{-2}, while Zhang et al. [22] noted that the SW model overestimated ET by 25.2% in a furrow-irrigated vineyard. For drip-irrigated vineyards (fc = 0.30), Ortega-Farías et al. [14] observed that the SW model was able to predict ET with a root-mean-square error (RMSE) = 0.51 mm d^{-1} and an MAE = 0.41 mm d^{-1}. In a furrow-irrigated vineyard (fc = 0.30), Zhao et al. [23] observed that the SW model estimated ET with a RMSE and MAE of 0.68 and 0.52 mm d^{-1}, respectively. In a drip-irrigated olive orchard, Ortega-Farías and Lopez-Olivari [24] observed that the SW model overestimated LE by approximately 2% of measured values, with a RMSE = 28 W m^{-2} and I_a = 0.98. This study also indicated that the SW model was very sensitive to errors in the values of stomatal resistance (r_{st}) and leaf area index (LAI). On this matter, Brenner and Incoll [25] indicated that the SW model overestimated crop transpiration of sparsely vegetated shrublands due to the over-estimation of the radiation that is absorbed by the canopy.

The traditional approach of SW model does not consider the effect of intra-orchard spatial variability of soil and canopy vigor on the estimation of ET. Canopy covers of commercial olive orchards are generally incomplete as a result of the canopy geometry that depends on canopy size, leaf area index, and plant density [26]. The canopy training system and associated canopy geometry may significantly affect the partitioning of Rn into LE, H, or G and thus orchard water requirements. To include the effect of intra-field spatial variability into the SW model, the instantaneous values of net radiation (Rn_i) and soil heat flux (G_i) over orchards can be estimated using remote sensing images. In this regard, several studies have suggested that Rn_i and G_i can be adequately estimated in heterogeneous canopies using satellite images. In a drip-irrigated olive orchard, Ortega-Farías et al. [27] indicated that Rn_i was estimated with an error of 3% of measured values when using thermal and multispectral images from the Landsat platform. In a drip-irrigated vineyard, Carrasco-Benavides et al. [28] using Landsat 7 images observed that the METRIC model estimated Rn_i and G_i with an error of 11% and 5% of measured values, respectively. However, the estimation of ET from the METRIC model requires the selection of the two "anchor" pixels, which is subjective and depends on the ability of the operator in search and isolate of the most appropriate hot and cold pixels [29–31]. For drip-irrigated olive orchards (fc < 0.3), the selection of the hot and cold pixels is critical because sensible heat flux (H), which is produced at the soil surface, is the main component of energy balance and plays a key role in the tree transpiration and stomatal closure [24].

Dhungel et al. [32] successfully used remote sensing tools to facilitate a comparison of different parameterizations to estimate LE over sparse canopies. This comparison was conducted using gridded weather and Landsat satellite data, such as instantaneous values of ET, surface roughness (Z_{om}), emissivity (ε_0), albedo (α), and leaf area index (LAI), to parameterize the Penman-Monteith (PM) equation. According to our knowledge, there is no available information about the estimation of the intra-orchard spatial variability of olive ET using ground-based weather measurements and remote sensing images as inputs to the SW model. Thus, the main objective of this study is to evaluate the potential use of the two-source SW model to compute the intra-orchard spatial variability of ET of a drip-irrigated olive orchard using satellite images and ground-based climate data. Moreover, the sub-models that were used to estimate the instantaneous values of latent heat flux (LE_i), available energy (A_i), net radiation (Rn_i), and soil heat flux (G_i) were also evaluated for the main phenological growth stages of olive trees.

2. Materials and Methods

2.1. Study Site Description

Measurements were collected in a drip-irrigated commercial olive orchard (*Olea europaea* L. cv Arbequina) that was located in Pencahue, Maule Region, Chile (35°23' S, 71°44' W, WGS84, 96 m above sea level) during the 2009–2010 season. The study area has a Mediterranean climate with a medium average temperature of 14.8 °C and an accumulated ET_0 of 1013 mm. The average annual rainfall in the region is approximately 602 mm, which is falling mainly during the winter (May to September). The summer period (December to March) is generally hot and dry with a high water vapor pressure deficit. The soil at the experimental site was classified as Quepo series with a clay loam soil. The trees were established using a hedgerow system with 5.0 m between rows and 1.5 m between trees (1333 trees ha^{-1}). The orchard was irrigated by two drippers per tree (2.0 L h^{-1}) spaced at intervals of 0.75 m along the rows. After irrigation, the percentage of wetted area (A_W) by the drippers located under the tree canopy was 4.5% of the total area.

To evaluate the irrigation management, the midday stem water potential (Ψ_{md}) was measured using a pressure chamber (PMS Instruments Co., Model 1000, Corvallis, OR, USA). A chosen shoot (two per tree, one tree per replicate) containing five to six pairs of leaves was encased in a plastic bag and then wrapped in aluminum foil for (at least) 2 h before being cut [33,34]. In addition, a LI-COR gas analyzer (Li-6400, LI-COR Inc., Lincoln, NE, USA) was used to measure stomatal resistance (r_{st}) on two leaves that were directly exposed to the sun and were located on the mid-section of the tree [35]. The Ψ_{md} and r_{st} measurements were collected from exposed leaves on 10 trees that were located on both sides of the olive rows. Finally, the LAI was measured weekly using a plant canopy analyzer (LAI-2000, LI-COR, Lincoln, NE, USA), which was calibrated according to Lopez-Olivari et al. [9]. Also, it is important to indicate that Ψ_{md} and r_{st} were measured at the time of satellite overpass.

2.2. Measurements of Climate and Energy Balance Data

Two towers (4.8 m height) were installed on a flat and homogeneous plot (6.45 ha) in an olive orchard to measure the surface energy balance components and micrometeorological variables at intervals of 30 min. Hobo sensors (Onset Computer, Inc., Bourne, MA, USA) were used to measure relative humidity (RH) and air temperature (Ta). An anemometer (03101-5, R. M. Young Co., Traverse City, MI, USA) and a pluviometer (A730RAIN, Adcon Telemetry, Klosterneuburg, Austria) were used to measure wind speed (u) and precipitation (Pp), respectively. The net radiation (Rn) was measured with a four-way net radiometer (CNR1, Kipp & Zonen, Delft, The Netherlands). Finally, the sensors to measure u, Pp, Ta, RH, and Rn were located at 1.9 m above the tree canopy.

Soil heat flux was estimated through the use of eight flux plates installed on either side of the rows (four plates in the inter row and four plates below the row). This arrangement considers the row shade effect throughout the day [36]. The flux plates with a constant thermal conductivity (HFT3,

Campbell Sci., Logan, UT, USA) were installed at a depth of 0.08 m. Additionally, two averaging thermocouple probes (TCAV, Campbell Sci., Logan, UT, USA) that were used to measure soil temperature were installed above each flux plate at depths of 0.02 and 0.06 m. All of the thermocouple probe signals were recorded on an electronic datalogger (CR3000, Campbell Sci, Logan, UT, USA) with a thirty-minute interval. Finally, soil heat flux was calculated at each position by adding the measured flux at 0.08 m to the heat that was stored in the layer above the heat flux plates [9,37].

The latent (LE) and sensible (H) heat fluxes were measured with an eddy covariance (EC) system, which is composed of a three-dimensional sonic anemometer (CSAT3, Campbell Scientific Inc., Logan, UT, USA) and an infrared gas analyzer (LI-7500, LI-COR Inc., Lincoln, NE, USA). The fluxes were registered at 1.6 m above the canopy at intervals of 10 Hz and were stored on a data logger (CR5000, Campbell Scientific Inc., Logan, UT, USA). Finally, a post-processing step was conducted to correct the sonic temperature due to crosswind influences [38] and water vapor density due to the influences of the fluctuations in temperature and humidity [39].

To reduce the uncertainty that is associated with the errors in the LE and H measurements, entire days were excluded from the study when the ratios of (H + LE) to (Rn − G) were outside the range between 0.8 and 1.2 [9]. Assuming that the measurements of Rn and G were representative of the available energy over the olive orchard, the fluxes of H and LE were forced to close the energy balance using the Bowen ratio approach (B = H/LE) [8,24,36,40]:

$$LE_B = \frac{(Rn - G)}{(1 + B)} \tag{1}$$

$$H_B = \frac{(Rn - G)}{\left(1 + B^{-1}\right)} \tag{2}$$

On a daily basis, the olive ET was calculated, as follows:

$$ET_{EC} = \frac{\sum_{n=1}^{24} LE_{B\,n}}{\lambda \rho_w} 1.8 \tag{3}$$

where ET_{EC} is the actual evapotranspiration of the olive orchard measured by the EC system (mm d^{-1}), 1.8 is a conversion factor, λ is the latent heat of vaporization (1013 MJ kg^{-1}), ρ_w is the water density (1000 kg m^{-3}), and n is the number of measurements during a 24-h period. The subscript B indicates that turbulent fluxes were recalculated using the Bowen ratio approach.

2.3. Shuttleworth and Wallace Model Description

At the time of satellite overpass, the SW model can estimate latent heat flux over olive orchards as the sum of the Penman-Monteith equation for evaporation and transpiration weighted by a set of coefficients that account for the combination of soil and canopy resistances. This can be done as follows [41]:

$$LE_i = T_i + E_i \tag{4}$$

$$T_i = C_c \frac{\Delta A_i + \left(\frac{\rho_a C_p D_i - \Delta r_a^c A_{si}}{r_a^a + r_a^c}\right)}{\Delta + \gamma\left(1 + \frac{r_s^c}{(r_a^a + r_a^c)}\right)} \tag{5}$$

$$E_i = C_s \frac{\Delta A_i + \left(\frac{\rho_a C_p D_i - \Delta r_a^s (A_i - A_{si})}{r_a^a + r_a^s}\right)}{\Delta + \gamma\left(1 + \frac{r_s^s}{(r_a^a + r_a^s)}\right)} \tag{6}$$

where LE_i is the instantaneous latent heat flux computed from the SW model (W m^{-2}), T_i is the latent heat flux from the tree transpiration (W m^{-2}), E_i is the latent heat flux from soil evaporation (W m^{-2}), C_c is the canopy resistance coefficient (dimensionless), C_s is the soil surface resistance coefficient (dimensionless), Δ is the slope of the saturation vapor pressure curve at the mean temperature (kPa $°C^{-1}$), A_i is the available energy leaving the complete canopy (W m^{-2}), A_{Si} is the available energy at the soil surface (W m^{-2}), C_P is the specific heat of the air at a constant pressure (1013 J kg^{-1} K^{-1}), D_i is the water vapor pressure deficit at the reference height (kPa), r_a^c is the bulk boundary layer resistance of the vegetative elements in the canopy (s m^{-1}), r_a^a is the aerodynamic resistance between the canopy source height and reference level (s m^{-1}), γ is the psychrometric constant (kPa $°K^{-1}$), r_s^c is the canopy resistance (s m^{-1}), r_a^s is the aerodynamic resistance between the soil and canopy source height (s m^{-1}), and r_s^s is the soil surface resistance (s m^{-1}). Subscripts "i" denote the values that are computed at the time of satellite overpass.

2.4. Estimation of Resistances

Values of r_s^c and r_a^c are calculated as:

$$r_s^c = \frac{r_{st}}{LAI} \tag{7}$$

$$r_a^c = \frac{r_b}{LAI} \tag{8}$$

where r_{st} is the mean stomatal resistance (s m^{-1}), r_b is the mean boundary layer resistance (s m^{-1}) and LAI is the leaf area index (m^2 m^{-2}). For this study, r_b was assumed to be equal to 25 s m^{-1}, and a constant value of r_s^s = 2000 sm^{-1} was used because of the dryness of the soil surface (between rows) during the study periods [41]. Additionally, a mean value of measured r_{st} = 235 (±60.84) s m^{-1} was used for the simulation period. Finally, general descriptions of r_a^a, r_a^s, C_s, and C_c can be found in Shuttleworth and Wallace [41] and Ortega-Farías et al., 2007 [21].

2.5. Estimation of Available Energy at the Time of the Satellite Overpass

The instantaneous values of available energy over the canopy (A_i) and soil surface (A_{Si}) can be calculated as:

$$A_i = Rn_i - G_i \tag{9}$$

$$A_{si} = Rn_{Si} - G_i \tag{10}$$

where Rn_i is the net radiation (W m^{-2}), G_i is the soil heat flux (W m^{-2}) and Rn_{Si} is the net radiation on the surface of the ground (W m^{-2}), which can be calculated using Beer's law, through the next expression:

$$Rn_{Si} = Rn_i \times \exp(-C \times LAI) \tag{11}$$

where C is the extinction coefficient of the net radiation of the crop (dimensionless). In this study, a mean value of measured LAI = 1.29 (±0.07) was used as a constant while C was assumed to be 0.66 [24]. Values of Rn_i were obtained, as follows [42]:

$$Rn_i = (1 - \alpha_i) \times R_{s\downarrow} + R_{L\downarrow} - R_{L\uparrow} - (1 - \varepsilon_0) \times R_{L\downarrow} \tag{12}$$

where α_i is the surface albedo, $R_{s\downarrow}$ is the short-wave incident radiation (W m^{-2}), $R_{L\downarrow}$ and $R_{L\uparrow}$ are the incoming and outgoing long-wave radiation, respectively, and ε_0 corresponds to the superficial emissivity. The values of $R_{L\downarrow}$ and $R_{L\uparrow}$ were obtained as:

$$R_{L\uparrow} = \varepsilon_0 \times \sigma \times T_s^4 \tag{13}$$

$$R_{L\downarrow} = \varepsilon_a \times \sigma \times T_a^4 \tag{14}$$

where σ is the Stefan-Boltzmann constant (5.67 × 10^{-8} W m^{-2} K^{-4}), T_s is the superficial temperature (°K), ε_0 is the "broad-band" surface emissivity (dimensionless), T_a is the near-surface air temperature (°K), and ε_a is the effective atmospheric emissivity (dimensionless). The methodologies to estimate ε_0, ε_a, and T_s are indicated by Allen et al. [42].

G_i was calculated using the following empirical relation [26]:

$$G_i = 0.3236 \times Rn_i - 51.52 \qquad (15)$$

2.6. Instantaneous to Daily Extrapolation of Actual Evapotranspiration

Using multispectral and thermal images from the Landsat satellite and ground-based weather data, the daily ET was computed for each image pixel using the following equations [28,42]:

$$ET_i = 3600 \frac{LE_i}{\rho_w \lambda} \qquad (16)$$

$$E_F = \frac{ET_i}{ET_{oi}} \qquad (17)$$

$$ET_{sw} = E_F \times ET_o \qquad (18)$$

where ET_i is the instantaneous ET computed from the SW model (mm h^{-1}), ρ_w is the density of water (1000 kg m^{-3}), λ is the latent heat of vaporization (J kg^{-1}), ET_{oi} is the instantaneous reference evapotranspiration (mm h^{-1}), ET_{sw} is the daily actual evapotranspiration estimated using the SW model (mm d^{-1}), ET_o is the daily reference evapotranspiration (mm d^{-1}), and E_F is the reference evapotranspiration fraction.

2.7. Image Processing

Seven Landsat (7 ETM+) satellite images were downloaded from the USGS GloVis web platform for free (http://glovis.usgs.gov) using the coordinates WRS-2 Path/Row: 233/85 (Table 1). The SLC instrument suffered a problem in 2003; thus, the images have lines of pixels without information. A method that was proposed by Storey et al. [43] was implemented to fill the pixels without data. This method estimates the value of the pixel based on the neighboring pixels in an image taken on a similar date. The estimations present good results in most of the images, but not in images with abrupt temporal transitions [44]. Also, the image processing for satellite information considered the methodology that was proposed by Allen et al. [42], aiming to generate radiometric and atmospheric corrections.

Table 1. Images selected for processing the Shuttleworth and Wallace model (SW) and main phenological stages of a drip-irrigated olive orchard.

Date	Day of Year	Overpass Time	Scene Cloud Cover	Phenological Stages
(dd-mm-yy)	(DOY)	(UTC)	(%)	
4/02/2009	35	2:24:02 p.m.	1	FC
20/02/2009	51	2:24:12 p.m.	2	FC
3/11/2009	307	2:24:42 p.m.	7	F
5/12/2009	339	2:25:05 p.m.	21	FS
21/12/2009	355	2:25:21 p.m.	1	PH
6/01/2010	6	2:25:38 p.m.	1	PH
22/01/2010	22	2:25:52 p.m.	0	PH

F = flowering; FS = fruit set; PH = pit hardening; FC = fruit colouring.

Multispectral and thermal data from each satellite image were processed pixel by pixel to estimate LE_i, Rn_i, G_i, and ET_i. In this regard, the satellite images were obtained from November to February where the main phenological periods were observed in the drip-irrigated olive orchard (Table 1). Finally, it is important to indicate that the number of days with a complete data set (satellite images and EC fluxes) was limited by cloudiness when the satellite overpassed the experimental site and persistent noise of the EC system (instrumental problems and flow distortion through the tower). For these reasons, several researchers have used a limited number of images to evaluate the satellite-based remote sensing (SBRS) models [6,45–49].

2.8. Statistical Analysis

The SW model used to compute the instantaneous latent heat flux (LE_i) and daily actual evapotranspiration (ET_{sw}) was validated using ground-based weather measurements. The validation was carried out using the ratio of the estimated to observed values (b), root-mean-square error (RMSE), mean absolute error (MAE), and index of agreement (I_a) [50,51]. The *t*-test was used to determine whether the b value was significantly different from unity at the 95% confidence level. In addition, the sub-models for estimating A_i, Rn_i, and G_i were included in the validation. Values of RMSE, MAE, and I_a were computed, as follows:

$$RMSE = \left[N^{-1} \sum_{i=1}^{N} (P_i - O_i)^2 \right]^{0.5} \tag{19}$$

$$MAE = N^{-1} \sum_{i=1}^{N} |P_i - O_i| \tag{20}$$

$$I_a = 1 - \left[\frac{\sum_{i=1}^{N} (P_i - O_i)^2}{\sum_{i=1}^{N} (|P_i - \overline{O}| + |O_i - \overline{O}|)^2} \right] \quad 0 \leq I_a \leq 1 \tag{21}$$

where N is the total number of observations, P_i and O_i are the estimated and observed values, respectively, and $\overline{O}$ is the mean of the observed values. Values of RMSE, MAE, P_i, O_i, and $\overline{O}$ are in W m^{-2} or mm d^{-1}

Finally, a sensitivity analysis of the inputs parameters (C, d, LAI, n) and resistances (r_b, r_{st}, r_a^a, r_a^c, r_a^s) was conducted to evaluate their effects on the ability of the SW model for estimating ET_i. In this case, the percent deviation of the mean ET_i was computed when the input values of the parameters and resistances individually varied by +/− 30%.

3. Results

The atmospheric conditions were very dry and hot during the main phenological periods (FS, PH, and FC) of the olive orchard, which were observed from December to February (Table 1). During this period, the values of u, Ta and D were between 0.3 and 4.5 m s^{-1}, 28.3–13.4 °C, and 0.2–2.9 kPa, respectively (Figure 1a,b). During the study, four rainfall events were observed with values of less than 5 mm. The accumulated value of ET_o from September to March (growing season) was 907 mm while that from December to March (summer) was 616 mm. During this study, the average values of ET_o and ET_{oi} were 6.95 (±0.99) mm d^{-1} and 0.58 (±0.06) mm h^{-1}, respectively (Table 2). These results indicate that the drip-irrigated olive orchard was under high atmospheric demand for water vapor (Figure 1). Under this atmospheric condition, values of Ψ_{md} ranged between −1.37 and −1.49 MPa, indicating that olive trees were well irrigated during the main phenological periods (F, FS, PH, and FD). Finally, Figure 1 shows that clear days were observed when the satellite passed over the experimental site.

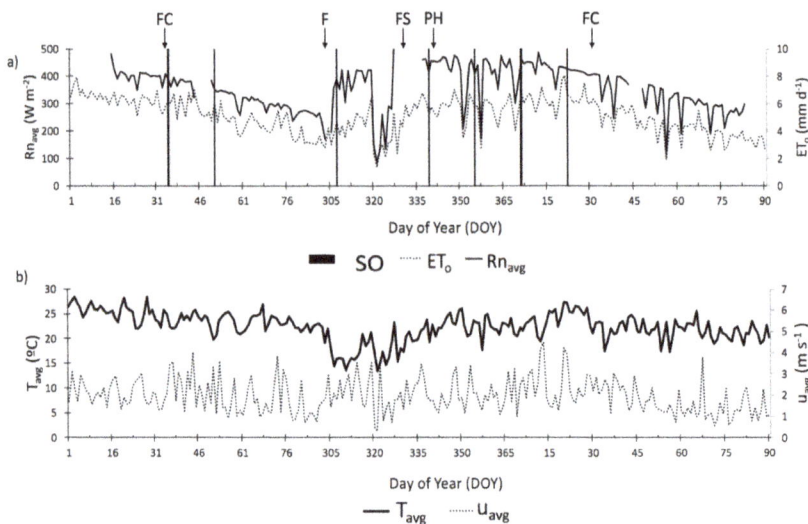

Figure 1. (**a**) Daily variation of net radiation (Rn$_{avg}$) and reference evapotranspiration (ET$_o$); (**b**) average daily values of temperature (T$_{avg}$) and average daily values of wind speed (u$_{avg}$). The vertical bars indicate the time of satellite overpass (SO). Also, phenological stages are indicated by an abbreviation followed by an arrow where F = flowering, FS = fruit Set, PH = pit hardening and FC = fruit colouring.

Table 2. Meteorological conditions during the study period.

DOY	Date	ET$_o$ (mm d^{-1})	ET$_{oi}$ (mm h^{-1})	RH (%)	Ta (°C)	U (m s^{-1})
35	4/02/2009	7.88	0.56	51.94	22.91	0.81
51	20/02/2009	6.88	0.52	54.16	17.92	4.11
307	3/11/2009	5.13	0.54	58.85	14.12	2.33
339	5/12/2009	6.47	0.57	56.12	16.36	1.17
355	21/12/2009	6.69	0.69	47.6	21.29	2.69
6	6/01/2010	7.84	0.6	53.53	21.33	2.69
22	22/01/2010	7.75	0.61	52.99	22.17	1.29
	Mean	6.95	0.58	53.6	19.44	2.16
	D.E.	0.99	0.06	3.5	3.33	1.15

where ET$_o$ = daily reference evapotranspiration, ET$_{oi}$ = instantaneous reference evapotranspiration, RH = Relative air humidity, Ta = Air Temperature and u = wind speed.

At 30 min time intervals, the ratio of (Rn − G) to (H + LE) was 0.89, suggesting that the orchard energy balance (SEB) was systematically imbalanced by approximately 11% (Figure 2). Several researchers reported a similar situation where the values of (LE + H) were less than those of (Rn − G) above olive orchards when the EC technique was used. In flood-irrigated olive orchards, Williams et al. [52], Ezzahar et al. [53], and Er-Raki et al. [4] reported imbalances ranging between 9% and 26%, while in drip-irrigated olive orchards, Villalobos et al. [54] and Testi et al. [55] indicated imbalances varying between 5% and 17%. The potential errors can be attributed to the uncertainties in the measurements of Rn and G [56–58], and energy storage within the olive tree biomass [9,53]. In this case, Twine et al. [40] suggested that LE and H values from de EC system can be corrected using the Bowen ratio (β = H/LE) because the different problems affect the measured values of H and LE in a similar proportion. Using this correction, however, Allen et al. [59] indicated that the lack of closure can be only attributed to errors in the measurements of H and LE without considering the potential bias in Rn-G. According to several researchers, the uncertainties observed in this study are modest and turbulent fluxes were recalculated using the Bowen-ratio approach [8,24,36,40,60].

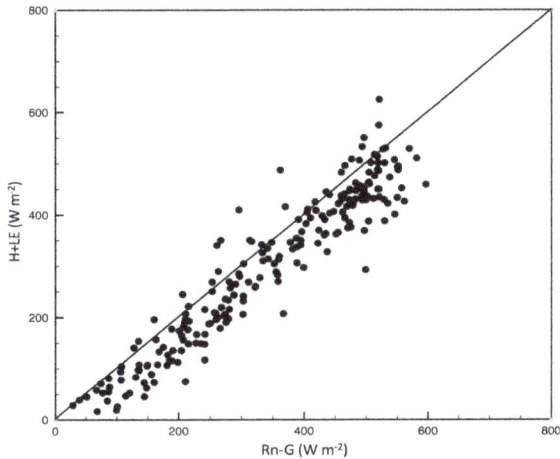

Figure 2. Energy balance closure at interval of 30 min for the days when the satellite overpassed the experimental site. H, LE, Rn, and G correspond to sensible heat flux (W m^{-2}), latent heat flux (W m^{-2}), net radiation (W m^{-2}), and soil heat flux (W m^{-2}), respectively. The coefficient of determination (R^2) was 0.88 and the (Rn $-$ G)/(H $+$ LE) ratio was 0.89.

3.1. Model Validation of Available Energy, Latent Heat Flux and Actual Evapotranspiration

The model validation of Rn$_i$, G$_i$, and A$_i$ is indicated in Table 3 when using ground-based weather measurements and satellite images. The results indicated that the sub-model was able to predict values of A$_i$ with a RMSE = 25 W m^{-2} and MAE = 21 W m^{-2}. The RMSE and MAE values for Rn$_i$ were between 39 and 32 W m^{-2}, while those for G$_i$ were between 33 and 27 W m^{-2}, respectively (Table 3). The *t*-test showed that the b values were different from unity, indicating that A$_i$, Rn$_i$, and G$_i$ were underestimated, with errors ranging between 4% and 8% of observed values in the drip-irrigated olive orchard. Finally, the comparison between the measured and modeled values in Figure 3 shows that all points were close to the 1:1 line, suggesting a good performance of the sub-models to estimate A$_i$, Rn$_i$, and G$_i$. For Rn$_i$ and G$_i$, the results that were observed in this study are similar to those that are found in literature for heterogeneous canopies, such as orchards and vineyard. For winegrape (drip irrigated) and table grape (micro sprinkler), Teixeira et al. [61] observed values of coefficient of determination (R^2) = 0.94 and RMSE = 17.5 W m^{-2} for Rn$_i$ when using satellite images (Landsat 5 and 7). For a drip-irrigated olive orchard (fc = 0.29), Ortega-Farías et al. [27] found that the METRIC model underestimated Rn$_i$ by approximately 3% of the observed values, with RMSE and MAE values of 40 and 33 W m^{-2}, respectively. In a super intensive olive orchard (fc = 0.3), Ortega-Farías et al. [26] indicated that the model was able to estimate Rn$_i$ and G$_i$, with errors of 5% and 2%, respectively, when using high-resolution thermal and multispectral data acquired with an unmanned aerial vehicle (UAV). In a drip-irrigated vineyard (fc = 0.3), Carrasco-Benavides et al. [28] noted that the METRIC model was able to estimate Rn$_i$ and G$_i$, with errors of 11% and 5%, respectively. In an experiment that was carried out using remote sensing techniques over heterogeneous landscapes, Liu et al. [62] found RMSE and MAE values of 51 and 25 W m^{-2}, respectively, when G$_i$ was evaluated.

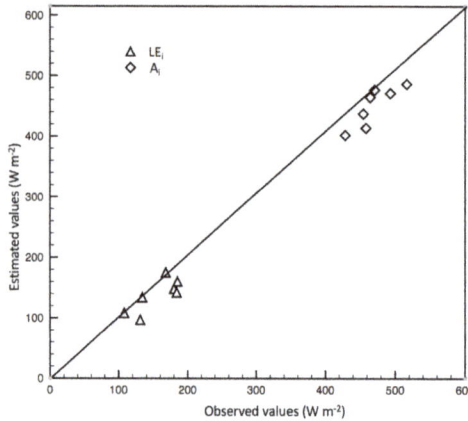

Figure 3. Comparison between estimated and observed values of instantaneous latent heat flux (LE$_i$) from the two-source model and instantaneous available energy (A$_i$) at the time of satellite overpass. Values of index of agreement (d) for LE$_i$ and A$_i$ were 0.8 and 0.82, respectively.

Table 3. Model validation of instantaneous values of net radiation (Rn$_i$), soil heat flux (G$_i$), available energy (A$_i$), and latent heat flux (LE$_i$) over a drip irrigated olive orchard. Also, daily actual evapotranspiration (ET$_{SW}$) is included in the analysis.

Variable	RMSE	MAE	I$_a$	b	*t*-Test
Rn$_i$ (W m^{-2})	39	32	0.79	0.96	F
G$_i$ (W m^{-2})	33	27	0.5	0.92	F
A$_i$ (W m^{-2})	25	21	0.82	0.96	F
LE$_i$ (W m^{-2})	26	20	0.8	0.88	F
ET$_{SW}$ (mm d^{-1})	0.31	0.28	0.95	0.90	F

where RMSE = Root mean square error, MAE = Mean absolute error, b = Ratio of estimated to observed values and I$_a$ = Index of agreement. T = null hypothesis (b = 1) True; F = alternative hypothesis (b ≠ 1).

Table 3 indicates that the SW model underestimated LE$_i$ by approximately 12% of observed values, with RMSE = 26 W m^{-2} and MAE = 20 W m^{-2} (Table 3). In addition, the comparisons between the observed and estimated values of LE at the time of satellite overpass show that most of the points are close to the 1:1 line (Figure 3). Furthermore, the model validation indicated the SW model was able to simulate the ET with RMSE and MAE values that are equal to 0.31 and 0.28 mm d^{-1}, respectively (Table 3). Additionally, the statistical analysis indicated that the ratio of ET$_{sw}$/ET$_{EC}$ was significantly different from unity, suggesting that the SW model underestimated ET with an error of 10% of observed values. In this regard, Figure 4 shows that all of the points were distributed below the 1:1 line. In a super intensive olive orchard (fc = 0.3, LAI = 1.3), Ortega-Farías and Lopez-Olivari [24] indicated that the SW model overestimated LE$_i$ and ET by approximately 2% and 6% of observed values, respectively, suggesting that the model was very sensitive to errors in the estimation of r$_{st}$ and LAI. Finally, in a furrow-irrigated vineyard (fc = 0.35, IAF = 2), Zhang et al. [20] observed MAE values about of 39 W m^{-2} when SW model was evaluated.

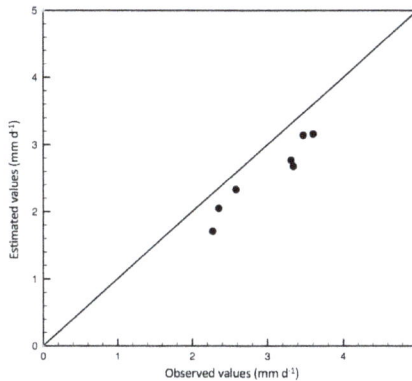

Figure 4. Comparison between estimated and observed values of daily actual evapotranspiration of a drip irrigated olive orchard. (ET_{sw}). Index of agreement (I_a) for ET_{sw} was equal to 0.95. This comparison was done using days when satellite images were available.

3.2. Sensitivity Analysis

Table 4 indicates that when Rn_i varied by ±30%, a variation of ±20% was observed in the estimation of ET_i. Also, the results indicated that the sensitivity of the predicted ET_i in respect to the uncertainties in C, d, n, and r_b was minimal. In addition, the sensitivity analysis indicated that the estimation of ET_i was significantly affected by variations of ±30% in the values of LAI and r_s^c. In this regard, the values of ET_i varied between +15% and −14% and between −13% and +20% when LAI and r_s^c varied by ±30%, respectively. (Table 4). This sensitivity analysis suggests that the ET_i that was estimated using the SW model was very sensitive to Rn_i, LAI, and r_{st}, but it was not sensitive to errors in the estimations of r_a^a, r_a^c, and r_a^s. Similar results in a drip-irrigated olive orchard were observed by Ortega-Farías and Lopez-Olivari [24], who indicated that the ET predicted by the SW model was sensitive to errors of ±30% in LAI and r_{st}, but was not significantly affected by errors in the estimation of aerodynamic resistances. Therefore, correct estimations of Rn_i, LAI and r_{st} become critical to increasing the accuracy of the SW model.

Table 4. Sensitivity analysis to instantaneous actual evapotranspiration (ET_i).

Variables	Symbol	Parameters	30%	−30%
Extinction coefficient of crop for net radiation (dimensionless)	C	0.66 *	0.09	0.53
Zero plane displacement of crop with complete canopy cover (LAI = 4) (m)	d	2.01 *	−0.26	3.29
Leaf Area Index (m^2 m^{-2})	LAI	1.29 *	13.93	−14.52
Eddy diffusivity decay constant in crop with complete canopy crop cover (dimensionless)	n	2.5 *	2.16	−0.95
Mean boundary layer resistance per unit area of vegetation (s m^{-1})	r_b	25 *	1.87	−1.44
Mean stomatal resistance (s m^{-1})	r_{st}	235 *	−13.06	19.96
Aerodynamic resistance between canopy source height and reference level (s m^{-1})	r_a^a	**	4.51	−4.45
Bulk boundary layer resistance of the vegetative elements in the canopy (s m^{-1})	r_a^c	**	1.87	−1.44
Aerodynamic resistance between the substrate and canopy source height (s m^{-1})	r_a^s	**	1.99	−1.61
Bulk stomatal resistance of the canopy (s m^{-1})	r_s^c	**	−12.82	20.3
Surface resistance of the substrate (s m^{-1})	r_s^s	2000 *	−3.19	5.93
Instantaneous net radiation (W m^{-2})	Rn_i	**	20.60	−20.03
Instantaneous Soil heat flux (W m^{-2})	G_i	**	−2.18	2.75
Instantaneous available energy (W m^{-2})	A_i	**	17.96	−17.39

* Constant values; ** Computed values.

3.3. Spatial Variability of Actual Evapotranspiration

For flowering, pit hardening and fruit colouring, maps of the actual evapotranspiration were generated using the temporal variability of ET_{sw} within the drip-irrigated olive orchard (Figure 5). While differences were observed in the ET_{sw} values over the entire time series (from 1.34 to 3.69 mm), the spatial variability for each of these days was low, with a maximum variation between 2.85 and 3.07 mm d^{-1} (DOY 22). These low variations of the spatially distributed ET_{sw} maps could be associated with the low intra-orchard spatial variability of A_i (Figure 6), which presented a standard deviation between 3 and 6 W m^{-2} for the entire study period. The highest value of A_i was observed on DOY 339 when the A_i ranged between 476 and 494 W m^{-2}. On the other hand, when we consider the daily variation, the results reflected a low spatial variability of the available energy. These results are logical when considering that the orchard presents similar management conditions (irrigation practices, training systems, canopy management, fertilization, among others), which are expressed by the low spatial variability at the experimental site.

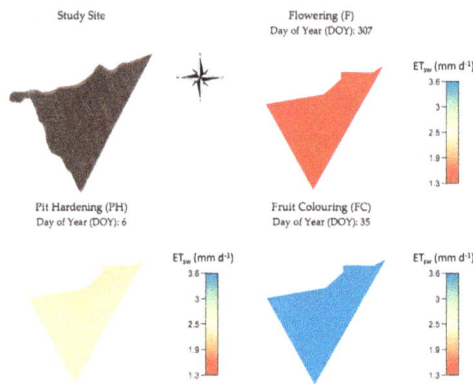

Figure 5. Maps of daily actual evapotranspiration (ET) computed from the SW model using remote sensing and meteorological data.

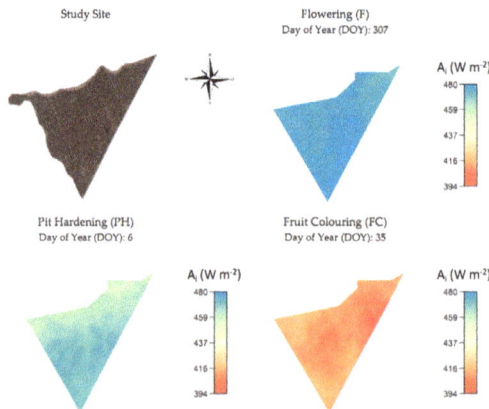

Figure 6. Maps of instantaneous Available energy (A_i) generated from satellite images and ground-based climate data.

4. Final Remarks

The results that were obtained in this study were similar to those that were reported in the literature for estimating olive ET when using the traditional approach of SW model [24,26,27]. This approach, however, does not include satellite images to account for the spatial variability of ET. For the main phenological stages, this study suggested that the SW model that uses satellite images and meteorological data could be suitable to estimate olive ET on a pixel-by-pixel basis. For practical application, however, the SW model requires good parameterizations of r_{st} and LAI, which are included in the formulation of r_s^c. According to Ortega-Farías and Lopez-Olivari [24], the parameterization of r_s^c depends on the canopy characteristics, which are mainly expressed by the training system and plant water status. In this study, the canopy architecture was maintained almost constant during the study with values of LAI and fc ranging between 1.16–1.38 m^2 m^{-2} and 0.26–0.3, respectively. Also, the drip-irrigated olive orchard was maintained under non-water stress conditions ($\Psi_{md} > -1.49$ MPa), indicating that the constant value of stomatal resistance (235 ± 60.84 s m^{-1}) used in the SW model was adequate. However, commercial olive orchards that possess different training systems, tree water status, and canopy sizes require the incorporation of spatially distributed values of LAI and stomatal conductance to increase the accuracy of the SW model to simulate ET. To accomplish this task, it is necessary to evaluate sub-models for predicting the spatial variability of A_i, LAI, and r_s^c of olive orchards with different canopy geometry and tree water status. Finally, the results of this study suggested that the SW model could be used to estimate olive ET as a complement to the satellite-based remote sensing (SBRS) models, like METRIC.

5. Conclusions

Results of this study suggested that the SW model could be a potential tool to compute spatial the variability of olive ET when using satellite images and ground-based climate data. In this case, Rn_i and G_i were obtained from satellite information and were introduced to the SW model through A_i. Simulated values of LE_i, Rn_i, G_i, and A_i were generally in good agreement with ground-based measurements within the olive orchard. In this case, the errors and RMSE ranged between 4% and 12% of observed values and 25–39 W m^{-2}, respectively. For daily water consumption, the results indicated that the SW model was able to predict olive ET with errors of 10% and RMSE = 0.3 mm d^{-1}. The observed results were consistent with the literature and are within the acceptable range for applications of remote sensing models in agriculture. Furthermore, the results also encourage the continued testing of this methodology in heterogeneous (vineyards and orchards) canopies with the aim of evaluating the effect of intra-orchard spatial variabilities of soil and canopy vigor on the estimation of ET. However, the practical application of the SW model requires adequate parameterizations of r_{st} and LAI of olive orchards under different training and system.

Acknowledgments: This study was supported by the Chilean government through the projects FONDECYT (No. 1130729) and FONDEF (No. D10I1157) and by the Universidad de Talca through the research program "Adaptation of Agriculture to Climate Change (A2C2)". The authors would like to thank Manuel Barrera and Alvaro Ried from the "Olivares de Quepu" Company for their technical support.

Author Contributions: Fernando Fuentes-Peñailillo and Samuel Ortega-Farías conceived and designed the experiments, supported the statistical analysis, and wrote the original manuscript draft. Cesar Acevedo-Opazo and David Fonseca-Luengo contributed extensively to results interpretation and discussions and editing of the manuscript.

Conflicts of Interest: The authors declare no conflict of interest.

List of Symbols

A_i	Instantaneous available energy leaving the complete canopy (W m^{-2})
A_{Si}	Available energy at the soil surface (W m^{-2})
A_W	Percentage of wetted area (%)

b	Ratio of estimated to observed values (dimensionless)
B	Bowen ratio (dimensionless)
C_C	Canopy resistance coefficient (dimensionless)
C_S	Soil surface resistance coefficient (dimensionless)
C	Extinction coefficient of crop for net radiation (dimensionless)
C_P	Specific heat of the air at constant pressure (1013 J kg^{-1} °K^{-1})
CV	Coefficient of variation (%)
d	Zero plane displacement of crop with complete canopy cover (LAI = 4) (m)
D_i	Water vapor pressure deficit at the reference height (kPa)
E_i	Latent heat flux from soil evaporation (W m^{-2})
E_F	Reference evapotranspiration fraction (dimensionless)
ET	Daily actual evapotranspiration (mm d^{-1})
ET_{EC}	Daily actual evapotranspiration obtained from de eddy covariance (EC) method (mm d^{-1})
ET_i	Instantaneous actual evapotranspiration computed from the Shuttleworth and Wallace (SW) model (mm h^{-1})
ET_o	Reference evapotranspiration (mm d^{-1})
ET_{oi}	Instantaneous reference evapotranspiration (mm h^{-1})
ET_{sw}	Daily actual evapotranspiration estimated using the SW model (mm d^{-1})
fc	Fractional crop cover (dimensionless)
G	Soil heat flux (W m^{-2})
G_i	Instantaneous soil heat flux (W m^{-2})
H	Sensible heat flux (W m^{-2})
H_B	H corrected using the Bowen ratio approach (W m^{-2})
I_a	Index of agreement (dimensionless)
Kc	Crop coefficient (dimensionless)
Kcb	basal crop coefficient (dimensionless)
Ke	soil evaporation coefficient (dimensionless)
LAI	Leaf area index (m^2 m^{-2})
LE_{EC}	Latent heat flux obtained from the EC method (W m^{-2})
LE_B	Instantaneous LE corrected using the Bowen ratio approach (W m^{-2})
LE_i	Instantaneous LE estimated using SW (W m^{-2})
MAE	Mean absolute error (dimensionless)
N	Total number of observations,
n	Eddy diffusivity decay constant in crop with complete canopy crop cover (dimensionless)
O_i	Observed values (W m^{-2} or mm d^{-1})
$\overline{O}$	Mean of the observed values (W m^{-2} or mm d^{-1})
P_i	Estimated values (W m^{-2} or mm d^{-1})
Pp	precipitation (mm)
RH	Relative air humidity (%)
Rn	Net radiation (W m^{-2})
Rn_{avg}	Average daily values of net radiation (W m^2)
Rn_i	Instantaneous net radiation (W m^2)
Rn_{Si}	Instantaneous net radiation on the surface of the ground (W m^{-2})
$R_{s\downarrow}$	Incident short-wave radiation (W m^{-2})
$R_{L\downarrow}$	Incoming long-wave radiation (W m^{-2})
$R_{L\uparrow}$	Outgoing long-wave radiation (W m^{-2})
R^2	Coefficient of determination (dimensionless)
r_b	Mean boundary layer resistance per unit area of vegetation (s m^{-1})
r_a^c	Bulk boundary layer resistance of the vegetative elements in the canopy (s m^{-1})
r_a^a	Aerodynamic resistance between the canopy source height and reference level (s m^{-1})
r_s^c	Canopy resistance (s m^{-1})
r_a^s	Aerodynamic resistance between the soil and canopy source height (s m^{-1})
r_s^s	Soil surface resistance (s m^{-1})
r_{st}	Mean stomatal resistance (s m^{-1})

Ta	Air temperature (°C)
T_{avg}	Average daily values of temperature (°C)
T_i	Latent heat flux from the olive transpiration (W m^{-2})
T_s	Superficial temperature (°K)
T_n	Near surface air temperature (°K)
u	Wind speed (m s^{-1})
u_{avg}	Average daily values of wind speed (m s^{-1})
Z_{om}	Surface roughness (m)
zo	Roughness length of crop with complete canopy cover (s m^{-1})
zó	Roughness length of soil surface (s m^{-1})
α	Surface albedo (dimensionless)
α_i	Instantaneous surface albedo (dimensionless)
ε_a	Effective atmospheric emissivity (dimensionless)
ε_0	Superficial emissivity (dimensionless)
λ	Latent heat of vaporization (1013 MJ kg^{-1}),
ρ_w	Water density (1000 kg m^{-3}),
σ	Stefan-Boltzmann constant (5.67×10^{-8} Wm^{-2} °K^{-4})
γ	Psychrometric constant (kPa °K^{-1})
τ_{sw}	Wide-band atmospheric transmissivity (dimensionless)
Ψ_{md}	Midday stem water potential (MPa)
Δ	Slope of the saturation vapor pressure curve at the mean temperature (kPa °C^{-1})

References

1. Ortega-Farías, S.; Irmak, S.; Cuenca, R.H. Special issue on evapotranspiration measurement and modeling. *Irrig. Sci.* **2009**, *28*, 1–3. [CrossRef]
2. Fereres, E.; Soriano, M.A. Deficit irrigation for reducing agricultural water use. *J. Exp. Bot.* **2007**, *58*, 147–159. [CrossRef] [PubMed]
3. Allen, R.G.; Pereira, L.S.; Raes, D.; Smith, M. *Crop Evapotranspiration—Guidelines for Computing Crop Water Requirements—FAO Irrigation and Drainage Paper 56*; Food and Agriculture Organization of the United Nations (FAO): Rome, Italy, 1998.
4. Er-Raki, S.; Chehbouni, A.; Boulet, G.; Williams, D.G. Using the dual approach of FAO-56 for partitioning ET into soil and plant components for olive orchards in a semi-arid region. *Agric. Water Manag.* **2010**, *97*, 1769–1778. [CrossRef]
5. Er-Raki, S.; Chehbouni, A.; Duchemin, B. Combining Satellite Remote Sensing Data with the FAO-56 Dual Approach for Water Use Mapping in Irrigated Wheat Fields of a Semi-Arid Region. *Remote Sens.* **2010**, *2*, 375–387. [CrossRef]
6. Cammalleri, C.; Ciraolo, G.; Minacapilli, M.; Rallo, G. Evapotranspiration from an Olive Orchard using Remote Sensing-Based Dual Crop Coefficient Approach. *Water Resour. Manag.* **2013**, *27*, 4877–4895. [CrossRef]
7. Paço, T.A.; Pocas, I.; Cunha, M.; Silvestre, J.C.; Santos, F.L.; Paredes, P.; Pereira, L.S. Evapotranspiration and crop coefficients for a super intensive olive orchard. An application of SIMDualKc and METRIC models using ground and satellite observations. *J. Hydrol.* **2014**, *519*, 2067–2080. [CrossRef]
8. Martínez-Cob, A.; Faci, J.M. Evapotranspiration of an hedge-pruned olive orchard in a semiarid area of NE Spain. *Agric. Water Manag.* **2010**, *97*, 410–418. [CrossRef]
9. Lopez-Olivari, R.; Ortega-Farías, S.; Poblete-Echeverría, C. Partitioning of net radiation and evapotranspiration over a superintensive drip-irrigated olive orchard. *Irrig. Sci.* **2016**, *34*, 17–31. [CrossRef]
10. Allen, R.G.; Pereira, L.S. Estimating crop coefficients from fraction of ground cover and height. *Irrig. Sci.* **2009**, *28*, 17–34. [CrossRef]
11. Pôças, I.; Paco, T.A.; Paredes, P.; Cunha, M.; Pereira, L.S. Estimation of Actual Crop Coefficients Using Remotely Sensed Vegetation Indices and Soil Water Balance Modelled Data. *Remote Sens.* **2015**, *7*, 2373–2400. [CrossRef]

12. Er-Raki, S.; Chehbouni, A.; Hoedjes, J.; Ezzahar, J.; Duchemin, B.; Jacob, F. Improvement of FAO-56 method for olive orchards through sequential assimilation of thermal infrared-based estimates of ET. *Agric. Water Manag.* **2008**, *95*, 309–321. [CrossRef]
13. Anadranistakis, M.; Liakatas, A.; Kerkides, P.; Rizos, S.; Gavanosis, J.; Poulovassilis, A. Crop water requirements model tested for crops grown in Greece. *Agric. Water Manag.* **2000**, *45*, 297–316. [CrossRef]
14. Ortega-Farías, S.; Poblete-Echeverría, C.; Brisson, N. Parameterization of a two-layer model for estimating vineyard evapotranspiration using meteorological measurements. *Agric. For. Meteorol.* **2010**, *150*, 276–286. [CrossRef]
15. Testi, L.; Villalobos, F.J.; Orgaz, F. Evapotranspiration of a young irrigated olive orchard in southern Spain. *Agric. For. Meteorol.* **2004**, *121*, 1–18. [CrossRef]
16. Were, A.; Villagarcía, L.; Domingo, F.; Moro, M.J.; Dolman, A.J. Aggregating spatial heterogeneity in a bush vegetation patch in semi-arid SE Spain: A multi-layer model versus a single-layer model. *J. Hydrol.* **2008**, *349*, 156–167. [CrossRef]
17. Zhou, M.C.; Ishidaira, H.; Hapuarachchi, H.P.; Magome, J.; Kiem, A.S.; Takeuchi, K. Estimating potential evapotranspiration using Shuttleworth–Wallace model and NOAA-AVHRR NDVI data to feed a distributed hydrological model over the Mekong River basin. *J. Hydrol.* **2006**, *327*, 151–173. [CrossRef]
18. Kato, T.; Kamichika, M. Determination of a crop coefficient for evapotranspiration in a sparse sorghum field. *Irrig. Drain.* **2006**, *55*, 165–175. [CrossRef]
19. Gardiol, J.M.; Serio, L.A.; Della Maggiora, A.I. Modelling evapotranspiration of corn (*Zea mays*) under different plant densities. *J. Hydrol.* **2003**, *271*, 188–196. [CrossRef]
20. Zhang, B.; Kang, S.; Li, F.; Zhang, L. Comparison of three evapotranspiration models to Bowen ratio-energy balance method for a vineyard in an arid desert region of northwest China. *Agric. For. Meteorol.* **2008**, *148*, 1629–1640. [CrossRef]
21. Ortega-Farías, S.; Carrasco, M.; Olioso, A.; Acevedo, C.; Poblete, C. Latent heat flux over Cabernet Sauvignon vineyard using the Shuttleworth and Wallace model. *Irrig. Sci.* **2006**, *25*, 161–170. [CrossRef]
22. Zhang, B.; Kang, S.; Zhang, L.; Tong, L.; Du, T.; Li, F.; Zhang, J. An evapotranspiration model for sparsely vegetated canopies under partial root-zone irrigation. *Agric. For. Meteorol.* **2009**, *149*, 2007–2011. [CrossRef]
23. Zhao, P.; Li, S.; Li, F.; Du, T.; Tong, L.; Kang, S. Comparison of dual crop coefficient method and Shuttleworth–Wallace model in evapotranspiration partitioning in a vineyard of northwest China. *Agric. Water Manag.* **2015**, *160*, 41–56. [CrossRef]
24. Ortega-Farías, S.; Lopez-Olivari, R. Validation of a Two-Layer Model to Estimate Latent Heat Flux and Evapotranspiration in a Drip-Irrigated Olive Orchard. *Trans. ASABE* **2012**, *55*, 1169–1178. [CrossRef]
25. Brenner, A.J.; Incoll, L.D. The effect of clumping and stomatal response on evaporation from sparsely vegetated shrublands. *Agric. For. Meteorol.* **1997**, *84*, 187–205. [CrossRef]
26. Ortega-Farías, S.; Ortega-Salazar, S.; Poblete, T.; Kilic, A.; Allen, R.; Poblete-Echeverría, C.; Ahumada-Orellana, L.; Zuniga, M.; Sepulveda, D. Estimation of energy balance components over a drip-irrigated olive orchard using thermal and multispectral cameras placed on a helicopter-based unmanned aerial vehicle (UAV). *Remote Sens.* **2016**, *8*, 638. [CrossRef]
27. Ortega-Farías, S.; Ortega-Salazar, S.; Aguilar, R.; De la Fuente, D.; Fuentes, F. Evaluation of a model to estimate net radiation over a drip-irrigated olive orchard using landsat satellite images. *Acta Hortic.* **2014**, *1057*, 309–314. [CrossRef]
28. Carrasco-Benavides, M.; Ortega-Farías, S.; Lagos, L.O.; Kleissl, J.; Morales-Salinas, L.; Kilic, A. Parameterization of the satellite-based model (METRIC) for the estimation of instantaneous surface energy balance components over a drip-irrigated vineyard. *Remote Sens.* **2011**, *6*, 11342–11371. [CrossRef]
29. Cuenca, R.H.; Ciotti, S.P.; Hagimoto, Y. Application of Landsat to Evaluate Effects of Irrigation Forbearance. *Remote Sens.* **2011**, *5*, 3776–3802. [CrossRef]
30. Long, D.; Singh, V.P. A modified surface energy balance algorithm for land (M-SEBAL) based on a trapezoidal framework. *Water Resour. Res.* **2012**, *48*. [CrossRef]
31. Long, D.; Singh, V.P. Assessing the impact of end-member selection on the accuracy of satellite-based spatial variability models for actual evapotranspiration estimation. *Water Resour. Res.* **2013**, *49*, 2601–2618. [CrossRef]

32. Dhungel, R.; Allen, R.G.; Trezza, R.; Robison, C.W. Comparison of Latent Heat Flux Using Aerodynamic Methods and Using the Penman-Monteith Method with Satellite-Based Surface Energy Balance. *Remote Sens.* **2014**, *6*, 8844–8877. [CrossRef]

33. Tognetti, R.; D'Andria, R.; Sacchi, R.; Lavini, A.; Morelli, G.; Alvino, A. Deficit irrigation affects seasonal changes in leaf physiology and oil quality of *Olea europaea* (cultivars Frantoio and Leccino). *Ann. Appl. Biol.* **2007**, *150*, 169–186. [CrossRef]

34. Pérez-López, D.; Gijón, M.C.; Marino, J.; Moriana, A. Water relation response to soil chilling of six olive (*Olea europaea* L.) cultivars with different frost resistance. *Span. J. Agric. Res.* **2010**, *8*, 780–789. [CrossRef]

35. Dichio, B.; Xiloyannis, C.; Sofo, A.; Montanaro, G. Osmotic regulation in leaves and roots of olive trees during a water deficit and rewatering. *Tree Physiol.* **2006**, *26*, 179–185. [CrossRef] [PubMed]

36. Poblete-Echeverría, C.; Sepúlveda-Reyes, D.; Ortega-Farías, S. Effect of height and time lag on the estimation of sensible heat flux over a drip-irrigated vineyard using the surface renewal (SR) method across distinct phenological stages. *Agric. Water Manag.* **2014**, *141*, 74–83. [CrossRef]

37. Shao, C.; Chen, J.; Li, L.; Xu, W.; Chen, S.; Gwen, T.; Xu, J.; Zhang, W. Spatial variability in soil heat flux at three Inner Mongolia steppe ecosystems. *Agric. For. Meteorol.* **2008**, *148*, 1433–1443. [CrossRef]

38. Schotanus, P.; Nieuwstadt, F.T.M.; de Bruin, H.A.R. Temperature measurement with a sonic anemometer and its application to heat and moisture fluxes. *Bound.-Layer Meteorol.* **1983**, *26*, 81–93. [CrossRef]

39. Webb, E.K.; Pearman, G.I.; Leuning, R. Correction of flux measurements for density effects due to heat and water vapour transfer. *Q. J. R. Meteorol. Soc.* **1980**, *106*, 85–100. [CrossRef]

40. Twine, T.E.; Kustas, W.P.; Norman, J.M.; Cook, D.R. Correcting eddy-covariance flux underestimates over a grassland. *Agric. For. Meteorol.* **2000**, *103*, 279–300. [CrossRef]

41. Shuttleworth, W.J.; Wallace, J.S. Evaporation from sparse crops—An energy combination theory. *Q. J. R. Meteorol. Soc.* **1985**, *111*, 839–855. [CrossRef]

42. Allen, R.G.; Tasumi, M.; Trezza, R. Satellite-based energy balance for mapping evapotranspiration with internalized calibration (METRIC)—Model. *J. Irrig. Drain. Eng.* **2007**, *133*, 380–394. [CrossRef]

43. Storey, J.; Scaramuzza, P.; Schmidt, G.; Barsi, J. Landsat 7 scan line corrector-off gap-filled product development. In Proceedings of the Pecora 16 Global Priorities in Land Remote Sensing, Sioux Falls, SD, USA, 23–27 October 2005.

44. Zeng, C.; Shen, H.; Zhang, L. Recovering missing pixels for Landsat ETM plus SLC-off imagery using multi-temporal regression analysis and a regularization method. *Remote Sens. Environ.* **2013**, *131*, 182–194. [CrossRef]

45. Spiliotopoulos, M.; Holden, N.M.; Loukas, A. Mapping evapotranspiration coefficients in a temperate maritime climate using the metric model and landsat TM. *Water* **2017**, *9*, 23. [CrossRef]

46. Reyes-González, A.; Kjaersgaard, J.; Trooien, T.; Hay, C.; Ahiablame, L. Comparative Analysis of METRIC Model and Atmometer Methods for Estimating Actual Evapotranspiration. *Int. J. Agron.* **2017**, *2017*, 3632501. [CrossRef]

47. Jimenez-Bello, M.A.; Castel, J.R.; Testi, L.; Intrigliolo, D.S. Assessment of a Remote Sensing Energy Balance Methodology (SEBAL) Using Different Interpolation Methods to Determine Evapotranspiration in a Citrus Orchard. *IEEE J. Sel. Top. Appl. Earth Obs. Remote Sens.* **2015**, *8*, 1465–1477. [CrossRef]

48. Song, L.; Liu, S.; Kustas, W.P.; Zhou, J.; Xu, Z.; Xia, T.; Li, M. Application of remote sensing-based two-source energy balance model for mapping field surface fluxes with composite and component surface temperatures. *Agric. For. Meteorol.* **2016**, *230*, 8–19. [CrossRef]

49. Serra, P.; Salvati, L.; Queralt, E.; Pin, C.; Gonzalez, O.; Pons, X. Estimating Water Consumption and Irrigation Requirements in a Long-Established Mediterranean Rural Community by Remote Sensing and Field Data. *Irrig. Drain.* **2016**, *65*, 578–588. [CrossRef]

50. Mayer, D.G.; Butler, D.G. Statistical validation. *Ecol. Model.* **1993**, *68*, 21–32. [CrossRef]

51. Willmott, C.J.; Ackleson, S.G.; Davis, R.E.; Feddema, J.J.; Klink, K.M.; Legates, D.R.; O'Donnell, J.; Rowe, C.M. Statistics for the Evaluation and Comparison of Models. *J. Geophys. Res. Oceans* **1985**, *90*, 8995–9005. [CrossRef]

52. Williams, D.G.; Cable, W.; Hultine, K.; Hoedjes, J.; Yepez, E.A.; Simonneaux, V.; Er-Raki, S.; Boulet, G.; de Bruin, H.; Chehbouni, A.; et al. Evapotranspiration components determined by stable isotope, sap flow and eddy covariance techniques. *Agric. For. Meteorol.* **2004**, *125*, 241–258. [CrossRef]

53. Ezzahar, J.; Chehbouni, A.; Hoedjes, J.C.B.; Er-Raki, S.; Chehbouni, A.; Boulet, G.; Bonnefond, J.M.; de Bruin, H.A.R. The use of the scintillation technique for monitoring seasonal water consumption of olive orchards in a semi-arid region. *Agric. Water Manag.* **2007**, *89*, 173–184. [CrossRef]

54. Villalobos, F.J.; Testi, L.; Orgaz, F.; García-Tejera, O.; López-Bernal, Á.; Victoria Gonzalez-Dugo, M.; Ballester-Lurbe, C.; Ramon Castel, J.; Jose Alarcon-Cabanero, J.; Nicolas-Nicolas, E.; et al. Modelling canopy conductance and transpiration of fruit trees in Mediterranean areas: A simplified approach. *Agric. For. Meteorol.* **2013**, *171*, 93–103. [CrossRef]

55. Testi, L.; Villalobos, F.J.; Orgaz, F.; Fereres, E. Water requirements of olive orchards: I simulation of daily evapotranspiration for scenario analysis. *Irrig. Sci.* **2006**, *24*, 69–76. [CrossRef]

56. Lee, X.; Black, T.A. Atmospheric turbulence within and above a douglas-fir stand. Part II: Eddy fluxes of sensible heat and water vapour. *Bound.-Layer Meteorol.* **1993**, *64*, 369–389. [CrossRef]

57. Wilson, K.; Goldstein, A.; Falge, E.; Aubinet, M.; Baldocchi, D.; Berbigier, P.; Bernhofer, C.; Ceulemans, R.; Dolman, H.; Field, C.; et al. Energy balance closure at FLUXNET sites. *Agric. For. Meteorol.* **2002**, *113*, 223–243. [CrossRef]

58. Leuning, R.; van Gorsel, E.; Massman, W.J.; Isaac, P.R. Reflections on the surface energy imbalance problem. *Agric. For. Meteorol.* **2012**, *156*, 65–74. [CrossRef]

59. Allen, R.G.; Pereira, L.S.; Howell, T.A.; Jensen, M.E. Evapotranspiration information reporting: I. Factors governing measurement accuracy. *Agric. Water Manag.* **2011**, *98*, 899–920. [CrossRef]

60. Liu, H.; Foken, T. A modified Bowen ratio method to determine sensible and latent heat fluxes. *Meteorol. Z.* **2001**, *10*, 71–80. [CrossRef]

61. De Teixeira, A.H.C.; Bastiaanssen, W.G.M.; Ahmad, M.D.; Bos, M.G. Reviewing SEBAL input parameters for assessing evapotranspiration and water productivity for the Low-Middle São Francisco River basin, Brazil. *Agric. For. Meteorol.* **2009**, *149*, 462–476. [CrossRef]

62. Liu, S.; Lu, L.; Mao, D.; Jia, L. Evaluating parameterizations of aerodynamic resistance to heat transfer using field measurements. *Hydrol. Earth Syst. Sci.* **2007**, *11*, 769–783. [CrossRef]

water

MDPI

Article

Assessing the Impact of LAI Data Assimilation on Simulations of the Soil Water Balance and Maize Development Using MOHID-Land

Tiago B. Ramos [1],*, Lucian Simionesei [1], Ana R. Oliveira [1], Hanaa Darouich [2] and Ramiro Neves [1]

[1]   Centro de Ciência e Tecnologia do Ambiente e do Mar (MARETEC), Instituto Superior Técnico, Universidade de Lisboa, Av. Rovisco Pais, 1, 1049-001 Lisboa, Portugal; lucian.simionesei@tecnico.ulisboa.pt (L.S.); anaramosoliveira@tecnico.ulisboa.pt (A.R.O.); ramiro.neves@tecnico.ulisboa.pt (R.N.)

[2]   Centro de Investigação em Agronomia, Alimentos, Ambiente e Paisagem (LEAF), Instituto Superior de Agronomia, Universidade de Lisboa, Tapada da Ajuda, 1349-017 Lisboa, Portugal; hanaa.darouich@gmail.com

*   Correspondence: tiagobramos@tecnico.ulisboa.pt; Tel.: +35-121-841-9428

Received: 8 August 2018; Accepted: 27 September 2018; Published: 30 September 2018

Abstract: Hydrological modeling at the catchment scale requires the upscaling of many input parameters for better characterizing landscape heterogeneity, including soil, land use and climate variability. In this sense, remote sensing is often considered as a practical solution. This study aimed to access the impact of assimilation of leaf area index (LAI) data derived from Landsat 8 imagery on MOHID-Land's simulations of the soil water balance and maize state variables (LAI, canopy height, aboveground dry biomass and yield). Data assimilation impacts on final model results were first assessed by comparing distinct modeling approaches to measured data. Then, the uncertainty related to assimilated LAI values was quantified on final model results using a Monte Carlo method. While LAI assimilation improved MOHID-Land's estimates of the soil water balance and simulations of crop state variables during early stages, it was never sufficient to overcome the absence of a local calibrated crop dataset. Final model estimates further showed great uncertainty for LAI assimilated values during earlier crop stages, decreasing then with season reaching its end. Thus, while model simulations can be improved using LAI data assimilation, additional data sources should be considered for complementing crop parameterization.

Keywords: biomass; crop transpiration; direct forcing; leaf area index; soil evaporation

---

1. Introduction

In recent decades, modeling has become an essential part of the decision-making process for improving irrigation water use [1–4], optimizing fertilization practices [5,6], predicting crop yields [7,8] and coping with climate change [9,10] at the field and regional scales. However, modeling tools require first a considerable time investment in calibration to provide feasible results to their users. This is often accomplished at the plot scale, where most variables influencing crop development (soil properties, plant physiology, groundwater levels and weather conditions) can be more easily monitored. The problem often arises when upscaling to the field or regional scales due to the difficulties in portraying landscape heterogeneity, including soil, land use and climate variability.

Remote sensing technology offers today a potential solution for accurately and reliably describing the spatial distribution of soil properties and canopy state variables (leaf area index, canopy height, biomass) at the field and regional scales [11]. A vast number of new satellite sensors (Landsat 8, Sentinel-2, Spot-6, RapidEye, Huanjing-1) together with versatile, light-weighted and low-cost sensors mounted in farm tractors or unmanned aerial vehicles are now available for providing information with high spatial and temporal resolution to farmers and technicians. This information can potentially be also assimilated into field or regional scale models, overcoming many constraints in terms of input parameterization.

The purpose of assimilation is to optimize model input parameters by integrating, both in space and time, soil or canopy state variables derived from remote sensing methods [12]. Accurate and up-to-date information has been increasingly available at low cost, which has led to numerous research studies focusing on assimilation of remote sensing measurements [13–19]. Some of these have estimated leaf area index (LAI) using different remote sensing data sources, then assimilating those values by directly replacing the simulated LAI to improve model estimates of the aboveground dry biomass, yield and crop transpiration [14,17]. Other studies have used more advanced assimilation techniques, usually based on the Extended Kalman Filter [20] and Ensemble Kalman Filter [21] assimilation methods, for integrating remote sensing LAI into model simulations [15,16,19]. Overall, regardless of the technique used, most of those studies concluded that remote sensing provides useful measurements which can then be used for improving model simulations.

While a wide variety of models exist capable of simulating crop growth processes at the regional scale, thus portraying landscape heterogeneity at some extent [22–24], fully distributed process-based models such as MIKE SHE [25], SHETRAN [26] and MOHID-Land [27] are often considered ideal for further studying distributed state variables (the spatiotemporal variability of soil moisture) and flow paths (sediment and nutrient transport) [28]. These fully distributed process-based models consider interactions between multiple components of the soil-water-atmosphere continuum, with fundamental process being formulated at fine spatial (plot) and temporal scales, contributing to the overall dynamics at a higher organizational level, such as the watershed [28]. For the case of MOHID-Land, the model has been used for improving irrigation practices at the plot and field scales [3,29,30], understanding the contribution of flood events to the eutrophication of water reservoirs [31,32] and forecasting fresh water quantity and quality in coastal rivers [33]. Extensive calibration has been normally required for characterizing soil, groundwater, crop and river flow properties. Thus, data assimilation may have here a decisive contribution for more accurately describing the spatial and temporal variability of many of the required input parameters. However, the impact of data assimilation on final model outputs needs to be first assessed.

The main objective of this study was thus to understand the impact of LAI assimilation on MOHID-Land's estimates of the soil water balance and crop state variables (LAI, canopy height, aboveground dry biomass and yield). The hypothesis addressed were that (i) the MOHID-Land model could accurately estimate the soil water balance and aboveground biomass growth in a one-dimensional domain; (ii) LAI assimilation could improve simulations of crop development and (iii) the related uncertainties could be assessed. Results from this study will help to improve hydrological modeling at the field and regional scales by quantifying the uncertainty related to data assimilation using the MOHID-Land model.

2. Materials and Methods

2.1. Field Site Description and Data

Field data used in this study was collected at Herdade do Zambujeiro (22 ha), Benavente, southern Portugal (38°58'0.97'' N, 8°44'46.63'' W, 6 m a.s.l.) (Figure 1). The climate in the region is semi-arid to dry sub-humid, with hot dry summers and mild winters with irregular rainfall. The mean annual temperature is 16.8 °C, with the mean daily temperatures at the coolest (January) and warmest (August) months reaching 11.4 and 22.7 °C, respectively. The mean annual precipitation is 668 mm, mostly occurring between October and May. The soil was a Haplic Fluvisol [34], with the main soil physical and chemical properties presented in Table 1. The bottom layers exhibited higher dry bulk density and lower measured saturated hydraulic conductivity values than the topsoil layers [3], evidencing some soil compaction due to tillage operations carried out throughout the years and the relatively high soil moisture that was constant along the seasons because of the shallow groundwater levels.

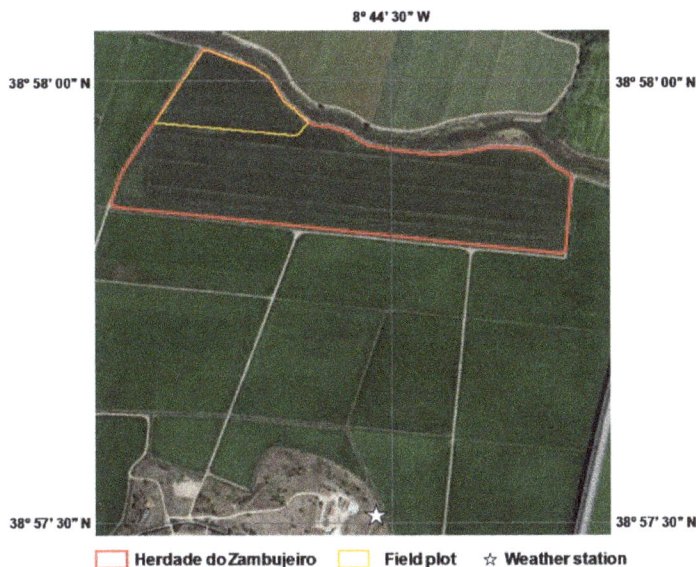

Figure 1. Location of the study site.

The MOHID-Land model was previously implemented in the area by Ramos et al. [3]. These authors evaluated the model's capacity in predicting soil water contents and fluxes and the evolution of different crop growth parameters, including the leaf area index (LAI), canopy height, aboveground dry biomass and yields during the 2014 and 2015 maize growing seasons. Details on the calibration/validation approach can be found in the cited reference. For that, the field was cropped with maize hybrid P1574 (FAO 600) with a density of approximately 89,000 plants ha^{-1}. Management practices, including fertilization and irrigation, were performed according to the standard practices in the region and were decided by the farmer. During 2014, maize was sown on May 24 and harvested on October 8; the net rainfall reached 163 mm, while the net irrigation amounted 365 mm (Figure 2). During 2015, maize was sown on April 16 and harvested on September 20; the net rainfall reached only 12 mm, while the net irrigation summed 620 mm (Figure 2). Irrigation was applied with the farmer's stationary sprinkler system. Groundwater depth (GWD) varied between approximately 1.5 m depth at the beginning of the growing season to 1.0 m depth during irrigation, further reaching 0.3 m depth during September 2014 after successive rain events (Figure 2). Crop stages were set as in Table 2 based on field observations.

One SM1 capacitance probe (Adcon Telemetry, Klosterneuburg, Austria) and one ECH2O-5 capacitance probe (Decagon Devices, Pullman, WA, USA) were installed at depths of 10, 30 and 50 cm to continuously measure soil water contents. One LEV1 level sensor (Adcon Telemetry, Klosterneuburg, Austria) was used to continuously monitor the groundwater level (Figure 2). One RG1 (Adcon Telemetry, Klosterneuburg, Austria) and two QMR101 (Vaisala, Helsinki, Finland) rain gauges were used to measure the amount of water applied per irrigation event.

LAI, canopy height and the aboveground dry biomass were further monitored by harvesting 3 random plants in four locations distributed randomly throughout the field plot, every 15 days, between May and September, during the 2014 and 2015 maize growing seasons (Table 3). The same crop parameters were measured at the end of each crop season, but by harvesting all plants in random areas of 1.5 m² (corresponding to approximately 12 plants). The length and width of crop leafs were measured in every harvested plant and then converted to LAI values as documented in Ramos et al. [4]. The aboveground dry biomass was determined by oven drying maize stems, leaves and grain at 70 °C to a constant weight. Maize yield was obtained from the grain's dry biomass measured at the end of each crop season.

Table 1. Main physical and chemical soil characteristics.

Depth (m)	Soil Layers			
	0–0.2	0.2–0.4	0.4–0.6	0.6–0.8
Coarse Sand, 2000–200 μm (%)	3.4	6.8	11.5	14.7
Fine Sand, 200–20 μm (%)	44.6	47.8	53.6	48.4
Silt, 20–2 μm (%)	33.3	28.1	20.6	23.2
Clay, <2 μm (%)	18.8	17.3	14.3	13.7
Texture	Silty Loam	Loam	Loam	Loam
Bulk Density (g cm^{-3})	1.57	1.52	1.66	1.66
Organic Matter (%)	1.73	0.96	0.57	0.59
θ_{FC} (cm^3 cm^{-3})	0.321	0.293	0.311	0.311
θ_{WP} (cm^3 cm^{-3})	0.209	0.235	0.223	0.223
Van Genuchten-Mualem Parameters:				
θ_r (cm^3 cm^{-3})	0.078	0.067	0.065	0.065
θ_s (cm^3 cm^{-3})	0.393	0.356	0.340	0.340
α (cm^{-1})	0.009	0.016	0.005	0.005
η (-)	1.75	1.31	1.80	1.80
ℓ (-)	−1.0	−1.0	−1.0	−1.0
K_s (cm d^{-1})	500.3	22.6	0.7	0.7

θ_{FC}, soil water content at field capacity; θ_{WP}, soil water content at the wilting point; θ_r, residual water content; θ_s, saturated water content; α and η, empirical shape parameters; ℓ, pore connectivity/tortuosity parameter; K_s, saturated hydraulic conductivity.

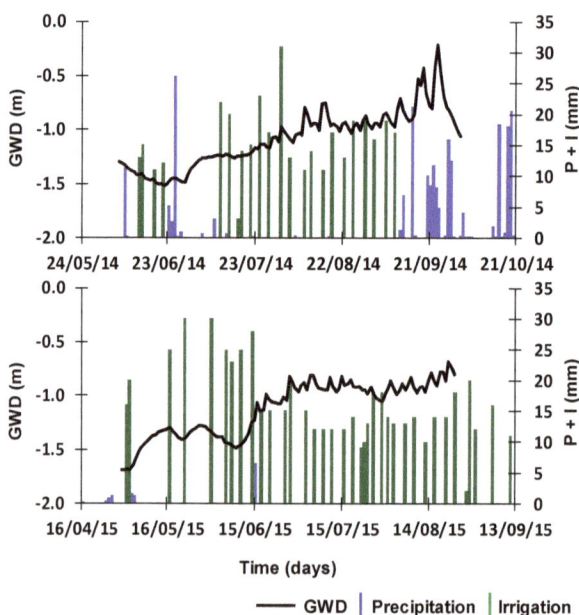

Figure 2. Daily values of precipitation, irrigation and groundwater depth (GWD) during the 2014 (**top**) and 2015 (**bottom**) crop seasons.

Table 2. Dates of crop growth stages.

Stage	2014		2015	
	Days	Date	Days	Date
Initial	31	24 May	37	16 April
Canopy Development	30	24 June	32	23 May
Mid-Season	52	24 July	53	24 June
Late Season	27	14 September	26	16 August
Harvest	-	8 October	-	20 September

Table 3. Measured values of Leaf Area Index (LAI), canopy height (h$_c$) and aboveground dry biomass during the 2014 and 2015 crop seasons.

Date	LAI (m^2 m^{-2})		h$_c$ (m)		Biomass (kg ha^{-1})	
	Mean	Standard Deviation	Mean	Standard Deviation	Mean	Standard Deviation
11/06/2014	0.1	0.0	0.14	0.02	48.3	15.3
25/06/2014	1.3	0.1	0.54	0.04	826.7	56.6
10/07/2014	3.2	0.4	1.02	0.10	2906.7	1083.2
24/07/2014	5.6	0.1	2.27	0.23	8279.4	339.2
11/08/2014	6.3	1.0	3.08	0.03	18,593.8	1881.7
27/08/2014	6.0	0.6	3.11	0.10	27,228.2	594.8
15/09/2014	5.3	0.4	3.03	0.15	34,945.6	982.8
08/10/2014	-	-	-	-	30,423.2	1735.6
15/05/2015	0.2	0.0	0.19	0.03	80.0	16.8
28/05/2015	1.3	0.3	0.49	0.01	1090.1	61.8
11/06/2015	4.0	0.3	1.45	0.06	5309.5	280.6
25/06/2015	5.9	0.6	2.65	0.09	11,550.0	958.9
14/07/2015	6.4	0.3	3.12	0.02	23,711.2	812.9
28/07/2015	6.2	0.6	3.12	0.01	27,373.9	229.2
13/08/2015	5.2	0.2	2.78	0.06	29,434.4	7624.1
14/09/2015	-	-	-	-	34,540.3	2670.5

2.2. Model Description

MOHID-Land is a distributed model capable of computing different physical and chemical processes in a three-dimensional domain using a finite-volume approach [27]. Variably-saturated water flow is described using the Richards equation, while the van Genuchten–Mualem functional relationships [35] are used for defining the unsaturated soil hydraulic properties, as follows:

$$S_e(h) = \frac{\theta(h) - \theta_r}{\theta_s - \theta_r} = \frac{1}{(1 + |\alpha\, h|^\eta)^m} \tag{1}$$

$$K(h) = K_s S_e^\ell \left[1 - \left(1 - S_e^{1/m}\right)^m\right]^2 \tag{2}$$

where S_e is the effective saturation (L^3 L^{-3}), θ_r and θ_s denote the residual and saturated water contents (L^3 L^{-3}), respectively, K_s is the saturated hydraulic conductivity (L T^{-1}), α (L^{-1}) and η (-) are empirical shape parameters, $m = 1 - 1/\eta$ and ℓ is a pore connectivity/tortuosity parameter (-).

Crop evapotranspiration (ET$_c$) is determined from reference evapotranspiration (ET$_0$) values computed with the FAO Penman-Monteith method using the single crop coefficient (K$_c$) approach [36]. ET$_c$ is then partitioned into potential soil evaporation (E$_p$) and potential crop transpiration (T$_p$) as a function of LAI [37]:

$$T_p = ET_c \left(1 - e^{(-\lambda\, \text{LAI})}\right) \tag{3}$$

$$E_p = ET_c - T_p \tag{4}$$

where λ is the extinction coefficient of radiation attenuation within the canopy (-).

Root water uptake is computed using the macroscopic approach introduced by Feddes et al. [38], meaning that T_p is distributed over the root zone and may be diminished by the presence of depth-varying root zone stressors, namely water stress. Root water uptake reductions (i.e., actual transpiration, T_a) are then described using the piecewise linear model proposed by Feddes et al. [38]. In this model, root water uptake is at the potential rate when the pressure head (h) is between h_2 and h_3, drops off linearly when $h > h_2$ or $h < h_3$ and becomes zero when $h < h_4$ or $h > h_1$ (subscripts 1 to 4 denote for different threshold pressure heads). E_p values are limited by a pressure head threshold value to obtain the actual soil evaporation rate (E_a) [39].

MOHID-Land further includes a modified version of the EPIC model [40,41] for simulating crop growth. This model is based on the heat unit theory, which considers that all heat above the base temperature will accelerate crop growth and development:

$$PHU = \sum_{i=1}^{m} HU = \sum_{i=1}^{m} (T_{av} - T_{base}) \qquad \text{when } T_{av} > T_{base} \tag{5}$$

where PHU is the total heat units required for plant maturity (°C), HU is the number of heat units accumulated on day i (°C), i = 1 corresponds to the sowing date (-), m is the number of days required for plant maturity (-), T_{av} is the mean daily temperature (°C) and T_{base} is the minimum temperature for plant growth (°C).

Crop growth is modelled by simulating light interception, conversion of intercepted light into biomass and LAI development. Total biomass is calculated from the solar radiation intercepted by the crop leaf area using the Beer's law [42]:

$$\sum_{i=1}^{m} \Delta Bio_{act,i} = \sum_{i=1}^{m} \Delta Bio_i \, \gamma_i = \sum_{i=1}^{m} RUE \left(0.5 \, PAR_i \, (1 - e^{-\lambda \, LAI})\right) \gamma_i \tag{6}$$

where $\Delta Bio_{act,i}$ and ΔBio_i are the actual and potential increase in total plant biomass on day i (kg ha^{-1}), RUE is the radiation-use efficiency of the plant ((kg ha^{-1}) (MJ m^{-2})$^{-1}$), PAR_i is the daily incident photosynthetically active radiation (MJ m^{-2}), λ is again the light extinction coefficient (-) and γ_i is the daily plant growth factor (0–1) which accounts for water and temperature stresses [41].

Leaf area index is computed as a function of heat units, crop stress and the development stage [41]. During early stages (initial and plant development stages), LAI increment on a given day is a function of the fraction of the plant's maximum LAI (LAI_{max}) that needs to be reached during those stages (fr_{LAImax}) and crop stress.

$$\Delta LAI_{act,i} = \Delta LAI_i \, \sqrt{\gamma_i} = (fr_{LAImax,i} - fr_{LAImax,i-1}) \, LAI_{max} \left(1 - e^{(5 \times (LAI_{i-1} - LAI_{max}))}\right) \sqrt{\gamma_i} \tag{7}$$

where $\Delta LAI_{act,i}$ and ΔLAI_i are the actual and potential LAI increment added on day i (m^2 m^{-2}), respectively and $fr_{LAImax,i}$ and $fr_{LAImax,i-1}$ are the fraction of the plant's maximum LAI (LAI_{max}, m^2 m^{-2}) on day i and i − 1 (-), respectively. During the mid-season stage, LAI is assumed to be constant. During the late-season stage, LAI declines as a function of LAI_{max}, heat units and crop stress.

Root depth is also computed as a function of heat units [41], while root biomass is assumed to decrease from 0.4 of the total biomass at emergence to 0.2 at maturity [43]. Finally, yield is obtained from the product of the aboveground dry biomass and the actual harvest index [41]. A more detailed description of the MOHID-Land model governing equations can be found in Trancoso et al. [27] and Ramos et al. [3].

2.3. Model Setup and Data Assimilation

The assimilation of LAI data into MOHID-Land simulations has a direct influence on the water balance through the partition of ET_c values into T_p (Equation (3)) and E_p (Equation (4)) and on the computation of the aboveground dry biomass (Equation (6)) Three distinct approaches were thus considered for better understanding the impact of LAI assimilation on model simulations:

A—The model was run as in Ramos et al. [3], which simulations of soil water contents, LAI, h_c, aboveground dry biomass and yields served as baseline for this study (Calibrated model). These authors followed a traditional calibration/validation approach, where a trial-and-error procedure was carried out to adjust soil hydraulic (Table 1) and crop parameters (Table 4) until deviations between the measured 2014 dataset and simulated values were minimized. The calibrated parameters were then validated using the 2015 dataset, with model simulations being compared to measured data.

B—The model was run using LAI values extracted from satellite data as inputs (LAI assimilation). LAI values were derived from the normalized difference vegetation index (NDVI) using the relationship shown in Figure 3. This relationship was found by comparing NDVI values computed from Landsat 8 satellite images (band 4 and 5) with LAI values measured in the study site, at multiple locations and over the 2014 and 2015 growing seasons. The calibrated soil hydraulic parameters were here also adopted (Table 1). However, the default crop parameters from the MOHID-Land's database (Table 4) were considered instead so that model performance in the absence of a calibrated dataset could be assessed.

C—The model was again run using LAI values extracted from satellite data as inputs. However, Ramos et al. [3] calibrated crop parameters (Table 4) were here considered (Calibration + LAI assimilation), as well as the calibrated soil hydraulic parameters (Table 1).

Table 4. Parameters of the crop growth model.

Crop Parameter	Default	Calibrated
Optimal Temperature for Plant Growth, T_{opt} (°C)	25.0	25.0
Minimum Temperature for Plant Growth, T_{base} (°C)	8.0	8.0
Plant Radiation-Use Efficiency, RUE [(kg ha^{-1}) (MJ m^{-2})$^{-1}$]	45.0	39.0
Total Heat Units Required for Plant Maturity, PHU (°C)	1700	1900
Fraction of PHU to Reach the End of Stage 1 (Initial Crop Stage), $fr_{PHU,init}$ (-)	0.15	0.20
Fraction of PHU to Reach the End of Stage 2 (Canopy Development Stage), $fr_{PHU,dev}$ (-)	0.50	0.43
Fraction of PHU after which LAI Starts to Decline, $fr_{PHU,sen}$ (-)	0.70	0.73
Maximum Leaf Area Index, LAI_{max} (m^2 m^{-2})	6.0	6.5
Fraction of LAI_{max} at the End of Stage 1 (Initial Crop Stage), $fr_{LAImax,ini}$ (-)	0.05	0.05
Fraction of LAI_{max} at the end of Stage 2 (Canopy Development Stage), $fr_{LAImax,dev}$ (-)	0.95	0.95
Maximum Canopy Height, $h_{c,max}$ (m)	2.5	3.1
Maximum Root Depth, $Z_{root,max}$ (m)	2.0	0.6
Potential Harvest Index for the Crop at Maturity, HI_{opt} (-)	0.50	0.49
Minimum Harvest Index Allowed, HI_{min} (-)	0.30	0.30

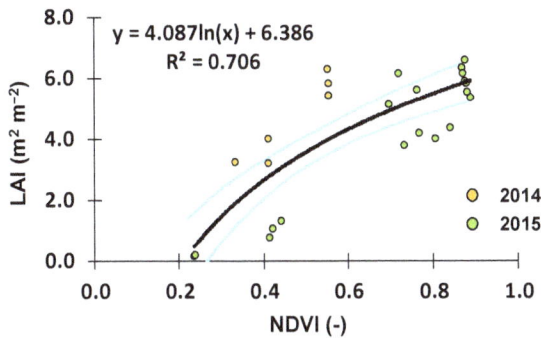

Figure 3. Relationship between measured Leaf Area Index (LAI) and the Normalized Difference Vegetation Index (NDVI). The blue lines correspond to the 95% confidence interval.

Landsat 8 images were first corrected to convert the TOA (Top of Atmosphere) planetary reflectance using reflectance rescaling coefficients provided in the Landsat 8 OLI metadata file and to correct the reflectance value with the sun angle. Two images were available from sowing to harvest during the 2014 growing season, while eight images were used during the 2015 growing season (Table 5). These images were used to extract the NDVI values corresponding to the multiple locations where LAI field observations were carried out, in a total of 26 measurements (Figure 3). LAI and NDVI values ranged from 0.41–5.85 and 0.23–0.88, respectively, in line with Pôças et al. [44].

Table 5. Leaf Area Index (LAI) assimilation data.

Date	Assimilated LAI (m² m⁻²)	NDVI (-)	LAI 95% Confidence Interval	
			Lower	Upper
09/07/2014	2.75	0.41	-	-
10/08/2014	3.97	0.55	-	-
23/04/2015	0.41	0.23	0.00	1.53
09/05/2015	0.48	0.24	0.00	1.58
25/05/2015	3.04	0.44	2.47	3.61
10/06/2015	5.50	0.81	4.89	6.11
26/06/2015	5.85	0.88	5.17	6.53
12/07/2015	5.79	0.86	5.12	6.46
28/07/2015	5.82	0.87	5.15	6.49
29/08/2015	5.31	0.77	4.73	5.89

Data assimilation in MOHID-Land was carried out using the forcing method [12]. The approach is relatively straightforward, with the model simply replacing the predicted value by a new input when an image becomes available, updating then fr_{LAImax} to account for what still needs to be reached during a specific crop stage and more interestingly, the water balance and the aboveground dry biomass estimates. From that date on and until another image becomes available model simulations follow the parameterization given in Table 4. Table 5 lists the assimilated LAI values and dates.

In all simulations (Approach A–C), the soil profile was specified with 2 m depth, divided into four soil layers according to observations (Table 1). The soil domain was represented using an Arakawa C-grid type [45], defined by one vertical column (one-dimensional domain) discretized into 100 grid cells with 1 m wide, 1 m long and 0.02 m thickness each (i.e., $1 \times 1 \times 0.02$ m³). The simulation periods covered from sowing to harvest. The upper boundary condition was determined by the actual evaporation and transpiration rates and the irrigation and precipitation fluxes (Figure 2). Weather data used in this study was taken from a meteorological station located 950 m from the study site (38°57'30.25'' N, 8°44'31.70'' W, 7 m a.s.l.; Figure 1) and included the average temperature (°C), wind speed (m s⁻¹), relative humidity (%), global solar radiation (W m⁻²) and precipitation (mm). ET_c values were computed from hourly ET_0 values and K_c values of 0.30, 1.20 and0.35 for the initial, mid-season and late season crop stages, respectively [31]. The K_c value for the initial crop stage was then adjusted for the frequency of the wetting events (precipitation and irrigation) and average infiltration depths, while the K_c values for mid-season and late season crop stages were adjusted for local climate conditions taking into consideration canopy height, wind speed and minimum relative humidity averages for the periods under consideration [36]. The following parameters of the Feddes et al. [38] model were used to compute T_p reductions due to water stress: $h_1 = -15$, $h_2 = -30$, $h_3 = -325$ to -600, $h_4 = -8000$ cm [46]. The bottom boundary condition was specified using the observed GWD (Figure 2). The initial soil water content conditions were set to field capacity.

2.4. Statistical and Uncertainty Analysis

Model calibration and validation was performed by comparing field measured values of soil water contents, LAI, h_c and aboveground dry biomass with the MOHID-Land simulations (Approaches A–C) using various quantitative measures of the uncertainty, such as, the coefficient of determination (R^2), the root mean square error (RMSE), the ratio of the RMSE to the standard deviation of observed data (NRMSE), the percent bias (PBIAS) and the model efficiency (EF). R^2 values close to 1 indicate that the model explains well the variance of observations. RMSE, NRMSE and PBIAS values close to zero indicate small errors of estimate and good model predictions [47–49]. Positive or negative PBIAS values refer to the occurrence of under- or over-estimation bias, respectively. Nash and Sutcliff [50] modelling efficiency EF values close to 1 indicate that the residuals variance is much smaller than the observed data variance, hence the model predictions are good; contrarily, when EF is very close to 0 or negative there is no gain in using the model.

Data assimilation is much dependent on the empirical relationship (Figure 3) established to derive LAI values from the NDVI measurements [14,18]. As such, the uncertainty related to that conversion was quantified on final model estimates of T_a, E_a and aboveground dry biomass using a Monte Carlo method. This evaluation was performed on modeling Approach C (Calibration + LAI assimilation) as the objective here was to assess if remote sensing data assimilation could further correct for simulation errors that result from model parameter uncertainty. The 2015 dataset was also considered as more satellite images were available during this season. A randomly population of 10,000 LAI values was first created for each available image date following a normal distribution with mean equal to the estimated parameter given by the LAI-NDVI regression equation and range defined by the 95% confidence intervals (Figure 3, Table 5). The model was then run following Approach C settings until reaching the dates of each of the eight available images (8 × 1 simulation). Afterwards, the 10,000 LAI randomly generated values were assimilated by the model, which then proceeded with simulations until the end season following Approach C settings again (8 × 10,000 simulations). In the end, the uncertainty of final model estimates of T_a, E_a and aboveground dry biomass were assessed for each assimilation date (8 dates) from 10,000 simulations (80,000 simulations in total). The Monte Carlo simulations were performed with a Python script.

3. Results and Discussion

3.1. LAI Evolution

Figure 4 shows the evolution of LAI estimated values using the calibrated model in Ramos et al. [3] (Approach A), direct LAI assimilation (Approach B) and the combination of the calibrated model and LAI assimilation (Approach C). Table 6 presents the statistical indicators used to evaluate the agreement between model simulations and measured values. Ramos et al. [3] showed that the MOHID-Land model could reasonably well simulate LAI evolution during the 2014 and 2015 growing seasons. In their study, the values of R^2 were very high (0.97), showing that the model could explain well the variability of the observed data. The errors of the estimates were quite small, resulting in RMSE values lower than 0.63 m^2 m^{-2} and NRMSE values lower than 0.16. The PBIAS values were lower than 6.40%, indicating some underestimation of the measured data. The modelling efficiency EF were also high ($\geq$0.93), meaning that the residual variance was much smaller than the measured data variance.

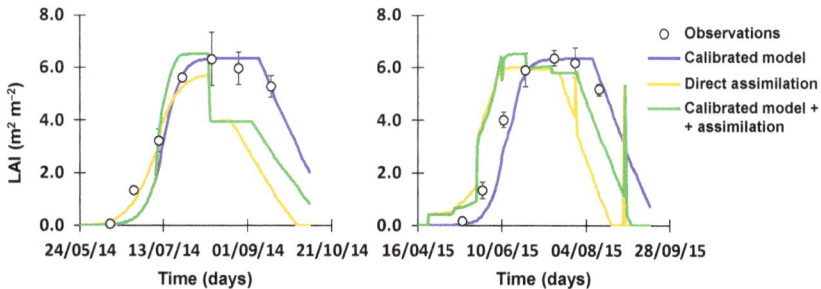

Figure 4. Measured and simulated leaf area index during 2014 (**left**) and 2015 (**right**) crop seasons. Vertical bars correspond to the standard deviation of measured data.

Table 6. Results of the statistical analysis between measured and simulated soil water contents, leaf area index (LAI), canopy height and aboveground dry biomass.

Statistic	R^2	RMSE	NRMSE	PBIAS	EF
2014					
Water Content ($cm^3 cm^{-3}$)					
Calibrated Model	0.73	0.018	0.061	−1.53	0.70
Direct Assimilation	0.85	0.012	0.041	−1.16	0.87
Calibrated Model + Assimilation	0.89	0.012	0.039	−0.95	0.88
LAI ($m^2 m^{-2}$)					
Calibrated Model	0.97	0.63	0.16	6.40	0.94
Direct Assimilation	0.60	2.13	0.54	33.82	0.26
Calibrated model + Assimilation	0.70	1.73	0.43	24.39	0.51
Canopy Height (m)					
Calibrated Model	0.93	0.42	0.22	−11.83	0.90
Direct Assimilation	0.86	0.58	0.31	0.96	0.80
Calibrated Model + Assimilation	0.90	0.50	0.26	−13.87	0.85
Dry Biomass ($kg ha^{-1}$)					
Calibrated Model	0.94	5128.3	0.39	19.20	0.87
Direct Assimilation	0.94	2183.7	0.24	4.60	0.95
Calibrated Model + Assimilation	0.96	2518.9	0.32	14.20	0.91
2015					
Water Content ($cm^3 cm^{-3}$)					
Calibrated Model	0.37	0.019	0.063	1.91	0.11
Direct Assimilation	0.40	0.017	0.057	0.90	0.28
Calibrated Model + Assimilation	0.39	0.018	0.060	1.58	0.18
LAI ($m^2 m^{-2}$)					
Calibrated Model	0.97	0.61	0.15	6.31	0.94
Direct Assimilation	0.35	2.16	0.52	−1.18	0.24
Calibrated Model + Assimilation	0.63	1.58	0.38	−10.34	0.59
Canopy Height (m)					
Calibrated model	0.96	0.33	0.17	−11.00	0.93
Direct Assimilation	0.85	0.65	0.33	−2.57	0.73
Calibrated Model + Assimilation	0.88	0.60	0.30	−19.13	0.77
Dry Biomass ($kg ha^{-1}$)					
Calibrated Model	0.93	4616.8	0.33	15.27	0.89
Direct Assimilation	0.98	6211.4	0.44	−31.49	0.80
Calibrated Model + Assimilation	0.96	6237.8	0.44	−28.52	0.79

R^2, coeficient of determination; RMSE, root mean square error; NRMSE, normalized RMSE; PBIAS, percent bias; EF, modeling efficiency.

The direct assimilation of LAI values into model simulations (Approach B) produced worse statistical indicators than when using the calibrated model (Approach A), with the R^2 values decreasing down to 0.35 and the RMSE and NRMSE values increasing up to 2.16 $m^2 m^{-2}$ and 0.53, respectively. The PBIAS showed contrasting results, while the EF values also decreased down to 0.24, indicating nonetheless that the model was still able to describe field measurements with relative success. The direct assimilation approach made that MOHID-Land's LAI simulated results were directly replaced by the remote sensing LAI values in the dates when satellite images were available. From that date on and until another image was available model simulations followed the default parameterization of the MOHID-Land crop database given in Table 4. As a result, assimilation of remote sensing LAI values using the forcing method available in the MOHID-Land model resulted in several unrealistic discontinuities in simulated LAI (Figure 4), a common feature when using this assimilation approach [14]. Also, LAI increased at a much faster pace during the initial and development crop stages, with maize also reaching senescence earlier. The difference of 200 °C in the total heat units required for plant maturity (PHU) considered between the default and calibrated crop parameters (Table 4) showed here to be critical for model performance. LAI assimilation was able to correct model simulations during the earlier crop stages but failed to counteract the end of the crop cycle as observed in the 2015 simulations (Figure 4). Here, despite assimilating higher LAI values, the model was obviously never able to extend the crop lifecycle longer than the allowed by the default PHU parameter.

The previous results show the importance of considering Approach C, where data assimilation forced simulations of the local calibrated model. Contrarily to the expected, LAI assimilation did not further improve Approach A results. The R^2 values still decreased down to 0.63, while the RMSE and NRMSE values increased up to 1.73 m^2 m^{-2} and 0.43, respectively. The EF values also decreased down to 0.51. However, these statistics were better than those obtained using only direct LAI assimilation (Approach B), showing the importance of local model calibration. Model simulations fully covered maize's lifecycle this time since no constraints in the PHU existed. However, results were still dependent on the quality of the assimilated data, with LAI evolution at the end of the 2015 season suggesting that some filtering would be needed during the assimilation process (Figure 4).

Despite the lower statistical indicators found when compared to those using only the calibrated model (Approach A), LAI evolution was also considered to be well represented when LAI data assimilation was included in the MOHID-Land model simulations (Approaches B and C), particularly during the earlier crop stages. Results further suggested that a higher time resolution of assimilated data would improve the agreement between model simulations and measured data. Nonetheless, more important than accurately predicting LAI evolution was to understand how data assimilation impacted the soil water balance, aboveground dry biomass and yield estimates during the 2014 and 2015 growing seasons, as shown below.

3.2. Soil Water Balance

Figure 5 presents the measured soil water contents at depths of 10, 30 and 50 cm during the 2014 and 2015 growing seasons and compares these values with model simulations following the approaches referred above. Contrarily to LAI results, forcing remote sensing LAI data into model simulations reduced deviations between measured and simulated soil water content values. During the 2014 growing season, the RMSE values decreased from 0.018 to 0.012 cm^3 cm^{-3}, while the NRMSE values reduced from 0.061 to 0.039 when considering LAI assimilation (Approaches B and C). Inversely, the EF values increased from 0.70 to 0.88 (Table 6). During the 2015 growing season, the positive impact of LAI data assimilation on soil water content simulation was more modest with only the EF values showing a relative improvement from 0.11 to 0.28. No noticeable differences were found between Approach B and Approach C statistical indicators. All simulations shared the same soil hydraulic parameters (Table 1) to better assess the actual impact of LAI assimilation on soil water content simulations, explaining thus the similarity of model results.

As LAI evolution was used in the partition of ET_c values into T_p (Equation (3)) and E_p (Equation (4)) [37], these two soil water balance components showed the greatest variation when considering LAI data assimilation (Table 7). The calibrated model (Approach A) produced estimates of T_a, E_a, capillary rise (CR) and deep percolation (DP), in line with other studies carried out in the region [51–53], some of which highlighting the importance of CR to the soil water balance in the Sorraia Valley region. Direct LAI assimilation produced always the lowest T_p and T_a values ($T_a/T_p = 1$), and, naturally, the highest E_p and E_a values during both seasons. The LAI data forcing on the calibrated model (Approach C) produced contrasting results when compared to Approach A, with T_p values decreasing in 2014 when LAI evolution was underestimated (PBIAS = 24.39%) and increasing in 2015 when the opposite occurred (PBIAS = −10.34%). Accurate LAI predictions were thus essential for simulating crop transpiration and soil evaporation, even though other important soil water balance components such as deep percolation and capillary rise were not significantly affected by less accurate LAI predictions. As a result, LAI assimilation in MOHID-Land may thus have a direct influence on biomass development, while estimates of groundwater recharge or solute leaching from the root zone may be impacted less significantly.

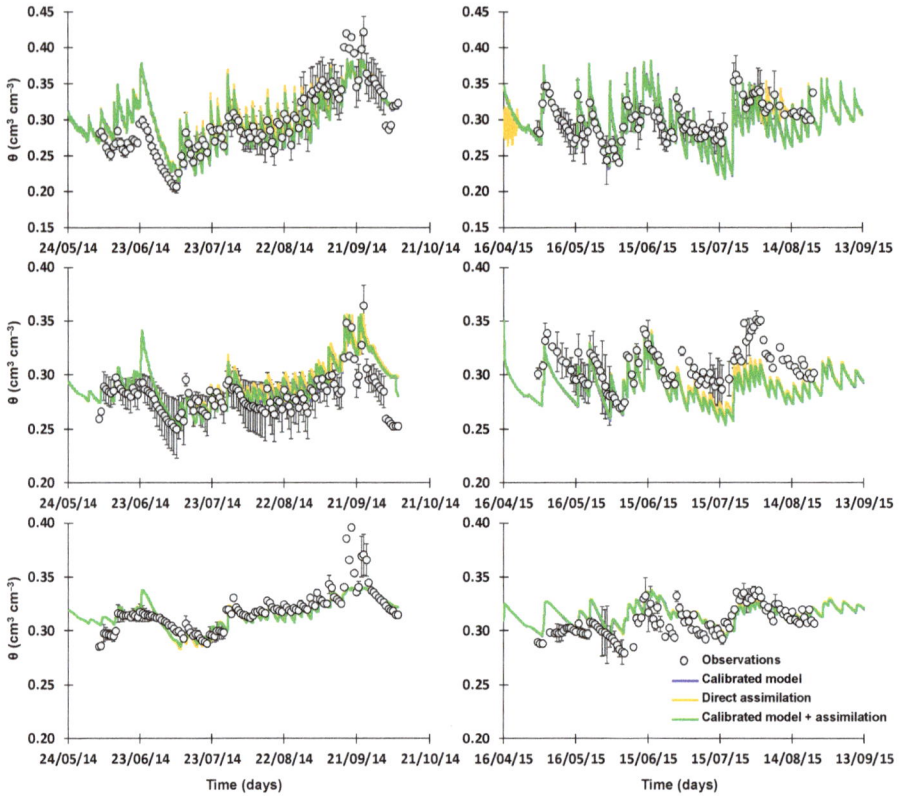

Figure 5. Measured and simulated soil water contents (θ) at depths of 10 (**top**), 30 (**middle**) and 50 cm (**bottom**) during 2014 (**left**) and 2015 (**right**) crop seasons. Vertical bars correspond to the standard deviation of measured data.

Table 7. Components of the soil water balance.

Approach	Inputs				Outputs			
	P (mm)	I (mm)	CR (mm)	ΔSS (mm)	E_a (mm)	T_a (mm)	T_a/T_p (-)	DP (mm)
2014								
Calibrated model	163	365	78	16	164	374	0.99	74
Direct assimilation	163	365	70	2	191	345	1.00	90
Calibrated model + Assimilation	163	365	78	2	183	355	1.00	84
2015								
Calibrated model	12	620	94	11	181	481	1.00	75
Direct assimilation	12	620	84	3	199	461	1.00	82
Calibrated model + Assimilation	12	620	95	3	150	512	1.00	75

P, precipitation; I, irrigation; CR, capillary rise; E_a, actual soil evaporation; T_a, actual crop transpiration; T_p, potential crop transpiration; DP, deep percolation.

3.3. Crop Height

The direct assimilation of LAI (Approach B) showed the maize canopy growing faster than that measured in the field or simulated by Ramos et al. [3] (Approach A), similarly to LAI predictions (Figure 6). Canopy height then assumed a default maximum value ($h_{c,max}$) of 2.5 m when the mid-season crop stage was reached (default value in Table 4), underestimating field values from that date onward and producing worse statistical indicators than those computed using the calibrated model (Table 6). The main problem here was thus the lack of a local calibrated dataset with the impact of LAI assimilation on canopy height simulations being only marginal as shown by the good indicators again obtained in Approach C.

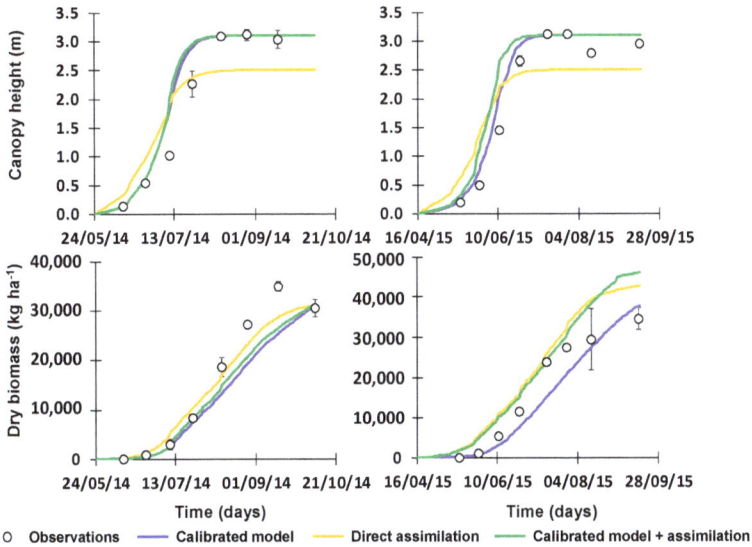

Figure 6. Measured and simulated canopy height (**top**) and aboveground dry biomass (**bottom**) during 2014 (**left**) and 2015 (**right**) crop seasons. Vertical bars correspond to the standard deviation of measured data.

3.4. Dry Biomass and Yields

Simulations of the aboveground dry biomass were concordant with the estimated LAI values during the 2014 and 2015 growing seasons. In 2014, the underestimation of LAI values led also to lower aboveground dry biomass estimates following the direct LAI assimilation approach (Approach B), with these being further closer to field measurements than the earlier results from the calibrated model (Approach A). As a result, the RMSE values decreased from 5128.3 to 2183.7 kg ha^{-1}, the NRMSE values reduced from 0.39 to 0.24 and the EF values increased from 0.87 to 0.95 when direct LAI assimilation was considered. Contrarily, the overestimation of LAI values produced larger errors in 2015 when compared to field measurements, with the RMSE values increasing from 4616.8 to 6211.4 kg ha^{-1}, the NRMSE values increasing from 0.33 to 0.44 and the EF values decreasing from 0.89 to 0.79. The local calibrated crop dataset (Approach C) did not improve aboveground dry biomass estimates.

Measured crop yields reached 16,093 and 17,300 kg ha^{-1} during the 2014 and 2015 seasons, respectively (Table 8). Yield predictions were in line with the same under- and overestimation tendencies observed in simulations of the aboveground dry biomass. Yield estimated from the three modeling approaches were relatively close (14,670–15,518 kg ha^{-1}) during the 2014 season, with Approach C producing the best estimates. On the other hand, during the 2015 season, yield estimates from that same approach produced the worse results (23,016 kg ha^{-1}), with all models diverging substantially from the measured value.

Table 8. Yield (kg ha^{-1}) estimates.

Season	Measured	Approach A	Approach B	Approach C
2014	16,093	14,670	15,196	15,518
2015	17,300	17,930	20,916	23,016

While many studies throughout the literature reported that precise knowledge of light interception and hence LAI, was critical for predicting biomass and yield accurately [12,17,54], results presented here were more in line with Linker and Ioslovich [19], who found that assimilation of easy-to-obtain canopy cover measurements did not always improve the predictions of biomass. They explained that by model choice, which in their case was a purely water-driven model in which solar radiation and light interception were not considered explicitly, likely resulting in underestimating the overall impact of canopy cover on crop development. A similar reasoning can be considered here. Ines et al. [16] also found that LAI assimilation could not always improve simulated aboveground dry biomass and yield predictions, particularly during dry conditions as the root zone soil moisture could not meet the increased water demand that resulted from improved canopy growth. Likewise, Nearing et al. [15] referred to the failure of LAI and soil moisture data assimilation in improving yield estimates, especially in water-limited environments, pointing out similar reasons as Ines et al. [16]. From a different perspective, Trombetta et al. [55] made use of remote sensed LAI data derived from the MODIS satellite images for calibrating/validating a hydrological model at the plot scale. Remote sensing LAI data, after being converted into canopy cover, was used as an alternative to field measurements during the calibration/validation process, with results suggesting this approach as a viable alternative for characterizing landscape heterogeneity (crop variability) at larger scales.

3.5. Uncertainty of Model Estimates

The previous sections showed that the impacts of LAI assimilation on MOHID-Land final estimates of T_a, E_a and aboveground biomass were substantial and that some filtering would be eventually needed for improving the quality of assimilated data. The uncertainty analysis carried out using a Monte Carlo method confirmed these early findings, with Figure 7 showing a relatively large uncertainty of final model estimates when LAI assimilation was performed during the first two dates (23/04/2015, 09/05/2015) of 2015 crop growing season. In these dates, T_a values ranged from 460 to 527 mm, while E_a values varied between 102 and 184 mm. Likewise, the aboveground dry biomass showed also considerable variation, ranging from 37,978 to 47,347 kg ha^{-1}. From those dates onwards, the uncertainty of final model estimates decreased, being relatively small when LAI assimilation was carried out at the end of the crop cycle. Filtering of the assimilation data would thus be important during the earlier crop stages, becoming irrelevant as the crop reaches the end of its life cycle.

The large uncertainty observed on model final estimates was already expected when LAI assimilation was performed during the earlier dates since the crop was still at the initial and development stages and thus is growth cycle was not yet well defined. Yet, the Monte Carlos simulation results showed that the assimilation of too low LAI values could lead the model to greatly underestimate crop transpiration and aboveground dry biomass, while soil evapotranspiration would be greatly overestimated. Hence, this can be quite problematic in the absence of additional information to update model simulations throughout the crop season if new satellite images are no longer available (e.g., due to cloud cover). In this case, the model will never be able to further update simulations of T_a, E_a and aboveground dry biomass, producing quite substantial errors.

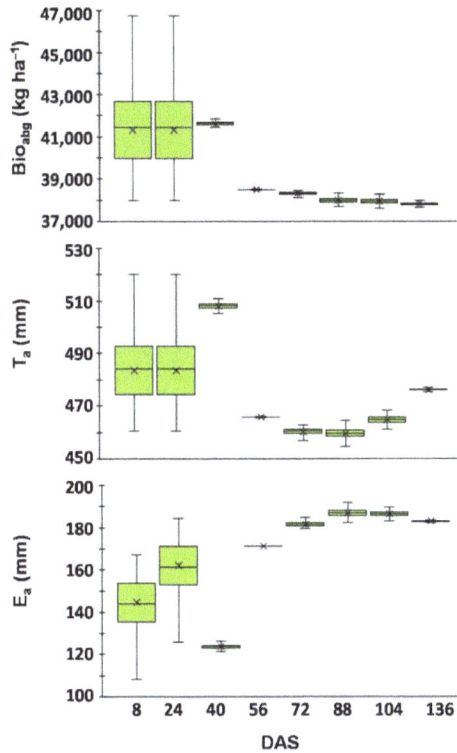

Figure 7. Uncertainty of data assimilation on aboveground dry biomass (**top**), actual transpiration T_a (**middle**) and actual soil evaporation E_a (**bottom**) final model estimates for the 2015 crop season on different dates (DAS, days after sowing). Box-plots indicate maximum and minimum values, median (–), first and third quartiles and average ($\times$) of 10,000 Monte Carlo simulations.

4. Conclusions

Remote sensing technology can provide valuable information for hydrological modeling at the field and regional scales by better characterizing the spatial and temporal variability of soils, land uses and climate, which otherwise are difficult to portray. This study showed that LAI assimilation from NDVI derived satellite data improved MOHID-Land estimates of the soil water balance and simulations of crop height and aboveground dry biomass during the early stages of the crop growing period. However, data assimilation was never sufficient to improve model simulations in the absence of a crop calibrated dataset, failing to simulate the entire growing season when calibrated potential head units (PHU) were missing or even crop maximum height when crop parameterization was misadjusted.

LAI data assimilation led also to great uncertainty on final estimates of crop transpiration, soil evaporation and the aboveground dry biomass when solely performed during the initial stages of the crop growing period. Although model uncertainty then decreased as LAI assimilation was being carried out closer to the end of the crop cycle, results showed that this approach may lead to complete erroneous estimates of the soil water balance and crop yields even when local calibrated soil and crop datasets are used. Therefore, while LAI remote sensing data can help defining MOHID-Land's input parameters, additional data sources should be accessed for complementing such characterization. The implementation of the MOHID-Land model at the regional scale cannot depend solely on inputs from the LAI data assimilation as estimates may diverge substantially from reality.

Author Contributions: Tiago Ramos conceived the experiment and wrote the paper; Lucian Simionesei set up the model and run simulations; Ana Oliveira performed the Monte Carlo analysis; Hanaa Darouich and Ramiro Neves made revisions and improvements to the draft version.

Funding: This study was funded by the Water JPI Project WATER4EVER (Optimizing water use efficiency in agriculture to preserve soil and water resources, Project WaterJPI/0010/2016, http://water4ever.eu). MARETEC acknowledges the national funds from the Foundation for Science and Technology (FCT) (Project UID/EEA/50009/2013). T. B. Ramos was supported by the FCT grant SFRH/BPD/110655/2015.

Conflicts of Interest: The authors declare no conflict of interest.

References

1. Paço, T.A.; Pôças, I.; Cunha, M.; Silvestre, J.C.; Santos, F.L.; Paredes, P.; Pereira, L.S. Evapotranspiration and crop coefficients for a super intensive olive orchard. An application of SIMDualKc and METRIC models using ground and satellite observations. *J. Hydrol.* **2014**, *519*, 2067–2080. [CrossRef]
2. González, M.G.; Ramos, T.B.; Carlesso, R.; Paredes, P.; Petry, M.T.; Martins, J.D.; Aires, N.P.; Pereira, L.S. Modelling soil water dynamics of full and deficit drip irrigated maize cultivated under a rain shelter. *Biosyst. Eng.* **2015**, *132*, 1–18. [CrossRef]
3. Ramos, T.B.; Simionesei, L.; Jauch, E.; Almeida, C.; Neves, R. Modelling soil water and maize growth dynamics influenced by shallow groundwater conditions in the Sorraia Valley region, Portugal. *Agric. Water Manag.* **2017**, *185*, 27–42. [CrossRef]
4. Paredes, P.; Rodrigues, G.J.; Petry, M.T.; Severo, P.O.; Carlesso, R.; Pereira, L.S. Evapotranspiration partition and crop coefficients of Tifton 85 Bermudagrass as affected by the frequency of cuttings. Application of the FAO56 dual K_c model. *Water* **2018**, *10*, 558. [CrossRef]
5. Cameira, M.R.; Fernando, R.M.; Ahuja, L.; Pereira, L.S. Simulating the fate of water in field soil–crop environment. *J. Hydrol.* **2005**, *315*, 1–24. [CrossRef]
6. Ramos, T.B.; Šimůnek, J.; Gonçalves, M.C.; Martins, J.C.; Prazeres, A.; Pereira, L.S. Two-dimensional modeling of water and nitrogen fate from sweet sorghum irrigated with fresh and blended saline waters. *Agric. Water Manag.* **2012**, *111*, 87–104. [CrossRef]
7. Jiang, Y.; Xu, X.; Huang, Q.; Huo, Z.; Huang, G. Assessment of irrigation performance and water productivity in irrigated areas of the middle Heihe River using a distributed agro-hydrological model. *Agric. Water Manag.* **2015**, *147*, 67–81. [CrossRef]
8. Vaghefi, S.A.; Abbaspour, K.C.; Faramarzi, M.; Srinivasan, R.; Arnold, J.G. Modeling crop water productivity using a coupled SWAT-MODSIM model. *Water* **2017**, *9*, 157. [CrossRef]
9. Valverde, P.; Serralheiro, R.; Carvalho, M.; Maia, R.; Oliveira, B.; Ramos, V. Climate change impacts on irrigated agriculture in the Guadiana river basin (Portugal). *Agric. Water Manag.* **2015**, *152*, 17–30. [CrossRef]
10. Fraga, H.; Atauri, I.G.C.; Santos, J.A. Viticultural irrigation demands under climate change scenarios in Portugal. *Agric. Water Manag.* **2018**, *196*, 66–74. [CrossRef]
11. Mulla, D.J. Twenty five years of remote sensing in precision agriculture: Key advances and remaining knowledge gaps. *Biosyst. Eng.* **2013**, *114*, 358–371. [CrossRef]
12. Jin, X.; Kumar, L.; Li, Z.; Feng, H.; Xu, X.; Yang, G.; Wang, J. A review of data assimilation of remote sensing and crop models. *Eur. J. Agron.* **2018**, *92*, 141–152. [CrossRef]
13. Vazifedoust, M.; van Dam, J.C.; Bastiaanssen, W.F.M.; Feddes, R.A. Assimilation of satellite data into agrohydrological models to improve crop yield forecasts. *Int. J. Remote Sens.* **2009**, *30*, 2523–2545. [CrossRef]
14. Thorp, K.R.; Hunsaker, D.J.; French, A.N. Assimilating leaf area index estimates from remote sensing into the simulations of a cropping systems model. *Trans. Am. Soc. Agric. Biol. Eng.* **2010**, *53*, 251–262.
15. Nearing, G.S.; Crow, W.T.; Moran, M.S.; Reichle, R.H.; Gupta, H.V. Assimilating remote sensing observations of leaf area index and soil moisture for wheat yield estimates: An observing system simulation experiment. *Water Resour. Res.* **2012**, *48*. [CrossRef]
16. Ines, A.V.; Das, N.N.; Hansen, J.W.; Njoku, E.G. Assimilation of remotely sensed soil moisture and vegetation with a crop simulation model for maize yield prediction. *Remote Sens. Environ.* **2013**, *138*, 149–164. [CrossRef]
17. Tripathy, R.; Chaudhari, K.N.; Mukherjee, J.; Ray, S.S.; Patel, N.; Panigrahy, S.; Parihar, J.S. Forecasting wheat yield in Punjab state of India by combining crop simulation model WOFOST and remotely sensed inputs. *Remote Sens. Lett.* **2013**, *4*, 19–28. [CrossRef]

18. Li, Y.; Zhou, G.Q.; Zhou, J.; Zhang, G.F.; Chen, C.; Wang, J. Assimilating remote sensing information into a coupled hydrology-crop growth model to estimate regional maize yield in arid regions. *Ecol. Model.* **2014**, *291*, 15–27. [CrossRef]
19. Linker, R.; Ioslovich, I. Assimilation of canopy cover and biomass measurements in the crop model AquaCrop. *Biosyst. Eng.* **2017**, *162*, 57–66. [CrossRef]
20. Evensen, G. Sequential data assimilation with a nonlinear quasi-geostrophic model using Monte Carlo methods to forecast error statistics. *J. Geophys. Res. Oceans* **1994**, *99*, 10143–10162. [CrossRef]
21. Evensen, G. The Ensemble Kalman Filter: Theoretical formulation and practical implementation. *Ocean Dyn.* **2003**, *53*, 343–367. [CrossRef]
22. Basso, B.; Liu, L.; Ritchie, J.T. A comprehensive review of the CERES-wheat, -maize and -rice models' performances. *Adv. Agron.* **2016**, *136*, 27–132.
23. Guzmán, S.M.; Paz, J.O.; Tagert, M.L.M.; Mercer, A.E.; Pote, J.W. An integrated SVR and crop model to estimate the impacts of irrigation on daily groundwater levels. *Agric. Syst.* **2018**, *159*, 248–259. [CrossRef]
24. Zhang, B.; Feng, G.; Ahuja, L.R.; Kong, X.; Ouyang, Y.; Adeli, A.; Jenkins, J.N. Soybean crop-water production functions in a humid region across years and soils determined with APEX model. *Agric. Water Manag.* **2018**, *204*, 180–191. [CrossRef]
25. Buts, M.; Drews, M.; Larsen, M.A.D.; Lerer, S.; Rasmussen, S.H.; Grooss, J.; Overgaard, J.; Refsgaard, J.C.; Christensen, O.B.; Christensen, J.H. Embedding complex hydrology in the regional climate system—Dynamic coupling across different modelling domains. *Adv. Water Resour.* **2014**, *74*, 166–184. [CrossRef]
26. Ewen, J.; Parkin, G.; O'Connell, P.E. SHETRAN: Distributed River Basin Flow and Transport Modelling System. *ASCE J. Hydrol. Eng.* **2000**, *5*, 250–258. [CrossRef]
27. Trancoso, A.R.; Braunschweig, F.; Chambel Leitão, P.; Obermann, M.; Neves, R. An advanced modelling tool for simulating complex river systems. *Sci. Total Environ.* **2009**, *407*, 3004–3016. [CrossRef] [PubMed]
28. Fatichi, S.; Vivoni, E.R.; Ogden, F.L.; Ivanov, V.Y.; Mirus, B.; Gochis, D.; Downer, C.W.; Camporese, M.; Davidson, J.H.; Ebel, B.; et al. An overview of current applications, challenges and future trends in distributed process-based models in hydrology. *J. Hydrol.* **2016**, *537*, 45–60. [CrossRef]
29. Simionesei, L.; Ramos, T.B.; Brito, D.; Jauch, E.; Leitão, P.C.; Almeida, C.; Neves, R. Numerical simulation of soil water dynamics under stationary sprinkler irrigation with MOHID-Land. *Irrig. Drain.* **2016**, *65*, 98–111. [CrossRef]
30. Simionesei, L.; Ramos, T.B.; Oliveira, A.R.; Jongen, M.; Darouich, H.; Weber, K.; Proença, V.; Domingos, T.; Neves, R. Modeling soil water dynamics and pasture growth in the montado ecosystem using MOHID-Land. *Water* **2018**, *10*, 489. [CrossRef]
31. Brito, D.; Neves, R.; Branco, M.C.; Gonçalves, M.C.; Ramos, T.B. Modeling flood dynamics in a temporary river draining to an eutrophic reservoir in southeast Portugal. *Environ. Earth Sci.* **2017**, *76*, 377. [CrossRef]
32. Brito, D.; Ramos, T.B.; Gonçalves, M.C.; Morais, M.; Neves, R. Integrated modelling for water quality management in a eutrophic reservoir in south-eastern Portugal. *Environ. Earth Sci.* **2018**, *77*, 40. [CrossRef]
33. Brito, D.; Campuzano, F.J.; Sobrinho, J.; Fernandes, R.; Neves, R. Integrating operational watershed and coastal models for the Iberian Coast: Watershed model implementation—A first approach. *Estuar. Coast. Shelf Sci.* **2015**, *167*, 138–146. [CrossRef]
34. IUSS Working Group. *World Reference Base for Soil Resources 2014: International Soil Classification System for Naming Soils and Creating Legends for Soil Maps*; World Soil Resources Reports No. 106; Food and Agriculture Organization of the United Nations (FAO): Rome, Italy, 2014.
35. Van Genuchten, M.T. A closed-form equation for predicting the hydraulic conductivity of unsaturated soils. *Soil Sci. Soc. Am. J.* **1980**, *44*, 892–898. [CrossRef]
36. Allen, R.G.; Pereira, L.S.; Raes, D.; Smith, M.; FAO. *Crop Evapotranspiration—Guidelines for Computing Crop Water Requirements*; Irrigation & Drainage Paper 56; Food and Agriculture Organization of the United Nations (FAO): Rome, Italy, 1998.
37. Ritchie, J.T. Model for predicting evaporation from a row crop with incomplete cover. *Water Resour. Res.* **1972**, *8*, 1204–1213. [CrossRef]
38. Feddes, R.A.; Kowalik, P.J.; Zaradny, H. *Simulation of Field Water Use and Crop Yield*; Wiley: Hoboken, NJ, USA, 1978.
39. American Society of Civil Engineers (ASCE). *Hydrology Handbook Task Committee on Hydrology Handbook*; II Series, GB 661.2. H93; ASCE: Reston, VA, USA, 1996; pp. 96–104.

40. Williams, J.R.; Jones, C.A.; Kiniry, J.R.; Spanel, D.A. The EPIC crop growth model. *Trans. Am. Soc. Agric. Biol. Eng.* **1989**, *32*, 497–511. [CrossRef]

41. Neitsch, S.L.; Arnold, J.G.; Kiniry, J.R.; Williams, J.R. *Soil and Water Assessment Tool: Theoretical Documentation*, 2009th ed.; Texas Water Resources Institute; Technical Report No. 406; Texas A&M University System: College Station, TX, USA, 2011.

42. Monsi, M.; Saeki, T. Uber den Lictfaktor in den Pflanzengesellschaften und sein Bedeutung fur die Stoffproduktion. *Jpn. J. Bot.* **1953**, *14*, 22–52. (In German)

43. Jones, C.A. *C4 Grasses and Cereals: Growth, Development and Stress Response*; John Wiley & Sons: New York, NY, USA, 1985; p. 419.

44. Pôças, I.; Paço, T.A.; Paredes, P.; Cunha, M.; Pereira, L.S. Estimation of actual crop coefficients using remotely sensed vegetation indices and soil water balance modelled data. *Remote Sens.* **2015**, *7*, 2373–2400. [CrossRef]

45. Purser, R.J.; Leslie, L.M. A semi-implicit, semi-lagrangian finite-difference scheme using high-order spatial differencing on a nonstaggered grid. *Mon. Weather Rev.* **1988**, *116*, 2069–2080. [CrossRef]

46. Wesseling, J.G.; Elbers, J.A.; Kabat, P.; van den Broek, B.J. *SWATRE: Instructions for Input Report*; Win and Staring Centre: Wageningen, The Netherlands, 1991.

47. Legates, D.R.; McCabe, G.J. Evaluating the use of "goodness-of-fit" measures in hydrologic and hydroclimatic model validation. *Water Resour. Res.* **1999**, *35*, 233–241. [CrossRef]

48. Moriasi, D.N.; Arnold, J.G.; van Liew, M.W.; Bingner, R.L.; Harmel, R.D.; Veith, T.L. Model Evaluation Guidelines for Systematic Quantification of Accuracy in Watershed Simulations. *Trans. Am. Soc. Agric. Biol. Eng.* **2007**, *50*, 885–900.

49. Wang, X.; Williams, J.R.; Gassman, P.W.; Baffaut, C.; Izaurralde, R.C.; Jeong, J.; Kiniry, J.R. EPIC and APEX: Model Use, Calibration and Validation. *Trans. Am. Soc. Agric. Biol. Eng.* **2012**, *55*, 1447–1462. [CrossRef]

50. Nash, J.E.; Sutcliffe, J.V. River flow forecasting through conceptual models part I—A discussion of principles. *J. Hydrol.* **1970**, *10*, 282–290. [CrossRef]

51. Cameira, M.R.; Fernando, R.M.; Pereira, L.S. Monitoring water and NO_3-N in irrigated maize fields in the Sorraia Watershed, Portugal. *Agric. Water Manag.* **2003**, *60*, 199–216. [CrossRef]

52. Paredes, P.; Rodrigues, G.C.; Alves, I.; Pereira, L.S. Partitioning evapotranspiration, yield prediction and economic returns of maize under various irrigation management strategies. *Agric. Water Manag.* **2014**, *135*, 27–39. [CrossRef]

53. Paredes, P.; Melo-Abreu, J.P.; Alves, I.; Pereira, L.S. Assessing the performance of the AquaCrop model to estimate maize yields and water use under full and deficit irrigation with focus on model parameterization. *Agric. Water Manag.* **2014**, *144*, 81–97. [CrossRef]

54. Yuping, M.; Shili, W.; Li, Z.; Yingyu, H.; Liwei, Z.; Yanbo, H.; Futang, W. Monitoring winter wheat growth in North China by combining a crop model and remote sensing data. *Int. J. Appl. Earth Obs. Geoinf.* **2008**, *10*, 426–437. [CrossRef]

55. Trombetta, A.; Iacobellis, V.; Tarantino, E.; Gentile, F. Calibration of the AquaCrop model for winter wheat using MODIS LAI images. *Agric. Water Manag.* **2016**, *164*, 304–316. [CrossRef]

water MDPI

Article

Effect of Alternative Irrigation Strategies on Yield and Quality of Fiesta Raisin Grapes Grown in California

Isabel Abrisqueta [1] and James E. Ayars [2,*]

[1]   Visiting Scientist, Irrigation Dpt., CEBAS-CSIC, P.O. Box 164, 30100 Murcia, Spain; iavillena@gmail.com
[2]   Research Agricultural Engineer, USDA Agricultural Research Service, San Joaquin Valley Agricultural Sciences Center, Parlier, CA 93648-9757, USA
*   Correspondence: james.ayars@gmail.com

Received: 14 March 2018; Accepted: 25 April 2018; Published: 30 April 2018

Abstract: Traditionally, grapes are fully irrigated, but alternative irrigation strategies to reduce applied irrigation water may be necessary in the future as occurrences of drought increase. This study was conducted in the San Joaquin Valley (SJV) of California from 2012 to 2014. Three irrigation treatments were used to study the effects on the yield and quality of Fiesta grapes. The treatments included: grower irrigation (GI) weekly irrigation lasting for approximately 65 h; sustained deficit (SD) equal to 80% of the GI treatment; and regulated deficit (RD) equal to 50% of the GI until fruit set when it was increased to 80% of the GI through harvest and reduced to 50% of the GI after harvest. Average water use across treatments was ≈489 mm. Average yield across all treatments was 7.9 t ha^{-1}, 9.1 t ha^{-1} and 11.8 t ha^{-1} in 2012, 2013, and 2014, respectively. Yield was sustained in SD and RD, with up to a 20% reduction in applied water use compared to GI. There were no differences in raisin quality and grade among any of the treatments in any year. The percentage of substandard grapes decreased from an average of 12.6% in 2012 to 3.6% in 2013 and 2014. Growers may use a sustained deficit approach during periods of limited water availability to minimize the effect on yield.

Keywords: deficit irrigation; Fiesta grapes; drip irrigation; dried on the vine; sustained deficit irrigation; regulated deficit irrigation

---

1. Introduction

Irrigation scheduling is a practice that determines both the depth and time of application of irrigation water. A wide range of scientific methods have been developed to facilitate scheduling. These methods require a detailed understanding of the changing water requirements during crop growth and a detailed understanding of climate. The Food and Agriculture Organization (FAO) method of scheduling [1], using reference evapotranspiration (ET$_o$) and a crop coefficient (K$_c$) is the classical scientific approach. This requires access to weather data and crop coefficients that describe the change in crop water use with time.

Eddy covariance and Bowen ratio systems have been used for direct measurement of crop water use. These systems are very expensive and used primarily for research but provide a detailed measurement of crop water use. Once crop water use is determined, it is integrated with available soil water and plant stress to determine the actual irrigation schedule.

Despite advances in science and technology and demonstrated success in improving water management, many farmers are reluctant to accept and incorporate scientific scheduling into their management practices. Instead, they rely on their experience and economic factors (cost of water, cost of electricity) for making irrigation decisions that typically result in over-irrigation.

California experienced five years of drought (2012–2017) that resulted in extensive fallowing of field crops throughout the state. Growers with perennial crops opted to remove older vineyards and orchards to extend available water supplies to the remainder of their farming operation. Due to the limited availability of surface water, groundwater became the primary source of irrigation water for many farmers resulting in an increased rate of over-drafting aquifers throughout California. The over-drafting resulted in legislation, the Sustainable Groundwater Management Act (SGMA) [2], to manage groundwater to achieve sustainable future withdrawals that will not restrict future water availability. This legislation will have a significant impact on the many growers in the San Joaquin Valley (SJV) who rely on groundwater as their primary source of irrigation water.

Deficit irrigation (DI) is a practice to reduce irrigation below the crop water requirement with minimal impact on yield and quality. There are several approaches, including: sustained deficit irrigation (SD), regulated deficit irrigation (RD), and partial root zone drying (PRD). SD is an approach that uniformly reduces the applied irrigation below the crop water requirement throughout the growing season. RD provides adequate water during critical growth periods that may affect yields and withholds water during those periods that will have minimal impact on the yield both during the current year and in subsequent years. PRD dries down half of the root zone and alternates between sides of the root zone when a threshold plant stress level in the canopy is reached on the side without irrigation. Studies have not demonstrated any advantage of the PRD approach relative to SD or RD [3].

RD has been successfully used in peach, pear, and apple [4–7]. However, recent results indicate the practice may be deleterious to long-term production in peach [7–9]. Early season varieties of plums [10] and grapes [11,12] are two examples of crops that can be deficit irrigated following harvest without impacting yield in subsequent years. Deficit irrigation is also routinely used as an irrigation strategy on wine grapes to promote quality parameters for wine production [13].

Nearly 80,000 ha of raisin grapes are produced in the SJV [14]. Traditionally, raisins are made by removing ripened grape clusters and laying them in the field on a paper "tray" to dry. New grape varieties have been bred to be dried on the vine (DOV) to reduce labor costs and minimize the impact of weather (rain) on the quality of the raisin [15]. Fiesta is a new variety developed for DOV production [16,17] that has little information available on its water requirements.

Using a weighing lysimeter, Williams et al. [18] demonstrated that Thompson seedless grapes could be irrigated with a sustained deficit equal to 80% of full crop evapotranspiration without impacting yield and quality. Thus, it may be possible to reduce irrigation on other table and raisin grape varieties without negative impacts on yield and quality. The objective of this study was to determine the water requirement and the effect of alternative irrigation strategies (SD and RD), on the yield and quality of Fiesta raisin grapes grown in the San Joaquin Valley of California.

2. Materials and Methods

2.1. Research Site

The research was conducted in a commercial vineyard located in Caruthers, California, USA (Lat. 36.54655° N, Long. −119.8791° W). The own-rooted vines (*Vitis vinifera* L. cv. Fiesta) were trained on a 45-cm-wide two-wire T trellis located ≈1.4 m above ground. The plant spacing was 3.6 m between rows and 2.1 m within the row, resulting in a plant density of ≈1280 vines per ha. The soil is a Hesperia fine sandy loam (Coarse-loamy, mixed, superactive, nonacid, thermic Xeric Torriorthents) with a bulk density ranging from 1.58 to 1.62 g/cm^3. Field capacity (FC) is 0.198 cm^3 cm^{-3} and the permanent wilting point (PWP) is 0.119 cm [19]. The total available water (TAW) in a 1.5 m profile is approximately 118 mm.

2.2. Irrigation Treatments and System

The irrigation schedule was determined by the grower based on his experience and the previous week's weather and the crop growth stage. Weekly irrigations started at 5 pm on Friday and ended at 12 pm on Monday, typically lasting 65 h which corresponded to a period of reduced rates for electricity. We assumed the grower irrigation was at least meeting 100% of the water requirement and developed our irrigation treatments as a percentage of the grower irrigation practice (GI).

The three irrigation treatments were: grower irrigation (GI), consisting of weekly irrigations beginning in March and continuing through October except for approximately 2 to 3 weeks while the fruit was drying beginning at cane cutting and ending after harvest; sustained deficit (SD), equal to 80% of the GI treatment throughout the growing season; and regulated deficit (RD), equal to 50% of GI until fruit set at which time it was increased to 80% of GI until cane cutting and following harvest it was reduced to 50% of GI. Deficit irrigation strategies were implemented in 2012 and continued through 2014.

Irrigation was applied using a drip system with $2 \, L \, h^{-1}$ emitters per plant spaced 0.5 m on either side of a vine, with an additional dripper $2 \, L \, h^{-1}$ mid-way between vines in a row. The application rate was approximately $0.8 \, mm \, hr^{-1}$. The drip line was suspended approximately 0.5 m above the ground and attached to the trellis posts. Irrigation water was supplied by a well located in the vineyard. The experimental block was isolated from the remainder of the field by installing sub-mains that connected all replications of a treatment. Each submain had a flow meter and valve installed and flow was recorded using a CS 3000 data logger (Campbell Scientific, Logan UT) that also controlled and monitored the irrigation in each treatment. The irrigation control program began operating at ≈ 5 pm on Friday afternoon and ended at 12 pm on the following Monday. All treatments began at the same time, but the SD and RD treatments ended before the GI finished irrigating based on their operational criteria. All fertilization was done through the drip system at the beginning of an irrigation. Disease, pest management, and fertilization were applied by the grower based on commercial practices.

2.3. Weather, Leaf Water Potential, and Soil Water Data

The experimental site was in an area of the SJV without any close coverage by a California Irrigation Management Information System (CIMIS) station. However, the CIMIS network provides an option for the spatial determination of the reference evapotranspiration (ET_o) and this option was used to estimate the ET_o at the research site. For spatial determination of ET_o CIMIS uses a combination of data derived from satellites and interpolated from CIMIS station measurements to estimate ET_o at a 2 kilometer (km) spatial resolution using the American Society of Civil Engineers version of the Penman-Monteith equation (ASCE-PM). Required input parameters are solar radiation, air temperature, relative humidity, and wind speed at two meters' height. The CIMIS weather station at the University of California West Side Research and Extension Center (WSREC) is located approximately 50 km west of the research site. The WSREC CIMIS station was used for reference evapotranspiration data and precipitation data prior to the installation of the Puresense Irrigation Manager (Jain Irrigation, Fresno, CA) system located in the research plot.

Leaf water potential (Ψ) was measured weekly throughout the summer using a pressure chamber (Soilmoisture Equipment Corp., Santa Barbara, CA). The measurements were taken at midday on 12 mature fully sunlit leaves in each treatment, following the recommendations of Hsiao [20]. Accumulated water stress over the season was characterized using a water stress integral (WSI) estimated over the entire growing season as represented by (t) in equation 1 from daily measurements of Ψ at an interval of n days between measurement using Equation (1),

$$\sum_{i=0}^{I=t} (\Psi(i, \, i+1) - c)n \tag{1}$$

where Ψ $(i, i + 1)$ is the mean of Ψ for any interval $(i, i + 1)$ and c is the least stress measured during the season [21].

Soil water content was measured in 30-cm increments to a depth of 1.5 m using a Puresense Irrigation Manager capacitance probe (Jain Irrigation, Fresno, CA, USA). One probe was in a single replication of each treatment. The probes were installed in June 2012 and were run continuously until the end of the project using the factory calibration. These are non-saline soils and the factory calibration was suitable for characterizing the changes in soil water content. Data from the probes were used to characterize the changes in soil water content. Crop water was calculated for the growing season and dormancy using precipitation (P), irrigation (I), and change in stored soil water (ΔSW). There was no water table in the region and thus no capillary rise and no surface runoff since the field was drip irrigated. Deep percolation was not considered significant because of the interval between irrigation and the crop water use dried down the profile and stored most of the applied water. Dormancy was defined as the period from harvest in one year to bud break in the following year and included November through March. The growing season was defined as running from April through October, which corresponded to approximately bud break to leaf drop. The ET_o was determined for the growing season from the spatial CIMIS data prior to the installation of the Puresense equipment.

2.4. Yield and Fruit Quality

Prior to cutting the canes, the number of clusters on the 10-experimental vines was counted, harvested into a bin, and weighed to determine an average fresh cluster weight. The fruit on 5 additional contiguous vines was partially dried by cutting the canes and after 3 to 5 days on the vine, the fruit was mechanically harvested and placed on a continuous paper tray to complete the drying process [16]. The dates for cane cutting, laying the raisins on the ground, and harvest are provided in Table 1.

Table 1. Dates for cutting canes, laying fruit on continuous paper tray and harvest (picking fruit off the continuous paper tray) at Caruthers, California field site.

Cut Canes	Fruit on Ground	Harvest
08/16/2012	08/31/2012	09/07/2012
08/20/2013	08/28/2013	09/04/2013
08/18/2014	08/29/2014	09/05/2014

Yield was measured by weighing the total fruit from the additional five contiguous vines in each plot, after being dried for approximately 10 days on the ground. A 1 kg subsample of fruit from each plot was taken to the U.S. Department of Agriculture (USDA) grading station located on the Sun Maid packing facility, Selma, CA. Raisins were graded using an airstream sorter and reported as Grade A, with 80% of well-matured grapes, Grade B, with 70% of reasonably well matured grapes, and Grade C, with 55% of fairly well matured grapes. Raisins failing to meet the Grade C maturity requirements were categorized as substandard grade (United States Standards for Grades of Processed Raisins, 1978). Grade A and Grade B were combined and reported as B or Better. Samples were tested for moisture, substandard percent, and % grade B or better. In terms of moisture, USDA dictates that raisins should contain no more than 16%.

2.5. Statistics

The treatments were distributed in a randomized complete block design with three replications. Each plot consisted of three rows with ten experimental vines identified in the center row of each plot. The 10 vines were subdivided into 2 groups of 3 contiguous vines and one group of 4 contiguous vines distributed along the row. Because of the machine harvest, five contiguous vines in the center row were used for yield and quality analysis. Raisin moisture content was adjusted to 14% prior to yield analysis [17]. Data were analyzed by one-way analysis of variance (ANOVA: SPSS, Chicago, IL, USA), and means were separated at $p \leq 0.05$ using Tukey's test.

3. Results

Climate in the SJV is characterized as Mediterranean, with most rainfall occurring during the winter months (November to March) and no rainfall occurring during the summer months (Figure 1). Rainfall in the winter of 2011–2012 was 140 mm, 96 mm in the winter of 2012–2013 and 65 mm in the winter of 2013 to 2014 compared the 27-year average of 218 mm. There was 183 mm, 3 mm, and 10 mm rainfall during 2012, 2013, and 2014 growing seasons (Table 2). The large difference in the total water use between the 2012 growing season and the 2013 and 2014 seasons was the rainfall that extended in April and May in 2012 and not in 2013 and 2014 which increased the total water availability and crop use.

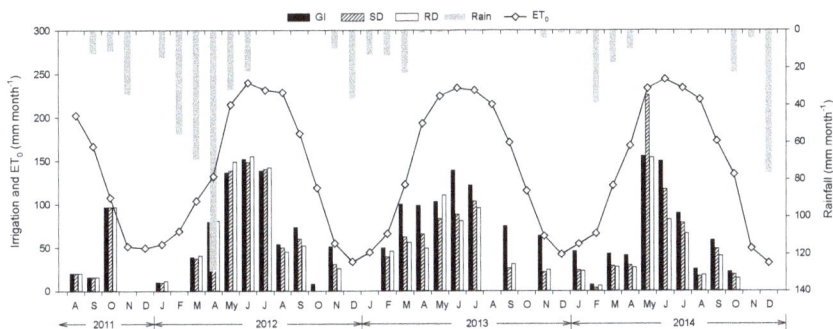

Figure 1. Monthly irrigation water applied in each treatment (GI—grower irrigation, SD—sustained deficit, RD—regulated deficit), rainfall and reference crop evapotranspiration (ET$_0$) from 2011 to 2014.

To ensure that the plant has adequate water for early season growth and development and delay the time of the first irrigation, growers of perennial crops in the SJV typically try to have a full soil water profile in the root zone prior to bud break in the spring. This is done using a combination of rainfall and irrigation. In this area, the rainfall period extends from dormancy of the previous year to bud break in the current year. Growers will estimate and apply the irrigation water needed to supplement rainfall and fill the soil water profile prior to bud break. To capture this practice, we divided the crop water use calculations in Table 2 into two periods; from harvest the previous year to bud break in the current year, and from bud break to dormancy in the current year (growing season). We found that the combined rainfall and irrigation data in Table 2 averaged across years and treatments is approximately 198 mm. This amount of applied water was adequate to fill the crop root zone in these soils prior to bud break.

The total crop water use in 2012 included 140 mm of rainfall plus 43 mm of irrigation prior to bud break and 183 mm of rainfall and irrigation in the GI treatment (47% of ET$_0$) during the growing season as well. There was significant crop water use from soil water in all treatments during this growing season as well. Applied irrigation in the GI treatment was reduced in 2013 and 2014 to 41% and 36% of ET$_0$ respectively, (Table 2) due to drought restrictions and concern by the grower for the pump operation and depth of available water in the irrigation well. The total irrigation for the growing season averaged across all treatments was 625 mm in 2012 mm 369 mm in 2013, and 426 mm in 2014. The average contribution of the stored soil water to the vines across all treatments also declined each year from a high of 207 mm in 2012 to 174 mm in 2013 to 67 mm in 2014 (Table 2). The total calculated water use for the GI treatment was 1070 mm in 2012, which was equal to 80% of ET$_0$, 527 mm in 2013 which was equal of 38% of ET$_0$, and 530 mm in 2014, which was equal to 40% of ET$_0$.

Table 2. Crop reference evapotranspiration (ET₀) based on California Irrigation Management Information System (CIMIS) and local data, precipitation (P) and applied water (I) and change in soil water content (ΔSW) data from harvest the previous year to bud break the current year and the growing season bud break to harvest in the current year and percentage of reduction (Red) in applied water compared to grower practice (GI) at Fiesta raisin project in Caruthers, CA. Grower Irrigation (GI) weekly beginning at bud break until cane cutting, halt until harvest then resume weekly until end of October, Sustained deficit (SD)—sustained deficit equal to 80% of GI, and Regulated deficit (RD)—50% of GI until fruit set, increased to 80% of GI, decreased to 50% of GI after harvest. Crop water use (ET$_c$) is sum of rainfall and irrigation and change in soil water.

Year	Treatment	Harvest—Bud Break (mm)			Growing Season (mm)						% Red
		P	I	I + P	ET₀	P	I	I + P	ΔSW	ET$_c$	
2012	GI	140	48.8	188.8	1345.6	183.5	641.2	824.7	246.8	1071.5	-
	SD		46.9	186.9			609.8	793.3	230.0	1023.3	5
	RD		52.2	192.2			624.5	808.0	144.2	952.2	11
2013	GI	96.3	172.5	268.8	1383.4	3.1	438.8	441.9	160.8	602.7	-
	SD		109.0	205.3			301.4	304.5	194.6	449.1	25
	RD		125.2	224.6			360.9	364.0	167.9	528.8	12
2014	GI	65.6	160.1	225.7	1270.6	10.2	460.7	470.9	60.0	530.9	-
	SD		78.8	144.4			468.7	478.9	75.4	554.3	4
	RD		82.1	147.7			348.3	358.5	66.6	426.1	20

The variation in the soil water content in 1.5 m of the soil profile in Figure 2. The data show that during the growing season there was a consistent change in the stored soil water that reflected the irrigation and crop water use. This ranged between 325 and 425 mm of water stored in the soil profile. At the end of the irrigation season before leaf drop each of the treatments extracted significant quantities of water from the soil profile. A total input of rainfall and irrigation of approximately 190 mm returned the stored water in the profile to 360 mm which was sufficient to meet the early season demand from bud break to early canopy development.

The crop water use for the Fiesta grapes in the SD treatment was 1023.3 mm in 2012, 449.1 mm in 2013 and 554.3 mm in 204. The water use for the RD treatment was 952.2 mm in 2012, 528.8 mm in 2013, and 426.1 mm in 2014. The 3-year crop water use averages for the GI, SD and RD treatments are 735 mm, 675 mm, and 635 mm, respectively. When just 2013 and 2014 are considered, the average water use across years was 566.8, 501.7 and 477.4 mm for the GI, SD and RD treatments respectively. This compares to the estimate of 560 and 710 mm of water for fully irrigated Thompson seedless grapes grown for raisins in the southern SJV [22].

Since ET$_c$ is proportional to the leaf area exposed to sunlight, the difference in well-watered Thompson Seedless and Fiesta may be the result of differences in the size of the cross arm used to support the crop. Williams et al. [18] using a weighing lysimeter found that ET$_c$ for high frequency irrigated Thompson Seedless grape ranged from 718 to 865 mm. Using the same weighing lysimeter, Williams et al. [23] demonstrated that crop water use of Thompson Seedless was a linear function of the shaded area. In the Williams, et al., study [18] the vines were trained on 0.6-m cross-arms rather than 0.45-m cross-arms used in the present study A proportional reduction the leaf area exposed to sunlight based on the cross-arm length would result in a 25% reduction in the ET$_c$ [23,24].

There is not a crop coefficient for Fiesta raisin grapes to use in scheduling irrigation or determining crop water use. Williams et al. [25] developed a K$_c$ for Thompson Seedless grapevine as a function of the day of the year. We applied that K$_c$ to ET₀ values measured during this experiment. The resulting values for ET$_c$ adjusted for the reduced leaf area are: 520.4 mm, 507.2 mm and 533.8 mm for 2012, 2013 and 2014, respectively. These values reflect the crop water use for the GI treatment and demonstrate that the 2012 irrigation was excessive. The 2013 GI exceeded the above estimates for well-watered grapevines.

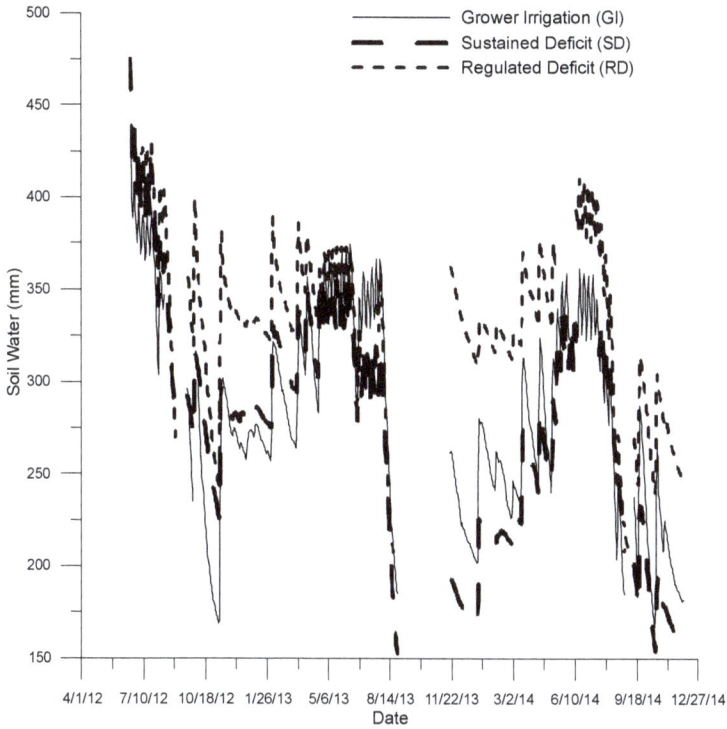

Figure 2. Soil water (mm) in 1.5 m of soil profile measured with capacitance probe. One measurement site in each treatment.

There continues to be a lack of acceptance in the grower community of a scientific approach to irrigation management. This project was developed using the grower methods as demonstration of a method to refine his current practice with the goal of improving water management. Williams et al. [18] had demonstrated with Thompson seedless that irrigation could be reduced by 80% without affecting yield and quality of grapes. This represented a starting point for the grower to modify his current practice. This was an effective technique and one that could be easily applied in other regions. Typically, this would be tested on a small block and expanded as the grower gained confidence. It also provides a management alternative during periods of limited water supply.

There were no differences in Ψ among the treatments early in the season each year, except in 2014 (Figure 3). In 2012, rain along with irrigation maintained high levels of soil water content (data not shown), minimizing plant water stress. Thus, Ψ ranged from −0.75 to −1.20 MPa early on in 2012 and dropped to ≈−1.4 MPa in August, at which point irrigation was withheld in order to dry the raisins; Ψ reached a minimum of −1.6 MPa just after harvest which would not affect the yield or quality of the fruit. In 2013, there was rain prior to bud break, and Ψ remained >−1.3 MPa until mid-June. The stress did not increase until after fruit development was completed and thus had no impact on the final yield.

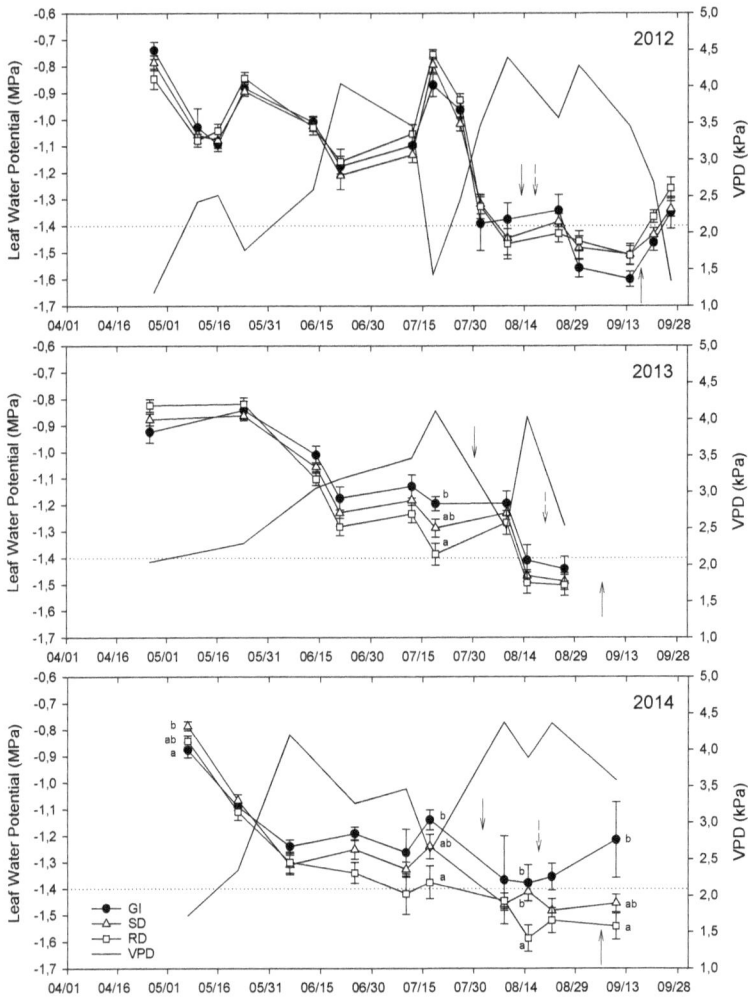

Figure 3. Leaf water potential for 2012, 2013, and 2014 for grower irrigation (GI), sustained deficit (SD) and regulated deficit (RD) vapor pressure deficit (VPD). The down arrow represents irrigation withhold start point. The dotted shape arrow down represents the date when the canes were cut. The arrow up represents the restoration of the irrigation after harvest. Each value is the mean ± SE of six individual measurements. Different letters at each time indicates significant differences according to the Tukey's Test HSD ($p \leq 0.05$).

With one exception in mid-July, there were no differences in Ψ among the treatments in 2013. Due to a problem with the pressure chamber, Ψ was not measured after the canes were cut in 2013. In 2014, Ψ reached −1.4 MPa by the beginning of July.

The stress integral as a measure of accumulated stress over a growing season has been used to characterize stress development to differentiate treatment effects. Overall, there was very little difference in accumulated water stress as characterized by the stress integral among the treatments in 2012 and 2013. In 2012 the GI had slightly greater accumulated stress and the RD treatment had the greatest accumulated stress in 2013. In 2014, the deficit treatments had higher values of accumulated stress than GI (Table 3). However, the accumulated stress in the SD and RD treatments in 2014 were like the values found in 2012.

Table 3. Stress integral in megapascal a day (MPa * Day) for grower irrigation (GI), sustained deficit (SD) and regulated deficit (RD) irrigation treatments in Fiesta raisin grape deficit irrigation study for the period from April to August at the Caruthers, CA. field site.

Irrigation Treatment	Stress Integral (MPa*Day) (April–August)		
	2012	2013	2014
GI	45.3	30.1	13.5
SD	41.6	32.0	45.6
RD	43.2	45.6	45.6

In 2013 and 2014 there were statistical differences in the number of clusters per vine between the SD and RD treatment (Table 4). In 2013 and 2014, the RD treatment had a greater number of clusters per vine than the GI treatment. The SD had a greater number of clusters per vine in 2014 than either of the other two and fewer in 2013 than the remaining treatments. There was no difference in the cluster weights in 2013 statistically, however, there was a statistical difference in the cluster weights between treatments in 2014.

In all three years, the total yield estimated on the weight per vine resulted in no yield differences statistically, however, there were differences in yield. In 2012 and 2013 the SD and RD treatments had higher yields than the GI treatment which was not the case in 2014 when the GI treatment had the highest of the three yields. The average yield across all treatments was 7.9 t ha^{-1} in 2012, 9.6 t ha^{-1} 2013 and 11.8 t ha^{-1} in 2014.

The SD treatment showed similar yield, fruit quality, and clusters per vine as the GI treatment but used less water, which increased water productivity. Except for 2012 when the SD treatment had significantly greater % substandard than the GI treatment. Fidelibus et al. [16] described an industry average volume of production for a Fiesta vineyard of 10.1 t ha^{-1}, with individual vineyards mostly averaging between 7.6 and 15.0 t ha^{-1}. Using the yield per vine data and the vines per ha estimate, the yields in this experiment are comparable to the industry average. During 2014, the SD treatment had slightly, although not significantly, lower yield than the rest of the treatments, but within the range described by Christensen [24]. In 2014 the GI treatment had fewer clusters per vine than the deficit irrigated treatments but this was offset by the increased cluster weight resulting in greater yield than the remaining treatments.

Even though there were no statistical differences in the yields between treatments, the actual yield data (t ha^{-1}) in Table 4 were divided by the applied water (I) data were used to determine the water productivity (WP). However, the WP data demonstrated that there were statistical differences in the water productivity between treatments in 2013 and 2014. In both years, RD had the highest water productivity of the three treatments. In 2013 there was a difference in water productivity between the GI and both SD and RD treatments but not between the SD and RD treatments. In 2014 the RD and GI were statistically different from the SD treatment but not from each other.

Raisin quality in the SD and RD treatments was comparable to the GI treatment in all three years. In 2012 there were statistical differences in the substandard treatment but not in the B and better categories. The percentage of substandard raisins in 2013 and 2014 was significantly less than in 2012, coming from an average of 12.6% to 3.6%. The average B and better percentages were 55.3%, 84.1%, and 75.4% in 2012, 2013 and 2014, respectively. This resulted in a reduction of the percentage of grade C fruit going from 33.9% in 2012 to 12.3% and 21.0% in 2013 and 2014, respectively. The major difference between 2012, 2013 and 2014 is the depth of applied irrigation water and the total estimated crop water use.

Table 4. Summary of the yield and quality components of the fiesta table grapes grown in Caruthers California. Water Productivity (WP) was calculated as kilogram of fruit/ha per mm of applied water for the grower irrigation (GI), sustained deficit (SD) and regulated deficit (RD) irrigation treatments. Moisture remaining in the fruit is given as % Moisture. Raisin grades include A, B and C and raisins graded C are considered Substandard and B or Better include both grades A and B (See Materials and Methods for Definition of Grades).

Year	Treatment	Clusters vine^{-1}	Cluster Weight (g)	Yield (kg vine^{-1})	Yield (t ha^{-1})	WP (kg ha^{-1} mm^{-1})	% Moisture	% Substandard	% B or Better
2012	GI	nd	nd	5.92	7.58	11.8	12.2	9.6 a	60.5
	SD	nd	nd	6.34	8.11	13.3	11.3	16.0 b	49.2
	RD	nd	nd	6.35	8.13	13.0	12.5	12.3 ab	50.7
2013	GI	44.5 a	137	6.02	7.71	17.5 a	14.6	3.8	83.6
	SD	42.4 a	156	7.31	9.36	31.0 b	13.3	3.7	82.3
	RD	52.4 b	150	8.01	10.25	28.4 b	13.5	3.4	86.3
2014	GI	27.9 a	271 b	10.14	12.97	28.1 ab	10.5	3.3	77.9
	SD	37.6 b	155 a	8.3	10.63	22.6 a	10.5	3.7	71.6
	RD	33.5 ab	211 ab	9.24	11.83	33.9 b	11.2	3.7	76.6

Values for each year and parameter followed by different letters are significantly different by Tukey's HDS ($p \leq 0.05$). nd—Data not collected.

Water **2018**, *10*, 583

4. Conclusions

The average crop water use for 2013 and 2014 was 567 mm, 502 mm, and 477 mm for the GI, SD and RD treatments respectively, when the soil water content is at field capacity at bud break. Approximately 190 mm of combined rainfall and irrigation are required to ensure that the soil profile is up to field capacity prior to bud break in this soil. Yield was sustained in SD and RD, with up to a 20% reduction in applied water use compared to the GI (566 mm) in 2013 and 2014. There were no statistical differences in yield between treatments in any of the three years of the project. The yield increased from an average of 7.9 tons per hectare to 11.8 tons per hectare in 2014. The accumulated stress as characterized by the stress integral over the season was similar across the treatments indicating that deficit irrigation was a viable alternative in situations with limited water supplies. There were no differences in the % B or Better raisins between any of the treatments in any year. The percent B or Better was 53.5, 84.0, and 75.4% in 2012, 2013, and 2014, respectively. There were statistical differences in the percent of substandard in 2012 but not in 2013 and 2014. The percentage of substandard grapes decreased from an average of 12.6% in 2012 to 3.6% in 2013 and 2014. This corresponded to a reduction in total average water use from 1015 mm in 2012 to 503 mm in 2014. There is little acceptance and implementation of scientific irrigation management by the grower community and this project demonstrated an approach that can be used to facilitate improving irrigation management. Use of a sustained deficit similar to the SD treatment would be recommended to a grower for ease of implementation. Previous research on fully irrigated grapevines has demonstrated that applied water could be reduced by 20% without impacting yield or quality. This represents a starting point.

Author Contributions: This research was conceptualized by James E. Ayars during the funding acquisition. The methodology for the field investigation was a joint effort between James E. Ayars and Isabel Abrisqueta. Isabel Abrisqueta was responsible for the field data collection, analysis and the original draft preparation under James E. Ayars' supervision. Ayars reviewed, edited and submitted the manuscript and has followed up on required revisions. James E. Ayars was responsible for administering the project.

Funding: This research was funded by National Institutes of Food and Agriculture – Specialty Crops Research Initiative (NIFA-SCRI) grant number 2010-01119.

Acknowledgments: The authors thank Richard Schoneman for his assistance in the installation and maintenance of the project. We thank Tim Rodrigues for the use of his vineyard for the project and the assistance and cooperation throughout the project. We are indebted to Sun Maid Growers for their support in particular, Rick Stark and Mike Moriyama. The authors also thank the two anonymous reviewers for their assistance and thoughtful suggestions and annotations used during the revision of this manuscript. This research was supported in part by a National Institutes of Food and Agricultural – Specialty Crops Research Initiative (NIFA-SCRI) grant "Vineyard water management strategies with Limited and Impaired Water Supplies". "The U.S. Department of Agriculture (USDA) prohibits discrimination in all its programs and activities on the basis of race, color, national origin, age, disability, and where applicable, sex, marital status, familial status, parental status, religion, sexual orientation, genetic information, political beliefs, reprisal, or because all or part of an individual's income is derived from any public assistance program. (Not all prohibited bases apply to all programs.) Persons with disabilities who require alternative means for communication of program information (Braille, large print, audiotape, etc.) should contact USDA's TARGET Center at (202) 720-2600 (voice and TDD). To file a complaint of discrimination, write to USDA, Director, Office of Civil Rights, 1400 Independence Avenue, S.W., Washington, D.C. 20250-9410, or call (800) 795-3272 (voice) or (202) 720-6382 (TDD). USDA is an equal opportunity provider and employer."

Conflicts of Interest: The authors declare no conflict of interest. The funding sponsors had no role in the design of the study; in the collection, analyses, or interpretation of data; in the writing of the manuscript, and in the decision to publish the results.

References

1. Allen, R.A.; Pereira, L.S.; Raes, D.; Smith, M. *Crop Evapotranspiration Guidelines for Computing Crop Water Requirements*; FAO Drainage and Irrigation Paper 56; FAO: Rome, Italy, 1998; 300p.
2. Sustainable Groundwater Management Act. Composed of Following Bills, AB 1739, SB 1319, and SB 1186, 2014. Available online: https://leginfo.legislature.ca.gov/faces/billNavClient.xhtml?bill_id=201320140AB1739 (accessed on 24 June 2016).
3. Sadras, V.O. Does partial root-zone drying improve irrigation water productivity in the field? A meta-analysis. *Irrig. Sci.* **2009**, *27*, 183–190. [CrossRef]

4. Chalmers, D.J.; Mitchell, P.D.; van Heek, L. Control of peach tree growth and productivity by regulated water supply, tree density, and summer pruning. *J. Am. Soc. Hortic. Sci.* **1981**, *106*, 307–312.
5. Chalmers, D.J.; Burge, G.; Jerie, P.H.; Mitchell, P.D. The mechanism of regulation of "Bartlett" pear fruit and vegetative growth by irrigation withholding and regulated deficit irrigation. *J. Am. Soc. Hortic. Sci.* **1986**, *6*, 904–907.
6. Girona, J.; Mata, J.; Arbones, A.; Alegre, S.; Rufat, J.; Marsal, J. Peach tree response to single and combined regulated deficit irrigation regimes under shallow soils. *J. Am. Soc. Hortic. Sci.* **2003**, *128*, 432–440.
7. Girona, J.; Gelly, M.; Mata, M.; Arbones, A.; Rufat, J.; Marsal, J. Peach tree response to single and combined deficit irrigation regimes in deep soils. *Agric. Water Manag.* **2005**, *72*, 97–108. [CrossRef]
8. Abrisqueta, I.; Vera, J.; Tapia, L.M.; Abrisqueta, J.M.; Ruiz-Sanchez, M.C. Soil water content criteria for peach trees water stress detection during the postharvest period. *Agric. Water Manag.* **2012**, *104*, 62–67. [CrossRef]
9. Vera, J.; Abrisqueta, I.; Abrisqueta, J.M.; Ruiz-sánchez, M.C. Effect of deficit irrigation on early-maturing peach tree performance. *Irrig. Sci.* **2013**, *31*, 747–757.
10. Johnson, R.S.; Handley, D.E.; Day, K.R. Postharvest water stress of an early maturing plum. *J. Hortic. Sci. Biotechnol.* **1994**, *69*, 1035–1041.
11. Netzer, Y.; Chongren, Y.; Shenker, M.; Bravdo, B.A.; Schwartz, A. Water use and the development of seasonal crop coefficients for Superior Seedless grapevines trained to an open-gable trellis system. *Irrig. Sci.* **2009**, *27*, 109–120. [CrossRef]
12. Serman, F.V.; Liotta, M.; Parrea, C. *Effects of Irrigation Deficit on Table Grapes cv. Superior Seedless Production*; International Society for Horticultural Science (ISHS): Leuven, Belgium, 2004.
13. Acevedo-Opazo, C.; Ortega-Farias, S.; Fuentes, S. Effects of grapevine (*Vitis vinifera* L.) water status on water consumption, vegetative growth and grape quality: An irrigation scheduling application to achieve regulated deficit irrigation. *Agric. Water Manag.* **2010**, *97*, 956–964. [CrossRef]
14. Staff, California Grape Acreage. *Report. Food and Agriculture*; Sacramento, California Department of Food and Agriculture: Sacramento, CA, USA, 2013; 63p.
15. Peacock, W.L.; Swanson, F.H. *The Future California Raisins is Drying on the Vine*; California Agriculture, University of California: Davis, CA, USA, 2005; Volume 59, pp. 70–74.
16. Fidelibus, M.W. Grapevine Cultivars, Trellis Systems, and Mechanization of the California Raisin Industry. *HortTechnology* **2014**, *24*, 285–289.
17. Fidelibus, M.W.; Christensen, P.L.; Katayama, D.G.; Ramming, D.W. Early-ripening Grapevine Cultivars for Dry-on-vine Raisins on an Open-gable Trellis. *HortTechnology* **2008**, *18*, 740–745.
18. Williams, L.; Phene, C.J.; Grimes, D.W.; Trout, T.J. Water use of young Thompson Seedless grapevines in California. *Irrig Sci.* **2003**, *22*, 1–9.
19. Isidoro, D.; Grattan, S.R. Predicting soil salinity in response to different irrigation practices, soil types and rainfall scenarios. *Irrig. Sci.* **2011**, *29*, 197–211. [CrossRef]
20. Hsiao, T.C. Measurements of plant water stress. In *Irrigation of Agricultural Crops*; Stewart, B.A., Nielsen, D.R., Eds.; Agronomy Monograph No. 30; American Society of Agronomy: Madison, WI, USA, 1990; pp. 243–279.
21. Myers, B.J. Water stress integral—A link between short-term and long-term growth. *Tree Physiology* **1988**, *4*, 315–322. [CrossRef] [PubMed]
22. Peacock, W. *Water Management for Grapevines*; IG1-95; UC Cooperative Extension Bulletin Pub.: Davis, CA, USA, 1995; p. 4.
23. Williams, L.; Ayars, J.E. Grapevine water use and the crop coefficient are linear functions of the shaded area measured beneath the canopy. *Agric. For. Meteorol.* **2005**, *132*, 201–211. [CrossRef]
24. Christensen, L.P. Raisin grape varieties. In *Raisin Production Manual*; University of California, Agricultural and Natural Resources Publication: Oakland, CA, USA, 2000; pp. 38–47.
25. Williams, L.E.; Phene, C.J.; Grimes, D.W.; Trout, T.J. Water use of mature Thompson Seedless grapevines in California. *Irrig. Sci.* **2003**, *22*, 11–18.

water

MDPI

Article

Assessment of a Smartphone Application for Real-Time Irrigation Scheduling in Mediterranean Environments

Marie Therese Abi Saab [1,*], Ihab Jomaa [2], Sleiman Skaf [2], Salim Fahed [1] and Mladen Todorovic [3]

[1] Climate and Water Unit, Lebanese Agricultural Research Institute, Fanar 90-1965, Lebanon; salimfahed@gmail.com

[2] Department of Irrigation and Agrometeorology, Lebanese Agricultural Research Institute, Tal Amara 278, Lebanon; ijomaa@lari.gov.lb (I.J.); sleimskaf@hotmail.com (S.S.)

[3] CIHEAM—Mediterranean Agronomic Institute of Bari, Via Ceglie 9, 70010 Valenzano (BA), Italy; mladen@iamb.it

* Correspondence: mtabisaab@lari.gov.lb or abisaabm@hotmail.com; Tel.: +961-3902994

Received: 26 November 2018; Accepted: 13 January 2019; Published: 1 February 2019

Abstract: The suitability of cloud-based irrigation technologies remains questionable due to limited information on their evaluation in the field. This study focussed on the on-field assessment of a smartphone irrigation scheduling tool—Bluleaf®—with respect to traditional water application practices. Bluleaf® uses weather, crop, soil, and irrigation system data to support a farmer's decision on the timing and amounts of irrigation. The smartphone application was tested in Bekaa Valley, Lebanon, on durum wheat, a strategic Mediterranean crop, during the 2017 and 2018 growing seasons. The simulation results on soil water balance were in "acceptable to very good" agreement with the measured soil moisture values, with a root mean square error (RMSE) between 15.1 and 26.6 mm and a modelling efficiency (NSE) that ranged from 0.77 to 0.92. The appropriateness of the adopted smartphone irrigation scheduling was confirmed also by leaf water potential measurements and the Crop Water Stress Index (CWSI). A water saving of more than 1000 m^3/ha (25.7%) was observed with Bluleaf® with respect to traditional irrigation scheduling. Therefore, new technologies could bring about substantial benefits to farmers and support water saving efforts in the Mediterranean region.

Keywords: irrigation scheduling; wheat; soil water balance; new technologies; smartphone application

1. Introduction

Sustainable irrigation management requires reliable and easy-to-use methods and tools to support real-time scheduling with respect to the availability of water, specific soil and weather conditions, a crop's water requirements, and a crop's response to stress. Studies, conducted in recent years in various parts of the world, have shown that the use of innovative technologies, management approaches, and modelling tools can improve irrigation scheduling, save water, enhance a farmer's income, and reduce the environmental burden [1–8]. In this context, real-time automatized irrigation scheduling, based on reliable low-cost sensors and simple water balance models, is receiving a growing amount of attention [9].

A plethora of innovative technological solutions in the agricultural sector is emerging worldwide. However, the commercialization of these products is intricate due to difficulties in demonstrating the on-field applicability and the effectiveness of proposed innovations and in creating direct contact with potential users. Over the past decade, numerous web-based irrigation scheduling tools have been developed that integrate real-time weather data and simple water balance models for irrigation scheduling [10–13]. However, their suitability in practice has been limited due to the need to use desktop computers (or laptops) and, therefore, scarce user–tool interaction. In contrast, the smartphone applications that have been developed for a new generation of mobile phones offer continuous user–tool interaction, increase operational flexibility, and permit the update of data in real-time during field inspection activities.

In the last few years, a huge number of studies have reported advances in ICT (Internet and Communication Tools) applications in agriculture, and, in particular, in irrigation at different scales [14–19]. They include the latest cloud-based technologies for on-field data acquisition, transmission, and management, monitoring of the soil-plant-atmosphere continuum and irrigation network performance, satellite- and ground-based remote-sensing applications, soil water balance and crop growth models, and the remote control of the irrigation process.

In Florida, smart irrigation applications (apps) were developed to provide real-time irrigation scheduling for selected crops (i.e., avocado, citrus, cotton, peanut, strawberry, and vegetables). Irrigation scheduling is based on crop evapotranspiration (ETc) or a water balance methodology using real-time weather data [15]. The apps were customized for different users considering the adopted irrigation systems, water conservations options, and other management practices.

In Italy, a new smartphone application, called Bluleaf®, was developed through collaboration between research institutions and the private sector [17]. Bluleaf® is based on a Decision Support System (DSS) platform that integrates weather and soil sensors with soil water balance and irrigation scheduling models that are fully adapted to the specific conditions of irrigated plots and describe in detail the crop's phenological stages and the characteristics of the on-farm irrigation systems. This permits the optimization of irrigation inputs and the enhancement of the irrigation application's efficiency. Bluleaf® was tested in southern Italy, and the results confirmed the robustness of the approach and its capacity to save water and energy with respect to traditional irrigation practices [17]. However, further investigations are needed to examine its performance under different soil and weather conditions and management practices.

The application of technological advances should be investigated more in the Mediterranean region in order to provide a step toward improved agricultural water use in terms of both increased economic benefits and a reduced environmental impact [20]. This is particularly relevant for the Middle East and North Africa (MENA) region, where water resources are scarce and agriculture is primarily based on irrigation. Therefore, it is important to consider the efficiency of adopted technological innovations in the context of specific hydrological realities and agronomic constraints [21,22] and to apply adequate indicators of water use performance and productivity for the sustainable conservation of resources [23].

The Bekaa valley is considered to be the food basket of Lebanon, where winter cereals are produced under supplemental irrigation and spring/summer vegetables are cultivated under full or deficit irrigation (depending on the availability of water). The increasing water scarcity in the valley constitutes the main driver threatening farmers to use less water on food production. In the valley, traditional irrigation scheduling, based on the farmer's knowhow, is the norm. This means that the irrigation scheduling is performed according to a time-set calendar schedule, the number of days that has elapsed since the last irrigation, visual detection of a change in crop color or wilting leaves, and/or according to how dry the soil feels. However, none of these traditional methods can provide information on 'how much' water to apply.

The main objective of the study was to assess the performance of the Bluleaf® smart irrigation scheduling application and to determine whether it could save water and improve yield water productivity over traditional irrigation scheduling practices in the Bekaa valley. The study focused on durum wheat, a strategic Mediterranean crop, cultivated during the 2016–2017 and 2017–2018 growing seasons.

This investigation is important at the local and regional scales to improve farmers' irrigation practices and to disseminate the benefits of use of innovative technologies that can increase water productivity, save resources, and reduce pressure on freshwater withdrawal.

2. Materials and Methods

2.1. Description of Smartphone-Based Decision Support System for Irrigation Management

2.1.1. General Overview of the Tool

Bluleaf® is a smartphone-based Decision Support System (DSS) (developed by Sysman Progetti & Servizi, Mesagne, BR, Bari, Italy) designed to provide integrated support to farmers, including, under a common shell, weather data (historical, real-time, and forecasted) and management tools for irrigation and nutrient application and plant disease risk-alert and protection. The system integrates scientific achievements and technological innovations in the fields of irrigation, agronomy, weather and soil moisture sensors, and data acquisition, transmission, and management and the application of web and app tools in agriculture.

The irrigation component uses weather, soil, crop, and irrigation system data to estimate water balance components during the crop-growing cycle and to elaborate irrigation scheduling (Figure 1). Soil water balance is modelled on a daily basis and it is based on the weather data collected from an agro-meteorological station close to the area of interest and weather forecasting data for a period of 7 days provided by the Meteoblue forecast system. The weather forecasting data are provided on an hourly and a daily basis and are available for each specific location, which permits the computation of the expected variation of water balance components and the optimization of irrigation scheduling 7 days ahead.

Soil data requirements are limited to the knowledge of the volumetric soil water content at field capacity, the wilting point and saturation, the electrical conductivity, the organic matter, and the effective soil depth. The user can select the soil characteristics from the soil database, insert his own data, or use a specific pedo-transfer function (PTF) to create and store new data according to a site-specific soil analysis. Moreover, the system can be equipped with capacitance sensors for the real-time monitoring of soil water content and testing/update of soil water balance modeling results.

The default crop database is arranged using the indicative crop growth and development parameters (e.g., crop coefficient, duration of growing cycles, minimum and maximum root depth, optimum yield threshold) as reported in the literature [24–26]. A more detailed description and the update of crop development and growth parameters is possible using the user's observations in the field during the crop-growing season.

The basic irrigation system data include the irrigation method, the application efficiency, the number of irrigation lines per row, and the distance between an emitter and the emitter's discharge. Additional data can be obtained from remote monitoring of the water supply network (the discharge and pressure at hydrants, the groundwater level in the wells) to support on-farm operational management of irrigation through a series of actuators (i.e., electro-valves) at pumping stations, hydrants, and valves. In this context, a specific multiplot and multicrop management module can be used for optimal water allocation considering all of a farm's irrigated plots, the water availability, and the economic parameters of cultivation.

2.1.2. Irrigation Scheduling: The Water Balance Method

The 'core' algorithm of Bluleaf® is designed to run simulations at the scale of a single 'irrigated plot', which is defined as 'the field-unit cultivated with the same crop (also in terms of variety type, planting date, density, etc.), with relatively homogeneous soil characteristics (average depth, texture, soil water holding capacity, etc.), equipped with a specific irrigation system and receiving the same irrigation applications (in terms of timing and amount)'. The crop-soil water balance and irrigation scheduling are computed by means of a specific model that was originally written in the MS Excel® programming language and previously tested and applied for similar applications [27]. The model estimates crop evapotranspiration, irrigation water requirements, and relative yield through the standard procedure proposed by the FAO 56 Paper [24].

The reference evapotranspiration (ET_o) was estimated by the FAO Penman–Monteith equation [24] as:

$$ET_o = \frac{0.408 \cdot \Delta \cdot (Rn - G) + \gamma \cdot \frac{900}{T+273} \cdot U_2 \cdot (e_s - e_a)}{\Delta + \gamma \cdot (1 + 0.34 \cdot U_2)} \tag{1}$$

where Rn is the net radiation available at the crop surface ($MJ/m^2/d$), G is the soil heat flux density ($MJ/m^2/d$), T is the mean air temperature at 2 m height (°C), U_2 is the wind speed at 2 m height (m/s), ($e_a - e_s$) is the vapour pressure deficit at 2 m height (kPa), Δ is the slope of the vapour pressure curve (kPa/°C), and γ is the psychometric constant (kPa/°C). The computations of the considered parameters were performed following the formulas reported in [24].

Crop evapotranspiration (ET_c) was calculated as a product of the reference evapotranspiration and the crop coefficient Kc:

$$ET_c = K_c ET_o \tag{2}$$

Soil water balance is estimated on a daily basis, and it is expressed in terms of water depletion in the effective root zone $D_{r,i}$ (mm) at the end of each day through the following equation:

$$D_{r,i} = D_{r,i-1} - P_i - IR_i + ET_{c,i} + DP_i \tag{3}$$

where $D_{r,i-1}$ represents the root zone depletion at the end of the previous day $i - 1$ (mm), P_i is the effective precipitation on day i (mm), IR_i is the net irrigation supply on day i (mm), $ET_{c,i}$ is the crop evapotranspiration (mm), and DP_i is the deep percolation on the same day (mm). Surface runoff (RO) is not considered since it occurs only when the precipitation and irrigation inputs are greater than the soil infiltration rate and refers to the water amount that does not enter into the soil. Capillary rise (CR) is usually a very small value that should be taken into consideration only in the case of a shallow groundwater level. Deep percolation is considered to be the excess of water above field capacity within the root zone. Therefore, it occurs when the effective rainfall is higher than the potential water storage of the root zone or in cases of excessive irrigation water supply. The effective precipitation was assumed to be a percentage of total precipitation; a default of 80% was adopted, and it can change during the growing season. Moreover, daily precipitation lower than 2 mm was not taken into account.

Total available water (TAW) for a crop is estimated as the difference between water content at field capacity and water content at wilting point and changes during the growing season as a function of root depth (Rd). Therefore,

$$TAW = (FC - WP)/100 \, Rd \tag{4}$$

where Rd is given in mm. Readily available water (RAW) represents the amount of water that can be depleted from the root zone without compromising crop growth. RAW is estimated as a fraction (p) of TAW as

$$RAW = p \times TAW \tag{5}$$

where p represents a threshold for maximum crop production (the optimum yield threshold). This threshold is crop-specific and can change during the growing season. It ranges between 0.3 (for water-stress-sensible crops) and 0.7 (for water-stress-tolerant crops).

Different irrigation strategies can be adopted, and they depend on the management allowable depletion (MAD), which can be lower than or equal to RAW (the optimal irrigation and maximum yield) and greater than RAW (a crop water stress and yield reduction).

When the soil water depletion in the root zone (D_r) is greater than RAW (i.e., it is below the optimal yield threshold), a dimensionless coefficient K_s (0–1) is used to account for the level of water stress. K_s is calculated by the following formula:

$$K_s = \frac{TAW - D_r}{TAW - RAW} \tag{6}$$

where D_r is the root zone's soil water depletion, TAW is the total available water, and RAW is readily available water (a predetermined fraction of the total available water). Hence, crop evapotranspiration is adjusted for water stress using the reduction coefficient K_s as:

$$ET_{c,adj} = K_s K_c ET_o \tag{7}$$

Yield reduction under water stress is estimated using the approach of [28] as:

$$\left(1 - \frac{Y_a}{Y_m}\right) = K_y \left(1 - \frac{ET_{c,adj}}{ET_c}\right) \tag{8}$$

where Y_a is the actual yield under water stress corresponding to the adjusted crop evapotranspiration ($ET_{c,adj}$), Y_m is the maximum yield corresponding to the optimal water supply, and K_y is a crop-specific yield response factor that can vary during the growing season.

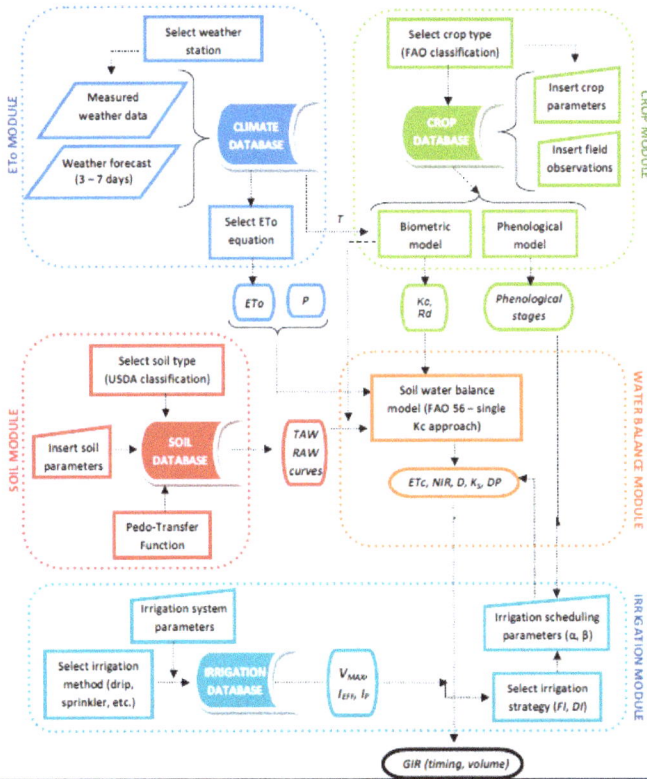

Figure 1. The Bluleaf® architecture. TAW, total available water; RAW, readily available water; NIR, net irrigation requirements; GIR, gross irrigation requirements; FI, full irrigation, DI, deficit irrigation.

2.2. Description of the Testing Site and Experimental Activities

2.2.1. Description of the Testing Site

The testing of the smartphone-based DSS was carried out at Tal Amara in the Bekaa valley (Lebanon) on the experimental field of the Lebanese Agricultural Research Institute (LARI). The Bluleaf® application was tested on durum wheat (cv. Icarasha) during two growing seasons: 2016–2017 and 2017–2018. The main weather parameters, including solar radiation, air temperature and relative humidity, wind speed, and precipitation, were taken from an agro-meteorological station located 400 m away from the experimental field. Monthly weather data, from November 2016 to June 2018, are given in Table 1 together with the reference evapotranspiration (ET_0) estimated by the FAO Penman–Monteith standard approach [24].

Table 1. Monthly climate data from November 2016 to June 2018 as recorded for the Tal Amara region.

Month	P (mm)	T_{max} (°C)	T_{min} (°C)	Rh_{mean} (%)	Rs (W/m^2)	Ws (m/s)	ET_0 (mm)
16 November	29.00	20.35	2.81	45.65	89.12	1.41	61.60
16 December	101.80	9.84	0.85	83.04	66.14	2.01	21.50
17 January	119.40	10.67	−0.88	78.20	68.20	1.69	35.70
17 February	29.80	14.13	−1.39	62.20	82.75	1.38	48.00
17 March	36.80	16.24	3.90	71.65	111.10	1.98	70.40
17 April	10.80	23.22	6.11	52.71	250.73	2.11	136.90
17 May	2.00	28.75	9.32	44.56	320.98	2.07	190.30
17 June	2.40	33.19	14.53	38.95	334.91	1.71	205.00
17 July	0.00	35.34	14.32	43.23	329.06	1.22	193.70
17 August	0.20	34.46	12.97	42.89	305.01	1.34	165.80
17 September	0.20	26.93	8.88	53.18	250.94	1.29	100.20
17 October	35.60	20.56	6.22	64.95	174.53	1.18	59.70
17 November	22.80	19.42	5.02	67.04	119.52	1.15	50.60
17 December	0.20	17.37	2.83	65.43	96.30	1.28	24.23
18 January	150.00	12.71	1.72	80.92	84.84	1.50	34.89
18 February	141.20	15.49	2.55	72.64	123.69	1.33	47.49
18 March	8.80	20.12	3.47	59.83	201.21	1.66	97.52
18 April	25.20	23.35	6.25	54.01	248.99	1.37	117.38
18 May	30.40	27.23	11.85	52.18	246.94	1.30	132.79
18 June	12.20	30.18	15.16	46.18	282.41	1.78	190.10

Soil samples were taken in both seasons at up to 50 cm depth at five locations within the experimental field. The soil texture was determined, and the corresponding texture class referred to the USDA (United States Department of Agriculture) classification. The basic soil hydrological properties for the two seasons were estimated in Bluleaf from the soil granulometric analysis, the organic matter, and the electrical conductivity by means of the pedo-transfer function as suggested by [29]. The results are presented in Table 2. In the first year, the soil water holding capacity was 180 mm/m, while in the second it was almost 25% lower (136 mm/m).

Table 2. The soil hydrological properties for the 2017 and 2018 growing seasons.

Parameters	2016–2017 Season	2017–2018 Season
Sand (%)	31	29
Silt (%)	25	25
Clay (%)	44	46
Soil texture (the USDA classification)	Clay	Clay
Soil water content at saturation (%vol)	52	48.7
Soil water content at field capacity (%vol)	40	40.3
Soil water content at wilting point (%vol)	22	26.7
Soil water holding capacity (mm/m)	180	136

The basic crop parameters were initially set according to [24,30]. However, both biometric measurements (the percentage of effective ground cover) and a phenological survey were done on a weekly basis to correct Bluleaf®'s specific crop parameters according to what was observed in the field.

2.2.2. Description of the Trial

The irrigation scheduling experiment consisted of nine plots, each 300 m² size (15 m × 20 m). The plots were randomized in a complete block design consisting of three irrigation scheduling treatments and three replicates: I-Bluleaf, irrigated according to Bluleaf®; I-farm, irrigated according to the farmer's knowhow; and I-rainfed.

For the I-Bluleaf treatment, irrigation was managed using the Bluleaf® smartphone application in such a way as to keep the soil moisture content above the threshold of readily available water starting from the booting until the grain-filling stage. For the I-farm treatment, irrigation was managed according to the common farmer's practice that consisted of providing the crop with water on a 15-day basis starting from early spring.

Irrigated plots were equipped with traditional impact sprinklers (with a discharge rate of 1.5 m³/h) spaced at 12 m on a lateral line overlaid on the middle of each plot. Each treatment had its own valve and flow meter. The adopted application efficiency was 75%. Durum wheat was sown in rows that were 0.18 m apart, with a density of 200 Kg of seeds per hectare. Nitrogen was applied at a rate of 60 kg/ha; in particular, ammonium sulfate (21% of N) was applied at the beginning of tillering, while ammonium nitrate (26–27%) was applied at the beginning of stem elongation.

The starting dates and duration of the main phenological stages observed during the two growing seasons are reported in Table 3.

Table 3. The starting dates and duration of the main phenological stages of durum wheat grown in the 2016–2017 and 2017–2018 growing seasons. In brackets are reported the days after sowing (DAS).

Growth Stage	Durum Wheat	
Season	2016–2017	2017–2018
Sowing	28 November 2016	4 December 2017
Tillering	7 February 2017 (71)	27 January 2018 (54)
Stem elongation	15 March 2017 (107)	1 March 2018 (87)
Booting	18 April 2017 (141)	29 March 2018 (115)
Flowering	8 May 2017 (161)	15 April 2018 (132)
Grain filling (milk stage)	10 May 2017 (163)	25 April 2018 (142)
Grain filling (dough stage)	30 May 2017 (183)	17 May 2018 (164)
Harvesting	5 July 2017 (219)	12 June 2018 (190)
Length of crop cycle (days)	219	190

2.2.3. Soil Moisture Content and Leaf Water Potential Measurements

Soil moisture readings were used to test the validity of Bluleaf® by checking the differences between simulated (model) and measured (sensor) data. For this purpose, an FDR (Frequency Domain Reflectometry) (or 'capacitance') sensor, the PR2 profile probe (DeltaT Devices Ltd, Cambridge, UK), was used for periodic monitoring of the soil water content along the profile in access tubes placed at a depth of 1 m in each of the nine plots of the trial.

Midday leaf water potential was measured three times in each growing season, during the irrigation season, on a two leaves per plot replicate using the Scholander pressure chamber (P3000, Soil Moisture Corp., Santa Barbara, CA, USA) in both trials. In addition, canopy temperature (Tc) was monitored by using an infrared thermometer (PhotoTemp™MXS™TD, Raytek®, Everett, WA, USA). The crop water stress index (CWSI) was estimated according to Idso's procedure [31] using air temperature (Ta) and vapor pressure deficit (VPD) data. In fact, the canopy temperature minus air temperature (Tc − Ta) values were plotted against the corresponding air VPD values, and delimited by the estimated upper and lower baselines. The CWSI values were then calculated as the relative distance of the measured point data between the upper and lower limits. A set of I-rainfed data under high-water-stress conditions was used to draw the upper baseline (UL) parallel to the VPD axis. Meanwhile, the lower baseline (LL) was derived from a set of I-Bluleaf and I-farm data.

2.2.4. Crop Yield, Crop Quality, and Yield Water Productivity

The final aboveground biomass, at the end of each crop cycle, was measured on a 1 m^2 (1 m × 1 m) surface in each plot replicate. The aboveground biomass was determined by oven drying samples at 70 °C until a constant weight was reached. In addition, at physiological maturity, the grain yield was measured by harvesting a sample area of 1 m^2 at the centre of each plot. Yield water productivity (Y-WP) was calculated as the ratio of yield over the sum of precipitation and irrigation water applied.

2.3. Statistical Analysis

The validation of a soil water balance estimate with Bluleaf® was done by comparing the simulated and measured values of soil water content over the whole growing cycle in both seasons. Goodness-of-fit parameters, mainly the root mean square error (RMSE), the coefficient of variation of the RMSE (CV(RMSE)), the Mean Bias Error (MBE), the Maximum Absolute Error (MAE), the index of agreement (d_{IA}), and the Nash–Sutcliffe coefficient (NSE), were applied to evaluate model performance against observed data.

The average difference between simulation outputs and experimental data was described by the root mean squared error (RMSE) as:

$$RMSE = \left[N \sum_{i=1}^{N} (P_i - O_i)^2 \right]^{0.5}$$

(9)

where N is the number of pairs of observed/measured (O_i) and predicted/simulated (P_i) data.

Then, the coefficient of variation of the RMSE (CV(RMSE)) was applied to normalize the RMSE to the mean of the observed/measured values ($\bar{O}$) as:

$$CV(RMSE) = \frac{RMSE}{\bar{O}}$$

(10)

The Mean Bias Error (MBE) was used to indicate the under/over estimations by the model as:

$$MBE = N^{-1} \sum_{i=1}^{N} (P_i - O_i)$$

(11)

The Maximum Absolute Error (MAE) was estimated as:

$$MAE = Max|P_i - O_i|_{i=1}^{N}$$

(12)

The index of agreement (d_{IA}), which represents the ratio between the mean square error and the "potential error", was calculated according as [32]:

$$d_{IA} = 1 - \frac{\sum_{i=1}^{n} (O_i - P_i)^2}{\sum_{i=1}^{n} {}^2(|P_i - \bar{O}| + |O_i - \bar{O}|)^2}$$

(13)

The d_{IA} is a descriptive parameter that varies between 0 and 1, with the value of 1.0 indicating excellent agreement.

In addition, the Modelling Efficiency (NSE), which represents a normalized statistic that determines the relative magnitude of the residual variance compared to the measured data variance [33,34], was defined as:

$$\text{NSE} = 1.0 - \frac{\sum\limits_{i=1}^{n} (O_i - P_i)^2}{\sum\limits_{i=1}^{n} (O_i - \overline{O})^2}. \tag{14}$$

According to [19], which worked on validating a smartphone application for avocado, the validity of the model performance was considered very good when the probability of fit showed an NSE = 0.9–1, good for NSE = 0.8–0.899, acceptable for NSE = 0.65–0.799, and unsatisfactory for NSE < 0.65.

All statistical analyses were carried out by using SAS v9.

3. Results and Discussion

3.1. Irrigation Water Supply and Saving

Irrigation dates and amounts of water supply considering two management approaches (Bluleaf® and the farmer's approach) are presented in Table 4 for both seasons. In the first season (2016–2017), the Bluleaf® irrigation was scheduled three times, each with 67.5 mm of net irrigation, which accounted for a total of 202.5 mm of net irrigation supply and 270 mm of gross irrigation input. The farmer's irrigation strategy assumed four irrigations of 67.5 mm each, with a total of 270 mm of net irrigation and 360 mm of gross irrigation. Therefore, a total water saving of 90 mm was obtained when the Bluleaf® application was used.

Table 4. The irrigation dates and amounts of water for the two seasons' trials.

Season 2017				
I-Bluleaf			**I-farm**	
Dates	Net Irrigation Amounts (mm)	Dates (DAS)	Net Irrigation Amounts (mm)	
22 April 2017 (145)	67.5	5 April 2017 (128)	67.5	
10 May 2017 (163)	67.5	20 April 2017 (143)	67.5	
23 May 2017 (176)	67.5	5 May 2017 (158)	67.5	
		20 May 2017 (173)	67.5	
Total seasonal irrigation (mm)	203		270	
Total seasonal rain (mm)	317		317	
Irrigation + rain (mm)	520		587	
Season 2018				
I-Bluleaf			**I-farm**	
Dates	Net Irrigation Amounts (mm)	Dates	Net Irrigation Amounts (mm)	
15 March 2018 (101)	53	15 March 2018 (101)	67.5	
5 April 2018 (122)	60	1 April 2018 (118)	67.5	
20 April 2018 (137)	67.5	16 April 2018 (133)	67.5	
8 May 2018 (155)	67.5	2 May 2018 (149)	67.5	
		17 May 2018 (164)	67.5	
Total seasonal irrigation (mm)	248		338	
Total seasonal rain (mm)	368		368	
Irrigation + rain (mm)	616		705	

Note: I-Bluleaf, irrigated according to Bluleaf®; I-farm, irrigated according to farmer know-how; I-rainfed, no irrigation.

In the second season (2017–2018), the Bluleaf® irrigation was scheduled four times with different water inputs, which accounted for 248 and 331 mm of net and gross irrigation, respectively. In the case of the farmer's irrigation strategy, five irrigation events occurred, each with 67.5 mm, which meant 337.5 mm of net irrigation supply and 450 mm of gross irrigation input. Thus, in the second year, a water saving of 119 mm was observed with the Bluleaf® application.

Overall, I-Bluleaf received 25.7% less water than I-farm independently on the growing season. In Lebanon, the area cropped with wheat represents about 30,000 ha, of which almost 50% is being irrigated [35]. Therefore, the annual water saving could reach about 15 million m^3 provided that, in all irrigated areas, a real-time irrigation scheduling tool is applied.

The difference in irrigation amounts between the two growing seasons was mainly due to the different soil characteristics of the two experimental seasons (the soil water holding capacity was about 25% greater in the first year with respect to the second year) and the distribution of precipitation throughout the growing season, which was more uniform in the first than in the second year. In fact, in the second growing season, the precipitation was higher than in the first year by about 51 mm; however, it was distributed mainly in January and February, when both the rooting depth and the crop evapotranspiration were low. Thus, in the second season, most of precipitation was lost mainly through deep percolation. The difference in net irrigation requirements estimated by Bluleaf® was about 45 mm. Nevertheless, the difference in net irrigation applications adopted by the farmer's approach was 67 mm. This confirms the suitability of smart irrigation technologies for irrigation scheduling, which permit us to have better knowledge of the soil-plant-atmosphere continuum and a more accurate irrigation supply.

3.2. Assessment of the Soil Water Content Estimation in the Root Zone

The results of soil water balance estimated on a daily basis are given by the values of soil water depletion in the root zone in Figures 2 and 3 for the three water treatments (I-Bluleaf, I-farm, and I-rainfed) in the 2016–2017 and 2017–2018 growing seasons. In both seasons, the plants were kept under optimal water conditions from the booting until the grain-filling stage. The irrigation was always applied before the root zone's soil water content went below the allowable depletion threshold. Consequently, no apparent water stress was experienced by the crops during the irrigation period. When irrigation was stopped, at the dough stage, the soil water content dropped below the readily available water threshold. The total net irrigation amounts of I-farm were 270 mm and 338 mm in the 2016–2017 and 2017–2018 growing seasons, respectively. Irrigation was performed every 15 days according to the common farmer's practice in the region. Accordingly, irrigation started at the early vegetative stage. For I-rainfed, water stress started at the booting stage and it was increased progressively until the ripening stage.

The testing of Bluleaf® to predict soil water content in the root zone demonstrated a good agreement with the measured values considering the soil water dynamics and the soil spatial variability and heterogeneity. The goodness-of-fit indicators are presented in Table 5. The RMSE ranged from 15.12 mm to 26.64 mm with a CV(RMSE) between 0.14 and 0.61 mm. The d_{IA} ranged from 0.77 to 0.98. The NSE showed that the model can be classified from acceptable to very good [19], with values ranging from 0.77 to 0.92. The differences between the simulated and observed soil water contents could be due to the fact that the soil water balance considers a one-dimensional flow of water through the soil, while ignoring lateral and preferential flow [36–38]. In addition, since soil properties are generally highly heterogeneous, the simulations were accepted as an average representation of soil water variations within the root zone. Trends in soil water content dynamics and not exact values of soil water content were reproduced by the model, as was also obtained in the study for the validation of a smartphone application for avocado [19].

Table 5. The goodness-of-fit indicators for the soil water balance simulation in the wheat trials.

Statistical Indicators	Season 2016–2017			Season 2017–2018		
	I-Bluleaf	I-farm	I-rainfed	I-Bluleaf	I-farm	I-rainfed
RMSE (mm)	26.64	25.57	20.63	17.59	24.86	15.12
CV(RMSE)	0.43	0.40	0.15	0.35	0.61	0.14
MBE (mm)	24.51	−12.14	16.43	4.74	−2.90	5.56
MAE (mm)	24.51	19.87	18.91	14.24	19.75	12.93
d_{IA}	0.77	0.91	0.98	0.93	0.92	0.97
NSE	0.85	0.88	0.90	0.92	0.77	0.92

Note: RMSE, root mean square error; CV(RMSE), coefficient of variation of the RMSE; MBE, mean bias error; MAE, mean absolute error; NSE, Nash–Sutcliffe Efficiency.

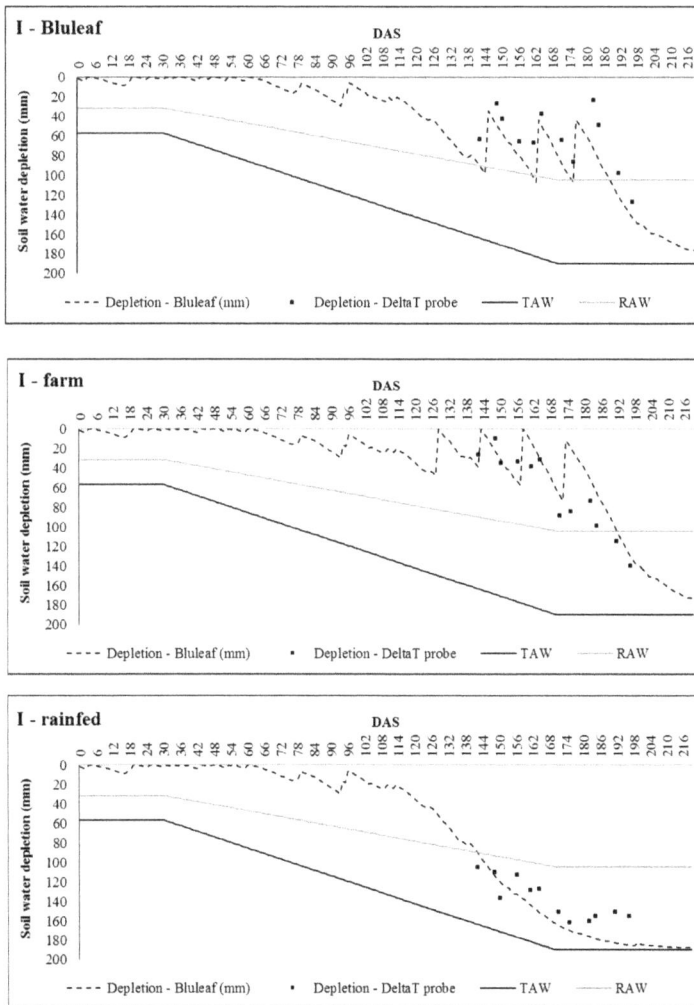

Figure 2. Simulated soil water depletion (Bluleaf®) versus measured soil water depletion (PR2 probes) for the three different treatments in the 2016–2017 wheat growing season. (I-Bluleaf, irrigated according to Bluleaf®; I-farm, irrigated according to farmer know-how; I-rainfed, no irrigation; TAW, total available water; RAW, readily available water).

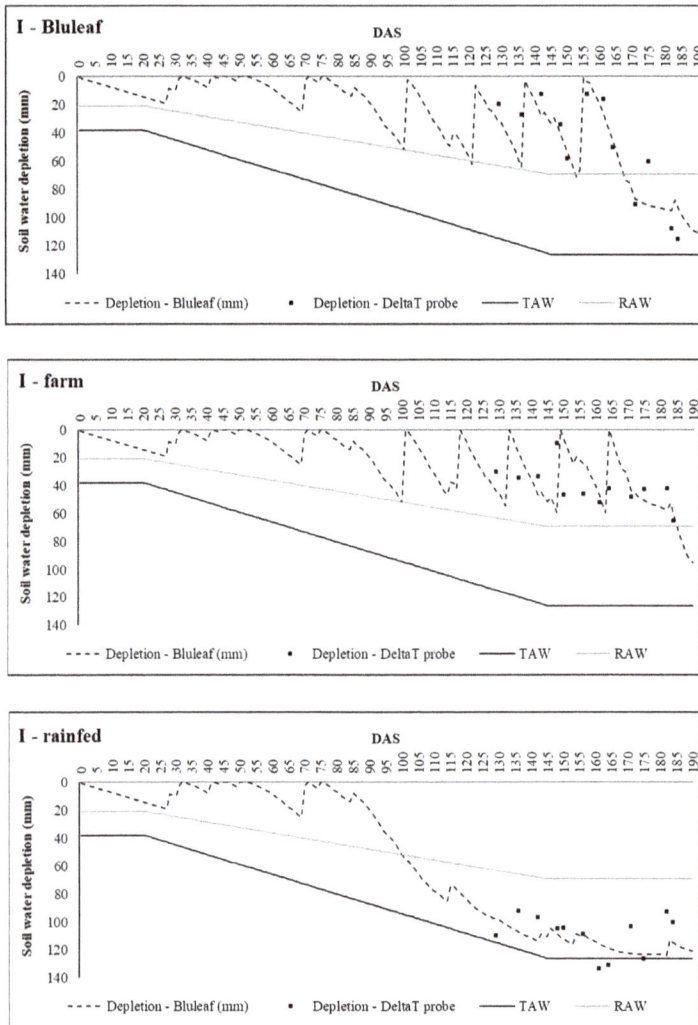

Figure 3. Simulated soil water depletion (Bluleaf®) versus measured soil water depletion (PR2 probes) for the three different treatments in the 2017–2018 wheat growing season. (I-Bluleaf, irrigated according to Bluleaf®; I-farm, irrigated according to farmer know-how; I-rainfed, no irrigation; TAW, total available water; RAW, readily available water).

3.3. Biomass, Yield, and Water Productivity

The main examined variables, particularly the final aboveground dry biomass (AGDB), the grain yield, and the yield water productivity (Y-WP), are reported in Table 6 for all treatments and both seasons. In addition, the grain yield, as affected by year (i.e., specific weather conditions) and treatment application, is presented in Figure 4.

Considering "year" as the source of variance, there was not a significant difference in terms of biomass production. However, the grain yield and the Y-WP were significantly different with higher mean values (4.81 t/ha and 1.01 kg/m^3, respectively, in the 2016–2017 growing season) than in 2017–2018 (3.22 t/ha and 0.59 kg/m^3). The reduction of yield in the second growing season could be explained by a higher air temperature during the crop development phase (an average of about 3.0 °C) in 2017–2018 than in 2016–2017, and a lack of precipitation in December (immediately after sowing) and in March. The high temperatures accelerated crop development and reduced biomass and yield in 2017–2018. Moreover, in the second season, the booting stage, which is sensitive to water stress, was anticipated in March when the precipitation was low and there was no irrigation. Thus, it could be an additional factor affecting the grain yield in that year.

Table 6. Aboveground dry biomass (AGDB), yield, and yield water productivity (Y-WP) as affected by year and irrigation treatment (Tr) application.

Source of Variation		AGDB (t/ha)	Grain Yield (t/ha)	Y-WP (kg/m^3)
Year (Y)		ns	*	*
	Y_{2017}	9.46 ± 2.05	4.81 ± 1.38 a	1.01 ± 0.11 a
	Y_{2018}	9.93 ± 1.43	3.22 ± 0.68 b	0.59 ± 0.10 b
Treatment (Tr)		**	***	ns
	I-Bluleaf	10.76 ± 0.89 a	4.63 ± 1.26 a	0.84 ± 0.29
	I-farm	10.73 ± 0.65 a	4.64 ± 1.22 a	0.74 ± 0.26
	I-rainfed	7.59 ± 0.99 b	2.77 ± 0.40 b	0.82 ± 0.18
Y × Tr		ns	*	ns

Note: ns, not significant; *, **, and ***, significant at P ≤ 0.05, P ≤ 0.01, and P ≤ 0.001, respectively. Means followed by a different letter in each column are significantly different according to the LSD test (P = 0.05).

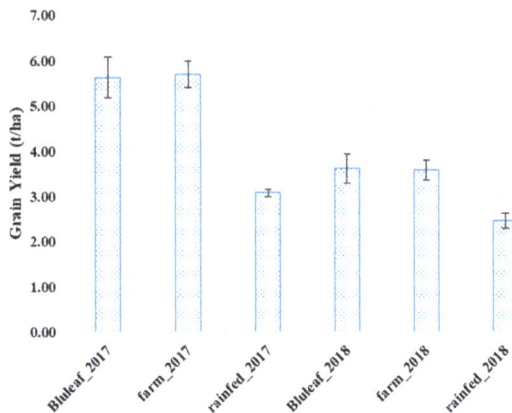

Figure 4. The Grain Yield in 2017 and 2018, as affected by Year and treatment application. The vertical bars indicate the standard error.

Considering "water" as the source of variance, the total dry biomass of wheat was significantly influenced by the water regime. In fact, I-Bluleaf and I-farm produced 41.76% and 41.37% higher biomass than I-rainfed, respectively. In addition, yield varied significantly in relation to water practice, with the values for I-Bluleaf and I-farm respectively 67.14% and 67.50% higher than I-rainfed. It should be mentioned that I-Bluleaf and I-farm were not significantly different in terms of final biomass and yield, and the values were very close.

Concerning the Y-WP that was expressed on the basis of grain yield, the results revealed that, although not significant, I-Bluleaf had a 13.5% higher Y-WP than I-farm. In fact, despite I-Bluleaf and I-farm presenting similar values of AGDB and grain yield, it is of great importance to emphasize that I-Bluleaf received 24.8% and 26.6% less water than I-farm, respectively, in seasons 2016–2017 and 2017–2018, which lead to a greater Y-WP.

The results confirmed that the application of supplemental irrigation and a limited amount of water increased the crop yield and water productivity of durum wheat [39–42].

3.4. Stress Indicators: Leaf Water Potential (LWP) and the Crop Water Stress Index (CWSI)

The midday leaf water potential was measured as illustrated in Figure 5, and corresponding to the time interval between the booting and grain-filling stages. In season 2016–2017, for the first measurement, the midday LWP presented almost similar values, between −24.5 and −25 bar in all treatments. However, for the second and third measurement, the LWP showed lower values (more negative), between −27.67 and −32 bar for rainfed treatments, which indicated water stress. In the case of irrigated treatments, the LWP was higher (around −20 bar) for the second measurement, and it was lower (around 26.5 bar) for the third measurement. The values of LWP were slightly higher for I-Bluleaf than for I-farm treatment. In season 2017–2018, the LWP was lower for rainfed treatment (between −28.67 and −33.17 bar) than for irrigated treatments (between −19.50 and −25.33 bar). There was a clear difference of midday LWP values between rainfed and irrigated treatments. Finally, the treatments I-Bluleaf and I-farm presented the same range of midday LWP, and no clear difference was noticed between the two treatments that mainly exhibited similar trends of LWP. The results obtained in this study are in agreement with the findings of [43], who reported similar LWP values for durum wheat grown under optimal and water-stress conditions in Avignon, France. These findings are also in agreement with the study of [42], who investigated the LWP variation under different water regimes for durum wheat and barley crops. This confirmed the validity of the Bluleaf® application for the irrigation scheduling of wheat, revealing that the plants were well-watered during the irrigation season.

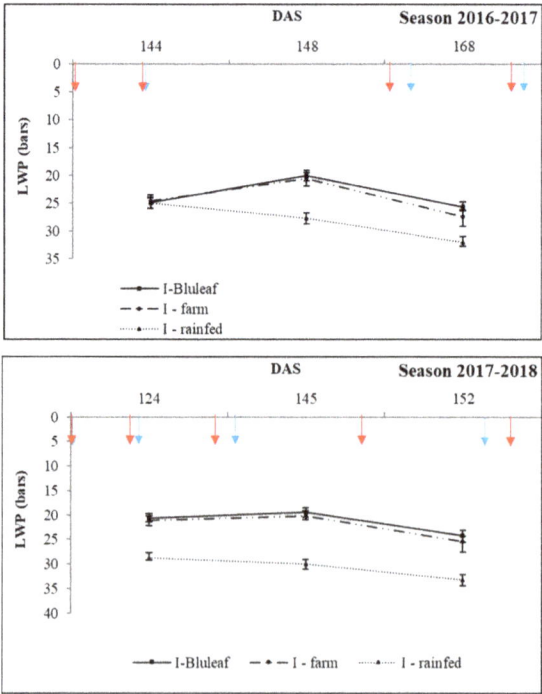

Figure 5. The midday leaf water potential (LWP) for the different treatments in the two wheat growing seasons. The blue arrows correspond to irrigation events for I-Bluleaf while the red arrows correspond to irrigation events for I-farm.

In Figure 6a, the canopy temperature minus air temperature (Tc – Ta) values are plotted against the corresponding values of air vapor pressure deficit (VPD), and delimited by the estimated upper and lower baselines.

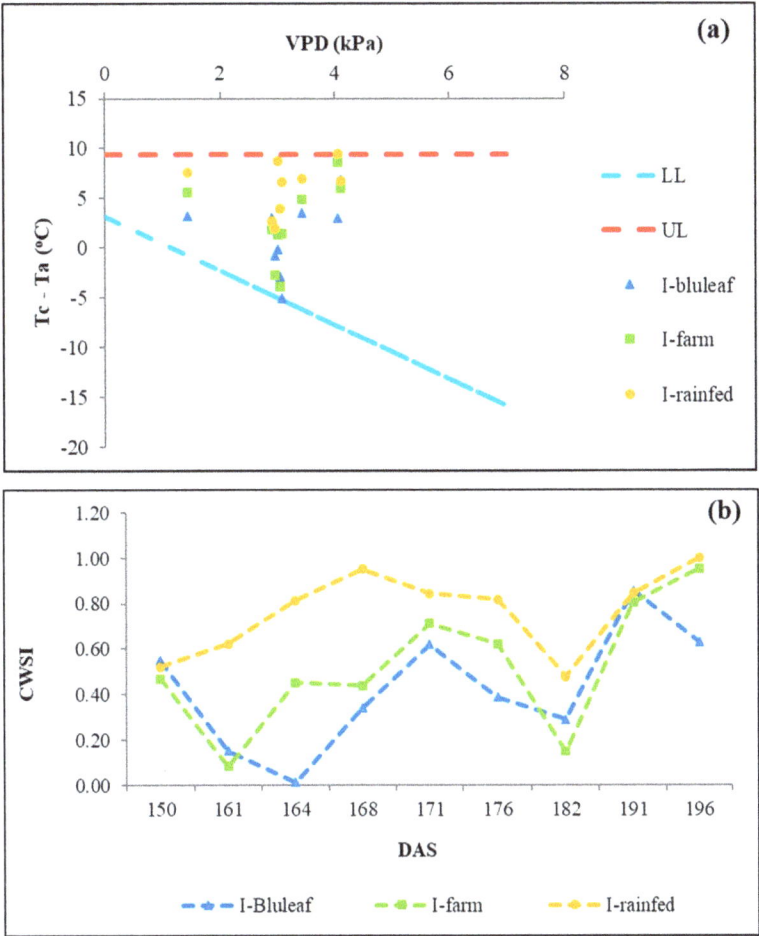

Figure 6. (**a**): A plot of canopy air temperature versus air vapor pressure deficit (VPD) with the corresponding upper and lower limits. (**b**): The Crop Water Stress Index (CWSI) variation for the three treatments during the wheat growing season.

The variation of the empirical CWSI under the three water regimes in 2016–2017 is shown in Figure 6b. Since the irrigation supply started late in the season (April), the first measurements showed a common stress (with a CWSI between 0.4 and 0.55). Later on, the CWSI decreased as a result of irrigation in I-Bluleaf and I-farm. The irrigated treatments followed a similar trend of CWSI in agreement with the adopted irrigation strategy. The CWSI threshold for irrigation can be adopted as 0.5, which is in agreement with other studies [44]. Therefore, the results on CWSI variation confirmed the validity of the Bluleaf® irrigation app for use in the irrigation management of durum wheat.

4. Conclusions and Perspectives

Field crops, such as wheat and other cereals, are important for stabilizing food security at the national level in the most of the MENA countries. In most cases, these crops are irrigated in order to reach a satisfactory level of production. Although water authorities are trying to develop regulations to limit water abstraction (especially from groundwater), their effective application requires support to help farmers in the implementation of new water-efficient technologies. For this reason, a user-friendly smartphone application for irrigation scheduling is of great importance in rationalizing water quantities.

The application of Bluleaf®, which was tested in this study, can provide daily customized irrigation scheduling for each farm at the irrigation sector scale using local meteorological data on a real-time basis, weather forecasting, soil, and crop data, and the hydraulic characteristics of the irrigation system. The results of the test indicated a considerable water saving of at least 1000 m^3/ha, which confirmed that the irrigation practices adopted by farmers are not efficient, cause a waste of water, nutrients, and energy, and trigger other environmental burdens. The assessment of Bluleaf®'s performance revealed that the presented tool could constitute a promising solution for irrigation scheduling with an "acceptable to very good" simulation of the soil water balance in the root zone.

Nowadays, there is an increasing demand for user-friendly platforms in the agricultural sector that should be able to provide relevant information for farmers by means of various types of sensors (local and remote) and modeling tools. Additionally, these applications could be used to produce reliably traceable records of farm activities and for the estimation of eco-efficiency [45], which is increasingly being demanded by the market.

The presented tool is one such platform, based on a smartphone application that allows for easy and instantaneous interaction with users. This provides additional insight into real-time irrigation management and permits more efficient water, nutrient, and energy use. Certainly, proper use of this and other similar tools depends on a concerted capacity-building effort and strong collaboration between researchers, extension service staff, and farmers. Further testing and calibration of other crop, soil, weather, and management conditions is needed.

Author Contributions: Data curation, M.T.A.S., I.J., S.S., and S.F.; Formal analysis, M.T.; Methodology, M.T.A.S.; Supervision, M.T.A.S., I.J., and M.T.; Writing (original draft), M.T.A.S.; Writing (review & editing), S.S., S.F., and M.T.

Funding: This research received no external funding.

Conflicts of Interest: The authors declare no conflict of interest.

References

1. Zotarelli, L.; Dukes, M.D.; Scholberg, J.M.S.; Femminella, K.; Munoz-Carpena, R. Irrigation scheduling for green bell peppers using capacitance soil moisture sensors. *J. Irrig. Drain. Eng.* **2011**, *137*, 73–81. [CrossRef]
2. Rosa, R.D.; Paredes, P.; Rodrigues, G.C.; Alves, I.; Fernando, R.M.; Pereira, L.S.; Allen, R.G. Implementing the dual crop coefficient approach in interactive software. 1. Background and computational strategy. *Agric. Water Manag.* **2012**, *103*, 8–24. [CrossRef]
3. Dobbs, N.A.; Migliaccio, K.W.; Li, Y.C.; Dukes, M.D.; Morgan, K.T. Evaluating irrigation applied and nitrogen leached using different smart irrigation technologies on bahiagrass (*Paspalum notatum*). *Irrig. Sci.* **2014**, *32*, 193–203. [CrossRef]
4. Pereira, L.S.; Paredes, P.; Rodrigues, G.C.; Neves, M. Modeling barley wateruse and evapotranspiration partitioning in two contrasting rainfall years. Assessing SIMDualKc and AquaCrop models. *Agric. Water Manag.* **2015**, *159*, 239–254. [CrossRef]
5. Todorovic, M.; Mehmeti, A.; Scardigno, A. Eco-efficiency of agricultural water systems: Methodological approach and assessment at meso-level scale. *J. Environ. Manag.* **2016**, *165*, 62–71. [CrossRef] [PubMed]
6. Mehmeti, A.; Todorovic, M.; Scardigno, A. Assessing the eco-efficiency improvement of Sinistra Ofanto irrigation scheme. *J. Clean. Prod.* **2016**, *138*, 208–216. [CrossRef]

7. Seidel, S.J.; Werisch, S.; Barfus, K.; Wagner, M.; Schutze, N.; Laber, H. Field Evaluation of Irrigation Scheduling Strategies using a Mechanistic Crop Growth Model. *Irrig. Drain.* **2016**, *65*, 214–223. [CrossRef]

8. Paredes, P.; D'Agostino, D.; Assif, M.; Todorovic, M.; Pereira, L.S. Assessing potato transpiration, yield and water productivity under various water regimes and planting dates using the FAO dual Kc approach. *Agric. Water Manag.* **2018**, *195*, 11–24. [CrossRef]

9. Adeyemi, O.; Grove, I.; Peets, S.; Domun, Y.; Norton, T. Dynamic Neural Network Modelling of Soil Moisture Content for Predictive Irrigation Scheduling. *Sensors* **2018**, *18*, 3408. [CrossRef]

10. Morgan, K.T.; Obreza, T.A.; Scholberg, J.M.; Parsons, L.R.; Wheaton, T.A. Citrus water uptake dynamics on a sandy Florida entisol. *SSSA J.* **2006**, *70*, 90–97. [CrossRef]

11. Kelley, L.; Miller, S. *Irrigation Scheduling Tools*; Irrigation Fact Sheet 3; East Lansing, M., Ed.; Michigan State University Extension: East Lansing, MI, USA, 2011. Available online: http://msue.anr.msu.edu/uploads/236/43605/resources/3_Scheduling_Tools.pdf (accessed on 22 January 2019).

12. Scherer, T. *Web-Based Irrigation Scheduler*; Fargo, N.D., Ed.; North Dakota State University: Fargo, ND, USA, 2014. Available online: www.ag.ndsu.edu/waterquality/documents/web-basedirrigation-scheduler (accessed on 22 January 2019).

13. Malamos, N.; Tsirogiannis, I.L.; Christofides, A.; Anastasiadis, S.; Vanino, S. Main Features and Application of a Web-based Irrigation Management Tool for the Plain of Arta. In Proceedings of the 7th International Conference on Information and Communication Technologies in Agriculture, Food and Environment (HAICTA 2015), Kavala, Greece, 17–20 September 2015; pp. 174–185.

14. Bartlett, A.C.; Andales, A.A.; Arabi, M.; Bauder, T.A. A smartphone app to extend use of a cloud-based irrigation scheduling tool. *Comput. Electron. Agric.* **2015**, *111*, 127–130. [CrossRef]

15. Migliaccio, K.W.; Morgan, K.T.; Vellidis, G.; Zotarelli, L.; Fraisse, C.; Zurweller, B.A.; Andreis, J.H.; Crane, J.H.; Rowland, D.L. Smartphone APPS for irrigation scheduling. *ASABE* **2016**, *59*, 291–301.

16. Vellidis, G.; Liakos, V.; Andreis, J.H.; Perry, C.D.; Porter, W.M.; Barnes, E.M.; Morgan, K.T.; Fraisse, C.; Migliaccio, K.W. Development and assessment of a smartphone application for irrigation scheduling in cotton. *Comput. Electron. Agric.* **2016**, *127*, 249–259. [CrossRef]

17. Todorovic, M.; Riezzo, E.E.; Buono, V.; Zippitelli, M.; Galiano, A.; Cantore, V. Hydro-Tech: An automated smart-tech Decision Support Tool for eco-efficient irrigation management. *Int. Agric. Eng. J.* **2016**, *25*, 44–56.

18. González Perea, R.; Fernández García, I.; Martin Arroyo, M.; Rodríguez Díaz, J.A.; Camacho Poyato, E.; Montesinos, P. Multiplatform application for precision irrigation scheduling in strawberries. *Agric. Water Manag.* **2017**, *183*, 194–201. [CrossRef]

19. Mbabazi, D.; Migliaccio, K.W.; Crane, J.H.; Fraisse, C.; Zotarelli, L.; Morgan, K.T.; Kiggundu, N. An irrigation schedule testing model for optimization of the Smartirrigation avocado app. *Agric. Water Manag.* **2017**, *179*, 390–400. [CrossRef]

20. Todorovic, M.; Mehmeti, A.; Cantore, V. Impact of different water and nitrogen inputs on the eco-efficiency of durum wheat cultivation in Mediterranean. *J. Clean. Prod.* **2018**, *183*, 1276–1288. [CrossRef]

21. Perry, C.; Steduto, P.; Allen, R.G.; Burt, C.M. Increasing productivity in irrigated agriculture: Agronomic constraints and hydrological realities. *Agric. Water Manag.* **2009**, *96*, 1517–1524. [CrossRef]

22. Molden, D.; Oweis, T.; Steduto, P.; Bidraban, P.; Hanjra, M.A.; Kijne, J. Improving agricultural water productivity: Between optimism and caution. *Agric. Water Manag.* **2010**, *97*, 528–535. [CrossRef]

23. Pereira, L.S.; Cordery, I.; Iacovides, I. Improved indicators of water use performance and productivity for sustainable water conservation and saving. *Agric. Water Manag.* **2012**, *108*, 39–51. [CrossRef]

24. Allen, R.G.; Pereira, L.S.; Raes, D.; Smith, M. *Crop Evapotranspiration*; Guidelines for Computing Crop Water Requirements; Irrigation and Drainage Paper 56; Food and Agriculture Organization: Rome, Italy, 1998.

25. Allen, R.G.; Pereira, L.S. Estimating crop coefficients from fraction ground cover and height. *Irrig. Sci.* **2009**, *28*, 17–34. [CrossRef]

26. Steduto, P.; Hsiao, T.C.; Fereres, E.; Raes, D. *Crop Yield Response to Water*; FAO Irrigation and Drainage Paper 66; FAO: Rome, Italy, 2012.

27. Todorovic, M. An Excel-based tool for real time irrigation management at field scale. In Proceedings of the International Symposium on Water and Land Management for Sustainable Irrigated Agriculture, Adana, Turkey, 4–8 April 2006; Çukurova University: Adana, Turkey, 2006.

28. Stewart, J.I.; Cuenca, R.H.; Pruit, W.O.; Hagan, R.M.; Tosso, J. *Determination and Utilisation of Water Production Functions for Principal California Crops*; W-67 California Contribution Project; University of California: Berkeley, CA, USA, 1977.

29. Saxton, K.E.; Rawls, W.J. Soil water characteristic estimates by texture and organic matter for hydrologic solutions. *Soil Sci. Soc. Am. J.* **2006**, *70*, 1569–1578. [CrossRef]

30. Asseng, S.; Milroy, S.; Bassu, S.; Abi Saab, M.T. Book Chapter: WHEAT. In *Crop Yield Response to Water. FAO Irrigation and Drainage Paper 66*; Steduto, P., Hsiao, T.C., Fereres, E., Raes, D., Eds.; Food and Agriculture Organization of UN: Rome, Italy, 2012; pp. 92–100.

31. Idso, S.B.; Jackson, R.D.; Pinter, P.J., Jr.; Reginato, R.J.; Hatfield, J.L. Normalizing the stress-degree-day parameter for environmental variability. *J. Agric. Meteorol.* **1981**, *24*, 45–55. [CrossRef]

32. Wilmot, C.J. Some comments on the evaluation of model performance. *Bull. Am. Meteorol. Soc.* **1982**, *64*, 1309–1313. [CrossRef]

33. Nash, J.E.; Sutcliffe, J.V. River flow forecasting through conceptual models: Part 1. A discussion of principles. *J. Hydrol.* **1970**, *10*, 282–290.

34. Moriasi, D.N.; Arnold, J.G.; Van Liew, M.W.; Bingner, R.L.; Harmel, R.D.; Veith, T.L. Model evaluation guidelines for systematic quantification of accuracy in watershed simulations. *Trans. ASABE* **2007**, *50*, 885–900. [CrossRef]

35. MoA/FAO. *Agricultural Census 2010*; MoA/FAO: Rome, Italy, 2010.

36. Beven, K.; Germann, P. Macropores and water flow in soils revisited. *Water Resour. Res.* **2013**, *49*, 3071–3092. [CrossRef]

37. Nimmo, J.R. Theory for source-responsive and free-surface film modeling ofunsaturated flow. *Vadose Zone J.* **2010**, *9*, 295–306. [CrossRef]

38. Nimmo, J.R.; Mitchell, L. Predicting vertically non sequential wetting patterns with a source-responsive model. *Vadose Zone J.* **2013**, *12*, 4. [CrossRef]

39. Oweis, T.; Pala, M.; Ryan, J. Stabilizing rainfed wheat yields with supplemental irrigation and nitrogen in a Mediterranean climate. *J. Agron.* **1998**, *90*, 672–681. [CrossRef]

40. Oweis, T.; Hachum, A.; Kijne, J. *Water Harvesting and Supplementary Irrigation for Improved Water Use Efficiency in Dry Areas*; SWIM Paper 7; Colombo, S.L., Ed.; International Water Management Institute: New Delhi, India, 1999.

41. Oweis, T.; Zhang, H.; Pala, M. Water use efficiency of rainfed and irrigated bread wheat in a Mediterranean environment. *J. Agron.* **2000**, *92*, 231–238. [CrossRef]

42. Albrizio, R.; Todorovic, M.; Matic, T.; Stellacci, A.M. Comparing the interactive effects of water and nitrogen on durum wheat and barley grown in a Mediterranean environment. *Field Crops Res.* **2010**, *115*, 179–190. [CrossRef]

43. Brisson, N.; Casals, M.-L. Leaf dynamics and crop water status throughout the growing cycle of durum wheat crops grown in two contrasted water budget conditions. *Agron. Sustain. Dev.* **2005**, *25*, 151–158. [CrossRef]

44. Zia, S.; Du, W.Y.; Spreer, W.; Spohrer, K.; He, X.K.; Müller, J. Assessing crop water stress of winter wheat by thermography under different irrigation regimes in North China Plain. *Int. J. Agric. Biol. Eng.* **2012**, *5*, 1–11.

45. Todorovic, M. Regional strategies in sustainable water management for irrigation: The eco-efficiency approach. In *Water Management for Sustainable Agriculture*; Oweis, T., Ed.; Burleigh Dodds Science Publishing Limited: Cambridge, UK, 2018; pp. 521–543.

water

MDPI

Article

Basin Irrigation Design with Multi-Criteria Analysis Focusing on Water Saving and Economic Returns: Application to Wheat in Hetao, Yellow River Basin

Qingfeng Miao [1], Haibin Shi [1,*], José M. Gonçalves [2,3] and Luis S. Pereira [2]

[1] College of Water Conservancy and Civil Engineering, Inner Mongolia Agricultural University, Hohhot 010018, China; imaumqf@imau.edu.cn

[2] Centro de Investigação em Agronomia, Alimentos, Ambiente e Paisagem (LEAF), Instituto Superior de Agronomia, Universidade de Lisboa, Tapada da Ajuda, 1349-017 Lisboa, Portugal; jmmg@esac.pt (J.M.G.); luis.santospereira@gmail.com (L.S.P.)

[3] Polytechnic of Coimbra, College of Agriculture, Bencanta, 3045-601 Coimbra, Portugal

* Correspondence: shi_haibin@sohu.com; Tel.: +86-0471-4300177

Received: 23 November 2017; Accepted: 5 January 2018; Published: 13 January 2018

Abstract: The sustainability of the Hetao Irrigation System, located in the water scarce upper Yellow River basin, is a priority considering the need for water saving, increased water productivity, and higher farmers' incomes. The upgrading of basin irrigation, the main irrigation method, is essential and includes the adoption of precise land levelling, cut-off management, improved water distribution uniformity, and adequate irrigation scheduling. With this objective, the current study focuses on upgrading wheat basin irrigation through improved design using a decision support system (DSS) model, which considers land parcels characteristics, crop irrigation scheduling, soil infiltration, hydraulic simulation, and environmental and economic impacts. Its use includes outlining water saving scenarios and ranking alternative designs through multi-criteria analysis considering the priorities of stakeholders. The best alternatives concern flat level basins with a 100 and 200 m length and inflow rates between 2 and 4 L s^{-1} m^{-1}. The total irrigation cost of designed projects, including the cost of the autumn irrigation, varies between 2400 and 3300 Yuan ha^{-1}; the major cost component is land levelling, corresponding to 33–46% of total irrigation costs. The economic land productivity is about 18,000 Yuan ha^{-1}. The DSS modelling defined guidelines to be applied by an extension service aimed at implementing better performing irrigation practices, and encouraged a good interaction between farmers and the Water Users Association, thus making easier the implementation of appropriate irrigation management programs.

Keywords: surface irrigation modelling; precise land levelling; irrigation systems design; beneficial water use; decision support systems (DSS); inflow rates; cut-off time

1. Introduction

The Yellow River basin is a water scarce region, with low water availability; about 500 m^3 per capita per year [1]. Agricultural irrigation corresponds to close to 90% of the total water use in the basin [2], and is particularly important in the Hetao Irrigation District. Climate change is likely a main cause for a decrease of water availability during the last decades [3–5], while increased water abstractions for industrial and domestic uses highly exacerbate water scarcity [2]. Forecasted scenarios on water resources allocation and use in the Yellow River basin point out the need to reduce irrigation water use [6].

The reduction of water resources allocation for irrigation due to the increased demand by non-agricultural sectors has unbalanced traditional irrigation management [6–9] and resulted in heavy challenges for the future use of water for irrigation. Thus, major priorities in the upper Yellow River basin refer to developing and implementing appropriate technologies aimed at water saving, improved water productivity, and increased farmer's incomes [7]. Since basin irrigation is the most used irrigation method in the 570,000 ha of irrigated land of Hetao, there is a requirement to focus on improving basin irrigation, which implies precise land levelling, appropriate inflow discharges and cut-off times, and adopting improved crop irrigation schedules [10–14], as well as improved supply management, namely modernizing canal conveyance and the distribution service aimed at upgrading water delivery and reducing runoff and seepage wastages [12,15–17]. When soils are saline [18], basin irrigation modernization also needs to consider salinity control practices [12,17,19], mainly adopting improved out of season autumn irrigation to appropriately leach the salts out of the soil's root zone. In addition, because water-saving practices impact groundwater dynamics [9,20], it is required that mutual influences of groundwater-irrigation are assessed and target groundwater depths are defined [21].

It is known that the performance of surface irrigation systems highly depends upon design and management [17,22–25]. Thus, appropriate design procedures and modelling are required because surface irrigation design based on simulation models produces results more easily, provides a better description of runoff and infiltration processes and the assessment of expected system performance, and results in the improved quality of design solutions [23–25]. In fact, there are a variety of factors that influence surface irrigation performance and shall be considered in the design: soil infiltration rates, hydraulic roughness, inflow discharge and duration, field length and slope, land shape, and surface micro-topography, as well as irrigation scheduling and control of salinity [17,22–29]. In addition, design must consider the negative impacts of irrigation, such as operational water losses by deep percolation and runoff out of the fields, water erosion due to surface flow, or relative to the control of fertilizer and chemical pollution and/or to control health impacts of irrigation with treated wastewater [29].

Decision support systems (DSS) aimed at the design of surface irrigation [30–32] may be the most adequate design tools because they may integrate data, models, and other calculation tools that focus on the various factors and impacts referred to above and, therefore, can be utilised for the easy creation of design alternatives. In addition, DSS integrate computational facilities that rank the considered design solutions, thus supporting design decision making. Ranking may be performed with multi-criteria analysis (MCA) [33], which identifies the compatibility among contradictory design criteria such as those relative to water saving and economic viability [34,35].

The application of DSS models for irrigation design easily associates issues relative to the hydraulics of the system with factors determining the irrigation performance and the environmental and economic results [30,36–39]. They are appropriate to be used in Hetao to assess solutions for water saving and economic returns for farmers because related design solutions depend upon numerous factors. However, design solutions cannot be field validated and model generated design alternatives have to be assessed and ranked to support the selection of the "best" solution, i.e., the alternative that better satisfies the design criteria. Thus, models and computational tools used by the DSS to create the design alternatives need to be parameterized using field data and validated models. Considering that good results were previously obtained with the DSS SADREG in surface irrigation design applied to wheat and cotton in Syria and Central Asia [31,34], this DSS model was selected for the current application to wheat in the Dengkou area of Hetao.

The objective of the present study was to assess and rank several design alternatives developed for basin irrigation applied to wheat in the experimental area of Dengkou, in the south-eastern part of Hetao. With this objective, the DSS model SADREG was used to create and rank various design alternatives. Ranking was performed with MCA considering two groups of design criteria, one relative to water saving and the other to economic returns. To appropriately parameterize SADREG, previous studies were developed during three years in the Dengkou area, one relative to basin irrigation [13], and the other to crop irrigation management [14], which provided field data for validating the simulation tools integrated in SADREG. Further objectives refer to preparing for extending the use of the DSS model for surface irrigation design to other areas of Hetao as a base for implementing irrigation water saving and modernization at the farm level.

2. Materials and Methods

2.1. The Study Area

The Hetao irrigation district is located in the upper reaches of the Yellow River and is one of the three largest irrigation districts of China, with 570,000 ha of irrigated land, and is 250 km long and 50 km wide (Figure 1). Hetao has an arid continental monsoon climate, with an average annual rainfall of 200 mm. According to the Köppen classification [40], the climate is BWk, with hot and dry summers and long, dry, and severely cold winters, which extend from November to March. Agriculture is only feasible during the spring-summer crop season and when irrigated.

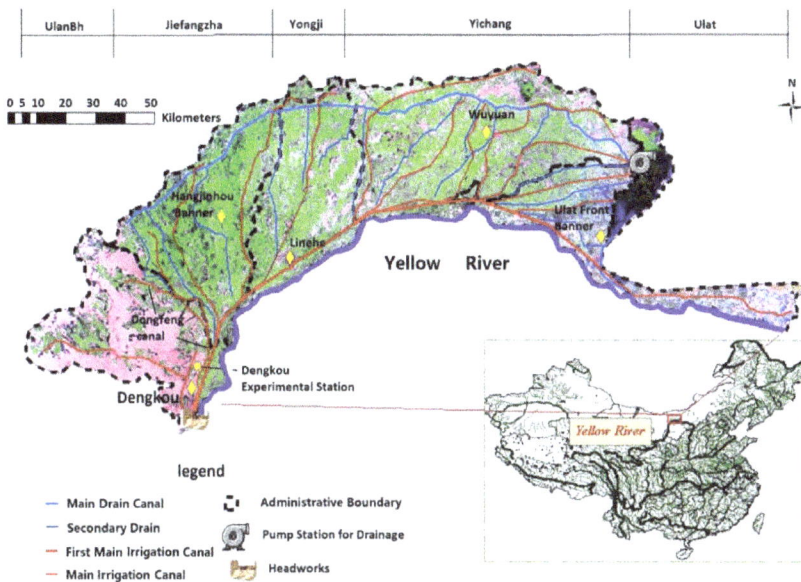

Figure 1. Map of the Hetao Irrigation District with representation of: (**a**) main irrigation drainage canals, (**b**) cropped areas, in green; (**c**) location of the Dengkou experimental area (Wu et al. [18]).

Water diverted from the Yellow River for irrigation totals about 5.2 billion m^3 year^{-1} [41,42]. To address water scarcity and the demand of non-irrigation water user' sectors, the Yellow River Water Conservancy Commission decided that diversions for irrigation in Hetao should be reduced to nearly 4.0 billion m^3 year^{-1}. However, a heavy reduction of water available for agriculture may have very important social impacts and a more flexible water allocation policy is advocated [43], limiting restrictions to irrigation water use to 10% in dry years. In addition, there are limitations in using groundwater due to salinity [18,44] and the presence of arsenic [45]. Only a small area is irrigated with groundwater and uses drip irrigation. Xu et al. [9,20] provided descriptions of the Hetao surface and groundwater systems and respective interactions as influenced by irrigation. Recent analysis of water use in Hetao includes a study on the water footprint of crop production [46] and an assessment of crop evapotranspiration dynamics [47].

The conveyance and distribution system of Hetao consists of seven levels of irrigation canals. The first main canal is gravity-fed from the Yellow River, with head-works located nearby Dengkou city and along the river (Figure 1). This first main canal supplies the main canals that flow South to North. Secondary drains also flow in the same direction into a main drain that flows West to East into a great lake. There are 61 areas served by the main and sub-main irrigation canals, named divisions, averaging ca. 9300 ha each. Branch and lower order canals of each division are managed by Water Users Associations (WUA), namely to deliver water to farms and to clean canals from deposited sediments carried by the irrigation water. Main and sub-main canals and drains, as well as the head-works and the drainage pumping station, are managed by the Hetao Administration, which is in charge of water allocation policies, water measurements, water fees, and the modernization of hydraulic structures.

An experimental area has been installed in Dengkou, in the upstream part of Hetao and where irrigation water is supplied by the Dongfeng canal (Figure 1). It is a main canal, 63 km long, designed for a discharge of 25 m^3 s^{-1}, and that supplies a division comprising 480 irrigation sectors and a total irrigated area of 16,300 ha. The branch and distributor canals that deliver water to the farms are currently being upgraded. A rotation delivery scheme is applied by the WUA, with fields supplied with a nearly constant discharge during each irrigation event. The application time is defined by the WUA depending upon the farmer's demand and the available water. The experimental area of Dengkou consists of a sector with 33.4 ha, with 394 land parcels and 210 farmers. The most common crops are maize, wheat, and sunflower, sometimes intercropped [14]. Experimentation is performed in the farmers' fields and respective irrigation management is agreed with the WUA.

2.2. Weather and Soils Data

Daily weather data, including precipitation, maximum and minimum air temperature (°C), maximum and minimum relative humidity (%), wind speed (m s^{-1}), and sunshine duration (h) were recorded in an automatic weather station (40°13′ N, 107°05′ E, and 1048 m elevation) located within the experimental area. Precipitation and grass reference evapotranspiration (ET$_0$) computed with the FAO-PM method [48] are shown in Figure 2, relative to the period of experimentation, 2010–2012. It may be noticed that rainfall is much smaller than ET$_0$ and highly varies with time. Differently, ET$_0$ varies little and its variability relates to the occurrence of rainfall in Summer.

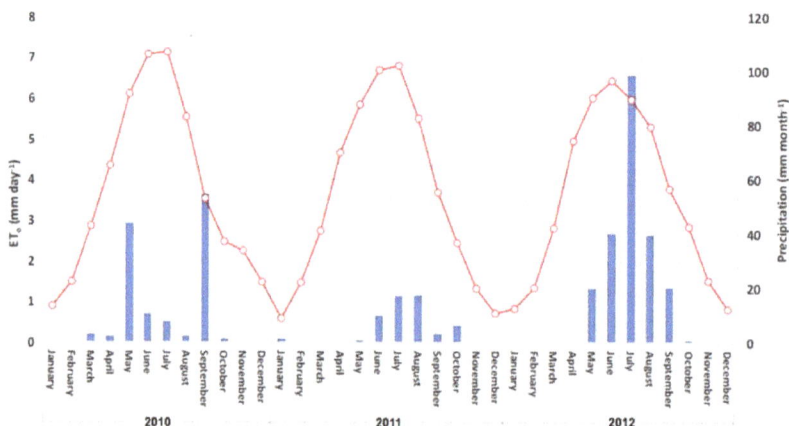

Figure 2. Monthly precipitation (❘) and daily reference evapotranspiration (–○–) observed in the Dengkou experimental area during the three years of experimentation, 2010–2012.

The soil in the experimental area is a siltic irragric Anthrosol [49] originated from sediments deposited by the Yellow River. Main soil textural and hydraulic properties were obtained from sampling in various locations within the study area. The texture of Dengkou soils is generally silt loamy in the upper layers, until a 0.60 m depth, and silt clay can be found below that depth. Soil textural and hydraulic properties were measured in a laboratory: texture was determined using a dry particle size analyser (HELOS RODOS, Sympatec, Clausthal-Zellerfeld, Germany); and the soil water retention curve was measured using a pressure plate extractor (model 1500F1, Soil Moisture Equipment Corp., Santa Barbara, CA, USA). Main soil physical characteristics are shown in Table 1. The total available soil water (TAW) ranges from 200 to 260 mm m^{-1}. Despite salinity occurring in large areas of Hetao, in the Dengkou area, the electrical conductivity of the saturation extract of the soil, EC_e, ranges from 0.11 to 1.58 dSm^{-1}. These values are smaller than the EC_e threshold relative to the referred main crops [48]. Moreover, such low salinity levels do not affect infiltration.

Table 1. Main soil textural and hydraulic properties of the soil in Dengkou (from [14]).

Depth (m)	Particle Size Distribution (%)			Soil Water Content (cm^3 cm^{-3})		
	Clay	Silt	Sand	At Saturation	At Field Capacity	At Wilting Point
0–0.20	23.0	76.7	0.3	0.47 ± 0.01	0.36 ± 0.01	0.16 ± 0.01
0.20–0.40	12.1	81.6	6.3	0.48 ± 0.01	0.37 ± 0.01	0.16 ± 0.01
0.40–0.60	14.6	84.2	1.2	0.49 ± 0.01	0.37 ± 0.01	0.16 ± 0.01
0.60–0.80	35.1	64.9	0.0	0.50 ± 0.02	0.39 ± 0.02	0.17 ± 0.02
0.80–1.00	42.5	57.5	0.0	0.52 ± 0.02	0.41 ± 0.02	0.18 ± 0.02

Following previous studies [13], infiltration is described by the Kostiakov equation [50]:

$$Z = K \cdot \tau^a \tag{1}$$

where Z is the cumulative infiltration depth (m), τ is the infiltration time (min), and K (m min^{-a}) and a (dimensionless) are empirically adjusted parameters. Because the duration of the water application in basin irrigation is small, the intake rate derived from Equation (1) does not significantly under-estimate infiltration at the end of irrigation [50]; thus, a third parameter representing the basic infiltration rate was not considered.

A large number of field measurements of irrigation events in Dengkou determined six standard infiltration curves [13]. Field basin infiltrometer tests [28] were performed, which provided a first estimation of the parameters K and a (Equation (1)). Later, these parameters were optimized using field advance and recession observations through the application of the inverse method [51,52] with the model SIRMOD [53]. This is a mechanistic surface irrigation simulation model aimed at the numerical solution of the Saint-Venant Equations for the conservation of mass and momentum [28].

Results of the infiltration tests performed have shown that the cumulative infiltration in silty soils increases with the precision of the adopted land levelling. Tests have also shown that infiltration rates decreased from the first to the following irrigation events, particularly for the precision levelled basins [13]. This behaviour was also observed in the nearby Huinong area [11] and by Bai et al. [26] in the North China Plain. It is likely due to the deposition of detached soil particles by the flowing water, which reduces infiltration due to the clogging of surface soil pores.

Six standard infiltration curves (SC-I to SC-VI) were obtained for the Dengkou silty soils from field observations [13]. For operational purposes, following the approach by Walker et al. [50], infiltration curves were clustered into three infiltration families (Figure 3) characterized by:

(i) High infiltration rates, when the first irrigation event is described by the observed curve SC-I, the second event to the curve SC-II, and the third and following events to the SC-III curve;

(ii) Medium infiltration rates, with the first event described by the curve SC-III, the second event by the curve SC-IV, and the third and following events by the curve SC-V; and

(iii) Low infiltration rates, where the first irrigation event is described by the curve SC-IV, the second event by the curve SC-V, and the third and later events by the curve SC-VI.

The K and a parameters relative to the infiltration curves are given in Figure 3. The distribution of high, medium, and low infiltration soils in the study area corresponds to 7–9%, 70–72%, and 20–22%, respectively. Further information on the methodologies applied is provided in Miao et al. [13,14].

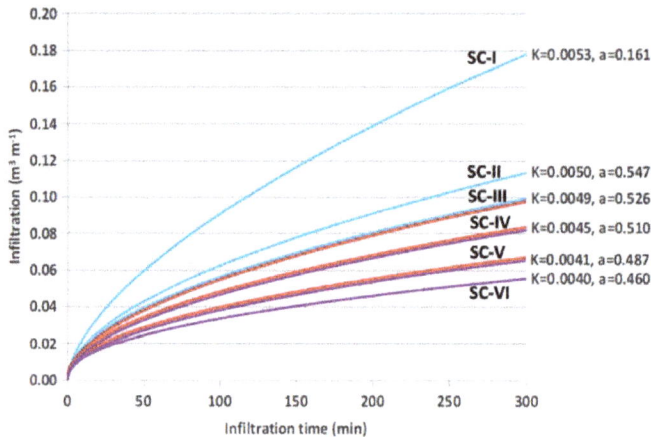

Figure 3. Cumulative infiltration curves SC-I to SC-VI characterizing the infiltration families relative to high, medium, and low infiltration rates represented with blue, red, and violet lines, respectively.

2.3. Irrigation and Yield Data

A survey on basin irrigation has been performed in the Dengkou experimental area [13] and the results have been used in the current design study. The typical sizes of field parcels and respective inflow discharges are summarized in Table 2. Irrigation basins commonly have a length of 50 m and widths ranging from 7 to 50 m. The wider fields often have more than one inlet. The field topography is flat but micro-topography is uneven.

Table 2. Field sizes and inflow discharges observed in Dengkou.

Field Sizes Length × Width (m)	Field Area (ha)	Inflow Discharge (Ls^{-1})	Field Sizes Occurrence in the Area (%)
50 × 10	0.05	10 ± 2	10
50 × 30	0.15	15 ± 3	30
65 × 20	0.30	20 ± 4	20
65 × 40	0.60	25 ± 5	10
100 × 25	0.25	15 ± 3	10
100 × 50	0.50	25 ± 5	20

A land levelling survey was also performed in several field parcels [13] with traditional and precise land levelling using the methodology described by Dedrick et al. [54]. The traditional land levelling (TL) consists of land smoothing using rudimentary equipment and practices and is performed by farmers without the support of topography surveys, hence resulting in a poor micro-topography and an uneven land surface. Differently, precise land levelling (PL) is performed with modern laser controlled levelling equipment, which provides a very regular soil surface with the target slopes. Precise land levelling is already well known in North China, including relative to its impacts on irrigation performance [25,26,55]. The latter were studied in Dengkou [13]. This study recognized the effects of inflow discharge control and irrigation scheduling on performance when aiming at water saving and higher crop yields.

The survey collected field data provided for calculating cut and fill volumes, and operation time and costs. Basin slopes were selected using terrain elevation data obtained by performing a field topography survey. It resulted in the following target slopes: zero cross slope for all cases, zero longitudinal slope (S_o) for level basins (LB), and S_o of 0.5‰ and 1.0‰ for graded basins (GB). The land levelling survey determined the following economic and technical parameters to be used in basin irrigation design:

(i) Operation time for maintenance: 3–4 and 4–5 h ha^{-1} for TL and PL basins, respectively, depending upon the distances between cut and fill sites, the power of the levelling equipment, the experience of the operator, and the soil conditions;

(ii) Hourly operation costs: 80 to 120 Yuan h^{-1} for TL basins and 200 to 240 Yuan h^{-1} for PL basins, with prices depending upon the equipment power and size;

(iii) Quality of land forming as expressed by the root mean of squared deviations between observed and target land elevations: 6 to 10 cm for TL basins and less than 4 cm for PL;

(iv) Frequency of land levelling maintenance: annual for both TL and PL basins.

According to observations [56], the spring wheat yield of 6000 kg ha^{-1} can be assumed for non-stressed conditions, i.e., full irrigation in a low salinity soil. The previous field and simulation study on the wheat crop irrigation scheduling [14] was used herein. The improved full irrigation scheduling implies a seasonal net irrigation depth of 300 mm with three irrigation events of 100 mm each. The irrigation practice includes, in addition to summer irrigation, out of season autumn irrigation, which is performed after the crop season and applies a high irrigation depth, usually close to 250 mm or larger, particularly when the soil salinity is high. The main objectives of autumn irrigation consist of: (a) controlling soil salinity through leaching the salts out of the root zone; (b) to improve soil structure, porosity, and permeability, due to the effect of successive soil water freezing and melting during winter; and (c) to store water in the soil to be available for cropping in early spring. Related processes are well known [57–59]. Following Li et al. [59,60], an irrigation depth of 230 mm was assumed adequate to leach the salts out of the root zone. Crop season and autumn irrigation data are summarized in Table 3.

Table 3. Water use components relative to current and improved wheat irrigation schedules (from [16]).

Irrigation Schedules	Number Irrigation Events	Net Target Irrigation Depth (mm)	Season Net Irrigation (mm)	Autumn Irrigation (mm)	Effective Rainfall (mm)	$ET_{c\,act}$ (mm)	$T_{c\,act}$ (mm)	Yield (kg ha^{-1})
Present	3	95	285	250	60	629	568	5880
Improved	3	100	300	230	60	644	574	6000

Notes: $ET_{c\,act}$—actual crop evapotranspiration; $T_{c\,act}$—actual crop transpiration.

To estimate the yield impacts of the various irrigation alternatives, the yield response curve proposed by Solomon [61] was adopted:

$$Y_a/Y_{max} = f(W_a/W_{max}) \tag{2}$$

where Y_a and Y_{max} are the actual and the maximum yield (kg ha^{-1}), respectively; W_a is the actual net irrigation water applied (mm); and W_{max} is the net water required to achieve Y_{max}. The respective parameterization is performed with the data in Table 4, which applies to Dengkou soils with low salinity and is based upon regionally observed data [56,62].

Table 4. Parameters used in the water-yield function.

W_a/W_{max}	0.5	0.75	1.0	1.5	2.0
Y_a/Y_{max}	0.40	0.70	1.0	0.95	0.90

2.4. Irrigation Performance

The irrigation performance indicators used consist of the distribution uniformity (DU, %) and the beneficial water use fraction (BWUF, %) [63]. DU is defined as:

$$DU = \frac{Z_{lq}}{Z_{avg}} \times 100 \tag{3}$$

where Z_{lq} is the average low quarter depth of water infiltrated (mm) and Z_{avg} is the average depth of water infiltrated in the whole irrigated field (mm). Two equations are used for BWUF to distinguish the cases of over-irrigation ($Z_{lq} > Z_{req}$) and under-irrigation ($Z_{lq} < Z_{req}$):

$$BWUF = \begin{cases} \frac{Z_{req}}{D} \times 100 & Z_{lq} > Z_{req} \\ \frac{Z_{lq}}{D} \times 100 & Z_{lq} < Z_{req} \end{cases} \tag{4}$$

where Z_{req} is the average depth (mm) required to refill the root zone in the quarter of the field having a higher soil water deficit, and D is the average water depth (mm) applied to the field. Z_{req} is estimated from measurements or using a soil water balance model. Z_{lq} and Z_{avg} are estimated from computing the depth of water infiltrated during the irrigation process with SIRMOD [53]. D is given by the product of the cut-off time (t_{co}) and the average inflow rate (Q_{in}).

The previous field basin irrigation evaluations [13] estimated DU and BWUF for both TL and PL basins. The results in Table 5 clearly show that traditional irrigation is not able to achieve water saving and salinity control since DU and BWUF indicators are far behind the potential values. Contrarily, precise levelling provides a high DU in modernized basins. However, BWUF values show a large gap between observed and potential values when irrigations follow traditional scheduling. High BWUF values are only attainable when adopting well-adjusted t_{co} and Q_{in}. Alternative values for t_{co} and Q_{in} were therefore used in model design simulations.

Table 5. DU and BWUF obtained from observations in traditional and precise levelled basins for various irrigation events and their potential values (from Miao et al. [13]).

Irrigation Event	DU (%)			BWUF (%)		
	Traditional	Improved Observed	Improved Potential	Traditional	Improved Observed	Improved Potential
1st	60	92	94	58	69	92
2nd	67	90	90	54	53	86
3rd	64	91	91	59	74	89

2.5. The DSS Model SADREG and Multi-Criteria Analysis

SADREG is a decision support system developed to assist the process of designing and planning improvements in farm surface irrigation systems as described by Gonçalves and Pereira [30]. Applications include those by Gonçalves et al. [31] to Fergana, Central Asia, and by Darouich et al. [34,64] to eastern Syria. The design component applies database information and produces a set of alternatives in agreement with the user options and field conditions. The hydraulic simulations are performed with the simulation model SIRMOD [53], which is incorporated in SADREG. The procedure for creating the required design alternatives and for their evaluation and ranking, follows various steps:

(i) Creating the "workspace" with main field data relative to soil water retention and soil infiltration rate characteristics, Manning's roughness coefficient, field length and width, longitudinal and cross slopes of the field, and land surface unevenness conditions;

(ii) Creating a "project" for selected combinations of workspace data, which characterizes the irrigation method, land levelling, crop data, field water supply, economic data, and number of units and outlets;

(iii) Grouping various projects to constitute a set of alternatives having different in-farm distribution systems and inflow rates Q_{in};

(iv) Application of associated model tools for land levelling design and for computing irrigation requirements, Z_{req} (mm);

(v) Performing the design simulation applying the SIRMOD model to every alternative, thus computing advance, wetting, and recession times, and infiltration depths, namely the average and the low quarter depths, Z_{avg} and Z_{lq} (mm), respectively;

(vi) Calculation of performance indicators for every alternative using the respective design data;

(vii) Application of multi-criteria analysis for ranking the alternatives according to the defined design criteria and user's priorities, based on the respective performance attributes.

The economic and labour input data reported for 2010 are presented in Table 6. At present, the farmer's irrigation fees in Hetao are not computed in terms of water use but just depend on the irrigated area. Fees vary from 600 to 800 Yuan ha^{-1} and cover WUA operation and maintenance (O&M) costs. The water price established by the Yellow River Commission for the water derived at the sector level ranges from 0.04 to 0.06 Yuan m^{-3}. In the current study, an irrigation cost averaging 700 Yuan ha^{-1} is considered, which is partitioned into a fixed cost of 420 Yuan ha^{-1} for O&M, and a variable cost for the gross water use was assumed with a water price of 0.05 Yuan m^{-3}.

Table 6. Economic and labour input data for wheat basin irrigation.

Type	Description	Value	Units
Distribution equipment	Non-lined canal cost (with field gate)	7 ± 1	Yuan m^{-1}
Irrigation water	Volumetric water cost	0.05 ± 0.01	Yuan m^{-3}
	Fixed cost per unit area	700 ± 100	Yuan ha^{-1}
Spring wheat crop	Yield price	3.0 ± 0.5	Yuan kg^{-1}
	Maximum yield	6000	kg ha^{-1}
	Production cost (excluding irrigation costs)	7.25 ± 0.2	10^3 Yuan ha^{-1}
Labour	Labour cost	11 ± 3	Yuan h^{-1}
Life-time	Building a non-lined distribution canal	1	year
Labour requirements	Operation of the non-lined canal	$t = t_{co}$	min
	Installing the non-lined canal	40	min 100 m^{-1}

The Manning's hydraulic roughness coefficient $n = 0.20$ m$^{-1/3}$ s was used for hydraulic simulations of basin irrigation when fields were cropped with wheat. That n value was obtained from a former field study in the same area [65]. Other studies [11,25,66] support the assumption that the parameter n essentially depends upon tillage and plants density, but not upon the land slope or land levelling precision. Pereira et al. [11] reported that n values slightly increase from the first to the last irrigation due to crop development. However, because impacts of n values on simulated basin irrigation performances are reported to be small [25,67], the constant value $n = 0.20$ m$^{-1/3}$ s was assumed in the current study.

The irrigation methods considered are the flat level basin (LB) and the flat graded basin (GB). Precise land levelling (PL) with a null cross slope was considered with three options for the longitudinal slope (S_o): zero level, 0.5‰, and 1.0‰.

The inflow rates (Q_{in}, L s^{-1} m^{-1} width) were defined in relation to the land parcel sizes (Table 7), i.e., the combination length-width, with a larger Q_{in} for longer basins.

Water 2018, 10, 67

Table 7. Basin sizes and related unit inflow rates for modernized design alternatives.

Inflow Rate Identifier	Length 50 m		Length 100 m		Length 200 m	
	Width (m)	Inflow Rate $(\text{L s}^{-1}\,\text{m}^{-1})$	Width (m)	Inflow Rate $(\text{L s}^{-1}\,\text{m}^{-1})$	Width (m)	Inflow Rate $(\text{L s}^{-1}\,\text{m}^{-1})$
(S)mall	30	0.5	30	1.0	30	2.0
(M)edium	15	1.0	15	2.0	15	3.0
(L)arge	7.5	2.0	7.5	3.0	7.5	4.0

The modernization scenarios are represented by projects and groups of alternatives as indicated in Figure 4. Projects refer to precision levelling (PL) and level basins (LB) and graded basins (GB) with S_o values of 0.5‰ and 1.0‰ slope. Groups refer to basin lengths, and alternatives are discriminated according to inflow rates S, M, and L, defined in Table 7 in combination with basin widths.

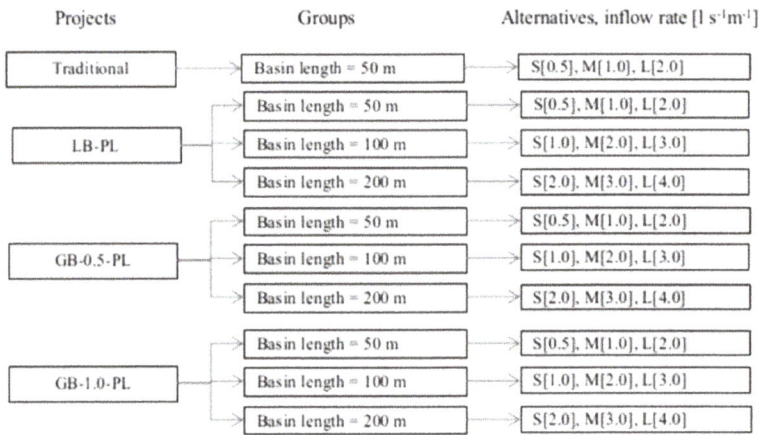

Figure 4. Structure of setting alternatives: projects (traditional, level basin LB and graded basins GB with slopes of 0.5‰ and 1.0‰), grouped for field lengths of 50, 100, and 200 m, and alternatives having different inflow rates and field widths.

The evaluation and selection of the design alternatives is the last task in the design decision making process. That selection is a multiple objective problem, for which a rational solution often requires multi-criteria analysis (MCA) to integrate different types of design attributes in a trade-off process, thus comparing adversative objectives or criteria [68,69]. In irrigation, adversative objectives generally refer to environmental, water saving, and economic criteria.

Linear utility functions were used for each criterion j:

$$U_j = \alpha_j x_j + \beta_j \tag{5}$$

which are normalized in the [0,1] interval, with zero for the most adverse and 1 for the most advantageous result. The slope parameter α is negative for criteria whose highest values are the worse, e.g., costs and water use, and is positive for criteria whose higher values are the best, e.g., water productivity. For each alternative, the linear weighted summation method [70,71] calculated the global utility that represents the integrated score performance of the considered alternative:

$$U_{glob} = \sum_{j=1}^{Nc} \lambda_j U_j \tag{6}$$

where U_{glob} is the global utility, scaled in the [0,1] interval; N_c is the number of criteria ($N_c = 7$ in this application); λ_j is the weight assigned to criterion j; and U_j is the utility relative to criterion j (Equation (5)). The decision criteria attributes, the respective weights, and the parameters of their linear utility functions (Equation (5)) are presented in Table 8. Overlapping or redundancy of criteria was checked and avoided.

Table 8. Criteria attributes, utility functions, and weights.

Decision Criteria Attributes	Symbol	Units	Weights (λ_j)	Utility Parameters (Equation (5))	
				α	β
Economic Productivity and Costs					
Economic land productivity	ELP	Yuan ha^{-1}	0.20	1.25×10^{-4}	-1.25
Fixed irrigation costs	FIC	Yuan ha^{-1}	0.10	-2.50×10^{-4}	1
Variable irrigation costs	VIC	Yuan ha^{-1}	0.10	-2.50×10^{-4}	1
Economic water productivity ratio	EWPR	ratio	0.10	0.25	1
Water Saving and Environment					
Total irrigation water use	IWU	m^3 ha^{-1}	0.20	-3.17×10^{-4}	1.95
Beneficial water use fraction	BWUF		0.15	1.818	-0.727
Irrigation water productivity	IWP	kg m^{-3}	0.15	-3.17×10^{-4}	1.95

The attributes relative to economic criteria are:

(i) Economic land productivity (ELP, €·ha^{-1}), the monetary yield value per unit of land;
(ii) Fixed irrigation costs (FIC, €·ha^{-1}), corresponding to investment costs per unit of land;
(iii) Variable irrigation costs (VIC, €·ha^{-1}), corresponding to the operation and maintenance costs per unit of land; and
(iv) Economic water productivity ratio (EWPR, dimensionless), defined as the ratio of total yield value to the total irrigation costs [63].

The attributes relative to water saving criteria consist of:

(i) Total irrigation water use (IWU, mm), corresponding to the seasonal gross irrigation depth (or irrigation volume, m^3);
(ii) Beneficial water use fraction (BWUF, dimensionless), defined with Equation (4); and
(iii) Irrigation water productivity (IWP, kg m^{-3}), ratio of total yield to IWU (in m^3).

Criteria are grouped into economic and water saving issues (Table 8); thus an economic utility (U_{EC}) and a water saving utility (U_{WS}) were defined:

$$U_{EC} = \sum_{i=1}^{Nc(EC)} \lambda_{ECi} U_{ECi} / \lambda_{EC} \tag{7}$$

$$U_{WS} = \sum_{i=1}^{Nc(WS)} \lambda_{WSi} U_{WSi} / \lambda_{WS} \tag{8}$$

where λ_{EC} and λ_{WS} are the sums of the weights relative to the economic and water saving criteria, respectively, with $\lambda_{EC} + \lambda_{WS} = 1.0$. The global utility corresponds to the sum of U_{EC} and U_{WS}:

$$U_{glob} = \sum_{j=1}^{Nc} \lambda_j U_j = \lambda_{EC} U_{EC} + \lambda_{WS} U_{WS}. \tag{9}$$

Solving Equation (9) in relation to U_{WS} results in

$$U_{WS} = \frac{U_{glob}}{\lambda_{WS}} - \frac{\lambda_{EC}}{\lambda_{WS}} U_{EC} \tag{10}$$

that allows a Cartesian representation of U_{glob} in the U_{EC}-U_{WS} Plane, and where the U_{glob} isolines are straight lines with slopes depending upon the values of λ_{EC} and λ_{WS}. That representation provides a better understanding of the impacts of water saving and economic results on the global utility.

To provide a sensitivity analysis of changes in the decision making priorities, several combinations of weights were used, starting when 20% of weights were assigned to farm economic results and 80% to water saving, and after considering pairs λ_{EC}-λ_{WS} of 40%-60%, 60%-40%, and 80%-20%. The weights λ_j used for the criteria attributes were consequently modified proportionally to those in Table 8, representing a balance between economic and water saving criteria (50% for each group).

3. Results and Discussion

3.1. Irrigation Water Use and Performance

Beneficial and non-beneficial water use (BWU and NBWU, $m^3\ ha^{-1}$) are compared in Figure 5 for 27 design alternatives. A smaller NBWU is achieved in level basin projects with a length of up to 200 m and for GB with 0.5‰ slopes when the length does not exceed 100 m. Naturally, a smaller NBWU corresponds to projects whose BWUF is higher and water productivity IWP is also higher (Table 9).

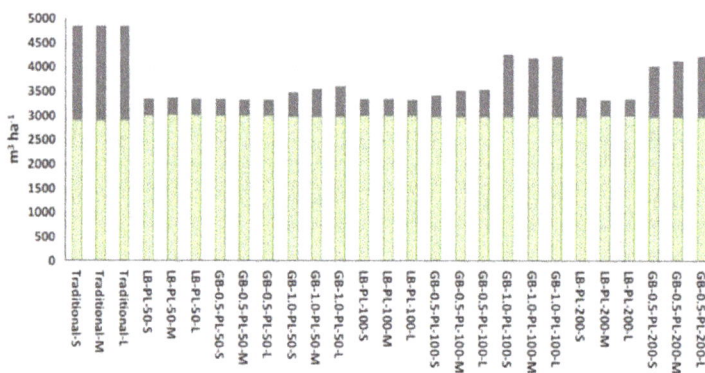

Figure 5. Beneficial and non-beneficial water use (BWU ▦ and NBWU ■, $m^3\ ha^{-1}$) relative to summer irrigation in a medium infiltration soil and considering level basins (LB) and graded basins (GB) with slopes of 1.0‰ and 0.5‰, with precise land levelling (PL); lengths of 50, 100, and 200 m; and inflow rates S, M, and L (defined in Table 7).

Main irrigation performance indicators relative to design alternatives for medium infiltration soils, the third irrigation event, and adopting an improved irrigation schedule, are presented in Table 9. These results indicate that:

(i) Modernization projects may achieve a BWUF of up to 90% when deep percolation (DP) is well controlled, thus when basin irrigation is highly improved relative to traditional systems, which have an average BWUF of 60% and DP of 40%.

(ii) Graded basin alternatives with a 200 m length and 1.0‰ slope are non-satisfactory (BWUF < 60% and DP > 40%), and hence were not considered further in the selection analysis.

(iii) The relationship between inflow rates (Q_{in}) and DU have been shown to be very weak, thus indicating that the magnitude of Q_{in} has small impacts on DU. However, as formerly observed for medium and low infiltration silty soils in China, inflow rates $Q_{in} \geq 2\ L\ s^{-1}\ m^{-1}$ are required [11,26,27]. This result also indicates that a high irrigation performance may be obtained with a flexible, varied inflow discharge. Differently, the cut-off time plays a crucial role in adjusting the applied depth to its target to avoid over-irrigation.

(iv) It was observed that S_o only slightly influences DU and that $S_o = 0‰$ (level basins) generally leads to a higher DU and BWUF. Poor results for long graded borders (L = 200 m) are likely due to the fact that $S_o > 0‰$ simultaneously favours advance and a long recession time resulting in high infiltration downstream, thus in a high DP and low DU.

(v) Irrigation water productivity is high (>1.8 kg m^{-3}) for LB and GB with $S_o = 0.5‰$ with a 50 m length, and for LB with a 100 m or 200 m length.

(vi) Relationships of field length (L) with DU are also quite weak, which may indicate that long basins are feasible but their performance depends on the combination S_o-Q_{in}.

Table 9. Irrigation performance and management indicators for various basin lengths, longitudinal slopes, and unit inflow rates in the case of medium infiltration soils.

Length (m)	Slope (‰)	Inflow Rate [1]	Project Alternatives	BWUF (%)	DU (%)	D (mm)	DP (mm)	t_{adv} (min)	t_{co} (min)	IWP [2] (kg m^{-3})
50	uneven	Small	Traditional-S	60.3	61.1	157	62	88	268	1.12
		Medium	Traditional-M	60.3	63.4	157	62	68	134	1.12
		Large	Traditional-L	60.2	67.0	157	62	48	67	1.12
	0	Small	LB-PL-50-S	90.1	90.0	111	11	96	186	1.80
		Medium	LB-PL-50-M	90.1	91.5	111	11	52	93	1.79
		Large	LB-PL-50-L	90.0	93.1	111	11	31	47	1.79
	0.5	Small	GB-0.5-PL-50-S	90.1	93.4	111	11	88	185	1.80
		Medium	GB-0.5-PL-50-M	90.0	92.7	111	11	48	93	1.80
		Large	GB-0.5-PL-50-L	89.9	92.5	111	11	29	46	1.80
	1.0	Small	GB-1.0-PL-50-S	84.4	85.8	118	18	83	197	1.72
		Medium	GB-1.0-PL-50-M	83.6	85.6	119	20	45	100	1.69
		Large	GB-1.0-PL-50-L	83.1	85.3	120	20	28	50	1.66
100	0	Small	LB-PL-100-S	90.1	93.3	111	11	136	186	1.79
		Medium	LB-PL-100-M	90.0	92.9	111	11	77	93	1.79
		Large	LB-PL-100-L	90.1	92.7	111	11	60	62	1.94
	0.5	Small	GB-0.5-PL-100-S	85.1	87.0	117	17	121	195	1.75
		Medium	GB-0.5-PL-100-M	83.9	86.1	119	19	71	99	1.70
		Large	GB-0.5-PL-100-L	84.0	85.5	119	19	54	66	1.69
	1.0	Small	GB-1.0-PL-100-S	65.1	70.5	153	53	111	255	1.35
		Medium	GB-1.0-PL-100-M	70.7	75.1	141	41	66	118	1.39
		Large	GB-1.0-PL-100-L	70.1	75.1	142	43	50	79	1.37
200	0	Small	LB-PL-200-S	90.1	93.0	111	11	198	186	1.77
		Medium	LB-PL-200-M	90.1	93.0	111	11	198	124	1.80
		Large	LB-PL-200-L	90.0	93.9	111	11	118	93	1.80
	0.5	Small	GB-0.05-PL-200-S	72.3	76.4	138	38	171	230	1.47
		Medium	GB-0.05-PL-200-M	71.7	75.8	139	39	128	154	1.42
		Large	GB0.05-PL-200-L	70.7	75.6	141	41	105	118	1.38
	1.0	Small	GB-0.10-PL-200-S	58.3	58.3	171	72	156	285	1.15
		Medium	GB-0.10-PL-200-M	46.5	55.3	214	115	117	238	0.77
		Large	GB-0.10-PL-200-L	51.1	59.7	195	95	97	163	0.83

Notes: BWUF—beneficial water use fraction; DU—distribution uniformity, gross irrigation depths; DP—deep percolation; t_{adv}—advance time; t_{co}—cut-off time; IWP—irrigation water productivity; LB and GB—level and graded basins; PL—precision land levelling; [1] inflow rates defined in Table 7; [2] IWP computed for irrigation water use during the crop season.

To assess the effects of soil infiltration on the irrigation performance, particularly on irrigation water use (IWU), the results relative to several alternatives applied to soils with high, medium, and low infiltration (Figure 3) are compared in Table 10. In general, IWU values for low and medium infiltration soils are similar, with differences not exceeding 8%. Differently, the IWU values of high infiltration soils are different of those for medium infiltration soils, particularly for basin lengths larger than 100 m. No feasible solutions were found for long basins in high infiltration soils due to excessive infiltration and very high percolation.

Table 10. Gross irrigation water use (IWU, m^3 ha^{-1}) during the crop season for several projects with medium inflow rates as influenced by soil infiltration—high, medium, and low (defined in Figure 3).

Projects	IWU (m^3 ha^{-1}) for Various Infiltration Rate Families		
	High	Medium	Low
LB-PL-50-M	3430	3360	3360
GB-0.5-PL-50-M	3410	3340	3330
GB-1.0-PL-50-M	3440	3550	3580
LB-PL-100-M	3530	3350	3350
GB-0.5-PL-100-M	3700	3520	3560
GB-1.0-PL-100-M	4010	4180	4160
LB-PL-200-M	4560	3330	3340
GB-0.5-PL-200-M	4670	4130	4160

Water saving, defined as the difference between the IWU of traditional irrigation (7350 m^3 ha^{-1}) and IWU relative to the retained alternatives, was estimated for the various design alternatives (Figure 6). IWU includes both the summer season and the autumn irrigation. The results in Figure 6 show that projects LB and GB with $S_o = 0.5‰$ provide annual water savings ranging from 1520 to 1740 m^3 ha^{-1}, i.e., 21% to 24% of IWU. LB perform slightly better than GB when the same basin length and inflow rate are considered. Water saving benefits of improved basin irrigation were reported in several studies carried out in China [7,11,13,27], Egypt [72], Portugal [30], Spain [73], and USA [74], supporting the assumption that water use decreases when the irrigation performance is improved.

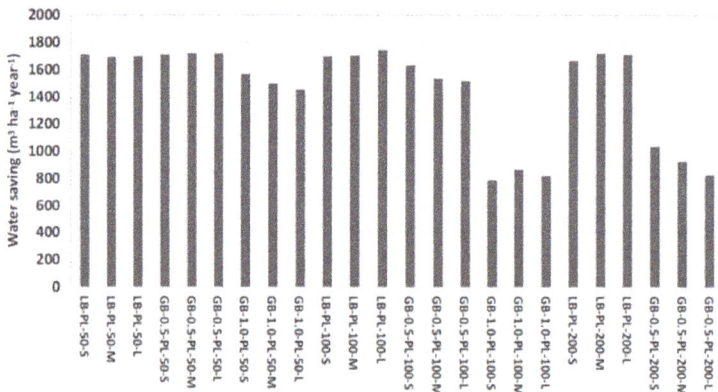

Figure 6. Water saving achievable by various design alternatives for level basins (LB) and graded basins (GB) with slopes of 1.0‰ and 0.5‰, with precise land levelling (PL); lengths of 50, 100, and 200 m; and inflow rates S, M, and L (as defined in Table 7) for a medium infiltration soil.

3.2. Economic Performance

The economic attributes relative to various design alternatives adopting a medium inflow rate are presented in Table 11. The total irrigation costs (TIC), relative to both the summer season and the autumn irrigation, vary between 2408 and 3292 Yuan ha^{-1}. Precision land levelling costs (1100 Yuan ha^{-1}) consist of the main component of TIC (33–46%). Considering that fixed water costs are 700 Yuan ha^{-1} (21–29% of TIC) and that variable water costs range from 269 to 328 Yuan ha^{-1}, i.e., only 8.8 to 13% of TIC, it can be inferred that the water operative costs are low and cannot play a large enough role as an incentive for water saving. Labour costs in traditional irrigation average 1330 Yuan ha^{-1}, about 45% of TIC, while for modernized systems, the labour costs are smaller, varying from 276 to 997 Yuan ha^{-1} (12–29% of TIC) because basin sizes are improved and processes of irrigation water supply require less manpower, and labour costs are lesser for level basins with a 200 m length. Nevertheless, further tests are required for long basins.

Table 11. Irrigation costs of various design alternatives and their cost components compared with yields, economic land productivity (ELP), and economic water productivity ratio (EWPR) of several projects for a medium infiltration soil and assuming medium inflow rates.

Design Alternatives	Components of the Irrigation Costs (Yuan ha^{-1})					Yield (kg ha^{-1})	ELP (Yuan ha^{-1})	EWPR
	Land Levelling	Supply System	Water	Labour	Total			
Traditional-M	350	200	1067	1335	2952	5447	16,342	5.54
LB-PL-50-M	1100	200	983	932	3215	6000	18,000	5.60
GB-0.5-PL-50-M	1100	200	982	925	3207	6000	18,000	5.61
GB-1.0-PL-50-M	1100	200	993	984	3277	5985	17,956	5.48
LB-PL-100-M	1100	100	982	555	2737	6000	18,000	6.58
GB-0.5-PL-100-M	1100	100	991	583	2774	5989	17,967	6.48
GB-1.0-PL-100-M	1100	100	1024	691	2915	5820	17,461	5.99
LB-PL-200-M	1100	50	982	276	2408	6000	18,000	7.47
GB-0.5-PL-200-M	1100	50	1021	341	2512	5850	17,549	6.99
GB-1.0-PL-200-M	1100	50	1106	479	2734	4494	13,483	4.93

Note: water and labour costs include summer and autumn irrigation.

The economic land productivity (ELP) and the economic water productivity ratio (EWPR) vary, as expected, from one project to another (Table 11). The minimal value for ELP refers to a 200 m long graded basin with $S_o = 1.0‰$ (GB-1.0-PL-200-M) because its irrigation performance is less good due to high percolation by downstream. The next poor performing alternative is the traditional one, with a low ELP value because yields are also less good. However, since the current ELP value is not much lower than for improved designs, it may be difficult to convince farmers to invest in modernization. Relative to EWPR, the best values (6.48 to 7.47) refer to basins of 100 or 200 m, both LB and GB, whose performance are good, yields are high, and total costs are low. It may be observed that issues relative to water saving (Figure 6) and to economic results (Table 12) are contradictory, so requiring the use of MCA to search for the best alternative designs, namely for other crops and different areas.

To evaluate the effects of increasing the irrigation costs, MCA was applied to rank the various projects under three scenarios of irrigation costs: (a) current costs; (b) the current costs increased by 20%; and (c) the current costs increased by 50%. The ranking of the alternatives was determined by the global utilities, U_{glob}. Results for the first 15 design alternatives for scenario (a) are presented in Table 12. It shows that the ranking based on U_{glob} values is different to the one that could result if considering economic results only, U_{EC}, due to the impact of water saving issues on U_{glob}, which evidences the need for associating U_{EC} and U_{WS} in the analysis. It is important to note that an increase of 20% of the irrigation costs does not produce a change of the ranking; contrarily, an increase of 50% produces a great change in ranking. For the current prices, the first six alternatives refer to level basins with lengths of 200 or 100 m without a great impact of the inflow rates. That ranking results from the fact that long basins have lower costs than the most common basins, with lengths of 50 m, which rank 8 to 14. However, adopting longer basins would lead to great changes in the structure of the irrigated fields when replacing the 50 m lengths with the 100 or 200 m long basins. The graded basins with a small slope ($S_o = 0.5‰$) rank 7 to 10; longer ones, of 200 m, are not included in the first 15 ranked projects.

Table 12. Impacts of increasing the total irrigation cost on the ranking of design alternatives determined by the global utilities U_{glob} considering an application to medium infiltration soils.

Design Alternatives	Current Total Irrigation Cost			Current Total Irrigation Cost Increased by 20%			Current Total Irrigation Cost Increased by 50%		
	U_{EC}	U_{glob}	Rank	U_{EC}	U_{glob}	Rank	U_{EC}	U_{glob}	Rank
LB-PL-200-M	0.86	0.87	1	0.83	0.85	1	0.64	0.76	9
LB-PL-200-L	0.86	0.87	2	0.83	0.85	2	0.64	0.75	10
LB-PL-200-S	0.86	0.85	3	0.83	0.84	3	0.64	0.76	8
LB-PL-100-L	0.81	0.85	4	0.77	0.83	4	0.72	0.81	5
LB-PL-100-M	0.80	0.84	5	0.76	0.82	5	0.71	0.75	12
LB-PL-100-S	0.80	0.83	6	0.76	0.81	6	0.71	0.75	15
GB-0.5-PL-100-S	0.80	0.82	7	0.76	0.80	7	0.73	0.84	1
GB-0.5-PL-50-L	0.74	0.80	8	0.69	0.78	8	0.79	0.82	4
GB-0.5-PL-50-M	0.74	0.80	9	0.69	0.78	9	0.79	0.83	2
GB-0.5-PL-50-S	0.74	0.80	10	0.69	0.78	10	0.79	0.83	3
LB-PL-50-S	0.74	0.80	11	0.69	0.78	11	0.73	0.63	21
LB-PL-50-L	0.74	0.80	12	0.69	0.78	12	0.75	0.68	19
LB-PL-50-M	0.74	0.80	13	0.69	0.78	13	0.74	0.65	20
GB-0.5-PL-100-M	0.80	0.79	14	0.76	0.77	14	0.72	0.80	6
GB-0.5-PL-100-L	0.80	0.79	15	0.75	0.77	15	0.72	0.79	7

It is interesting to note that level basins consist of the most commonly considered highly performing systems worldwide [17,23,25,29,75] and in China [7,11,13,26,27,56]. This common behaviour justifies why studies referring to graded basins are rare and point to solutions having small slopes [76].

Ranking is greatly modified if irrigation costs increase by 50%. The first six ranked projects are now graded basins with S_o of 0.5‰ and lengths of 50 m and 100 m. Level basins become low ranked and all basins 200 m long also fall in their ranking. The explanation for this is found when looking at the costs and the benefits, namely the economic land productivity (ELP) and the economic water productivity ratio (EWPR). These results indicate that the design approaches used may not be appropriate if large changes in irrigation costs occur, particularly if those increases are not well balanced with the economic benefits.

3.3. Ranking of Design Alternatives

The global utility, the economic utilities relative to the criteria attributes ELP, FIC, VIC, and EWPR, and the water saving utilities referring to the criteria attributes IWU, BWUF, and IWP are compared in Figure 7 for the best alternative of each project when applied to a medium infiltration soil. The results show that U_{glob} values relative to all design alternatives are significantly higher than U_{glob} characterizing the traditional systems. Nevertheless, the U values relative to costs FIC and VIC are similar for traditional and modernization systems, and the same occurs for ELP, particularly for short basin lengths (50 m). Differently, the U values referring to water use and saving, attributes IWU and BWUF, are much smaller than the corresponding U values for the modernization projects. These results evidence that modernization projects respond well to the need for adopting water saving irrigation but, simultaneously, make it clear that economic results are not advantageous enough for farmers to invest in modernization and water saving. These results show the need for economic incentives for farmers if the common attitude of "business as usual" is to be overcome.

Comparing the U_{glob} of the best modernization design alternatives, it can be observed that higher U_{glob} values are seen for level basins with a 100 and 200 m length. These high U_{glob} values result from high ELP and EWPR values and low costs, FIC and VIC, as indicated by the high U scores for these criteria attributes, particularly for the 200 m long basins. High U scores are also observed for criteria attributes IWU, BWUF, and IWP; the highest scores are for the 100 m length basins. The next ranking alternatives are for level and small slope (0.5‰) graded basins with 100 m lengths, whose utilities relative to the referred criteria are quite similar to those previously referred to. LB and GB basins of 50 m rank next because U scores relative to economic criteria are lower than the former. The last ranked alternative is GB with 1.0‰ slope, whose performance is affected by low U scores relative to IWU, BWUF, and IWP, i.e., to water saving.

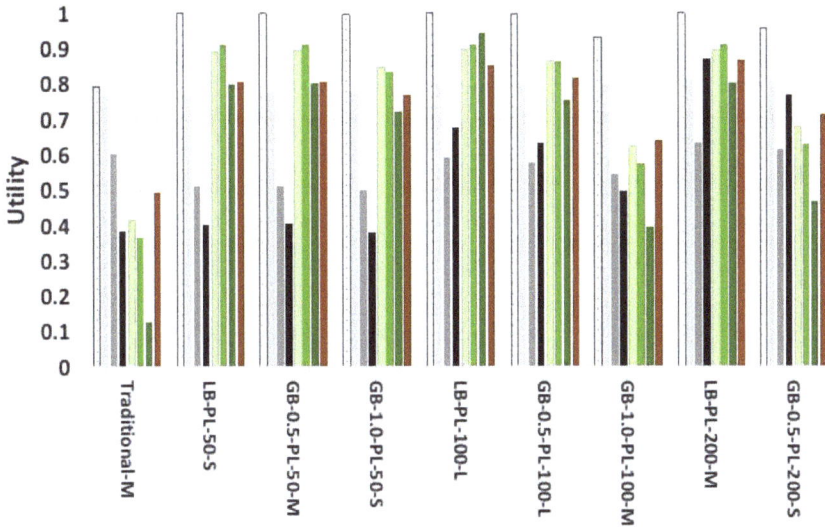

Figure 7. Comparison of the utilities relative to the considered criteria attributes ELP (⬜), FIC ( ), VIC (▦), EWPR (◼), IWU (▢), BWUF (◼), IWP (◼), and of the global utilities (◼) for the best project alternatives and the traditional one when applied to a medium infiltration soil.

A graphical evaluation of the best alternatives relative to a medium infiltration soil and adopting a medium inflow rate is presented in Figure 8. Using the representation of U_{glob} in the U_{EC}-U_{WS} Plane (Equation (10)), it is easy to understand through observing the contribution of the economic utilities (U_{EC}) and water saving utilities (U_{WS}) to the global utility (U_{glob}) of the considered design alternatives. The 0.60, 0.80, and 0.90 U_{glob} isolines were computed with Equation (10) for $\lambda_{EC} = \lambda_{WS} = 0.50$. The results in Figure 8 show that the best four ranked design alternatives have about the same U_{WS}, close to 0.87 (level basins with lengths of 50, 100, and 200 m, and a graded basin with $S_o = 0.5$‰ and length of 50 m), but have quite different U_{EC}, ranging from 0.75 to 0.87. Hence, the economic results dictated the ranking of those four design alternatives, with the 200 m long basins ranking first and the 50 m basins ranking fourth due to irrigation costs. It may also be observed that GB with $S_o = 1.0$‰ are the last ranked in terms of economic results, but the GB-1.0‰ for L = 50 m ranks high in terms of water saving, with a U_{WS} value of 0.78. These results indicate that using the Cartesian representation as in Figure 8 provides a good explanation on ranking, thus making easier the selection of alternatives by a decision maker.

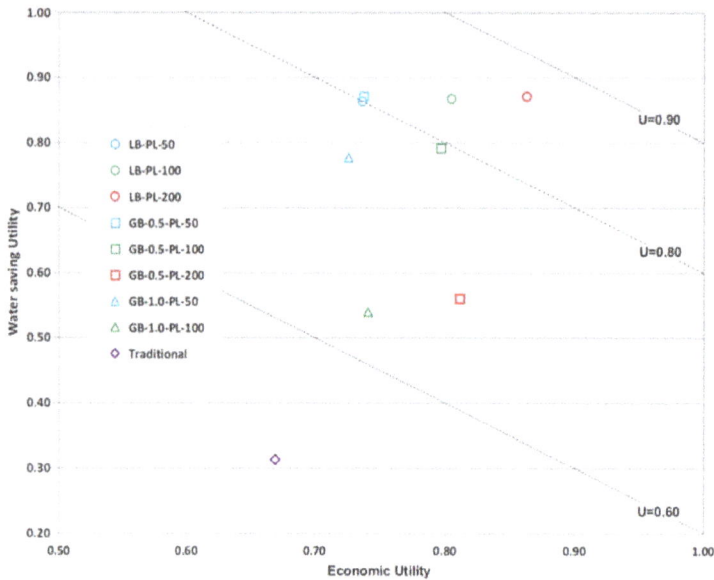

Figure 8. Evaluation of alternatives considering the joint effects of the economic and water saving utilities on the global utility of the best project alternatives for a medium infiltration soil and for medium inflow rates. Isolines of the global utility 0.60, 0.80, and 0.90 are included.

The results in Figures 7 and 8 make it evident that water saving and farm economics are contradictory, particularly when observing that the best rankings for U_{WS} are affected by the economic results when U_{glob} values are considered. This was also observed in former studies using MCA [31,34]. To better shed light on this behaviour, the rankings of the best 15 project alternatives were determined adopting weights attributed to economic issues and water saving different from 50%, which were adopted in the previous analysis. Various combinations of weights were therefore used (Table 13), starting when 20% of weights were assigned to farm economic results and 80% to water saving, and later considering different pairs λ_{EC}-λ_{WS} of 40%-60%, 60%-40%, and 80%-20%. The results clearly show that level basins have the best rankings for all priority combinations for basins with L = 100 m or 200 m. The length L = 100 m has a slight advantage when the priority is assigned to water saving and the length L = 200 m is more advantageous when prioritizing economic results. The graded border with a slope S_o = 0.5‰ and length L = 50 m is the following design alternative in the ranking when a higher priority is for water saving, followed by the LB projects of 50 m. If the priorities relative to farm economics increase, then longer GB are selected, always with the small slope of 0.5‰. A few cases are highlighted in Table 13 to give better visibility to changes in the ranking of selected project alternatives when the assigned priority weights change. These results indicate that the level basin is in general the best choice, that graded borders with S_o = 0.5‰ are feasible, and that basin lengths of 50 m, as at present, are also feasible, but have lower ranks than the 100 m long basins.

Table 13. Changes in ranking of the first 15 ranked design alternatives for a medium infiltration soil when priority scenarios relative to water saving and to farm economic results are modified. Some LB and GB alternatives are highlighted for easily observing changes in ranking when priorities are modified.

Rank	Large Priority for Water Saving	Small Priority for Water Saving	Small Priority for Economic Issues	Large Priority for Economic Issues
	Weights 20%-80%	Weights 40%-60%	Weights 60%-40%	Weights 80%-20%
1	LB-PL-100-L	LB-PL-200-M	LB-PL-200-M	LB-PL-200-M
2	LB-PL-200-M	LB-PL-200-L	LB-PL-200-L	LB-PL-200-L
3	LB-PL-200-L	LB-PL-100-L	LB-PL-200-S	LB-PL-200-S
4	LB-PL-100-M	LB-PL-200-S	LB-PL-100-L	LB-PL-100-L
5	LB-PL-100-S	LB-PL-100-M	LB-PL-100-M	LB-PL-100-M
6	LB-PL-200-S	LB-PL-100-S	LB-PL-100-S	LB-PL-100-S
7	GB-0.5-PL-50-L	GB-0.5-PL-100-S	GB-0.5-PL-100-S	GB-0.5-PL-100-S
8	GB-0.5-PL-50-M	GB-0.5-PL-50-L	GB-0.5-PL-100-M	GB-0.5-PL-100-M
9	GB-0.5-PL-50-S	GB-0.5-PL-50-M	GB-0.5-PL-100-L	GB-0.5-PL-100-L
10	LB-PL-50-S	GB-0.5-PL-50-S	GB-0.5-PL-50-L	GB-0.5-PL-200-S
11	LB-PL-50-L	LB-PL-50-S	GB-0.5-PL-50-M	GB-0.5-PL-50-L
12	LB-PL-50-M	LB-PL-50-L	GB-0.5-PL-50-S	GB-0.5-PL-50-M
13	GB-0.5-PL-100-S	LB-PL-50-M	LB-PL-50-S	GB-0.5-PL-50-S
14	GB-0.5-PL-100-M	GB-0.5-PL-100-M	LB-PL-50-L	LB-PL-50-S
15	GB-1.0-PL-50-S	GB-0.5-PL-100-L	LB-PL-50-M	LB-PL-50-L

4. Conclusions

This study aimed at the application of a DSS with multi-criteria analysis to design and rank alternative design solutions for water-saving basin irrigation of spring wheat in Hetao, currently focusing on its upstream area represented by the Dengkou experimental area. The DSS SADREG was successfully used, thus providing appropriate design information for implementing the modernization of basin irrigation in the area. It was able to generate and rank multiple design alternatives with a consideration of both water saving and economic returns. The adoption of a linear weighted sum MCA, where criteria weights can be changed to modify the priorities attributed to the criteria, was revealed to be appropriate for involving stakeholders in the decision process relative to the future implementation of best design alternatives. To support that implementation throughout Hetao, irrigation design alternatives must be assessed considering different crops and environmental conditions occurring in Hetao, namely relative to salinity.

The results of the study have shown a clear preference for level basins with a 100 m and 50 m length, particularly when priorities are assigned to water saving criteria because less water is then used and yields are high. Differently, project alternatives for longer basins, of 200 m, are highly ranked if the priorities are assigned to economic criteria because costs of modernized irrigation are reduced for long basins. It was evidenced that ranking for water saving or for farm economic results is contradictory, but MCA was able to rank project alternatives with a consideration of and associating both types of criteria, i.e., preferring one or another type of criteria does not imply that the other has to be excluded. Apparently, the best decision is to adopt level basins with a 50 m length, or graded basins with the same length and a small slope of 0.5‰ because these sizes would not require changes in the structure of the fields contrarily to adopt lengths of 100 or 200 m. In addition, selecting 50 m lengths agrees with the experience of the irrigators. Despite the fact that inflow rates do not play a major role, the results indicate that medium to large Q_{in} values should be selected taking into consideration the size of the irrigated fields.

This study provides an insight on the adequacy of modern basin irrigation in Hetao aimed at reducing/controlling the demand for irrigation water, which is a major requirement for the sustainability of irrigated agriculture. However, in addition to improving farm irrigation systems, it is definitely required to improve irrigation management, mainly irrigation scheduling. Yields and water use considered in the current study were determined for conditions of modern, rational irrigation scheduling; otherwise, results considered herein are not achievable. It is also required that the canal system operation is modernized to provide adequate delivery scheduling, i.e., that matches the irrigation demand of modernized irrigation scheduling. Considering the great pressure by the drip irrigation market, future studies are also required to appropriately compare surface and drip irrigation considering both water saving and economic criteria; otherwise, directions for change may be unclear.

The implementation of modern basin irrigation in Hetao, which implies a combination of surface irrigation design and management, definitely requires appropriate extension and training services for farmers and local irrigation canal operators, as well as institutional and economic incentives for farmers to invest in upgrading their irrigation systems. To support that implementation, irrigation design alternatives must be assessed considering different cropping systems and environmental conditions occurring in Hetao, namely relative to salinity.

Acknowledgments: This study was funded by the Key Project of National Natural Science Foundation, No. 51539005; The project of Inner Mongolia Agricultural University, No. 2017XQG-4, NDYB2016-23. National Natural Science Foundation, No. 51769024; National Thirteenth Project Five-Year Scientific and Technical Support Plan, No. 2016YFC0400205, contracted with the Ministry of Science and Technology, China; and The support of FCT through the research unit LEAF-Linking Landscape, Environment, Agriculture and Food (UID/AGR/04129/2013) is acknowledged.

Author Contributions: Qingfeng Miao, Haibin Shi, and José M. Gonçalves conceived and designed the experiments. Qingfeng Miao performed the field experiments under the supervision of the second author. Qingfeng Miao and José M. Gonçalves analysed the data and performed modelling with the advice of Luis S. Pereira. Qingfeng Miao and José M. Gonçalves wrote the paper with contributions of the other authors, and the final revision was performed by Luis S. Pereira.

Conflicts of Interest: The authors declare no conflict of interest.

References

1. Wang, H.; Yang, Z.; Saito, Y.; Liu, J.P.; Sun, X. Interannual and seasonal variation of the Huanghe (Yellow River) water discharge over the past 50 years: Connections to impacts from ENSO events and dams. *Glob. Planet. Chang.* **2006**, *50*, 212–225. [CrossRef]
2. Liu, C.M.; Xia, J. Water problems and hydrological research in the Yellow River and the Huai and Hai River basins of China. *Hydrol. Process* **2004**, *18*, 2197–2210. [CrossRef]
3. Fu, G.; Chen, S.; Liu, C.; Shepard, D. Hydro-climatic trends of the Yellow River Basin for the last 50 years. *Clim. Chang.* **2004**, *65*, 149–178. [CrossRef]
4. Yang, D.; Li, C.; Hu, H.; Lei, Z.; Yang, S.; Kusuda, T.; Koike, T.; Musiake, K. Analysis of water resources variability in the Yellow River of China during the last half century using historical data. *Water Resour. Res.* **2004**, *40*. [CrossRef]
5. Zhang, Q.; Peng, J.; Singh, V.P.; Li, J.; Chen, Y.D. Spatio-temporal variations of precipitation in arid and semiarid regions of China: The Yellow River basin as a case study. *Glob. Planet. Chang.* **2014**, *114*, 38–49. [CrossRef]
6. Yu, L. The Huanghe (Yellow) River: Recent changes and its countermeasures. *Cont. Shelf Res.* **2006**, *26*, 2281–2298. [CrossRef]
7. Pereira, L.S.; Cai, L.G.; Musy, A.; Minhas, P.S. (Eds.) *Water Savings in the Yellow River Basin. Issues and Decision Support Tools in Irrigation*; China Agriculture Press: Beijing, China, 2003; 290p.
8. Qu, Z.; Yang, X.; Huang, Y. Analysis and assessment of water-saving project of Hetao Irrigation District in Inner Mongolia. *Trans. Chin. Soc. Agric. Mach.* **2015**, *46*, 70–76. (In Chinese)
9. Xu, X.; Huang, G.H.; Qu, Z.Y.; Pereira, L.S. Assessing the groundwater dynamics and predicting impacts of water saving in the Hetao Irrigation District, Yellow River basin. *Agric. Water Manag.* **2010**, *98*, 301–313. [CrossRef]
10. Deng, X.-P.; Shan, L.; Zhang, H.; Turner, N.C. Improving agricultural water use efficiency in arid and semiarid areas of China. *Agric. Water Manag.* **2006**, *80*, 23–40. [CrossRef]

11. Pereira, L.S.; Gonçalves, J.M.; Dong, B.; Mao, Z.; Fang, S.X. Assessing basin irrigation and scheduling strategies for saving irrigation water and controlling salinity in the Upper Yellow River Basin, China. *Agric. Water Manag.* **2007**, *93*, 109–122. [CrossRef]

12. Pereira, L.S.; Cordery, I.; Iacovides, I. *Coping with Water Scarcity. Addressing the Challenges*; Springer: Dordrecht, The Netherlands, 2009; p. 382.

13. Miao, Q.; Shi, H.; Gonçalves, J.M.; Pereira, L.S. Field assessment of basin irrigation performance and water saving in Hetao, Yellow River basin: Issues to support irrigation systems modernisation. *Biosyst. Eng.* **2015**, *136*, 102–116. [CrossRef]

14. Miao, Q.; Rosa, R.D.; Shi, H.; Paredes, P.; Zhu, L.; Dai, J.; Gonçalves, J.M.; Pereira, L.S. Modeling water use, transpiration and soil evaporation of spring wheat-maize and spring wheat-sunflower relay intercropping using the dual crop coefficient approach. *Agric. Water Manag.* **2016**, *165*, 211–229. [CrossRef]

15. Roost, N.; Cui, Y.L.; Xie, C.B.; Huang, B. Water supply simulation for improved allocation and management. In *Water Savings in the Yellow River Basin. Issues and Decision Support Tools in Irrigation*; Pereira, L.S., Cai, L.G., Musy, A., Minhas, P.S., Eds.; China Agriculture Press: Beijing, China, 2003; pp. 275–290.

16. Gonçalves, J.M.; Pereira, L.S.; Fang, S.X.; Dong, B. Modelling and multicriteria analysis of water saving scenarios for an irrigation district in the Upper Yellow River Basin. *Agric. Water Manag.* **2007**, *94*, 93–108. [CrossRef]

17. Pereira, L.S.; Oweis, T.; Zairi, A. Irrigation management under water scarcity. *Agric. Water Manag.* **2002**, *57*, 175–206. [CrossRef]

18. Wu, J.W.; Vincent, B.; Yang, J.Z.; Bouarfa, S.; Vidal, A. Remote sensing monitoring of changes in soil salinity: A case study in Inner Mongolia, China. *Sensors* **2008**, *8*, 7035–7049. [CrossRef] [PubMed]

19. Hoffman, G.J.; Shalhevet, J. Controlling salinity. In *Design and Operation of Farm Irrigation Systems*, 2nd ed.; Hoffman, G.J., Evans, R.G., Jensen, M.E., Martin, D.L., Elliot, R.L., Eds.; American Society of Agricultural and Biosystems Engineers: St. Joseph, MI, USA, 2007; pp. 160–207.

20. Xu, X.; Huang, G.H.; Qu, Z.Y.; Pereira, L.S. Using MODFLOW and GIS to assess changes in groundwater dynamics in response to water saving measures in irrigation districts of the upper Yellow River basin. *Water Resour. Manag.* **2011**, *25*, 2035–2059. [CrossRef]

21. Xu, X.; Huang, G.H.; Sun, C.; Pereira, L.S.; Ramos, T.B.; Huang, Q.Z.; Hao, Y.Y. Assessing the effects of water table depth on water use, soil salinity and wheat yield: Searching for a target depth for irrigated areas in the upper Yellow River basin. *Agric. Water Manag.* **2013**, *125*, 46–60. [CrossRef]

22. Pereira, L.S. Higher performances through combined improvements in irrigation methods and scheduling: A discussion. *Agric. Water Manag.* **1999**, *40*, 153–169. [CrossRef]

23. Clemmens, A.J.; Walker, W.R.; Fangmeier, D.D.; Hardy, L.A. Design of surface systems. In *Design and Operation of Farm Irrigation Systems*, 2nd ed.; Hoffman, G.J., Evans, R.G., Jensen, M.E., Martin, D.L., Elliot, R.L., Eds.; American Society of Agricultural and Biosystems Engineers: St. Joseph, MI, USA, 2007; pp. 499–531.

24. Strelkoff, T.; Clemmens, A.J. Hydraulics of surface systems. In *Design and Operation of Farm Irrigation Systems*, 2nd ed.; Hoffman, G.J., Evans, R.G., Jensen, M.E., Martin, D.L., Elliot, R.L., Eds.; American Society of Agricultural and Biosystems Engineers: St. Joseph, MI, USA, 2007; pp. 436–498.

25. Reddy, J.M. Design of level basin irrigation systems for robust performance. *J. Irrig. Drain. Eng.* **2013**, *139*, 254–260. [CrossRef]

26. Bai, M.; Xu, D.; Li, Y.; Pereira, L.S. Stochastic modeling of basins microtopography: Analysis of spatial variability and model testing. *Irrig. Sci.* **2010**, *28*, 157–172. [CrossRef]

27. Bai, M.J.; Xu, D.; Li, Y.N.; Pereira, L.S. Impacts of spatial variability of basins microtopography on irrigation performance. *Irrig. Sci.* **2011**, *29*, 359–368. [CrossRef]

28. Walker, W.R.; Skogerboe, G.V. *Surface Irrigation. Theory and Practice*; Prentice-Hall: Englewood Cliffs, NJ, USA, 1987.

29. Pereira, L.S.; Gonçalves, J.M. Surface irrigation. In *Oxford Encyclopaedia of Agriculture and the Environment*; Oxford University Press: Oxford, UK, 2017.

30. Gonçalves, J.M.; Pereira, L.S. Decision support system for surface irrigation design. *J. Irrig. Drain. Eng.* **2009**, *135*, 343–356. [CrossRef]

31. Gonçalves, J.M.; Muga, A.P.; Horst, M.G.; Pereira, L.S. Furrow irrigation design with multicriteria analysis. *Biosyst. Eng.* **2011**, *109*, 266–275. [CrossRef]

32. Bazzani, G.M. An integrated decision support system for irrigation and water policy design: DSIRR. *Environ. Model. Softw.* **2005**, *20*, 153–163. [CrossRef]
33. Roy, B.; Bouyssou, D. *Aide Multicritère: Méthodes et Cas*; Economica: Paris, France, 1993.
34. Darouich, H.; Gonçalves, J.M.; Muga, A.; Pereira, L.S. Water saving vs. farm economics in cotton surface irrigation: An application of multicriteria analysis. *Agric. Water Manag.* **2012**, *115*, 223–231. [CrossRef]
35. Darouich, H.; Pedras, C.M.G.; Gonçalves, J.M.; Pereira, L.S. Drip vs. surface irrigation: A comparison focusing water saving and economic returns using multicriteria analysis applied to cotton. *Biosyst. Eng.* **2014**, *122*, 74–90. [CrossRef]
36. Oad, R.; Garcia, L.; Kinzli, K.-D.; Patterson, D.; Shafike, N. Decision support systems for efficient irrigation in the Middle Rio Grande Valley. *J. Irrig. Drain. Eng.* **2009**, *135*, 177–185. [CrossRef]
37. Pedras, C.M.G.; Pereira, L.S. Multicriteria analysis for design of microirrigation systems. Application and sensitivity analysis. *Agric. Water Manag.* **2009**, *96*, 702–710. [CrossRef]
38. Khadra, R.; Lamaddalena, N. Development of a decision support system for irrigation systems analysis. *Water Resour. Manag.* **2010**, *24*, 3279–3297. [CrossRef]
39. Barradas, J.M.M.; Matula, S.; Dolezal, F. A Decision Support System-Fertigation Simulator (DSS-FS) for design and optimization of sprinkler and drip irrigation systems. *Comput. Electron. Agric.* **2012**, *86*, 111–119. [CrossRef]
40. Kottek, M.; Grieser, J.; Beck, C.; Rudolf, B.; Rubel, F. World Map of the Köppen-Geiger climate classification updated. *Meteorol. Z.* **2006**, *15*, 259–263. [CrossRef]
41. IWC-IM. *Construction and Rehabilitation Planning Project for Water-Saving in Hetao Irrigation District of the Yellow River Basin, Inner Mongolia*; Institute of Water Conservancy and Hydropower of Inner Mongolia: Hohhot, China, 1999. (In Chinese)
42. Hetao Administration. *The Datasets of Evaluation of Irrigation Water Use Efficiency in Hetao Irrigation District of Inner Mongolia*; Hetao Irrigation Administration Bureau: Hohhot, China, 2003. (In Chinese)
43. Shao, W.W.; Yang, D.W.; Hu, H.P.; Sanbongi, K. Water resources allocation considering the water use flexible limit to water shortage—A case study in the Yellow River Basin of China. *Water Resour. Manag.* **2009**, *23*, 869–880. [CrossRef]
44. Yu, R.; Liu, T.; Xu, Y.; Zhu, C.; Zhang, Q.; Qu, Z.; Liu, X.; Li, C. Analysis of salinization dynamics by remote sensing in Hetao Irrigation District of North China. *Agric. Water Manag.* **2010**, *97*, 1952–1960. [CrossRef]
45. Guo, H.; Zhang, Y.; Jia, Y.; Zhao, K.; Li, Y.; Tang, X. Dynamic behaviors of water levels and arsenic concentration in shallow groundwater from the Hetao Basin, Inner Mongolia. *J. Geochem. Explor.* **2013**, *135*, 130–140. [CrossRef]
46. Sun, S.; Wu, P.; Wang, Y.; Zhao, X.; Liu, J.; Zhang, X. The impacts of interannual climate variability and agricultural inputs on water footprint of crop production in an irrigation district of China. *Sci. Total Environ.* **2013**, *444*, 498–507. [CrossRef] [PubMed]
47. Bai, L.; Cai, J.; Liu, Y.; Chen, H.; Zhang, B.; Huang, L. Responses of field evapotranspiration to the changes of cropping pattern and groundwater depth in large irrigation district of Yellow River basin. *Agric. Water Manag.* **2017**, *188*, 1–11. [CrossRef]
48. Allen, R.G.; Pereira, L.S.; Raes, D.; Smith, M. *Crop Evapotranspiration. Guidelines for Computing Crop Water Requirements*; Paper 56; FAO: Rome, Italy, 1998; p. 300.
49. IUSS Working Group WRB. *World Reference Base for Soil Resources 2014, Update 2015. International Soil Classification System for Naming Soils and Creating Legends for Soil Maps*; World Soil Resources Reports No. 106; FAO: Rome, Italy, 2015.
50. Walker, W.R.; Prestwich, C.; Spofford, T. Development of the revised USDA–NRCS intake families for surface irrigation. *Agric. Water Manag.* **2006**, *85*, 157–164. [CrossRef]
51. Katopodes, N.D.; Tang, J.H.; Clemmens, A.J. Estimation of surface irrigation parameters. *J. Irrig. Drain. Eng.* **1990**, *116*, 676–696. [CrossRef]
52. Horst, M.G.; Shamutalov, S.S.; Pereira, L.S.; Gonçalves, J.M. Field assessment of the water saving potential with furrow irrigation in Fergana, Aral Sea Basin. *Agric. Water Manag.* **2005**, *77*, 210–231. [CrossRef]
53. Walker, W.R. *SIRMOD—Surface Irrigation Modeling Software*; Utah State University: Logan, UT, USA, 1998.
54. Dedrick, A.R.; Gaddis, R.J.; Clark, A.W.; Moore, A.W. Land forming for irrigation. In *Design and Operation of Farm Irrigation Systems*, 2nd ed.; Hoffman, G.J., Evans, R.G., Jensen, M.E., Martin, D.L., Elliot, R.L., Eds.; American Society of Agricultural and Biosystems Engineers: St. Joseph, MI, USA, 2007; pp. 320–346.

55. Zheng, H.X.; Shi, H.; Guo, K.Z.; Hao, W.L. Study on influence of field surface slope on irrigation efficiency in combination with different irrigation parameters. *Agric. Res. Arid Areas* **2011**, *29*, 43–48. (In Chinese)

56. Bai, G.; Du, S.; Yu, J.; Zhang, P. Laser land leveling improve distribution of soil moisture and soil salinity and enhance spring wheat yield. *Trans. Chin. Soc. Agric. Eng.* **2013**, *29*, 125–134. (In Chinese)

57. Chamberlain, E.J.; Gow, A.J. Effect of freezing and thawing on the permeability and structure of soils. *Eng. Geol.* **1979**, *13*, 73–92. [CrossRef]

58. Wang, L.P.; Akae, T. Analysis of ground freezing process by unfrozen water content obtained from TDR data in Hetao Irrigation District of China. *J. Jpn. Soc. Soil Phys.* **2004**, *98*, 11–19. [CrossRef]

59. Li, R.; Shi, H.; Flerchinger, G.N.; Zou, C.; Li, Z. Modeling the effect of antecedent soil water storage on water and heat status in seasonally freezing and thawing agricultural soils. *Geoderma* **2013**, *206*, 70–74. [CrossRef]

60. Li, R.; Shi, H.; Akae, T.; Zhang, Y.; Zhang, X.; Flerchinger, G.N. Scheme of water saving irrigation in autumn based on SHAW model in Inner Mongolia Hetao irrigation district. *Trans. Chin. Soc. Agric. Eng.* **2010**, *26*, 31–36. (In Chinese)

61. Solomon, K.H. Yield related interpretations of irrigation uniformity and efficiency measures. *Irrig. Sci.* **1984**, *5*, 161–172. [CrossRef]

62. Tong, W.J.; Liu, Q.; Chen, F.; Wen, X.Y.; Li, Z.H.; Gao, H.Y. Salt tolerance of wheat in Hetao irrigation district and its ecological adaptable region. *Acta Agron. Sin.* **2012**, *38*, 909–913. (In Chinese) [CrossRef]

63. Pereira, L.S.; Cordery, I.; Iacovides, I. Improved indicators of water use performance and productivity for sustainable water conservation and saving. *Agric. Water Manag.* **2012**, *108*, 39–51. [CrossRef]

64. Darouich, H.; Cameira, M.R.; Gonçalves, J.M.; Paredes, P.; Pereira, L.S. Comparing sprinkler and surface irrigation for wheat using multi-criteria analysis: Water saving vs. economic returns. *Water* **2017**, *9*, 50. [CrossRef]

65. Zheng, H.; Shi, H.; Zhu, M.; Liu, H.; Gonçalves, J.M. Estimation of infiltration parameters for border irrigation based on SIRMOD method and modelling of border irrigation. *Trans. Chin. Soc. Agric. Eng.* **2009**, *25*, 29–34. (In Chinese)

66. Strelkoff, T.S.; Clemmens, A.J.; Bautista, E. Estimation of soil and crop hydraulic properties. *J. Irrig. Drain. Eng.* **2009**, *135*, 537–555. [CrossRef]

67. Nie, W.; Fei, L.; Ma, X. Estimated infiltration parameters and Manning roughness in border irrigation. *Irrig. Drain.* **2012**, *61*, 231–239.

68. Pomerol, J.C.; Barba-Romero, S. *Multicriterion Decision in Management: Principles and Practice*; Springer Science and Business Media: Berlin, Germany, 2000; Volume 25.

69. Ishizaka, A.; Nemery, P. *Multi-Criteria Decision Analysis: Methods and Software*; Wiley: Chichester, UK, 2013.

70. Stanimirovic, I.P.; Zlatanovic, M.L.; Petkovic, M.D. On the linear weighted sum method for multi-objective optimization. *Facta Acta Univ.* **2011**, *26*, 49–63.

71. Yan, H.B.; Huynh, V.N.; Nakamori, Y.; Murai, T. On prioritized weighted aggregation in multi-criteria decision making. *Expert Syst. Appl.* **2011**, *38*, 812–823. [CrossRef]

72. Clemmens, A.J.; El-Haddad, Z.; Strelkoff, T.S. Assessing the potential for modern surface irrigation in Egypt. *Trans. ASAE* **1999**, *42*, 995–1008. [CrossRef]

73. Playán, E.; Mateos, L. Modernization and optimization of irrigation systems to increase water productivity. *Agric. Water Manag.* **2006**, *80*, 100–116. [CrossRef]

74. Clemmens, A.J.; Allen, R.G.; Burt, C.M. Technical concepts related to conservation of irrigation and rainwater in agricultural systems. *Water Resour. Res.* **2008**, *44*, W00E03. [CrossRef]

75. Playán, E.; Faci, J.M.; Serreta, A. Modeling microtopography in basin irrigation. *J. Irrig. Drain Eng.* **1996**, *122*, 339–347. [CrossRef]

76. González, C.; Cervera, L.; Moret-Fernández, D. Basin irrigation design with longitudinal slope. *Agric. Water Manag.* **2011**, *98*, 1516–1522. [CrossRef]

water

MDPI

Article

Influence of Straw Amendment on Soil Physicochemical Properties and Crop Yield on a Consecutive Mollisol Slope in Northeastern China

Shaoliang Zhang *, Yao Wang and Qingsong Shen

College of Resources and Environment, Northeast Agricultural University, 600 Changjiang Street, XiangFang District, Harbin 150030, China; wangyao.1993@foxmail.com (Y.W.); shenqingsong0130@163.com (Q.S.)
* Correspondence: shaoliang.zhang@neau.edu.cn

Received: 8 March 2018; Accepted: 24 April 2018; Published: 26 April 2018

Abstract: Straw amendment (SA) can be used to increase soil organic matter and decrease dioxide carbon emissions. However, the impact of SA on the crop yield is still subject to debate in different areas. In this study, soil temperature (ST), soil moisture (SM), soil bulk density, soil-available-nitrogen (AN), soil-available-phosphorus (AP), crop growth and yield were measured in SA and NSA (no straw amendment) at slope positions of a 130-m-long consecutive Mollisol slope during the maize (*Zea mays*) growth stages in the North Temperate Zone of China. Compared with NSA, the influence of SA on ST and SM was not consistent, while AN typically increased on the top slope. However, SA conventionally increased AP, increased daily ST and monthly ST (2.4–7.9%), and increased daily SM and monthly SM (2.1–12.5%) on the back slope. SA increased crop yield by 1–9.8% and 55.6–105.1% on the top and back slopes, respectively. At the bottom, SA conventionally decreased ST (0.20–1.48 °C in July and August), SM (3.5–29.6% from May to August), AN and AP, and decreased crop yield (4.1–30.6%). In conclusion, SA changed the equilibrium of ST and SM, influenced the dynamics of AN and AP on the consecutive slopes, and increased yield on both the top and back slopes but decreased yield at the bottom.

Keywords: soil moisture; soil temperature; soil nutrient; crop yield; Corn; Black soil

1. Introduction

Crop residues are the main by-products in agriculture, especially in maize cultivation [1]. The management of crop residues significantly influences soil quantity and environment [2–4]. In terms of management strategies, several options are available, including burning, incorporation, direct drilling-in of surface residues, under-sowing crops, and baling and removing of crop residues for use as stock feed, building material, fuel, livestock bedding, composting for mushroom cultivation, bedding for plants, and sources of chemicals [5]. Straw amendment (residue mulch or residue incorporation) added to soils is usually considered the most efficient method to increase soil organic carbon (SOC) and decrease salinity, gas emissions, and soil loss [3,5–11]. However, the impacts of straw amendment (SA) on crop yield are still subject to debate in different areas [9–15], especially considering different regions under different soil types and climatic conditions.

Effects of SA on the dynamics of soil water, runoff, infiltration, soil temperature, nutrients and soil water use and productivity are contradictory because of soil characteristics, climate and crop and soil management practices which vary enormously [9,12–14]. Generally, residue incorporation reduces surface runoff, increases SM, crop transpiration and WUE (crop water use efficiency), and decreases ST; the extent to which this happens is mainly determined by the growth stage and the amount of residue returned to the soil [9,16–19]. Furthermore, intensified straw application significantly reduces evapotranspiration at the grain-filling to the maturity stages, and significantly increases surface SM at the grain-filling stage and considerably improves rainfall-use efficiency (RUE) during the whole growth period [20]. However, in another study, ditch-buried straw return decreases SM, but increases mean ST in a humid, mid-subtropical monsoon climate and increases soil microbial activity [21]. In addition, the positive effects of residue incorporation on water balance and crop yield are more pronounced at dry sites than at wet sites. Residue incorporation can be an effective adaptation option for mitigating the impacts of climate change on winter crops by improving WUE, and it is particularly effective in narrow-leaf cropping systems in hot and dry environments [11]. Compared with NSA (no straw amendment), SA typically decreases the uptake of soil nitrogen, while higher levels of mineral N addition can mitigate the harmful effects of SA [3,15,22]. It is also reported that plant growth can be immediately enhanced by plant residues with a low C/N ratio, even under drought-stress conditions [23]. Furthermore, SA increases soil organic matter, decreases soil bulk density, reduces the stability of aggregates, water infiltration, saturated/unsaturated hydraulic conductivity, and air permeability, and thereby increases soil and water loss; SA changes the dynamics of SM and ST, and influences the activity of microbe and nutrient transformation [3]. However, most studies evaluating the impacts of straw amendment on soil physicochemical properties have been mainly carried out on flat or sloping plots [12,21,24], while the precise impacts of straw amendment in fields with consecutive slopes remain unknown.

In Northeast China, crop residues are typically burned after harvest, a practice which has been applied for several hundred years. In recent years, SA has been widely adopted in this Mollisol region, and SA typically decreases the crop yield in the flat area [12]. However, it is not clear how straw amendment influences crop yield in a sloped field. It is the premise of this study that crop yield influenced by SA was different between a sloped and flat field. In the study, crop yield and its main driving factors, e.g., SM, ST, soil available nutrients, were monitored in a field with consecutive slopes over a period of two years. We aim to determine how SA changes soil physicochemical properties and influences crop growth and yield in the Mollisol region of Northeast China. The information from the results of the impacts of straw amendment on crop yield can be used for the development of local residue management and can also be transferred to other Mollisol regions, such as the mid-latitudes of North America, Eurasia, and South America with larger Mollisol areas [25].

2. Materials and Methods

2.1. Research Location

Our research site (45°45′35.82″ N, 126°54′34.35″ E) is in the Xiangyang experimental station of Northeast Agricultural University (45°45′27″–45°46′33″ N, 126°35′44″–126°55′54″ E), Harbin city, Heilongjiang province in Northeast China (Figure 1). The experimental station is located in the North Temperate Zone of China. This area has a continental monsoon climate with hot and rainy conditions in the summer, and cold and arid weather in the winter. The average annual temperature is 3.5 °C and annual sunshine duration averages between 2400 and 2500 h. Total annual solar radiation is 4718.7 MJ m^{-2}, and the annual average available accumulated temperature ($\geq$10 °C) is 2558 °C. Average annual precipitation (rainfall and snow) is 534 mm (coefficient of variation = 2%) and annual rainfall is 497 mm (coefficient of variation = 2%) over a 20-year period. Average annual rainfall is 457.3 mm (coefficient of variation = 3.8%), with 65% of rainfall falling in June, July and August over the 10 years from 2003 to 2013 (Figure 2).

Formation of soils in the study area began during the Quaternary period on loess deposits under natural grasses and now have a rich, dark organic layer (mean depth of 30 cm) and are classified as Black soil by the Chinese Soil Taxonomy (CST), Phaeozems by the World Reference Base for Soil Resources (WRB), and Mollisols by the US Soil Taxonomy (USST) [26]. Generally, the most slopes are inclined at less than 5°, but are more 100 meters in length. These soils have a silty clay-loam texture with high productivity. The mean soil physicochemical properties of the research sites are shown in Table 1.

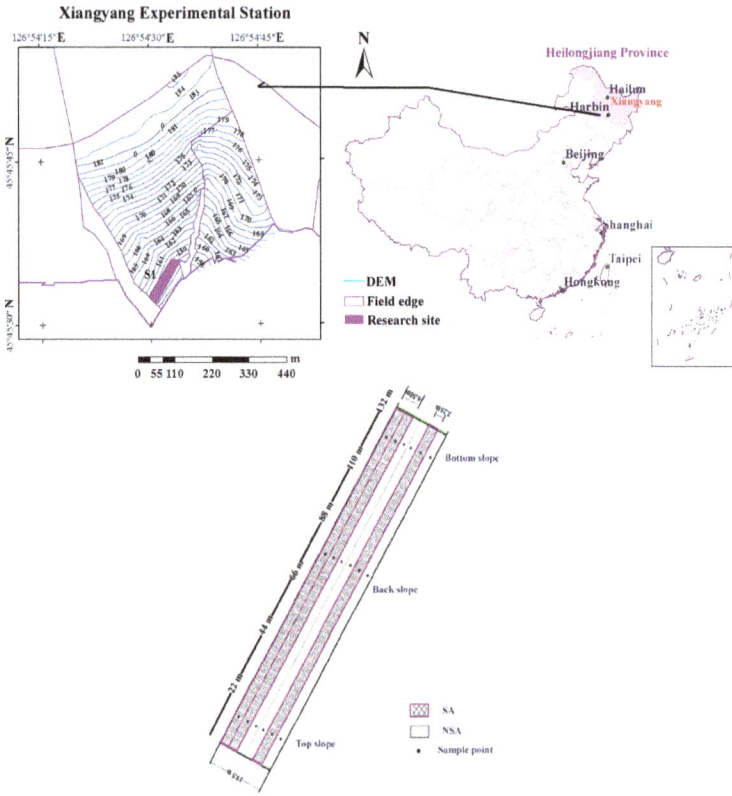

Figure 1. Location of the research area in China, and experimental plot design. Straw Amendment and No Straw Amendment are represented as SA and NSA, respectively.

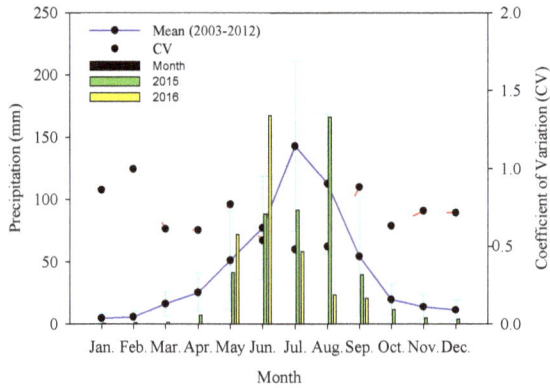

Figure 2. Monthly mean precipitation in 2015 and 2016, and monthly mean precipitation and associated variation from January to December in 2003–2013 in Harbin city, Heilongjiang province, China.

Table 1. Soil physicochemical properties before planting in the experimental fields.

Slope Position	Soil Depth (cm)	Organic Matter (g kg^{-1})	Total Nitrogen (g kg^{-1})	Total Phosphorus (g kg^{-1})	Bulk Density (g cm^{-3})	Total Porosity (%)	Field Capacity (cm^3 cm^{-3})	Wilting Point (cm^3 cm^{-3})	Saturated Water (cm^3 cm^{-3})	pH
Top ($n = 12$)	0–10	36.7 ± 4.2	1.6 ± 0.3	0.29 ± 0.04	0.98 ± 0.11	63.02 ± 5.01	27.5 ± 0.4	8.2 ± 0.9	54.2 ± 0.5	6.15 ± 0.04
	10–20	36.7 ± 4.2	1.6 ± 0.3	0.29 ± 0.04	0.99 ± 0.10	62.64 ± 4.90	28.5 ± 0.4	10.0 ± 1.0	50.3 ± 0.4	6.38 ± 0.03
Back slope ($n = 12$)	0–10	32.3 ± 3.1	1.2 ± 0.2	0.11 ± 0.03	1.21 ± 0.14	54.34 ± 6.1	30.6 ± 0.3	10.8 ± 1.5	61.0 ± 0.6	6.51 ± 0.03
	10–20	32.3 ± 3.1	1.2 ± 0.2	0.11 ± 0.03	1.26 ± 0.11	52.45 ± 5.1	32.8 ± 0.3	13.4 ± 1.5	56.8 ± 0.5	6.62 ± 0.05
Bottom ($n = 12$)	0–10	34.2 ± 3.9	1.4 ± 0.3	0.32 ± 0.02	1.02 ± 0.09	61.51 ± 6.1	27.5 ± 0.5	9.0 ± 0.8	52.2 ± 0.6	6.06 ± 0.02
	10–20	34.2 ± 3.9	1.4 ± 0.3	0.32 ± 0.02	1.05 ± 0.07	60.38 ± 4.6	27.8 ± 0.4	9.5 ± 0.7	53.7 ± 0.6	6.17 ± 0.04

Note: Data represent the mean of all soil samples.

2.2. Experimental Design

This experimental station was built in 2014, while most of the experiments were run from 2010. Corn (*Zea mays*) was planted without organic fertilizers (only inorganic fertilizers were used), and all crop residues were burned in the autumn from 2010 to 2014. The field is on a northeastern-facing slope (steepness of 5°) that was classified into three slope positions (top, back slope and bottom) at intervals of 44 m down the length of the slope (132 m). All consecutive slope positions from top to bottom were considered to reflect the effect of the straw amendment on the dynamics of SM and ST along the whole slope. Before the experiment was initiated, soils at a depth of 0–30 cm in the same slope position (22 m wide across the altitude) were mixed evenly by tractors in the autumn of 2014. A randomized complete block experimental design with three replications was established on the slope at the end of 2014, and the new ridges were formed in the spring of 2015. Each plot of the experimental site was 3.25 m wide across the slope and 130 m long up and down the slope; crop rows were planted 0.65 m apart. Treatment systems included SA and NSA. In the SA treatment, all crop residues were cut into lengths less than 20 cm by the harvester (John Deere), and then all crop residues were directly, evenly, and completely incorporated into 0–30 cm soil depth after harvest, and new ridges (0.65 m wide) were formed at the same time in the autumn. In the NSA treatment, all crop residues were removed from the field at harvest, and then the plots were rotated (30 cm deep), and new ridges (0.65 m wide) were formed at the same time in the autumn. Corn cultivars Xianyu 335 (*Zea mays*) and Pengcheng 216 (*Zea mays*) were planted in spring of 2015 and 2016, respectively (Table 2). All seeds were coated with an insecticide consisting of Sanmate (15%), Thiram (10%), and Carbofuran (10%) (40 kg of seed coated with 1 kg insecticide). Seeding and fertilization were done mechanically in the same operation. No other insect and disease control methods were used during the rest of the crop growing season.

Table 2. Crop management in 2015 and 2016.

Locations	Crop	Crop Variety	Seeding Date	Plant Density (Plant ha^{-1})	Fertilizer (kg ha^{-1})	Weed Control (L ha^{-1})
2016	Corn	Pengcheng 216	8 May	60,000	8 May N:58.5; P$_2$O$_5$:54.0 28 Jule N:150.0	4 May Xiongdilian (2,4-D butylate 15%, Metribuzin 5%, Acetochlor 40%) 2 28 July Nicosulfuron 1
2015	Corn	Xianyu 335	4 May	60,000	4 May N:117.8 P$_2$O$_5$:121.1	4 May Acetochlor 2 29 July Atrazine 1.2

2.3. Measurements and Calculations

Precipitation, SM and ST were all monitored once per hour by TBR (tipping bucket rain gauge), FDR (Frequency Domain Reflectomestry) and a thermistor probe (Made by Qingshen electronic equipment limitation company of China) from May to October. Only three transects 0–22 m (top slope), 66–88 m (back slope) and 110–132 m (bottom) on the slope field were used for sampling and monitoring (Figure 1). FDR and thermistor probes were buried in the soil at a depth of 10 cm in the center of the ridges, at the top slope, back slope and bottom slope. Consecutive 96-h data (four days) collected before and after precipitation were used to reflect the dynamics of daily ST and SM in May, June, July and August in this study. The mean values of ST and SM per hour for a whole month were used to represent the monthly ST and SM.

Soil samples were collected from the 0–22 m (top), 66–88 m (back slope) and 110–132 m (bottom) slope positions and used for nutrient analysis in July (growing stage) and September (mature stage) in 2015 and 2016. Each soil sample at the 0–20 cm depth was comprised of a mixture of five cores taken randomly from within a 1 m^2 plot across slope positions. Soil samples were transported to the laboratory as soon as possible after collection from the field, and then were measured at once. Ammonium nitrogen (NH$_4$$^+$-N) was analyzed using the colorimetric method (625 nm) of indophenol blue-KCl extraction, and nitrate nitrogen (NO_3^--N) was detected using the colorimetric method (220 nm minus 275 nm) after 2 mol L^{-1} KCl extraction [27]. AN (available nitrogen) was calculated as the sum of NO_3^--N and NH_4^+-N. AP was determined using the molybdenum-blue method after extraction with 0.5 mol L^{-1} NaHCO$_3$ [27]. Soil bulk density was measured on undisturbed soil samples taken from the center of the ridge with 100 cm^3 cylinders at 0–5 cm and 10–15 cm depth at slope positions before planting in spring 2016.

The Kolmogorov–Smirnov (K–S) test was used to test the normality of the distribution of various variables across slope position and months. The results showed that site variables (ST and SM) were all normally distributed. One-way analyses of variance (ANOVA) were performed on ST and SM among slope positions and months. Multiple comparisons using the Least Significant Difference (LSD) method were carried out using SPSS 19 statistical software (IBM SPSS Statistics for Windows, Version 19.0. Armonk, NY: IBM Corp.). Variance component estimates and 95% confidence intervals were calculated with restricted maximum likelihood (REML) and unbounded variance components (version 19.0. SPSS Inc.). All figures were drawn by SigmaPlot 10.

3. Results

3.1. Influence of Straw Amendment on Daily Soil Temperature and Soil Moisture in Early Crop Growth Stages

In the early crop growth stages (May and June), precipitation was lower in 2015 than in 2016, while ST was generally higher in 2015. In the early crop growth stages of 2015, ST exceeded values of 10 °C both during the day and at night and before and after precipitation. ST was lower in the SA treatment than in the NSA treatment during the day both before and after precipitation, irrespective of the position (top or bottom of the slope), while it was higher in the SA treatment during the night (Figure 3). In particular, the difference between SA and NSA increased in the middle of the afternoon (12:00–16:00) or after precipitation. On the back slope, ST was higher under SA than under NSA during the day time, while it decreased after precipitation and reached values close to those under NSA in the night.

In May 2016, ST values exceeded 10 °C during the day and at night before precipitation but were lower than 10 °C at night after precipitation across all slope positions. At the top slope, ST differed from that measured in 2015 and was higher in the SA than in the NSA treatment during the daytime and at night, as well as before and after precipitation. On the back slope, ST was lower in the SA than in the NSA treatment for most of the daytime, but was mostly higher in the NSA treatment, especially after precipitation and during the night. The difference in ST between SA and NSA increased during the morning (3:00–5:00) and afternoon (12:00–16:00). In May and June 2016, ST was also lower in the SA than in the NSA treatment during the day and at night both before and after precipitation, while it reached values close to those of SA after precipitation at the bottom of the slope.

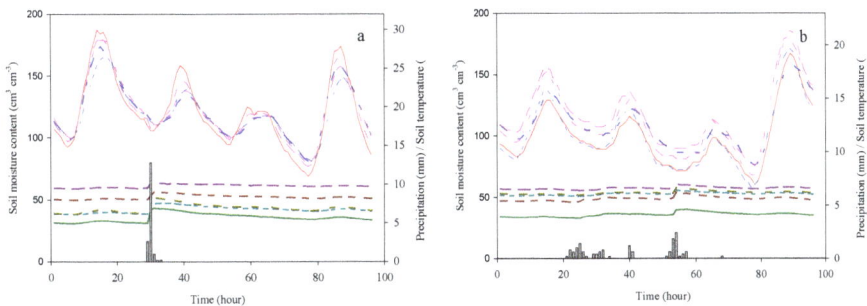

Figure 3. *Cont.*

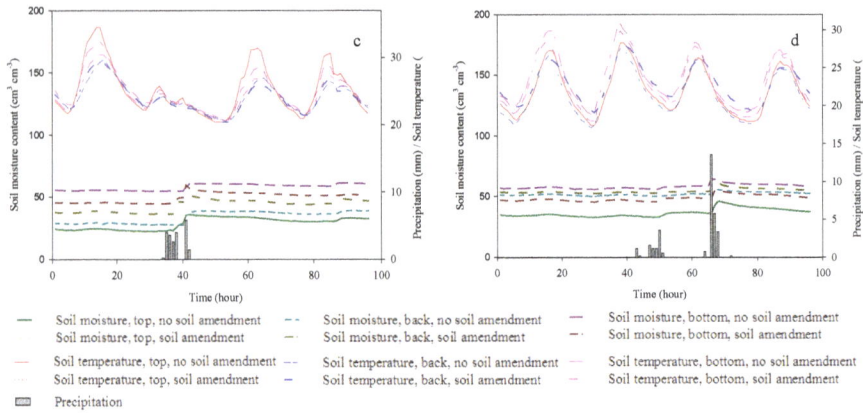

Figure 3. (a–d) represent the dynamics of soil temperature and soil moisture during the consecutive 96-h period from 27–30 May 2015, 14–17 May 2016, 24–27 June 2015 and 16–19 June 2016, respectively.

In the early crop growth stages of both 2015 and 2016, SM was lower in the SA than in the NSA treatment at the top and bottom of the slope both before and after precipitation, although this difference decreased from May to June. SM under SA was typically slightly higher than under NSA on the back slope in both 2015 and 2016, but lower at the bottom of SA than on the back slope of SA and NSA in 2016.

3.2. Influence of Straw Amendment on Daily Soil Temperature and Soil Moisture in Late Crop Growth Stages

In the late growth stages (July and August), ST exceeded 15 °C in both 2015 and 2016. In 2015, before and after precipitation and at the top and the bottom of the slope, ST was also lower in the SA than in the NSA treatment during the day, but increased during the night (Figure 4). The difference between SA and NSA was only slightly changed by precipitation. On the back slope, ST was generally lower in the SA than in the NSA treatment during the day, but at night, reached values similar to those in the NSA treatment.

In the late growth stages of 2016, at the top slope, ST differed from the values measured in 2015 and was higher in the SA than in the NSA treatment during both the daytime and at night before and after precipitation. On the back slope, ST was higher under SA than under NSA for most of the daytime, and the difference increased at night. ST at the bottom of the slope was also lower under SA compared with NSA during the day and at night both before and after precipitation, while the relationship was hardly changed by precipitation.

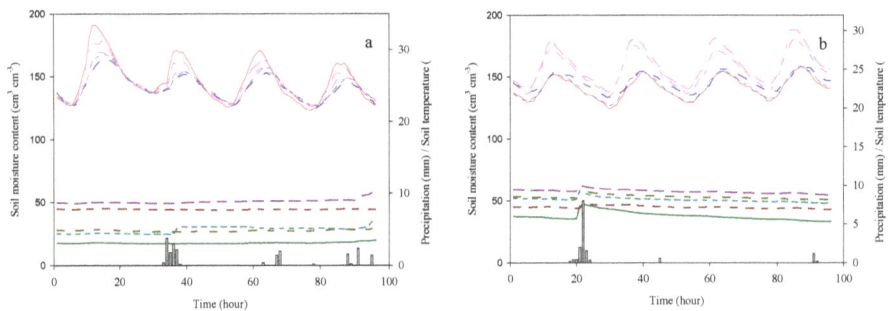

Figure 4. *Cont.*

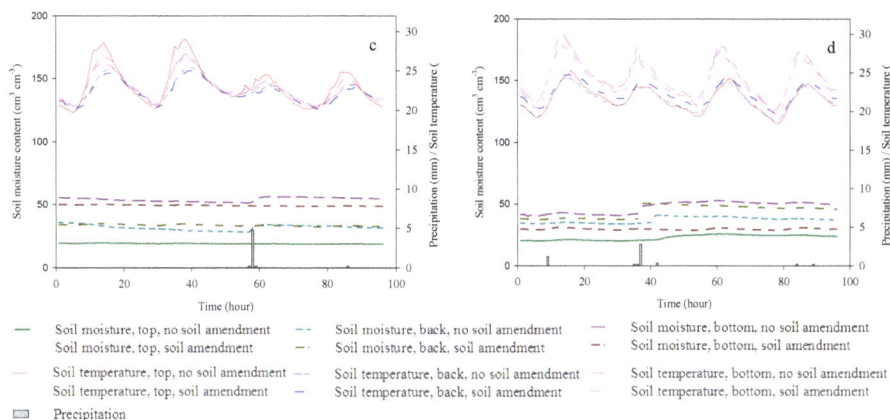

Figure 4. (a–d) represent the dynamics of soil temperature and soil moisture during the consecutive 96-h period from 20–23 July 2015, 25–28 July 2016, 13–16 August 2015 and 20–23 August 2016, respectively.

ST dynamics in the late growth stages of 2016 were comparable to those in 2015, while the difference between the bottom and the top slope or between the bottom and back slope increased in both July and August 2016. Compared with the early growth stage, the variation of ST on the back slope was lower than that at the top of the slope in 2015 and increased at the bottom in 2016.

In the late growth stages of both 2015 and 2016, SM dynamics were more consistent in SA compared with NSA for most slope positions. At the top of the slope, SM was higher under SA than under NSA in 2015 and lower under SA than under NSA in July 2016; in August 2016, there was no difference prior to precipitation. Generally, SM was slightly higher under SA than under NSA on the back slope in both 2015 and 2016, and this relationship changed slightly after precipitation. SM in the SA treatment at the bottom of the slope was also lower than that in the NSA treatment and even lower than that in both SA and NSA treatments on the back slope in 2016.

3.3. Influence of Straw Amendment on Monthly Soil Temperature across Different Slope Positions

Except in May, monthly ST was highest at the top slope, followed by the back and the bottom of the slope in both SA and NSA treatments in 2015, while it was highest at the bottom in the SA treatment, followed by the back and the top slopes in both SA and NSA treatments in 2016. Compared with NSA, monthly ST at the top slope was generally lower under SA in May, July, and August of 2015, but higher from May to August in 2016 (Table 3). On the back slope, ST under SA was significantly higher, i.e., 7.2%, 7.0%, 4.9%, 3.0%, and 2.4% than that in the NSA treatment in May, June, July, and August 2016, respectively, and similar to the values measured under NSA in 2015 (Table 3). In contrast, ST at the bottom of the slope was significantly lower under SA than under NSA from July to August 2015 and from May to September 2016; in particular, it was 2.2% (0.50 °C), 3.7% (0.98 °C), 2.0% (0.50 °C), and 7.6% (1.48 °C) lower in June, July, August, and September 2016, respectively, than under NSA.

The variance of ST (reflected by the CV) was lower under SA than under NSA at the top of the slope from May to September in both 2015 and 2016 and lower under SA at the bottom of the slope from May to September in 2015, but this trend was not consistent in 2016. On the back slope, the variance of ST was higher under SA than under NSA and similar from June to September in 2015. However, in 2016, the variance was lower under SA than under NSA from May to July on the back slope. Generally, in 2016, ST variance was higher, both at the top and the bottom of the slope, except under NSA in May 2016.

Table 3. Monthly soil temperature in the treatments with straw amendments (SA) and without straw amendments (NSA) in 2015 and 2016 ($n = 1448$).

	Straw Amendments (SA) (°C)						No Straw Amendments (NSA) (°C)					
	Top		Back Slope		Bottom		Top		Back Slope		Bottom	
	Mean	CV	Mean	CV	Mean	CV	Mean	CV	Mean	CV	Mean	CV
2015												
May	16.64 a [1]	0.31	16.56 ab	0.30	16.56 ab	0.31	16.06 b	0.37	16.51 ab	0.27	16.45 ab	0.31
Jun.	22.62 ab	0.24	22.20 cd	0.22	22.58 bc	0.24	23.01 a	0.29	22.02 d	0.22	22.49 dc	0.25
Jul.	23.23 a	0.12	22.54 b	0.10	22.84 c	0.11	23.53 d	0.15	22.78 e	0.10	23.24 f	0.12
Aug.	22.78 bc	0.07	22.29 e	0.06	22.70 c	0.07	23.09 a	0.10	22.50 d	0.06	22.93 ab	0.08
2016												
May	17.28 b	0.25	17.02 c	0.23	17.67 a	0.27	16.03 d	0.27	15.87 d	0.28	17.74 a	0.24
Jun.	21.64 d	0.15	21.92 c	0.12	22.11 b	0.16	20.87 e	0.15	20.49 f	0.15	22.61 a	0.16
Jul.	23.74 d	0.09	24.43 c	0.08	25.68 b	0.12	23.14 f	0.10	23.29 e	0.10	26.66 a	0.13
Aug.	22.43 d	0.11	22.64 c	0.11	24.36 b	0.13	22.13 e	0.12	21.98 f	0.11	24.86 a	0.13
Sept.	17.30 e	0.12	18.00 bc	0.11	18.06 b	0.16	17.91 c	0.13	17.57 d	0.12	19.54 a	0.14

Note: [1] Values followed by the same letter within the same rows (lowercase letter) are not significantly different based on the LSD multiple range test ($p \leq 0.05$).

3.4. Influence of Straw Amendments on Monthly Soil Moisture at Different Slope Positions

Monthly SM was highest at the bottom of the slope, followed by the back and the top for SA and NSA in 2015 and NSA in 2016. Monthly mean SM at the top was higher under SA than under NSA from June to September in 2015, but lower under SA than under NSA from May to September 2016 (Table 4). Compared with NSA, SM was significantly higher, i.e., by 1.4%, 32.3%, 37.4%, and 24.5% under SA in June, July, August, and September 2015, respectively, while it was significantly lower by 11.5%, 11.2%, 8.8%, 8.3%, and 10.0% in May, June, July, August, and September in 2016, respectively, at top of the slope. On the back slope, SM was significantly higher under SA than under NSA, i.e., by 2.9%, 10.5%, 8.8%, 8.5%, and 2.0% in May, June, July, August, and September in 2015, respectively, and significantly higher under SA than under NSA, i.e., by 2.1%, 5.7%, 4.2%, 12.5%, and 5.9% in May, June, July, August, and September in 2016, respectively. However, at the bottom of the slope, except in September, SM was significantly lower, i.e., by 14.7%, 12.9%, 6.7%, and 3.5% under SA compared with NSA in May, June, July, and August 2015, respectively, and was significantly higher, i.e., by 15.9%, 15.9%, 22.1%, 29.6%, and 28.3% under SA than under NSA in May, June, July, August, and September 2016, respectively.

Table 4. Monthly soil moisture in the SA and NSA treatments in 2015 and 2016 ($n = 1448$).

	Straw Amendments (SA) (cm^3 cm^{-3})						No Straw Amendments (NSA) (cm^3 cm^{-3})					
	Top		Back slope		Bottom		Top		Back slope		Bottom	
	Mean	CV	Mean	CV	Mean	CV	Mean	CV	Mean	CV	Mean	CV
2015												
May	30.16 f [1]	0.10	42.02 c	0.08	51.52 b	0.04	34.05 e	0.09	40.85 d	0.06	60.37 a	0.02
Jun.	33.23 e	0.14	44.55 c	0.08	52.44 b	0.07	32.77 f	0.14	40.32 d	0.16	60.22 a	0.04
Jul.	32.20 e	0.22	35.73 c	0.16	50.87 b	0.11	24.34 f	0.23	32.84 d	0.21	54.53 a	0.08
Aug.	34.21 e	0.20	39.83 c	0.15	54.39 b	0.10	24.90 f	0.25	36.72 d	0.20	56.34 a	0.08
Sept.	24.76 e	0.20	29.84 c	0.17	43.79 a	0.15	19.88 f	0.26	29.26 d	0.21	40.51 b	0.12
2016												
May	30.69 f	0.07	52.89 b	0.04	47.55 d	0.05	34.66 e	0.09	51.79 c	0.03	56.57 a	0.03
Jun.	33.43 f	0.10	55.46 b	0.05	49.49 d	0.06	37.64 e	0.10	52.47 c	0.03	58.85 a	0.03
Jul.	25.75 f	0.19	47.31 b	0.11	40.64 d	0.14	28.25 e	0.22	45.42 c	0.10	52.14 a	0.10
Aug.	21.77 f	0.07	43.12 b	0.07	33.06 d	0.10	23.75 e	0.12	38.34 c	0.09	46.95 a	0.06
Sept.	26.56 f	0.13	45.04 b	0.07	37.24 d	0.14	29.51 e	0.16	42.52 c	0.08	51.96 a	0.05

Note: [1] Values followed by the same letter within the same rows (lowercase letters) are not significantly different based on the LSD multiple range test ($p \leq 0.05$).

The variance of SM (reflected by the CV) was lower under SA at the top of the slope from July to September and lower under SA on the back slope in August and September; it was consistently higher under SA at the bottom of the slope from May to September. The variance decreased from the top to the bottom of the slope for both NSA and SA in 2015 and for NSA in 2016, while it deceased towards the back of the slope and increased towards the bottom under SA in 2016.

3.5. Influence of Straw Amendment on Soil Bulk Density

Soil bulk density was measured under SA and NSA before planting in spring 2016 (between 2015 and 2016). Soil bulk density was compared between SA and NSA across soil depths at the same slope position. At the top, soil bulk density was significantly higher (35%) at the depth of 10–15 cm compared with that at the 0–5 cm soil depth under NSA and significantly higher (16%) at 10–15 cm compared with that at 0–5 cm under SA; compared with NSA, SA decreased soil bulk density at the 10–15 cm soil depth, but this was not significant. At both the back of the slope and at the bottom, soil bulk density was significantly higher, i.e., 19% (back of the slope) and 31% (bottom) at the 10–15 cm soil depth compared with the 0–5 cm soil depth under NSA, but no significant differences were observed between the soil depths of 0–5 cm and 10–15 cm under SA; SA significantly decreased soil bulk density by 7% at the back of the slope and by 13% at the bottom at the 10–15 cm soil depth compared with NSA.

3.6. Influence of Straw Amendment on Soil Nutrients

AN and AP were compared between SA and NSA in July (growing stage) and September (mature stage). Except for the back of the slope of 2016, SA typically decreased AN compared with NSA at both the top and bottom of the slope. In 2015, SA decreased AN in all slope positions in July and significantly decreased (13%) AN compared with NSA on the back slope; this effect was not significant in September. In 2016, SA decreased AN at both the top and bottom positions and significantly decreased (16%) AN compared with NSA at the top in September. In 2015, SA only decreased AP on the back slope; the difference was not significant in July. However, SA decreased AP at all slope positions and significantly decreased 56% at the back of the slope in September. In 2016, SA decreased AP at both the back of the slope and at the bottom and significantly decreased (44%) AP at the bottom of the slope in July. However, SA only decreased AP at the bottom; this effect was not significant in September.

3.7. Influence of Straw Amendment on Crop Growth and Yield

At the top of the slope, under SA, crop height and yield were generally higher (1–9.8%) compared to NSA, but this difference was not significant in both 2015 and 2016 (Table 5). At the back of the slope, crop height was lower under SA in July and was higher than or close to the value under NSA in September. Compared with NSA at the back of the slope, crop yield under SA was significantly increased by 55.6% and 105.1% in 2015 and 2016, respectively. At the bottom of the slope, crop height under SA was higher than under NSA from July to September in 2015 and in September 2016, while crop yield under SA was 4.1–30.6% lower than under NSA; this difference was only statistically significant in 2016.

Table 5. Mean crop height and yield in the SA and NSA treatments in 2015 and 2016 (*n* = 3).

		Straw Amendments (SA)			No straw Amendments (NSA, Control)		
	Height/Yield	Top	Back Slope	Bottom	Top	Back Slope	Bottom
	Jul. (cm)	197.8 ± 3.0 a [1]	157.2 ± 6.2 c	177.7 ± 0.6 b	197.7 ± 5.3 a	166.2 ± 10.6 bc	170 ± 0.9 b
2015	Sept. (cm)	318.8 ± 3.2 ab	311.5 ± 6.6 ab	321.2 ± 3.7 a	311.2 ± 4.9 ab	302 ± 4.9 b	315.8 ± 7.0 ab
	Yield (t)	10.1 ± 1.4 a	8.4 ± 0.8 a	9.3 ± 0.7 a	9.2 ± 0.4 a	5.4 ± 0.1b	9.7 ± 1.3 a
	Jul. (cm)	125.9 ± 9.0 a	119.2 ± 1.5 a	120.1 ± 3.3 a	130.9 ± 4.0 a	133.6 ± 10.8 a	127.1 ± 7.2 a
2016	Sept. (cm)	300.8 ± 5.1 a	267.2 ± 4.1 b	280.3 ± 5.6 ab	286.8 ± 15.3 ab	269.3 ± 8.0 b	270.8 ± 1.3 b
	Yield (t)	13.4 ± 1.6 a	12.1 ± 1.6 ab	10.0 ± 1.4 b	13.3 ± 1.3 ab	5.9 ± 1.0 c	14.4 ± 0.1 a

Note: [1] Values followed by the same letter within the same rows (lowercase letters) are not significantly different based on the LSD multiple range test ($p \leq 0.05$).

4. Discussion

ST and SM are highly correlated with nutrient cycling processes in soil ecosystems [24,28,29]. The activity of most soil microorganisms increases when ST exceed 10 °C and reaches maximum levels at 25–35 °C. In addition, soil microbial activity increases rapidly at a soil water potential greater than −30 MPa and is highest at −0.01 MPa [30–32]. In this study, daily ST was greater than 22 °C, and SM was generally higher than 33% (v/v) from June to August at the bottom of the slope. In early crop growth stages, SM on the top slope and back slope was slightly higher than FC (Field Capacity) due to low temperature, high precipitation, and low transpiration, and increased especially after precipitation in this Mollisol soil with heavy texture. However, SM decreased in late crop growth stages, especially on the top slope and back slope. This was mainly due to both temperature and transpiration being higher during these stages, and most of the water moved from the top to the bottom of the slope. Relatively low levels of SM and ST impacted crop growth and significantly influenced crop growth at the top and the back of the slope.

Yang et al. (2016) reported decreased SM, increased ST, and increased crop yield in an SA treatment, mainly influenced by straw ditches and straw burial depth in a flat area. These results can be compared to the findings of our study, where daily and monthly average ST was effectively increased at the top (relative flat) of the slope under SA, but daily and monthly average SM decreased in 2016. Our findings for 2015 were largely different compared to those for 2016. Therefore, our results suggest that ST and SM are not consistently affected by SA in top slope positions (relative flat). This was mainly due to the fact that the available water capacity and hydraulic conductivity of saturated soil typically increased with straw amendment [33], and straw amendment increased infiltration and the slope facilitated water movement towards the bottom [11,18,19,34]. The different results between 2015 and 2016 may have also been caused by variations in precipitation and the history of residue return (one year vs. two years) between 2015 and 2016, and may also be influenced by straw burial because it was difficult to evenly distribute the straw amendment using a tractor. At the same time, straw amendment changed the soil bulk density and increased the capillary porosity in the deeper soil layers (Figure 5), which could also effectively balance the equilibrium of SM dynamics and influence the dynamics of ST at the top of the slope [12,21]. Thus, in this study, the SA treatment with improved soil structure effectively prevented low SM levels under low precipitation conditions in the early crop growth phase of 2015 and prevented low ST under high precipitation conditions in the early crop growth phase in 2016. Furthermore, SA changed the equilibrium of SM and ST and changed soil nutrient cycling. In the process of SA decomposition, microorganisms compete for soil N. Also, a considerable amount of N can easily be transferred from the top of the slope, while P is strongly adsorbed in soil [35]. Thus, AN was typically decreased, while AP was typically increased in both 2015 and 2016 at the top of the slope (Figure 6a). Therefore, compared with NSA, crop residue return (SA) at the top of the slope can effectively balance the equilibrium of SM dynamics and positively influence ST, thereby promoting crop growth and increasing crop yield.

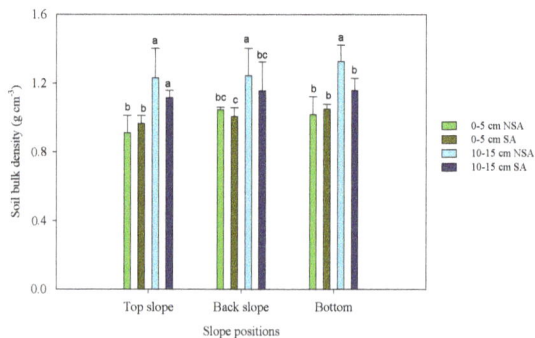

Figure 5. Soil bulk density before planting in spring of 2016. NSA and SA represent no straw amendment and amendment, respectively.

At the back of the slope, compared with NSA, SA notably increased daily ST. Monthly ST under SA was generally higher than under NSA from May to August 2016 and was similar in May and June 2015. Compared with NSA, higher ST under SA is suitable for crop growth. Daily SM was generally higher under SA than under NSA at the back of the slope, although this trend changed with precipitation, and SM was typically higher under SA than under NSA after precipitation. Our results are similar to the findings of a previous study which found that SA increased the response rate of SM to rainfall in the wet Yangtze River delta agricultural region [21]. Monthly SM was also generally higher under SA than under NSA at the back of the slope, in both 2015 and 2016. This is most likely because SA increases infiltration and reduces evaporation [34], and infiltration and subsurface runoff increase with an increase in straw amendment [18,19]. However, ST usually increases with decreasing SM levels [12,21], which is not in agreement with our results, possibly because straw amendment decreases soil bulk density, increases infiltration and decreases heat capacity [35]. This is mainly because the soil bulk density was higher than 1.2 g cm^{-3} under NSA, and was higher than the water density (1.0 g cm^{-3} at 101.325 kPa), while the soil bulk density was close to 1.0 g cm^{-3} under SA. In our study, monthly SM levels in the SA treatment in 2016 were highest at the back of the slope, followed by the bottom and the top of the slope; this could also be attributed to the effect of long-term crop residue return, especially SA, as soils with relatively low bulk density (Figure 5) and high infiltration capacity can absorb more water from the top of the slope [34]. Thus, crop residue return can also effectively balance the equilibrium of SM dynamics and positively influence ST; in particular, SM is increased. Furthermore, SA changed the equilibrium of SM and ST and changed the soil nutrient cycling. Also, most of the N that moved from the top of the slope was readily deposited at the back of the slope under SA, while P does not move easily in soil [35]. Thus, AP was typically decreased, while AN was typically increased in both 2015 and 2016 on the back of the slope (Figure 6a), thereby increasing crop yield at the back of the slope [9]. However, compared with NSA, crop height under SA was lower in the early crop growth stages. This may be due to the microbial decomposition of crop residues, which exhausted soil nitrogen pools and deprived the crops of nitrogen, although this difference was lower in August.

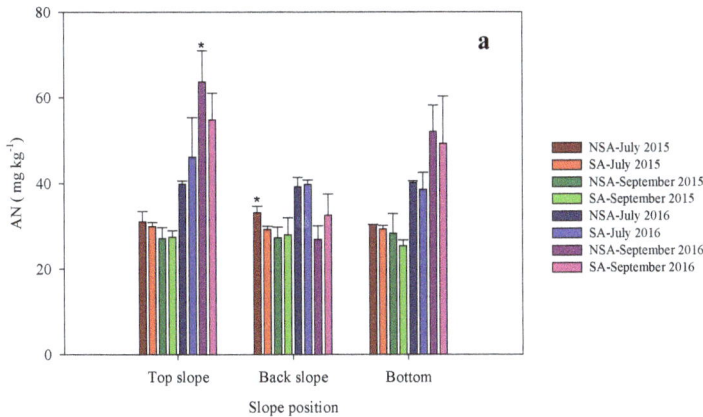

Figure 6. *Cont.*

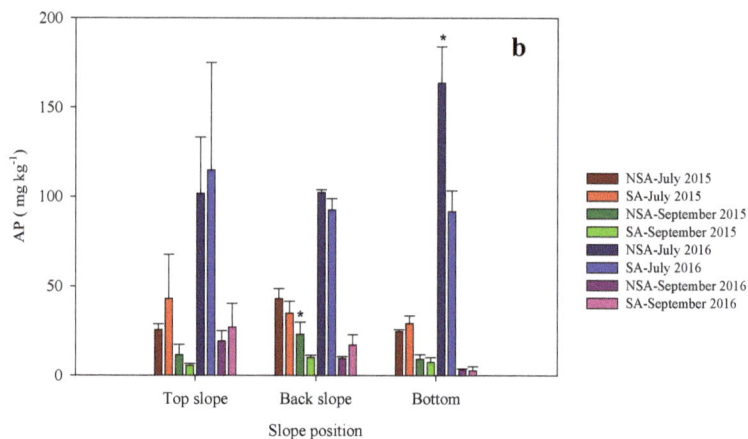

Figure 6. Soil available nitrogen (AN) (a) and soil available phosphorus (AP) (b)changed between the treatments in 2015 and 2016. Significantly differences between amendment and no straw amendment was marked with an "*" for the same slope position and year.

At the bottom of the slope, SM was generally higher compared to the top due to the "trickling down" effect. Daily ST at the bottom of the slope was lower under SA compared with NSA during the daytime and at night both before and after precipitation. Except in September, monthly ST at the bottom was also generally lower under SA than under NSA from May to August in both 2015 and 2016. However, low ST is not beneficial for soil microbial activities and the release of nutrients for crop growth. Daily SM at the bottom of the slope was lower under SA than under NSA (both before and after precipitation), and monthly SM at the bottom of the slope was generally lower under SA than under NSA, in both 2015 and 2016. This may be due to the fact that over time, soil amendment increases water infiltration and subsurface runoff at the bottom, especially as residue return increases annually [19,34]. Generally, the high SM at the bottom under both SA and NSA was not the factor that limited nutrient cycling and crop growth. However, residue return increases soil organic carbon amounts, decrease C:N ratios, and leads to competition between microbial communities and crops in terms of nitrogen. Furthermore, because SA changed the equilibrium of SM and ST at the bottom of the slope, soil available nutrients were not readily released from soil, which decreased both AP and AN in 2015 and 2016 (Figure 6a). Therefore, straw amendment significantly decreased crop yields at the bottom. Generally, crop residues with lower C:N ratios and relatively high SM levels (95% field capacity) can increase the numbers and the total dry weight of earth worms, thereby improving soil properties and fertility [36]. Furthermore, SA not only increases soil total C and N contents but also increases microbial biomass C and N contents and reduces N fertilizer use, especially when combined with manure [37,38]. However, excessive straw amendment, especially using straw with a high C:N ratio, could exhaust soil nitrogen pools [35]. This was also proved in this present study. Thus, in order to obtain higher crop yields, additional fertilizer, especially nitrogen application at the bottom of the slope is advisable. However, our study was carried out on a northeastern-facing slope, and the results could be different for other slope aspects [12,39], which should be investigated in further studies.

Wang et al. (2014) indicated that straw incorporation increased evapotranspiration from tasseling to the grain-filling stages of maize, while it reduced evapotranspiration from the ten-leaf collar to the tasseling stage and from the grain-filling to the maturity stages of maize. Liu et al. (2017) reported that residue incorporation also increased crop transpiration and increased WUE (crop water use efficiency) due to reduced soil evaporation and surface runoff [11]. However, in our study, SM in the SA treatment was consistently higher at the back of the slope and lower at the bottom of the slope, while not consistently higher at the top. This may also be influenced by crop growth, landscape, and vigorous crop growth, which significantly decreased or increased soil water levels [25]. In this study, straw amendment increased crop yield at the top and the back of the slope with relatively low SM, but decreased crop yield at the bottom. In that regard, our results are similar to the findings of other studies describing the positive effects of residue incorporation on water balance and crop yield, especially in narrow-leaf cropping systems in hot and dry environments [11]. Thus, straw amendment represents an effective method to increase crop yields at the top and the back of slope with relatively low SM, while it slightly negatively affected the yield at the bottom of the slope with higher SM. Previous studies have indicated that straw incorporation into deeper soil layers may improve crop yields, even in the face of global climate change, but potentially slows down cultivation influence, costs more time, labor, and energy [11,40]. In this study, we only investigated the effects of straw incorporation into the 0–30 cm soil layer on a consecutively sloping farmland in the Mollisol region. Further work should therefore focus on the impacts of straw incorporation into the deeper soil layers on SM, ST, crop yields and profit.

5. Conclusions

Soil temperature, soil moisture, soil nutrients and soil bulk density were changed after straw amendment. Straw amendment with the improved soil structure effectively prevented low soil moisture levels under low precipitation conditions and prevented low soil moisture under high precipitation conditions in the early crop growth phase on the top of the slope. Furthermore, SA changed soil nutrient cycling and typically decreased AN but increased AP on the top of the slope. Generally, straw amendment increased both soil temperature and moisture, increased AN, and decreased AP at the back of the slope. However, straw amendment decreased soil temperature and decreased AN and AP at the bottom of the slope.

Straw amendment can effectively change soil structure and balance the equilibrium of soil moisture and temperature in areas with relatively low soil moisture contents on the top and back of the slope in the Mollisol region of Northeast China, thereby not decreasing maize yields at the top of the slope, while increasing maize yields on the back of the slope. However, straw amendment decreased soil moisture and temperature, resulting in reducing soil available nutrients, and typically reduced crop yields at the bottom of the slope. In order to better manage soil temperature, moisture and nutrients, slope positions should be fully considered when straw amendment is adopted in sloped fields.

Author Contributions: The experimental field has been managed by Shaoliang Zhang since 2010, including seeding, ploughing, fertilizing, weeding and harvesting. Shaoliang Zhang, Yao Wang and Qingsong Shen buried FDR and thermistor probes, and collected the data. Shaoliang Zhang analyzed data, and wrote the manuscript. Yao Wang and Qingsong Shen checked the results, facilitated discussions, and reviewed the manuscript.

Acknowledgments: We thank the editors and the two anonymous reviewers for their comments. Part of the paper is sponsored by the project of Science and Technology Research Projects of the Education Department of Heilongjiang Province, China (No. 1254G012).

Conflicts of Interest: The authors declare no conflict of interest.

References

1. Recous, S.; Robin, D.; Darwis, D.; Mary, B. Soil inorganic N availability: Effect on maize residue decomposition. *Soil Biol. Biochem.* **1995**, *27*, 1529–1538. [CrossRef]
2. Kim, S.; Dale, B.E. Global potential bioethanol production from wasted crops and crop residues. *Biomass Bioenergy* **2004**, *26*, 361–375. [CrossRef]
3. Blanco-Canqui, H.; Lal, R. Crop Residue Removal Impacts on Soil Productivity and Environmental Quality. *Crit. Rev. Plant Sci.* **2009**, *28*, 139–163. [CrossRef]
4. Ghimire, R.; Machado, S.; Rhinhart, K. Long-Term Crop Residue and Nitrogen Management Effects on Soil Profile Carbon and Nitrogen in Wheat-Fallow Systems. *Agron. J.* **2015**, *107*, 2230–2240. [CrossRef]

5. Kumar, K.; Goh, K.M. Crop residues and management practices: Effects on soil quality, soil nitrogen dynamics, crop yield, and nitrogen recovery. In *Advances in Agronomy*; Sparks, D.L., Ed.; Elsevier: New York, NY, USA, 2000; Volume 68, pp. 197–319.

6. Poeplau, C.; Reiter, L.; Berti, A.; Katterer, T. Qualitative and quantitative response of soil organic carbon to 40 years of crop residue incorporation under contrasting nitrogen fertilisation regimes. *Soil Res.* **2017**, *55*, 1–9. [CrossRef]

7. Schmatz, R.; Recous, S.; Aita, C.; Tahir, M.M.; Schu, A.L.; Chaves, B.; Giacomini, S.J. Crop residue quality and soil type influence the priming effect but not the fate of crop residue C. *Plant Soil* **2017**, *414*, 229–245. [CrossRef]

8. Novelli, L.E.; Caviglia, O.P.; Pineiro, G. Increased cropping intensity improves crop residue inputs to the soil and aggregate-associated soil organic carbon stocks. *Soil Tillage Res.* **2017**, *165*, 128–136. [CrossRef]

9. Zhao, Y.G.; Pang, H.C.; Wang, J.; Huo, L.; Li, Y.Y. Effects of straw mulch and buried straw on soil moisture and salinity in relation to sunflower growth and yield. *Field Crops Res.* **2014**, *161*, 16–25. [CrossRef]

10. Saha, R.; Ghosh, P.K. Soil Organic Carbon Stock, Moisture Availability and Crop Yield as Influenced by Residue Management and Tillage Practices in Maize-Mustard Cropping System Under Hill Agro-Ecosystem. *Nat. Acad. Sci. Lett. India* **2013**, *36*, 461–468. [CrossRef]

11. Liu, D.L.; Zeleke, K.T.; Wang, B.; Macadam, I.; Scott, F.; Martin, R.J. Crop residue incorporation can mitigate negative climate change impacts on crop yield and improve water use efficiency in a semiarid environment. *Eur. J. Agron.* **2017**, *85*, 51–68. [CrossRef]

12. Zhang, S.L.; Zhang, X.Y.; Huffman, T.; Liu, X.B.; Yang, J.Y. Soil Loss, Crop Growth, and Economic Margins under Different Management Systems on a Sloping Field in the Black Soil Area of Northeast China. *J. Sustain. Agric.* **2011**, *35*, 293–311. [CrossRef]

13. Chen, S.Y.; Zhang, X.Y.; Pei, D.; Sun, H.Y.; Chen, S.L. Effects of straw mulching on soil temperature, evaporation and yield of winter wheat: Field experiments on the North China Plain. *Ann. Appl. Biol.* **2007**, *150*, 261–268. [CrossRef]

14. Singh, V.; Srivastava, A.; Singh, R.K.; Savita, U.S. Effect of tillage practices and residue management on soil quality and crop yield under maize (*Zea mays*)-based cropping system in Mollisol. *Indian J. Agric. Sci.* **2011**, *81*, 1019–1025.

15. Azam, F.; Lodhi, A.; Ashraf, M. Availability of soil and fertilizer nitrogen to wetland rice following wheat straw amendment. *Biol. Fertil. Soils* **1991**, *11*, 97–100. [CrossRef]

16. McCalla, T.M.; Duley, F.L. Effect of crop residues on soil temperature. *J. Am. Soc. Agron.* **1946**, *38*, 75–89. [CrossRef]

17. Mooers, C.A.; Washko, J.B.; Young, J.B. Effects of wheat straw, lespedeza sericea hay, and farmyard manure, as soil mulches, on the conservation of moisture and the production of nitrates. *Soil Sci.* **1948**, *66*, 307–315. [CrossRef]

18. Xing, X.G.; Li, Y.B.; Ma, X.Y. Effects on infiltration and evaporation when adding rapeseed-oil residue or wheat straw to a Loam soil. *Water* **2017**, *9*, 12. [CrossRef]

19. Lu, P.R.; Zhang, Z.Y.; Feng, G.X.; Huang, M.Y.; Shi, X.F. Experimental study on the potential use of bundled crop straws as subsurface drainage material in the newly reclaimed coastal land in Eastern China. *Water* **2018**, *10*, 15. [CrossRef]

20. Wang, X.; Jia, Z.; Liang, L. Effect of straw incorporation on soil moisture, evapotranspiration, and rainfall-use efficiency of maize under dryland farming. *J. Soil Water Conserv.* **2014**, *69*, 449–455. [CrossRef]

21. Yang, H.S.; Feng, J.X.; Zhai, S.L.; Dai, Y.J.; Xu, M.M.; Wu, J.S.; Shen, M.X.; Bian, X.M.; Koide, R.T.; Liu, J. Long-term ditch-buried straw return alters soil water potential, temperature, and microbial communities in a rice-wheat rotation system. *Soil Tillage Res.* **2016**, *163*, 21–31. [CrossRef]

22. Smika, D.E.; Ellis, R. Soil temperature and wheat straw mulch effects on wheat plant development and nutrient concentration. *Agron. J.* **1971**, *63*, 388–391. [CrossRef]

23. Abera, G.; Wolde-meskel, E.; Bakken, L.R. Effect of organic residue amendments and soil moisture on N mineralization, maize (*Zea mays* L.) dry biomass and nutrient concentration. *Arch. Agron. Soil Sci.* **2013**, *59*, 1263–1277. [CrossRef]

24. Andren, O.; Rajkai, K.; Katterer, T. Water and temperature dynamics in a clay soil under winter-wheat—Influence on straw decomposition and n-immobilization. *Biol. Fertil. Soils* **1993**, *15*, 1–8. [CrossRef]

25. Zhang, S.L.; Huang, J.; Wang, Y.; Shen, Q.S.; Mu, L.L.; Liu, Z.H. Spatiotemporal heterogeneity of soil available nitrogen during crop growth stages on mollisol slopes of Northeast China. *Land Degrad. Dev.* **2017**, *28*, 856–869. [CrossRef]

26. Zhang, S.L.; Yan, L.L.; Huang, J.; Mu, L.L.; Huang, Y.Q.; Zhang, X.Y.; Sun, Y.K. Spatial heterogeneity of soil C:N ratio in a mollisol watershed of Northeast China. *Land Degrad. Dev.* **2016**, *27*, 295–304. [CrossRef]

27. Bao, S. *Soil Agricultural Chemical Elements Analysis*; China Agriculture Press: Beijing, China, 2000; p. 3.

28. Bussiere, F.; Cellier, P. Modification of the soil-temperature and water-content regimes by a crop residue mulch—Experiment and modeling. *Agric. For. Meteorol.* **1994**, *68*, 1–28. [CrossRef]

29. Tian, Y.Q.; Liu, J.; Zhang, X.Y.; Gao, L.H. Effects of summer catch crop, residue management, soil temperature and water on the succeeding cucumber rhizosphere nitrogen mineralization in intensive production systems. *Nutr. Cycl. Agroecosyst.* **2010**, *88*, 429–446. [CrossRef]

30. Reth, S.; Hentschel, K.; Drosler, M.; Falge, E. DenNit—Experimental analysis and modelling of soil N_2O efflux in response on changes of soil water content, soil temperature, soil pH, nutrient availability and the time after rain event. *Plant Soil* **2005**, *272*, 349–363. [CrossRef]

31. Walker, A. *Soil Microbiology: A Critical Review*; Butterworth & Co.: London, UK, 1975.

32. Paul, E.A.; Clark, F.E. *Soil Microbiology and Biochemistry*; Academic Press: San Diego, CA, USA, 1989.

33. Bhagat, R.M.; Verma, T.S. Impact of rice straw management on soil physical-properties and wheat yield. *Soil Sci.* **1991**, *152*, 108–115. [CrossRef]

34. Qiao, H.L.; Liu, X.J.; Li, W.Q.; Huang, W. Effects of straw deep mulching on soil moisture infiltration and evaporation. *Sci. Soil Water Conserv.* **2006**, *2*, 34–39.

35. Brady, N.C.; Weil, R.R. *Nature and Properties of Soils*; Macmillan Publishing Company: New York, NY, USA, 2000.

36. Chen, J.; Gu, W.; Tao, J.; Xu, Y.J.; Wang, Y.; Gu, J.Y.; Du, S.Y. The effects of organic residue quality on growth and reproduction of Aporrectodea trapezoides under different moisture conditions in a salt-affected agricultural soil. *Biol. Fertil. Soils* **2017**, *53*, 103–113. [CrossRef]

37. Zhou, X.Q.; Wu, H.W.; Li, G.D.; Chen, C.R. Short-term contributions of cover crop surface residue return to soil carbon and nitrogen contents in temperate Australia. *Environ. Sci. Pollut. Res.* **2016**, *23*, 23175–23183. [CrossRef] [PubMed]

38. Zhao, B.Z.; Zhang, J.B.; Yu, Y.Y.; Karlen, D.L.; Hao, X.Y. Crop residue management and fertilization effects on soil organic matter and associated biological properties. *Environ. Sci. Pollut. Res.* **2016**, *23*, 17581–17591. [CrossRef] [PubMed]

39. Zhang, S.L.; Zhang, X.Y.; Huffman, T.; Liu, X.B.; Yang, J.Y. Influence of topography and land management on soil nutrients variability in Northeast China. *Nutr. Cycl. Agroecosyst.* **2011**, *89*, 427–438. [CrossRef]

40. Chamen, T. The effect of straw incorporation on diesel fuel use and the emission of pollutants. *Agric. Eng.* **1994**, *49*, 89–92.

water

MDPI

Article

Organic Amendments Influence Soil Water Depletion, Root Distribution, and Water Productivity of Summer Maize in the Guanzhong Plain of Northwest China

Li-Li Zhao [1,2,3], Lu-Sheng Li [4], Huan-Jie Cai [1,2,3,*], Xiao-Hu Shi [1,2,3] and Chao Zhang [1,2,3]

[1] College of Water Resources and Architectural Engineering, Northwest A&F University, Yangling 712100, China; sdytdxzll@163.com (L.-L.Z.); shixiaohu2006@126.com (X.-H.S.); chaozhang13@163.com (C.Z.)
[2] Key Laboratory for Agricultural Soil and Water Engineering in Arid Area of Ministry of Education, Northwest A&F University, Yangling 712100, China
[3] Institute of Water Saving Agriculture in Arid Areas of China, Northwest A&F University, Yangling 712100, China
[4] School of Water Conservancy, North China University of Water Resources and Electric Power, Zhengzhou 450046, China; lilusheng0715@163.com
*   Correspondence: caihj@nwsuaf.edu.cn; Tel.: +86-29-8708-2133

Received: 11 October 2018; Accepted: 9 November 2018; Published: 13 November 2018

Abstract: Organic amendments improve general soil conditions and stabilize crop production, but their effects on the soil hydrothermal regime, root distribution, and their contributions to water productivity (WP) of maize have not been fully studied. A two-year field experiment was conducted to investigate the impacts of organic amendments on soil temperature, water storage depletion (SWSD), root distribution, grain yield, and the WP of summer maize (*Zea mays* L.) in the Guanzhong Plain of Northwest China. The control treatment (CO) applied mineral fertilizer without amendments, and the three amended treatments applied mineral fertilizer with 20 Mg ha^{-1} of wheat straw (MWS), farmyard manure (MFM), and bioorganic fertilizer (MBF), respectively. Organic amendments decreased SWSD compared to CO, and the lowest value was obtained in MBF, followed by MWS and MFM. Meanwhile, the lowest mean topsoil (0–10 cm) temperature was registered in MWS. Compared to CO, organic amendments generally improved the root length density (RLD) and root weight density (RWD) of maize. MBF showed the highest RLD across the whole soil profile, while MWS yielded the greatest RWD to 20 cm soil depth. Consequently, organic amendments increased grain yield by 9.9–40.3% and WP by 8.6–47.1% compared to CO, and the best performance was attained in MWS and MBF. We suggest that MWS and MBF can benefit the maize agriculture in semi-arid regions for higher yield, and WP through regulating soil hydrothermal conditions and improving root growth.

Keywords: soil temperature; soil water storage depletion; root growth; maize yield; semi-arid region

1. Introduction

The intensification of crop cultivation and livestock leads to a large increase in the production of straw and animal manure in semi-arid regions of China and elsewhere [1–3]. Recycling crop straw and animal manure by land application has been regarded as a practical method to improve general soil conditions and stabilize crop production [4–6]. However, more recently, farmers in the semi-arid regions of China have used mineral fertilizer instead of traditional organic materials due to low costs [2,7]. Consequently, mismanagement of crop straw and animal manure not only leads to poor soil quality and low water productivity [8,9], but also causes adverse environmental consequences, such as water and air pollution [10,11].

The Guanzhong Plain is one of the most important grain-production areas in the semi-arid region of China [8,12]. Summer maize (*Zea mays* L.) is a widely cultivated crop in this area, with an annual production of 3.4 billion kg of grains (Shaanxi Provincial Bureau of Statistics, http://www.shaanxitj.gov.cn/). In this region, organic amendments from plant residues or livestock manure have been used, with the primary goal of improving soil water conditions and crop water productivity. For instance, Yu et al. [7] showed that a three-year application of ammoniated straw (ammonification of crop straw through adding urea) improved several aspects of soil quality, and significantly increased soil water storage and crop water productivity in a summer maize, winter wheat rotation. Wang et al. [13] confirmed that maize water productivity increased by 3–8% based on the improvement of grain yield in organic manured soils over four years, although soil water content decreased in the 50–150 cm soil layer. However, Yang et al. [1] reported that 8-year manure amendments increased soil water content in the 0–10 cm soil layer compared with the sole application of mineral fertilizer, while it depressed the crop yield of the wheat-soybean cropping system. The discrepancy in observed effects may be due to the variations in environmental duration, cropping system, and the characteristics of applied organic materials. The differences in the compositions of various organic amendments have different effects on soil water status and crop water productivity by the changes in soil's physical, chemical, and biological functions [5,14,15]. Therefore, it is essential to assess the effects of various organic amendments on soil water conditions and crop water productivity for adjusting the optimum organic amendment practices to local conditions.

Roots are important organs for plant growth and mediators of associations between the effects of organic amendments on soils, shoots, and grain yields [16]. Root growth is crucial in optimizing crop yield and water productivity in semi-arid regions [17,18]. Previous studies have provided important insights into organic amendments in terms of regulating plant root growth. The nutrients released from decomposing organic materials play a key role in the penetration and establishment of plant roots [19–21], which helps the crop to utilize water and nutrients from deeper layers and to maintain high water productivity under soil water stress conditions [18,22]. The type of applied organic amendments is also an important factor [14,23]. For example, Gaiotti et al. [23] found that adding compost from vine-pruning waste strongly stimulated vines' root development, while compost derived from cattle manure waste had no significant effect on the root systems. However, the relationships between root distribution and both yield and water productivity remain key issues of discussion.

This study aims to evaluate the effects of organic amendments on soil hydrothermal conditions and root growth to ensure the stable increase in grain yield and water productivity of summer maize. For this purpose, a two-year study was undertaken in the Guanzhong Plain of Northwest China to evaluate the effects of three common organic amendments (wheat straw, farmyard manure, and bioorganic fertilizer) on soil temperature, soil water storage depletion, root growth, and their contributions to grain yield and water productivity of summer maize. We hypothesized that: (i) organic amendments could alter soil hydrothermal conditions and root growth, and therefore favor yield and water productivity (WP) increase; (ii) these organic amendments-induced effects would be variable due to the differences in the types of added organic materials.

2. Materials and Methods

2.1. Study Site

A field experiment was conducted through two consecutive summer maize-growing seasons (June–October) from 2014 to 2015 in a maize–wheat cropping system. The experimental site is located at the Key Laboratory of Agricultural Soil and Water Engineering in Arid Area (34°17′ N, 108°04′ E, 506 m a.s.l.) in the Guanzhong Plain, a semi-arid region of northwest China. This region has a semi-arid to sub-humid climate, with 632 mm annual rainfall and 13 °C mean annual temperature. The soil at the site is classified as Eum-Orthic Anthrosols [8] with basic physicochemical properties presented in Table 1 (sampled on 9 June 2014).

Table 1. Initial characteristics of the soil sampled in different soil layers in the experimental fields in June 2014.

Parameters	Depth (cm)		
	0–20	20–60	60–100
Texture (international system)	Clay loam	Silty clay loam	Silty clay loam
Sand (0.02~2 mm) (g 100 g^{-1})	37.44	29.44	24.11
Silt (0.002~0.02 mm) (g 100 g^{-1})	44.01	48.48	54.52
Clay (<0.002 mm) (g 100 g^{-1})	18.59	23.10	21.37
Bulk density (g cm^{-3})	1.38	1.58	1.39
Field capacity (cm^3 cm^{-3})	0.33	0.31	0.30
Organic matter (g kg^{-1})	14.57	10.03	8.04
Total N (g kg^{-1})	0.92	0.71	0.66
Total P$_2$O$_5$ (g kg^{-1})	0.58	0.47	0.38
Total K$_2$O (g kg^{-1})	10.12	9.47	8.67
Soil pH	8.57	8.53	8.36

2.2. Treatments and Crop Management

The experiment was conducted in a randomized complete block design with four treatments replicated three times in 7.5 m × 4.0 m plots (fixed plot test). Four treatments consisted of a control (CO) with additions of mineral fertilizer (N:P = 170:170 kg ha^{-1}) without organic matter, and the same doses of mineral fertilizer along with wheat straw (MWS), farmyard manure (MFM), and bioorganic fertilizer (MBF). Three types of organic materials were applied in the field at the rate of 20 Mg ha^{-1} (dry weight). The wheat straw was cut into 2 cm segments. The farmyard manure used was from a mixture of 80% decomposed sheep manure and 20% soil by weight, following the production method of local farmers. The bioorganic fertilizer consisted of farmyard manure and microbial agent (60 kg ha^{-1}, 2 × 10^8 cfu g^{-1} living bacteria, supplied by Sino Green Agri-Biotech Company, Beijing, China). Basic physico-chemical properties of the organic materials are listed in Table 2.

Table 2. Physico-chemical characteristics of the applied organic materials.

Material	Organic Carbon (g kg^{-1})	Total N (g kg^{-1})	Total P (g kg^{-1})	Total K (g kg^{-1})	Dry Bulk Density (g cm^{-3})
Wheat straw	432.44a	10.84b	1.65b	9.71a	0.074c
Farmyard manure	190.30b	11.63ab	3.02a	6.49b	0.474a
Bioorganic fertilizer	184.19b	12.98a	3.19a	6.27b	0.325b

Note: Different letters in the same column indicate significant differences at the 5% probability level.

Summer maize (Zhengdan 958) was planted manually on 18 June 2014 and 15 June 2015, and was harvested manually on 15 October 2014 and 13 October 2015. The plant population was approximately 51,892 plant ha^{-1} with a row spacing of 60 cm and a plant spacing of 30 cm. Maize was fertilized with a uniform dose of 170 kg N ha^{-1} and 170 kg P$_2$O$_5$ ha^{-1} in the form of urea (46% N) and diammonium phosphate (18% N, 46% P$_2$O$_5$). 60% of the mineral fertilizer was applied as a basal fertilizer, and the remaining 40% was top-dressed at the 12-leaf stage. Before planting, 60% of mineral fertilizer and all of the organic materials were mixed up with the soils by a moldboard plow to a depth of around 25 cm as per the treatment. Organic insecticide (15 kg ha^{-1}) was also incorporated into the soil to control pest insects before sowing. Hand weeding was performed at the 6-leaf stage and 12-leaf stage during the maize growing seasons. All experimental plots were irrigated using flood irrigation with the same amount of underground water (75 mm) at 42 days after sowing (DAS). Crop residues were manually removed after harvest.

2.3. Sampling and Measurements

Rectangular geothermometers (Jingda Thermal Instruments, Hengshui, China) were installed at depths of 0, 5, 10 and 20 cm in the middle of each plot. The soil temperature was recorded at 10-day intervals after sowing, and the values recorded at 8:00, 14:00 and 20:00 were averaged and treated as mean diurnal soil temperature.

Soil water content was measured gravimetrically at 10 cm intervals, down to 100 cm depth, at the times of 0, 40, 60, 80, 100 and 120 DAS in 2014 and 2015. Seasonal soil water storage depletion (SWSD) and evapotranspiration (ET) was determined according to the following equations [7,13]:

$$SWS = \sum SWC_i \times SBD_i \times SD_i, \tag{1}$$

$$SWSD_{j\sim j+1} = SWS_j - SWS_{j+1}, \tag{2}$$

$$ET = P + I + SWSD, \tag{3}$$

where SWS is the soil water storage in the 0–100 cm soil layer (mm), SWC is soil gravimetric water content (g g^{-1}), SBD is soil bulk density (g cm^{-3}), SD is soil depth (cm), $SWSD_{j\sim j+1}$ is the change in SWS form stage j (SWS_j) to stage $j + 1$ (SWS_{j+1}) (mm), ET is the evapotranspiration from the sowing to maturity stage (mm), P is rainfall (mm), and I is irrigation (mm). Water losses to runoff and deep drainage were assumed to be negligible as no heavy rains or irrigations occurred during the maize growing seasons in this flat experiment field.

On five sampling occasions during the growing seasons (40, 60, 80, 100, and 120 DAS), soil-root columns were taken at 10 cm increments to a depth of 100 cm using a root auger (7.3 cm internal diameter). The columns were randomly collected at three positions located at the planting spot (0 cm), 15 cm to one side (−15 cm) and 15 cm to the opposite side (+15 cm) of three tagged plants in each plot, and the mean value was used. Roots were gently washed by swirling water through a 400 μm sieve to remove the surrounding soil. Washed root samples (live root) were scanned with a HP scanjet 8200 scanner and analyzed using Delta-T Scan image analysis software (Delta-T Devices Company, Cambridge, UK) to determine root length. These root samples were then oven-dried at 75 °C to constant weight, and root dry weight was recorded. The root length (cm) and root weight (g) were divided by the soil-root core volume (cm^3) to calculate root length density (RLD, cm cm^{-3}) and root weight density (RWD, g cm^{-3}), respectively.

At maturity, the two central rows were manually harvested to determine the aboveground biomass and grain yields of summer maize (standardized to 12% moisture). Yield components, including numbers of kernel spike^{-1}, grain weight spike^{-1} and 100-grain weight were also derived from measurements on these harvested plants. Water productivity (WP, kg ha^{-1} mm^{-1}) was calculated by dividing the grain yield (kg ha^{-1}) by ET (mm).

2.4. Statistical Analysis

All treatment effects were subjected to univariate analysis using the SPSS 17.0 software (SPSS, Inc., Chicago, IL, USA). The treatment effects at different time intervals were tested by the analysis of variances (ANOVA). Significant differences among treatments were separated using least significant difference (LSD) tests at $p < 0.05$. The influence of all treatments on soil hydrothermal, maize root, and yield parameters and their relationships among each other were analyzed by the principal component analysis (PCA). The RLD and RWD distributions were visualized as two-dimensional graphs generated using the SURFER 8.0 software package (Golden Software, LLC, Golden, CO, USA).

3. Results

3.1. Weather Conditions during the Study Period

The precipitation during the maize growing season was 381.3 mm in 2014 and 278.2 mm in 2015, and this accounted for 56.3% and 46.5% of the annual precipitation, respectively (Figure 1). A large proportion of precipitation occurred in August and September, compared with the other months of the vegetative season over the two years. Based on a 30-year (1986–2015) average precipitation (322.1 mm) during the maize-growing season, 2014 and 2015 can be considered as a wet year and a normal year, respectively. The trends of average daily temperature were similar during the two maize-growing seasons. The highest daily mean temperature was 32.3 °C and 31.3 °C in 2014 and 2015, respectively, in both cases in July.

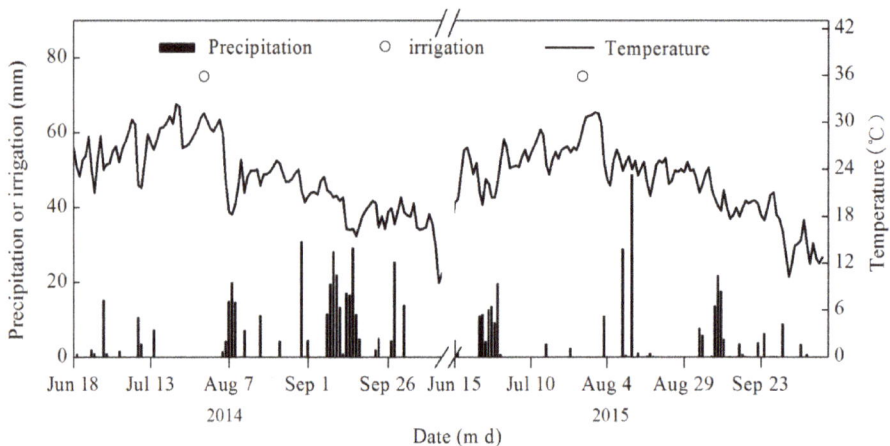

Figure 1. Daily rainfall, irrigation, and average daily temperature recorded during the summer maize growing seasons in 2014 and 2015.

3.2. Soil Temperature

The soil temperature trends were similar in the 2014 and 2015 growth periods (Table 3), as shown by the temporal and spatial variations in daily mean soil temperature under all treatments presented in Figure 2. Soil temperature at 0, 5, 10 and 20 cm depths followed a downward quadratic curve from sowing to maturation across all treatments, peaking at 40 DAS (July), following the similar pattern to the observed changes in air temperatures (Figure 1). Generally, organic amendments resulted in significant differences in the mean soil temperature at 0–10 cm depth, where the significantly ($p < 0.05$) lowest mean soil temperature was registered in MWS among all treatments. Nonetheless, the soil temperature remained almost invariable among the treatments at 20 cm depth, because it was weakly affected by solar radiation energy in this layer.

The effect of organic amendments on soil temperature was large at the early growth stages of maize (0–40 DAS), when the daily mean soil temperatures in MWS at 0, 5, and 10 cm depth were 1.3, 3.2, and 1.9 °C lower than those in CO, respectively ($p < 0.05$) (Figure 2). Conversely, the daily mean soil temperatures in MFM and MBF at 0 cm depth were 4.0 and 3.7 °C higher than those in CO, respectively ($p < 0.05$). This was probably due to the dark color of the farmyard manure and bioorganic fertilizer. The dark color organic materials showed high absorptivity and low reflectivity. However, at the middle growth stages (40–100 DAS), the full establishment of maize canopy and the spell of hot weather led to a minimal impact on soil temperature with organic amendments when compared to CO. At the later growth stages (100–120 DAS), organic amendments increased ($p < 0.05$) the daily mean soil temperatures at 0–10 cm depth by 0.6–1.6 °C compared to CO, due to the improvement of topsoil heat storage with the consecutive decomposition of organic materials, especially in cool conditions.

Table 3. Significance of the effects of organic amendment treatments, year, and their interactions on the soil and crop parameters.

Parameters	Tests of Normality	Tests of Homogeneity of Variances	ANOVA		
			Y	T	Y × T
Soil temperature	ns	ns	ns	*	ns
Soil water storage depletion	ns	ns	*	*	ns
Root length density	ns	ns	ns	**	*
Root weight density	ns	ns	ns	**	*
Grain yield	ns	ns	**	**	**
Water productivity	ns	ns	**	**	**

Note: Y, year; T, treatment; ns, not significant. *, significant differences at the 5% probability level; **, significant differences at the 1% probability level.

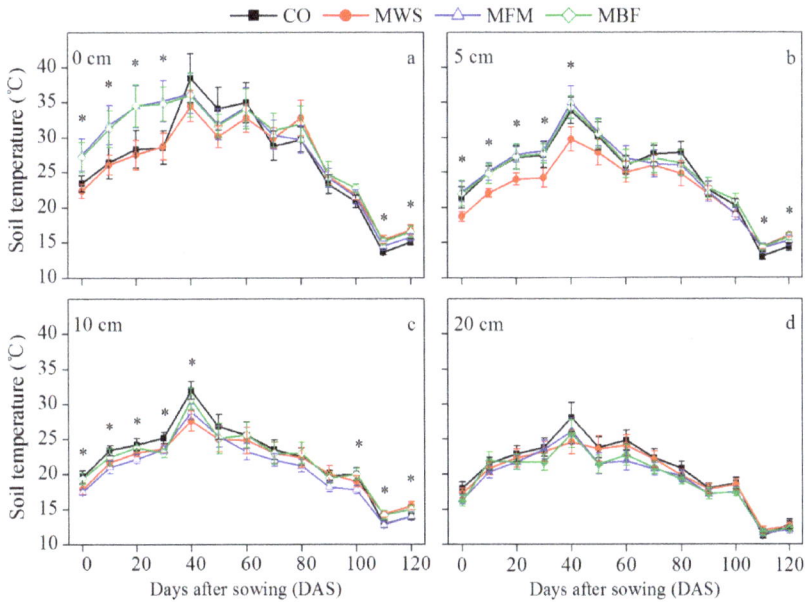

Figure 2. Effects of the treatments on average soil temperatures at indicated depths during the summer maize-growing seasons in 2014 and 2015. CO, mineral fertilizer without organic amendments; MWS, mineral fertilizer with wheat straw; MFM, mineral fertilizer with farmyard manure; MBF, mineral fertilizer with bioorganic fertilizer. *, $p < 0.05$.

3.3. Soil Water Depletion

As shown in Figure 3, organic amendments generally reduced the ET in the first 100 cm of the soil profile from sowing to harvesting over the two maize-growing seasons. More specifically, MBF significantly decreased the SWSD in the 0–100 cm soil layer by 18.3 and 29.5 mm compared to CO in 2014 and 2015 ($p < 0.05$), respectively. Meanwhile, MWS and MFM also decreased the SWSD by 0.8 and 4.1 mm in 2014 ($p > 0.05$) and 24.6 and 15.3 mm in 2015 ($p < 0.05$) compared to CO, respectively. Notably, the middle layer (20–60 cm) contributed to the major variations of the SWSD from the entire profile.

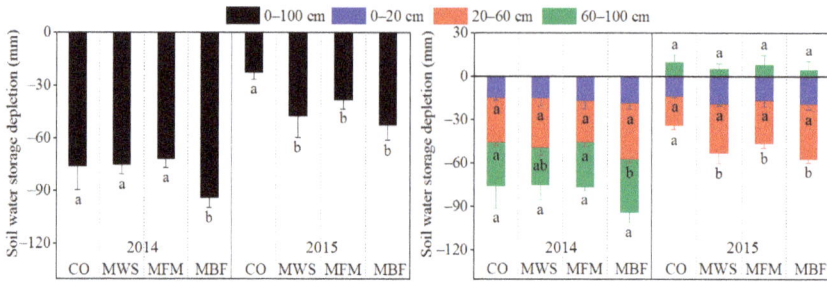

Figure 3. Total soil water storage depletion (SWSD) under indicated treatments from different soil layers in the 2014 and 2015 growing seasons. The abbreviations of the treatment names are the same as those described in Figure 2. Means of the SWSD within the same soil layer sharing the different lowercase letters are significantly different at $p < 0.05$.

There were clear variations in the SWSD across treatments at different growth stages (Figure 4), which depended on the intensity and distribution of rainfall and evapotranspiration. At the early growth stages, the maize plants were small and water depletion mainly resulted from soil evaporation. MWS and MBF reduced the SWSD compared to CO ($p < 0.05$) by decreasing soil evaporation loss in 2014, while these between-treatments differences were negligible in 2015. The reason for this phenomenon may be related to more replenished rainfall in 2015 (93.6 mm) than in 2014 (42.7 mm) during this stage. However, crop transpiration was the major pathway for water depletion due to the fast development of the maize canopy. This was especially true in MWS, which resulted in a significantly ($p < 0.05$) higher SWSD due to the best maize growth (Table 4) at 40–80 DAS over the two-year period. At 80–120 DAS, there was a desirable amount of rainfall in 2014 (224.6 mm) compared to the less rainfall (91.3 mm) in 2015. Therefore, there were no differences in the SWSD among the treatments in 2014, while MWS, MFM, and MBF resulted in significantly lower SWSD compared to CO in 2015 during this period ($p < 0.05$). The result clearly implied that the positive effects of organic amendments on SWSD mainly occurred in the early/middle–late growth stages of the dry spell.

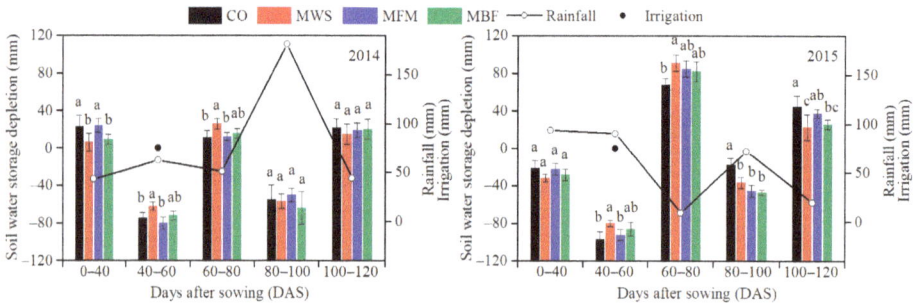

Figure 4. The soil water storage depletion (SWSD) in the 0–100 cm layers under indicated treatments during the different maize growth stages in 2014 and 2015. The abbreviations of the treatment names are the same as those described in Figure 2. Means of the SWSD within the same growth stage sharing the different lowercase letters are significantly different at $p < 0.05$.

Table 4. Maize grain yield, yield components, aboveground biomass, evapotranspiration (ET), and water productivity (WP) under indicated treatments in the 2014 and 2015 growing seasons.

Year	Treatments	Kernels Spike^{-1}	Grain Weight Spike^{-1} (g)	100-Grain Weight (g)	Grain Yield (kg ha^{-1})	Aboveground Biomass (kg ha^{-1})	ET (mm)	WP (kg ha^{-1} mm^{-1})
2014	CO	418.4c	106.4c	24.9c	5626.4d	11013.8b	379.7a	14.8c
	MWS	451.2b	136.4a	30.7a	7895.0a	13582.1a	380.5a	20.8a
	MFM	443.9bc	112.6bc	25.8c	6184.3c	11177.4b	383.8a	16.1b
	MBF	477.6a	117.3b	28.6b	7122.1b	12818.0a	361.4b	19.7a
2015	CO	424.1c	129.1b	29.8c	7204.8c	13434.6b	336.1a	21.4c
	MWS	489.4a	150.4a	31.8a	9376.1a	16631.3a	310.7b	30.2a
	MFM	461.1b	143.3a	31.4ab	8687.2b	14718.4b	320.9b	27.1b
	MBF	491.8a	151.3a	30.7bc	9662.6a	16266.9a	306.6b	31.5a

Note: The abbreviations of the treatment names are the same as those described in Figure 2. Different lowercase letters within columns in each year indicate significant differences at the 5% probability level.

3.4. Maize Root Systems

3.4.1. Vertical Distribution of RLD and RWD

There was no significant difference between the root densities at maturity in 2014 and 2015 (Table 3). The vertical distributions of RLD and RWD shown in Figures 5 and 6 were the average results over the two years.

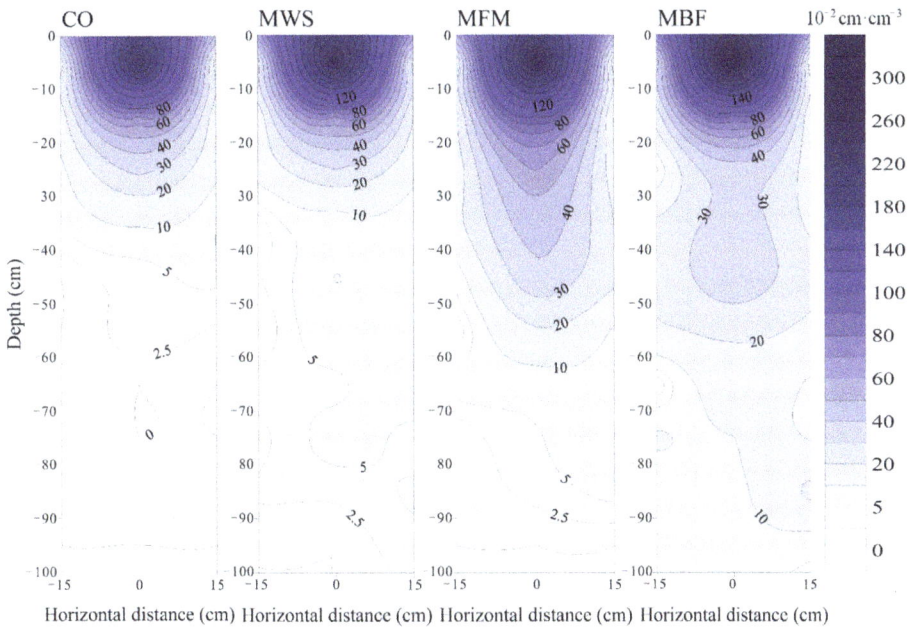

Figure 5. Two-dimensional distributions of the mean root-length density in the 0–100 cm layer under indicated treatments at maturity averaged over two years. The abbreviations of the treatment names are the same as those described in Figure 2.

The RLD in MBF was the highest among all treatments across all the soil profiles, and the ratio of the RLD between MBF and CO increased with increasing soil depth, which was 1.3, 3.1, and 20.5 times in the 0–20, 20–60, and 60–100 cm soil layers, respectively ($p < 0.05$) (Figure 5). A similar trend was observed in MFM, with 2.0 and 7.2 times higher RLD than that in CO at 20–60 cm and 60–100 cm depth, respectively ($p < 0.05$). However, this increase was only observed at 60–100 cm depth in MWS, where the RLD was 7.2 times higher in MWS than in CO ($p < 0.05$).

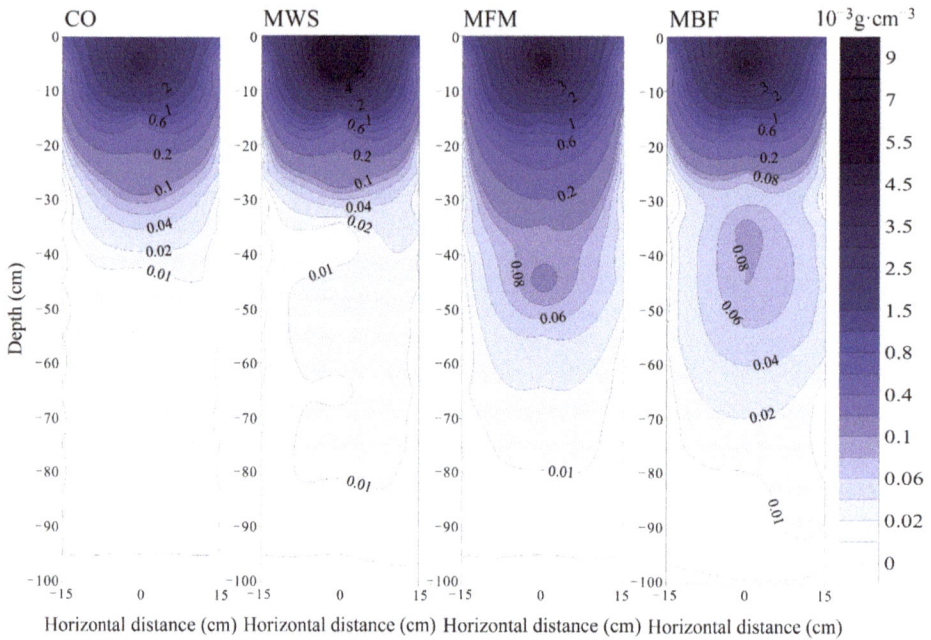

Figure 6. Two-dimensional distributions of the mean root-weight density in the 0–100 cm layer under indicated treatments at maturity averaged over two years. The abbreviations of the treatment names are the same as those described in Figure 2.

The RWD followed similar patterns to the observed RLD trends across the profile, except for the 0–20 cm depth, where the highest value of RWD occurred in MWS followed by MFM and MBF, and CO yielded the lowest value ($p < 0.05$) (Figure 6). At 20–60 cm depth, the RWD in MFM and MBF were significantly ($p < 0.05$) higher by 1.7 and 0.5 times than that in CO, respectively. At 60–100 cm depth, the RWD in MWS, MFM, and MBF were significantly ($p < 0.05$) higher by 4.4, 4.2, and 7.9 times that in CO, respectively. As expected, organic amendments improved the ability of the maize root system to grow down and stretch into the deep layer compared to CO, which was especially true in MBF.

3.4.2. Changes in RLD and RWD with Crop Development

Generally, different organic amendments had a significant impact on the root densities of maize at various growth stages in 2015 (Figure 7). MBF and MFM had significantly ($p < 0.05$) greater mean RLD compared to MWS and CO. This was especially true at 80–120 DAS, when the RLD in MFM and MBF were 10.8–65.1% significantly ($p < 0.05$) higher compared to MWS and CO (Figure 7a). Additionally, MWS also had positive changes in RLD compared to CO, but these changes were only significant ($p < 0.05$) at 40 DAS (35.2%).

There were notable between-treatment differences in RWD at 40 DAS and 80–120 DAS, when the mean RWD in MWS, MFM, and MBF was 106, 83.2, and 47.6% higher than that in CO, respectively ($p < 0.05$) (Figure 7b). Moreover, the RWD in MWS was 31.7–60.5% greater than that in other treatments at 40 DAS ($p < 0.05$), and these differences became more significant at 100 and 120 DAS.

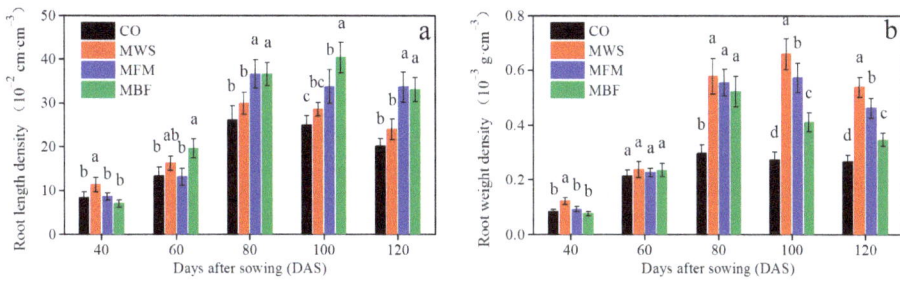

Figure 7. The root-length density (RLD) and root-weight density (RWD) under indicated treatments on different days after sowing during the 2015 growing season. The abbreviations of the treatment names are the same as those described in Figure 2. Means of the RLD or RWD within the same days after sowing sharing the different lowercase letters are significantly different at $p < 0.05$.

3.5. Maize Grain Yield, Yield Components, and WP

The grain yield, yield components, and WP of maize were considerably ($p < 0.05$) affected by the organic amendments, and the values were significantly ($p < 0.05$) higher in 2015 than in 2014 (Tables 3 and 4). Over the two years, organic treatments significantly ($p < 0.05$) increased grain yield (9.9–40.3%) and WP (8.6–47.1%) compared to CO, and MWS and MBF generally had the highest values. Similarly, MWS and MBF also had significantly ($p < 0.05$) higher aboveground biomass by 10.5–23.8% compared to CO and MFM. In addition, kernel spike^{-1}, grain weight spike^{-1} and 100-grain weight in MWS and MBF were significantly ($p < 0.05$) higher compared to CO over the two years, while MFM only had significantly ($p < 0.05$) higher values relative to CO in 2015.

3.6. Impact Factors for Yield and WP

The PCA of all treatments, soil hydrothermal characteristics, and crop growth parameters indicated that the first two principal components explained 70.4% of the experimental variance (Figure 8). Direction and size of the line revealed that a number of yield-related parameters were mainly associated with MBF and MWS treatments. However, these yield parameters, such as grain yield, WP, aboveground biomass, kernel spike^{-1}, grain weight spike^{-1}, and 100-grain weight were negatively related to MFM and CO treatments. Among the soil hydrothermal variables, soil temperature at 100–120 DAS was positively linked to crop yield and WP. The SWSD at 0–40 and 80–120 DAS were negatively correlated with yield-related parameters, while the SWSD at 40–80 DAS was positively related with the aforementioned parameters. Additionally, total SWSD from sowing to harvesting at 20–60 cm depth was negatively associated with yield-related parameters.

Among the crop root parameters, the RLD and RWD at 80–120 DAS were significantly related to grain yield and WP, and significant correlations were also detected between RWD at 60 DAS and both grain yield and WP (Table 5). The results indicated that root densities significantly affected grain yield and WP from the middle of the growing seasons onwards. With regard to soil depth, both RLD and RWD in the 0–20 and 60–100 cm layers at maturity provided a great contribution to crop yield and WP (Figure 8).

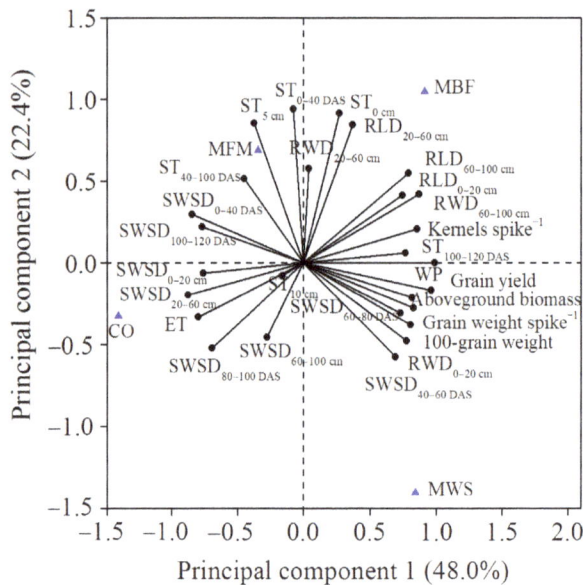

Figure 8. Principal component analysis (PCA) for various treatments as a function of soil and crop parameters. Treatment abbreviations are the same as those described in Figure 2. $ST_{0-40\ DAS}$, $ST_{40-100\ DAS}$, and $ST_{100-120\ DAS}$: soil temperature at 0–40, 40–100, and 100–120 DAS in the 0–10 cm soil layer; $ST_{0\ cm}$, $ST_{5\ cm}$, and $ST_{10\ cm}$: soil temperature at 0, 5, and 10 cm depth during the whole maize growth seasons; $SWSD_{0-40\ DAS}$-$SWSD_{100-120\ DAS}$: soil water storage depletion at 0–40, 40–60, 60–80, 80–100, and 100–120 DAS; $SWSD_{0-20\ cm}$, $SWSD_{20-60\ cm}$, and $SWSD_{60-100\ cm}$: soil water storage depletion from sowing to harvest at 0–20, 20–60, and 60–100 cm depths. $RLD_{0-20\ cm}$, $RLD_{20-60\ cm}$ and $RLD_{60-100\ cm}$: root length density at 0–20, 20–60, and 60–100 cm depths at maturity; $RWD_{0-20\ cm}$, $RWD_{20-60\ cm}$, and $RWD_{60-100\ cm}$: root weight density at 0–20, 20–60, and 60–100 cm depths at maturity; ET: evapotranspiration; WP: water productivity.

Table 5. Correlations between both grain yield and WP and root densities measured at different days after sowing of summer maize in various organic amendment management systems.

Root Densities		Days after Sowing (DAS)				
		40	60	80	100	120
Grain yield	RLD	0.024	0.649 **	0.506 *	0.584 *	0.577 *
	RWD	0.198	0.436	0.808 **	0.648 **	0.623 *
WP	RLD	−0.062	0.693 **	0.595 *	0.672 **	0.568 *
	RWD	0.109	0.487	0.818 **	0.598 *	0.562 *

Note: WP, water productivity; RLD, root length density; RWD, root weight density. *, $p < 0.05$. **, $p < 0.01$.

4. Discussions

4.1. Soil Temperature Responses to the Organic Amendments

Soil temperature is an important ecological factor for determining water and nutrient uptake by crop root systems [24]. Previous studies have shown the beneficial regulating effects of organic amendments on soil temperature for high soil organic carbon, soil available water content, and soil water holding capacity [4,25,26]. Consistent with these findings, MWS reduced the maximum soil temperature and raised the minimum soil temperature, while it significantly reduced the mean soil temperature at 0–10 cm depth compared to CO in our study (Figure 2). Hence, the practice of MWS can help reduce the adverse effects of terminal heat on the summer maize production in semi-arid regions [27]. By contrast, such beneficial effects did not present themselves in MFM and MBF. Previous studies on organic amendments have also reported that the incorporation of corn straw residue and farmyard manure failed to influence soil temperature [6,28]. Inconsistent results obtained here might be related to the type and characteristics of applied organic materials.

4.2. Soil Water Depletion Responses to the Organic Amendments

Understanding the process of soil water depletion of a cropping system in a semi-arid region helps maintain a soil hydrological balance and promote crop productivity. Many studies have demonstrated that adding manure and compost increased the available water content and reduced the soil water depletion by improving the soil's physical properties, such as porosity, infiltration, and water retention capacity [29,30]. Similarly, organic amendments generally decreased the SWSD in our experiment (Figure 3). The SWSD was significantly low in MBF plots (Figure 3), possibly because the bioorganic fertilizer with living bacteria strongly and rapidly improved the soil's physical environment and thus alleviated soil water stress [13,31]. In addition, MWS and MFM slightly reduced the SWSD compared with CO in 2014, whereas a significant decrease of SWSD under MWS and MFM occurred in 2015 (Figure 3). One probable reason for this may be the accumulative effects of the continuous application of organic materials [3], and another reason may be that nutrient management was more evident during the dry-spell period [20,29].

Furthermore, we observed that the effects of organic amendments on SWSD were mainly significant in the middle soil layer (20–60 cm) in our study (Figure 3). This phenomenon was partly consistent with previous studies in similar environments [13,32]. The likely reason for these effects was that loosened topsoil and good crop-root development affected by adding organic amendments promoted water storage and water extraction in the middle soil layer [19]. However, Wiesmeier et al. [33] observed that the application of organic wastes most strongly impacted soil water content in the topsoil, where organic amendments are mostly mixed by cultivation.

4.3. Maize Root Responses to the Organic Amendments

Organic amendments are important soil management practices affecting plant root growth. Previous researches have reported the beneficial roles of earthworm compost, green manuring, and farmyard manure on the penetration and establishment of plant roots [20,23,34]. Consistent with these findings, organic amendments generally exerted a clear improvement on the RLD and RWD of maize relative to CO in our experiment (Figures 5–7), because both soil temperature and water, being substantially influenced by organic amendments, affected plant root growth [17]. By contrast, Espinosa et al. [35] found that manure incorporation severely impaired the total root biomass of grasses when organic amendments were applied in clay subsoil to overcome nutrient deficiencies.

Not only did the maize produce a greater root system with organic amendments, but different types of amendments also had different effectivity on maize root densities—for example, MWS yielded the highest RWD, but the significantly highest RLD was recorded in MBF (Figures 5–7). This discrepancy may be attributed to two causes. Firstly, MWS exerted more favorable heat and water regulation capacities at the early growth stages (Figures 2 and 4), which were beneficial for primary roots' development (Figure 7). Normally, a more vigorous rooting system developed at the early growth stages can give the crop a better chance to survive and grow in semi-arid environments [17], which could lead to the highest RWD in the ensuing growth stages. Secondly, the high decomposition and mineralization turnover under MBF [36] could increase the supply of readily-available nutrients [19] and ultimately promote the development of 'finer' roots [21,23] relative to the other treatments, especially in the deeper layers (Figures 5 and 6). However, the related mechanisms still need to be explored in further studies.

4.4. Grain Yield and WP Responses to the Organic Amendments

Organic amendments had significantly beneficial effects on maize grain yield and WP over two growing seasons (Table 4), and similar findings have also been obtained in other crop species [8,11,20]. It is likely that organic amendments generally promote optimum root physiological function [20], which is favorable for the efficient utilization of soil water and nutrients [9,37] for higher yield and WP. Indeed, our results also demonstrated that grain yield and WP had significantly positive correlations with root densities from the middle of the growing seasons onwards (Table 5). Conversely, Yang et al. [1] found that the high C/N ratio of organic manure was responsible for its slightly negative effect on crop production in a wheat-soybean cropping system in loess soil, while Edmeades [38] found no significant influence of manure on crop production, with adequate fertilization in all plots.

With regard to the type of organic amendments, the maize grain yield and WP were higher under MWS and MBF than under MFM in both years (Table 4), presumably because they resulted in the closest optimum soil temperatures and water depletion process for promoting greater root density and ultimately boosting higher yield and WP (Figure 8) [10,20]. Interestingly, yield and WP were clearly higher in 2015 (Table 4) with lower cumulative rainfall (Figure 1) than in 2014, which might be attributed to the closer optimal rainfall distribution in 2015 [39].

5. Conclusions

Organic amendments resulted in a significant improvement in grain yield and WP of summer maize. Most importantly, MWS and MBF gained significantly higher values compared to MFM. The advantages under MWS and MBF were attributed not only to the optimized soil temperature in the top 10 cm layer and water depletion in the middle soil layer, but also to the improved root density. Overall, both MWS and MBF could be viable management practices used to alleviate production risks in maize agriculture in semi-arid regions. However, further research in this area is needed, including detailed analyses on the balance between crop nutrient requirements and the chemical composition of organic materials, focusing on long-term effects for relevant soil and crop parameters for sustainable agriculture.

Author Contributions: L.-L.Z. and H.-J.C. conceived and designed the experiments; L.-L.Z., L.-S.L. and X.-H.S. performed the experiments; L.-L.Z., L.-S.L., H.-J.C., X.-H.S. and C.Z. contributed reagents/materials/analysis tools; L.-L.Z. and L.-S.L. analyzed the data; L.-L.Z. wrote the paper; L.-S.L. and H.-J.C. revised the paper.

Funding: This work was supported by the National Key Research and Development Program of China [grant no. 2016YFC0400201] and 111 Project [grant no. B12007].

Acknowledgments: The authors thank Junliang Fan, Shaoping Xue and Jian Wang for their valuable assistance in the experimental work.

Conflicts of Interest: The authors declare no conflict interests.

References

1. Yang, X.Y.; Li, P.R.; Zhang, S.L.; Sun, B.H.; Chen, X.P. Long-term-fertilization effects on soil organic carbon, physical properties, and wheat yield of a loess soil. *J. Plant Nutr. Soil Sci.* **2011**, *174*, 775–784. [CrossRef]
2. Zhang, Y.L.; Li, C.H.; Wang, Y.W.; Hu, Y.M.; Christie, P.; Zhang, J.L.; Li, X.L. Maize yield and soil fertility with combined use of compost and inorganic fertilizers on a calcareous soil on the North China Plain. *Soil Tillage Res.* **2016**, *155*, 85–94. [CrossRef]
3. Kowaljow, E.; Gonzalez-Polo, M.; Mazzarino, M.J. Understanding compost effects on water availability in a degraded sandy soil of Patagonia. *Environ. Earth Sci.* **2017**, *76*, 255. [CrossRef]
4. Yang, B.J.; Huang, G.Q.; Qian, H.Y. Effects of straw incorporation plus chemical fertilizer on soil temperature, root micro-organisms and enzyme activities. *Acta Pedol. Sin.* **2014**, *51*, 150–157. (In Chinese) [CrossRef]
5. Yazdanpanah, N.; Mahmoodabadi, M.; Cerdà, A. The impact of organic amendments on soil hydrology, structure and microbial respiration in semiarid lands. *Geoderma* **2016**, *266*, 58–65. [CrossRef]

6. Lai, R.; Arca, P.; Lagomarsino, A.; Cappai, C.; Seddaiu, G.; Demurtas, C.E.; Roggero, P.P. Manure fertilization increases soil respiration and creates a negative carbon budget in a Mediterranean maize (*Zea mays* L.)-based cropping system. *Catena* **2017**, *151*, 202–212. [CrossRef]

7. Yu, K.; Dong, Q.G.; Chen, H.X.; Feng, H.; Zhao, Y.; Si, B.C.; Li, Y.; Hopkins, D.W. Incorporation of pre-treated straw improves soil aggregate stability and increases crop productivity. *Agron. J.* **2017**, *109*, 2253–2265. [CrossRef]

8. Li, S.; Li, Y.B.; Li, X.S.; Tian, X.H.; Zhao, A.Q.; Wang, S.J.; Wang, S.X.; Shi, J.L. Effect of straw management on carbon sequestration and grain production in a maize–Wheat cropping system in Anthrosol of the Guanzhong Plain. *Soil Tillage Res.* **2016**, *157*, 43–51. [CrossRef]

9. Das, A.; Patelb, D.P.; Kumara, M.; Ramkrushnaa, G.I.; Mukherjeec, A.; Layeka, J.; Ngachana, S.V.; Buragohain, J. Impact of seven years of organic farming on soil and produce quality and crop yields in eastern Himalayas, India. *Agric. Ecosyst. Environ.* **2017**, *236*, 142–153. [CrossRef]

10. Zhang, P.; Wei, T.; Jia, Z.K.; Han, Q.F.; Ren, X.L. Soil aggregate and crop yield changes with different rates of straw incorporation in semiarid areas of Northwest China. *Geoderma* **2014**, *230–231*, 41–49. [CrossRef]

11. Agegnehu, G.; Nelson, P.N.; Bird, M.I. Crop yield, plant nutrient uptake and soil physicochemical properties under organic soil amendments and nitrogen fertilization on Nitisols. *Soil Tillage Res.* **2016**, *543*, 295–306. [CrossRef]

12. Zheng, J.; Fan, J.L.; Zhang, F.C.; Yan, S.C.; Xiang, Y.Z. Rainfall partitioning into throughfall, stemflow and interception loss by maize canopy on the semi-arid Loess Plateau of China. *Agric. Water Manag.* **2018**, *195*, 25–36. [CrossRef]

13. Wang, X.L.; Ren, Y.Y.; Zhang, S.Q.; Chen, Y.L.; Wang, N. Applications of organic manure increased maize (*Zea mays* L.) yield and water productivity in a semi-arid region. *Agric. Water Manag.* **2017**, *187*, 88–98. [CrossRef]

14. Barzegar, A.R.; Yousefi, A.; Daryashenas, A. The effect of addition of different amounts and types of organic materials on soil physical properties and yield of wheat. *Plant Soil* **2002**, *247*, 295–301. [CrossRef]

15. Hossain, M.Z.; Niemsdorff, P.V.F.U.; Heß, J. Effect of different organic wastes on soil properties and plant growth and yield: A review. *Environ. Sci.* **2017**, *48*, 224–237. [CrossRef]

16. Liu, Z.; Zhu, K.L.; Dong, S.T.; Liu, P.; Zhao, B.; Zhang, J.W. Effects of integrated agronomic practices management on root growth and development of summer maize. *Eur. J. Agron.* **2017**, *84*, 140–151. [CrossRef]

17. Guan, D.H.; Zhang, Y.S.; Al-Kaisi, M.M.; Wang, Q.Y.; Zhang, M.C.; Li, Z.H. Tillage practices effect on root distribution and water use efficiency of winter wheat under rain-fed condition in the North China Plain. *Soil Tillage Res.* **2015**, *146*, 286–295. [CrossRef]

18. Chilundo, M.; Joel, A.; Wesström, I.; Brito, R.; Messing, I. Response of maize root growth to irrigation and nitrogen management strategies in semi-arid loamy sandy soil. *Field Crop. Res.* **2017**, *200*, 143–162. [CrossRef]

19. Gill, J.S.; Sale, P.W.G.; Peries, R.R.; Tang, C. Changes in soil physical properties and crop root growth in dense sodic subsoil following incorporation of organic amendments. *Field Crop. Res.* **2009**, *114*, 137–146. [CrossRef]

20. Bandyopadhyay, K.K.; Misra, A.K.; Ghosh, P.K.; Hati, K.M. Effect of integrated use of farmyard manure and chemical fertilizers on soil physical properties and productivity of soybean. *Soil Tillage Res.* **2010**, *110*, 115–125. [CrossRef]

21. Zhang, X.Y.; Cao, Y.N.; Tian, Y.Q.; Li, J.S. Short-term compost application increases rhizosphere soil carbon mineralization and stimulates root growth in long-term continuously cropped cucumber. *Sci. Hortic.* **2014**, *175*, 269–277. [CrossRef]

22. Mandal, U.K.; Singh, G.; Victor, U.S.; Sharma, K.L. Green manuring: Its effect on soil properties and crop growth under rice–wheat cropping system. *Eur. J. Agron.* **2003**, *19*, 225–237. [CrossRef]

23. Gaiottia, F.; Marcuzzoa, P.; Belfiorea, N.; Lovata, L.; Fornasierb, F.; Tomasia, D. Influence of compost addition on soil properties, root growth and vine performances of *Vitis vinifera* cv *Cabernet sauvignon*. *Sci. Hortic.* **2017**, *225*, 88–95. [CrossRef]

24. Yang, H.S.; Feng, J.X.; Zhai, S.L.; Dai, Y.J.; Xu, M.M.; Wu, J.S.; Shen, M.X.; Bian, X.M.; Koidec, R.T.; Liu, J. Long-term ditch-buried straw return alters soil water potential, temperature, and microbial communities in a rice-wheat rotation system. *Soil Tillage Res.* **2016**, *163*, 21–31. [CrossRef]

25. Mahanta, D.; Bhattacharyya, R.; Gopinath, K.A.; Tuti, M.D.; Jeevanandan, K.; Chandrashekara, C.; Arunkumar, R.; Mina, B.L.; Pandey, B.M.; Mishra, P.K.; et al. Influence of farmyard manure application and mineral fertilization on yield sustainability, carbon sequestration potential and soil property of gardenpea–french bean cropping system in the Indian Himalayas. *Sci. Hortic.* **2013**, *164*, 414–427. [CrossRef]

26. Zhang, S.L.; Wang, Y.; Shen, Q.S. Influence of straw amendment on soil physicochemical properties and crop yield on a consecutive mollisol slope in Northeastern China. *Water* **2018**, *10*, 559. [CrossRef]

27. Singh, V.K.; Singh, Y.; Dwivedi, B.S.; Singh, S.K.; Majumdar, K.; Jat, M.L.; Mishra, R.P.; Rani, M. Soil physical properties, yield trends and economics after five years of conservation agriculture based rice-maize system in North-Western India. *Soil Tillage Res.* **2015**, *155*, 133–148. [CrossRef]

28. Song, Z.W.; Gao, H.J.; Zhu, P.; Chang, P.; Deng, A.X.; Zheng, C.Y.; Mannaf, M.A.; Islam, M.N.; Zhang, W.J. Organic amendments increase corn yield by enhancing soil resilience to climate change. *Crop J.* **2015**, *3*, 110–117. [CrossRef]

29. Wang, X.J.; Jia, Z.K.; Liang, L.Y. Effect of straw incorporation on soil moisture, evapotranspiration, and rainfall-use effciency of maize under dryland farming. *J. Soil Water Conserv.* **2014**, *69*, 449–455. [CrossRef]

30. Ramos, M.C. Effects of compost amendment on the available soil water and grape yield in vineyards planted after land levelling. *Agric. Water Manag.* **2017**, *191*, 67–76. [CrossRef]

31. Eldardiry, E.; Hellal, F.; Mansour, H.; Elhady, M.A. Assessment cultivated period and farm yard manure addition on some soil properties, nutrient content and wheat yield under sprinkler irrigation system. *Agric. Sci.* **2013**, *4*, 14–22. [CrossRef]

32. Liu, C.A.; Li, F.R.; Zhou, L.M.; Zhang, R.H.; Jia, Y.; Lin, S.L.; Wang, L.J.; Siddique, K.H.M.; Li, F.M. Effect of organic manure and fertilizer on soil water and crop yields in newly-built terraces with loess soils in a semi-arid environment. *Agric. Water Manag.* **2013**, *117*, 123–132. [CrossRef]

33. Wiesmeier, M.; Spörlein, P.; Geuß, U.; Hangen, E.; Haug, S.; Reischl, A.; Schilling, B.; Lützow, M.; Kögel, K.I. Soil organic carbon stocks in Southeast Germany (Bavaria) as affected by land use, soil type and sampling depth. *Glob. Chang. Biol.* **2012**, *18*, 2233–2245. [CrossRef]

34. Canellas, L.P.; Olivares, F.L.; Okorokova-Façanha, A.L.; Façanha, A.R. Humic acids isolated from earthworm compost enhance root elongation, lateral root emergence, and plasma membrane H^+-ATPase activity in maize roots. *Plant Physiol.* **2002**, *130*, 1951–1957. [CrossRef] [PubMed]

35. Espinosa, D.; Peter, W.G.S.; Tang, C.X. Changes in pasture root growth and transpiration efficiency following the incorporation of organic manures into a clay subsoil. *Plant Soil* **2011**, *348*, 329–343. [CrossRef]

36. Khaliq, A.; Abbasi, M.K. Improvements in the physical and chemical characteristics of degraded soils supplemented with organic-inorganic amendments in the Himalayan region of Kashmir, Pakistan. *Catena* **2015**, *126*, 209–219. [CrossRef]

37. Meade, G.; Lalor, S.T.J.; Cabe, T.M. An evaluation of the combined usage of separated liquid pig manure and inorganic fertiliser in nutrient programmes for winter wheat production. *Eur. J. Agron.* **2011**, *34*, 62–70. [CrossRef]

38. Edmeades, D.C. The long-term effects of manures and fertilisers on soil productivity and quality: A review. *Nutr. Cycl. Agroecosys.* **2003**, *66*, 165–180. [CrossRef]

39. Wen, Z.H.; Shen, J.B.; Martin, B.; Li, H.G.; Zhao, B.Q.; Yuan, H.M. Combined applications of nitrogen and phosphorus fertilizers with manure increase maize yield and nutrient uptake via stimulating root growth in a long-term experiment. *Pedosphere* **2016**, *26*, 62–73. [CrossRef]

Article

Perturbation Indicators for On-Demand Pressurized Irrigation Systems

Bilal Derardja [1,2], Nicola Lamaddalena [2,*] and Umberto Fratino [1]

[1] Department of Civil, Environmental, Land, Building Engineering and Chemistry, Polytechnic of Bari, 70126 Bari, Italy; bilal.derardja@poliba.it (B.D.); umberto.fratino@poliba.it (U.F.)

[2] Land and Water Resources Management Dept., Centre International de Hautes Etudes Agronomiques Méditerranéennes (CIHEAM)—Mediterranean Agronomic Institute of Bari, 70010 Valenzano, Bari, Italy

* Correspondence: lamaddalena@iamb.it

Received: 18 January 2019; Accepted: 13 March 2019; Published: 18 March 2019

Abstract: The perturbation in hydraulic networks for irrigation systems is often created when sudden changes in flow rates occur in the pipes. This is essentially due to the manipulation of hydrants and depends mainly on the gate closure time. Such a perturbation may lead to a significant pressure variation that may cause a pipe breakage. In a recent study, computer code simulating unsteady flow in pressurized irrigation systems—generated by the farmers' behavior—was developed and the obtained results led to the introduction of an indicator called the relative pressure variation (RPV) to evaluate the pressure variation occurring into the system, with respect to the steady-state pressure. In the present study, two indicators have been set up: The hydrant risk indicator (HRI), defined as the ratio between the participation of the hydrant in the riskiest configurations and its total number of participations; and the relative pressure exceedance (RPE), which provides the variation of the unsteady state pressure with respect to the nominal pressure. The two indicators could help managers better understand the network behavior with respect to the perturbation by defining the riskiest hydrants and the potentially affected pipes. The present study was applied to an on-demand pressurized irrigation system in Southern Italy.

Keywords: pressurized irrigation systems; on-demand operation; perturbation; unsteady flow; hydrant risk indicator; relative pressure exceedance

---

1. Introduction

Pressurized irrigation systems have been developed during the last decades with considerable advantages as compared to open canals. They guarantee better services to the users and a higher distribution efficiency [1]. On-demand systems are designed to deliver water at the flow rates and pressures required by on-farm irrigation systems, considering the time, duration, and frequency as defined by the farmers [2].

In pressurized systems operating on-demand, a group of hydrants operating at the same time is known as a configuration. Changing from a configuration to another is the main origin of perturbation in pressurized irrigation systems [3].

Many authors [4,5] have emphasized that improving the performance of existing irrigation schemes is a critical topic to decrease excessive water use and enhance system efficiency.

To evaluate obstacles, constraints and opportunities, to set up a consistent modernization strategy and improve the irrigation service to users, Food and Agriculture Organization (FAO) has developed a methodology called MASSCOTE (mapping system and services for canal operation techniques, [6]). FAO is currently working with CIHEAM-Bari (Centre International de Hautes Etudes Agronomiques Méditerranéennes-Mediterranean Agronomic Institute of Bari) to adapt the MASSCOTE approach to pressurized irrigation systems. This new approach is called MASSPRES, or mapping system and services for pressurized irrigation systems. One of the main steps is to map the perturbation. The main objective of the present study is to define effective indicators that can estimate the unsteady flow effects in an on-demand pressurized system.

Fluid transient analysis is one of the most challenging and complicated flow problems in the design and operation of water pipeline systems. Transient flow control has become an essential requirement for ensuring the safe operation of water pipeline systems [7]. The computer modeling tools for the hydraulic unsteady flow simulation have been widely used in simple pipeline systems. Little is known about the behavior of the unsteady flow in complex pipe network systems. This phenomenon could be analyzed by several numerical methods. One of them is the method of characteristics, which could be used for very complex systems [8].

In a recent study [3], a user-friendly tool was developed to simulate the unsteady flow in a pressurized irrigation network using an indicator called relative pressure variation (RPV). In addition, a specific numerical analysis was carried out to select the appropriate gate-valve closing time to avoid potential pipe damages.

In the present work, two new indicators have been set up, the hydrant risk indicator (HRI), that describes the degree of risk of each hydrant on the system by causing pressure waves propagating through the system pipes, and the relative pressure exceedance (RPE), that represents the variation of the pressure in the system with respect to the pipe nominal pressure. The latter is interpreted as a warning signal that a pipe in the system may collapse.

The present study intends to provide both designers and managers with adequate analysis on the hydraulic behavior of the system under unsteady flow conditions, only generated by the opening and closing of hydrants (i.e., gate-valves). Other conditions are out of the scope of this work. In this perspective, the above-mentioned user-friendly tool was updated.

2. Methodology

2.1. Unsteady Flow Assumptions

Possible mechanisms that may significantly affect pressure waveforms include unsteady friction, cavitation, a number of fluid–structure interaction effects, and viscoelastic behavior of the pipe-wall material, leakages and blockages. These are usually not included in standard water hammer software packages and are often hidden in practical systems [9].

The usual assumptions (Reference [10]) have been considered to develop the software code:

- The flow in the pipeline is considered to be one-dimensional with the mean velocity and pressure values in each section.
- The unsteady friction losses are approximated to be equal to the losses for the steady-state losses.
- The pipes are full of water during all the transient flow and no water column separation phenomenon occurs.
- The wave speed is considered constant.
- The pipe wall and the liquid behave linearly elastically.

The Euler and the conservation of mass equations are:

$$\frac{dV}{dt} + \frac{1}{\rho}\frac{\partial P}{\partial s} + g\frac{dz}{ds} + \frac{f}{2D}V|V| = 0 \tag{1}$$

$$a^2\frac{\partial V}{\partial s} + \frac{1}{\rho}\frac{dP}{dt} = 0 \tag{2}$$

where, g is the gravitational acceleration (ms^{-2}), D (m) is the pipe diameter, V (ms^{-1}) is the mean velocity, P (Nm^{-2}) is the pressure, z is the pipe elevation (m), f is the Darcy–Weisbach friction factor and a (ms^{-1}) is the celerity. t (s) and s (m) represent the independent variables.

The equation includes V and its module for preserving the shear stress force direction on the pipe wall according to the flow direction.

The characteristic method makes it possible to replace the two partial differential Equation (1) and Equation (2) with a set of ordinary differential equations. The resulting equations will be expressed in terms of the piezometric head H (m). These equations are deeply described in any hydraulic textbook discussing the water hammer phenomenon (e.g., Reference [11]).

It is important to mention that the slope of the characteristic curves on the space–time planes is a function of V (s, t). This is introduced in the numerical solution procedure as explained hereafter.

$$C^+ : \frac{dV}{dt} + \frac{g}{a}\frac{dH}{dt} - \frac{g}{a}V\frac{dz}{ds} + \frac{f}{2D}V|V| = 0 \text{ only when } \frac{ds}{dt} = V + a \tag{3}$$

$$C^- : \frac{dV}{dt} - \frac{g}{a}\frac{dH}{dt} + \frac{g}{a}V\frac{dz}{ds} + \frac{f}{2D}V|V| = 0 \text{ only when } \frac{ds}{dt} = V - a \tag{4}$$

The equations $\frac{ds}{dt} = V + a$ and $\frac{ds}{dt} = V - a$ are the characteristics of the Equation (3) and Equation (4), respectively. The integration of ($\frac{ds}{dt} = V + a$) gives ($t = \frac{1}{V+a} \times s + $ constant), that is represented by the curve C^+. Similarly, for ($\frac{ds}{dt} = V - a$), ($t = -\frac{1}{a-V} \times s + $ constant) is determined and represented by the curve C^-, shown in Figure 1.

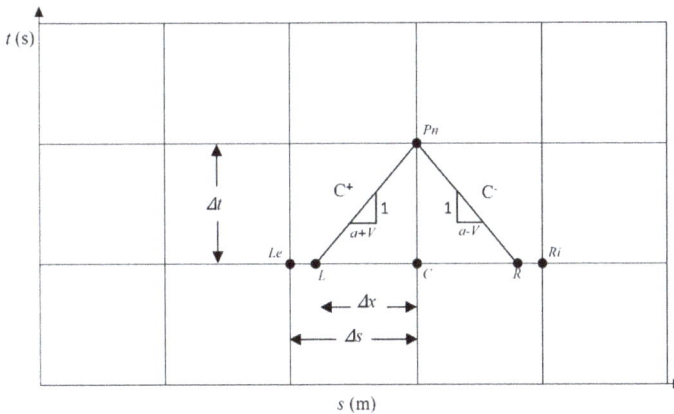

Figure 1. Characteristic curves on the space–time plane.

2.2. The Numerical Solution for the Ordinary Differential Equations

The characteristic curves can be approximated to straight lines over each single Δt interval. In fact: (i) Δt may be made as small as one wishes, and (ii) usually a >> V, causing $\frac{ds}{dt}$ to be nearly constant [12]. We seek to find the values of V and H at point *Pn*. They are calculated based on V and H at the points C, Le and Ri of the previous time following the characteristic curves C^+ and C^-. The velocity and the head at *Pn* become the known values for the subsequent time calculation, shown in Figure 1.

The characteristic curves passing through *Pn* intersect the earlier time (t is constant) at the points L and R. Consequently, the finite difference approximations to Equations (3) and (4) become

$$C^+ : \frac{V_P - V_L}{\Delta t} + \frac{g}{a}\frac{H_P - H_L}{\Delta t} - \frac{g}{a}V_L\frac{dz}{ds} + \frac{f}{2D}V_L|V_L| = 0 \tag{5}$$

$$C^- : \frac{V_P - V_R}{\Delta t} - \frac{g}{a}\frac{H_P - H_R}{\Delta t} + \frac{g}{a}V_R\frac{dz}{ds} + \frac{f}{2D}V_R|V_R| = 0 \tag{6}$$

The last two equations include six unknown terms: V_P, H_P, V_L, H_L, V_R and H_R. In the earlier time, values of P and V are known only at the points C, Le and Ri. Using linear interpolation, as shown in Figure 1, V_L, H_L, V_R and H_R are to be expressed as a function of V_C, H_C, V_{Le}, H_{Le}, V_{Ri} and H_{Ri}. In detail, along the C^+ characteristic, we assume:

$$\frac{\Delta x}{\Delta s} = \frac{V_L - V_c}{V_{Le} - V_c} = \frac{H_L - H_c}{H_{Le} - H_c} \tag{7}$$

solving the above equations for V_L and H_L, we obtain:

$$V_L = V_c + a\frac{\Delta t}{\Delta s}(V_{Le} - V_c) \tag{8}$$

$$H_L = H_c + a\frac{\Delta t}{\Delta s}(H_{Le} - H_c) \tag{9}$$

An analogous approach can be applied along the C^- characteristic. This leads to solving Equation (5) and Equation (6) simultaneously for V_{Pn} and H_{Pn}, as follows:

$$V_{Pn} = \frac{1}{2}[(V_L + V_R) + \frac{g}{a}(H_L - H_R) + \frac{g}{a}\Delta t(V_L - V_R)\sin\theta - \frac{f\Delta t}{2D}(V_L|V_L| + V_R|V_R|)] \tag{10}$$

$$H_{Pn} = \frac{1}{2}[\frac{a}{g}(V_L - V_R) + (H_L + H_R) + \Delta t(V_L + V_R)\sin\theta - \frac{a}{g}\frac{f\Delta t}{2D}(V_L|V_L| - V_R|V_R|) \tag{11}$$

Usually, the slope term ($\frac{dz}{ds} = \sin\theta$) is small and may be neglected [13].

The complexity of irrigation systems is the non-uniformity of pipe materials and pipe sizes, which requires a pipe discretization where each elementary section has constant geometrical and physical properties. Each elementary section is divided into an integer number of elements NS_i, with length Δs_i, whose value is calculated, to have the same Δt in all the system [14].

A steady-state simulation is executed for each configuration of hydrants simultaneously operating. The obtained results (H and V) will constitute the initial conditions for running the transient simulation. Assuming an instantaneous closure of valves, the computer code calculates the water hammer process until the simulation time reaches a predefined observation time (T_{max}), generally assumed large enough to reach again the new steady-state flow conditions.

The application of the differential equations assumes the boundary conditions described hereafter. The variables V and H are indexed with P_i corresponding to the points, one on each side of the boundary section, which is nearly superposed (Figure 2). For all the other parameters, only the number of pipes is used as an index to prevent any complication in naming. In both cases of upstream and downstream end boundaries of the systems, only one point exists following C^- and C^+, respectively.

2.3. The Boundary Conditions

The boundary conditions at each end of the pipes describe externally imposed conditions on velocity and/or pressure head. The used and the common ones are examined hereafter.

The potency of the characteristics method is the adequacy of analyzing each boundary and each conduit section separately along the unsteady flow time occurrence.

The boundary conditions described below were considered.

2.3.1. Reservoir

If a reservoir with constant pressure head H_0 is located upstream of the network, then:

$$H_{P_1} = H_0 \tag{12}$$

2.3.2. Valve

Being located at the downstream end of the pipes, the valve closure is assumed to induce a linear flow velocity variation at the cross-section according to the following equation:

$$V_{P_1} = V_0 \times \left(1 - \frac{t}{T_c}\right) \tag{13}$$

where V_0 (ms^{-1}) is the initial flow velocity and T_c (s) is the valve closure time.

In the case of a sudden closure ($T_c = 0$), V_{P_1} becomes equal to zero directly after the perturbation creation.

A detailed analysis with different gate-valves' closing time (from Tc = 0 to Tc = 6 s) was carried out and published in a previous study by the same authors [3]. The sudden closure was considered to clearly show the effect of the phenomenon.

2.3.3. Internal Boundary Conditions

Junctions with two and three pipes are considered:

1. Two-pipe junction:

    A two-pipe junction is shown in Figure 2a.

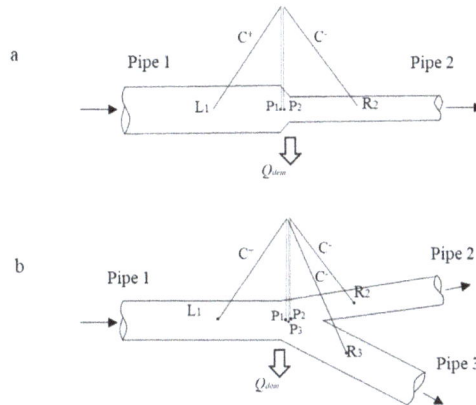

Figure 2. Boundary conditions at a typical series of (**a**) two and (**b**) three pipes junction.

In the case of no external demand, the values of the four unknowns can be found by solving the set of equations below:

- Following the C^+ equation (Equation (5)):

$$V_{P_1} = \left(V_{L_1} + \frac{g}{a_1}H_{L_1} - \frac{f_1 \Delta t}{2D_1}V_{L_1}|V_{L_1}|\right) - \left(\frac{g}{a_1}\right)H_{P_1} \tag{14}$$

- Following the C^- equation (Equation (6)):

$$V_{P_2} = \left(V_{R_2} - \frac{g}{a_2}H_{R_2} - \frac{f_2 \Delta t}{2D_2}V_{R_2}|V_{R_2}|\right) + \left(\frac{g}{a_2}\right)H_{P_2} \tag{15}$$

- The conservation of mass equation:

$$V_{P_1}A_1 = V_{P_2}A_2 \tag{16}$$

- The energy equation at the points P_1 and P_2, neglecting the difference in velocity heads and any local losses:

$$H_{P_1} = H_{P_2} \tag{17}$$

Solving the above system of equations, the head value H at the junction can be calculated as follows:

$$H_{P_1} = H_{P_2} = \frac{C_3 A_1 - C_1 A_2}{C_2 A_2 + C_4 A_1} \tag{18}$$

where C_1, C_2, C_3 and C_4 are the function of the known values obtained from the earlier time. By means of a back-substitution, also the flow velocities can be found.

In the case of a series of two pipes with an external constant demand Q_{dem} ($m^3 \ s^{-1}$) (that is delivered by one hydrant), a similar system of equations can be used, modifying Equation (16) only, as follows:

$$V_{P_1} A_1 = V_{P_2} A_2 + Q_{dem} \tag{19}$$

which makes Equation (18) become:

$$H_{P_1} = H_{P_2} = \frac{C_3 A_1 - C_1 A_2 - Q_{dem}}{C_2 A_2 + C_4 A_1} \tag{20}$$

2. Three-pipe junction

A three-pipe junction is shown in Figure 2b.

In the case of a pipe junction with one inflow and two outflows, the following equations are used to find the six unknowns:

$$\text{Pipe 1, } C^+: \ V_{P_1} = C_1 - C_2 H_{P_1} \tag{21}$$

$$\text{Pipe 2, } C^-: \ V_{P_2} = C_3 + C_4 H_{P_2} \tag{22}$$

$$\text{Pipe 3, } C^-: \ V_{P_3} = C_5 + C_6 H_{P_3} \tag{23}$$

$$\text{Conservation of mass}: \ V_{P_1} A_1 = V_{P_2} A_2 + V_{P_3} A_3 \tag{24}$$

$$\text{The energy balance, neglecting local losses between 1 and 2}: \ H_{P_1} = H_{P_2} \tag{25}$$

$$\text{The energy balance, neglecting local losses between 1 and 3}: \ H_{P_1} = H_{P_3} \tag{26}$$

Solving the previous set of equations leads to:

$$H_{P_1} = H_{P_2} = H_{P_3} = \frac{C_1 A_1 - C_3 A_2 - C_5 A_3}{C_2 A_1 + C_4 A_2 + C_6 A_3} \tag{27}$$

In the case of a three-pipe junction with an outlet, in the previous set of equations, only Equation (24) has to be modified, as follows:

$$V_{P_1} A_1 = V_{P_2} A_2 + V_{P_3} A_3 + Q_{dem} \tag{28}$$

while Equation (27) becomes:

$$H_{P_1} = H_{P_2} = H_{P_3} = \frac{C_1 A_1 - C_3 A_2 - C_5 A_3 - Q_{dem}}{C_2 A_1 + C_4 A_2 + C_6 A_3} \tag{29}$$

2.4. Calculation Process

In order to define the initial conditions for the unsteady flow analysis, a steady-state simulation was executed for each configuration. Starting from the upstream reservoir water level, by substituting the head losses calculated using the Darcy–Weisbach equation, the piezometric elevation (H) and the velocity (V) are defined in each section of the system.

Starting with the pre-computed H and V (calculated from the steady-state conditions), calculations of the new values H_{Pn} and V_{Pn} are carried out for each grid point with an increment of ΔT (Figure 1). Therefore, new values of H and V are obtained, and they replace the previous ones. The process continues for the pre-fixed simulation time. The software selects the maximum and the minimum pressure occurring at each section through the time of simulation (selection through time). A second selection through the pipe sections for P_{max} and P_{min} is performed (selection through space). The analysis results are tabulated as the maximum and minimum pressure head occurred for each pipe, which will be the basis of the calculation of the indicators.

As it was mentioned before, in this study T_{max} has been chosen to be equal to 30 s, where the water hammer magnitude variation was no more significant. The calculation process is summarized in the flow chart of the software (Figure 3).

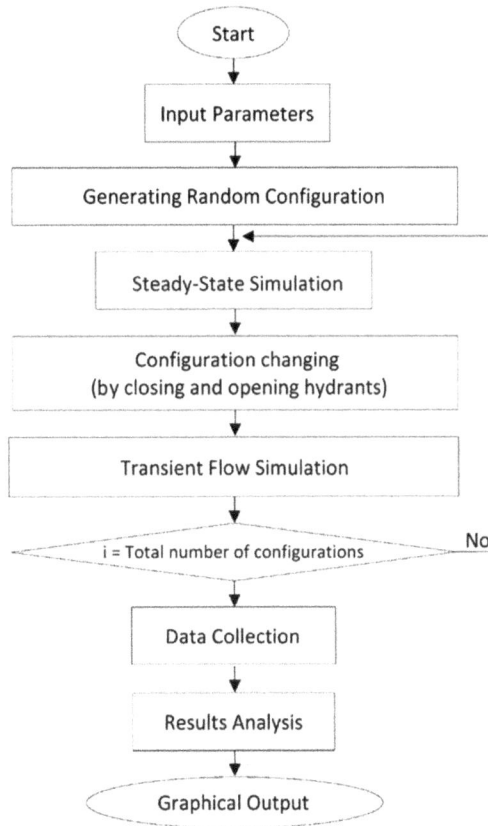

Figure 3. Computer code flowchart.

2.5. The Hydrant Risk Indicator (HRI)

By analyzing the impacts of different configurations on the hydraulic behavior of the irrigation system, the hydrant risk indicator evaluates the sensitivity of the network in terms of pressure to each hydrant manipulation. It is defined as the ratio between the participation of the hydrant in the riskiest configurations and its total number of participations.

A chosen percentage of the riskiest configurations will yield an upper and lower pressure envelope. The upper envelope represents the maximum pressure magnitude recorded through all the pipes, while the lower envelope represents the minimum recorded pressure magnitude values.

The HRI reflects the potential risk created by each hydrant. Hydrants significantly impacting the overall performance in terms of pressure are expected to appear more frequently in the risky configurations.

$$\text{HRI} = \frac{\text{RPN}}{\text{TPN}} \tag{30}$$

where TPN is the total participation number, and RPN is the risky participation number. A hydrant will be considered when it is being closed or opened, which is the main reason for the perturbation.

Knowing that the maximum and the minimum pressures are separately treated and presented:

- The HRI_{Pmin} indicates the ability of each hydrant to create a negative wave (P_{min}). RPN takes into consideration only the opening mode.
- The HRI_{Pmax} indicates the ability of each hydrant to create a positive wave (P_{max}). RPN takes into consideration the closing mode.

It is worth mentioning that the total number of configurations has to be taken to ensure that the indicator achieves the stabilization stage, as explained hereafter. The developed software does not identify the configuration as riskiest, it only identifies the probability of risk.

2.6. Relative Pressure Exceedance (RPE)

This numerically represents the pressure variation and the created risk with respect to the nominal pressure. The RPE was introduced to help both the designer and manager analyze the irrigation systems operating on-demand and illustrate the weak points of the system where any pipe damage may occur. The RPE is defined by:

$$\text{RPE} = 100 \times \frac{P_{max} - \text{NP}}{\text{NP}} \tag{31}$$

The RPE is the relative pressure exceedance (percentage), P_{max} (bar) is the maximum pressure recorded throughout the simulation time at each section, and NP (bar) is the nominal pressure. In the present study, NP was assumed equal 10 bars for a possible representation of the indicator. The RPE is presented as 10% equiprobability curves, where each curve represents a probability of occurrence.

The user introduces a percentage in the model that represents an envelope percentage of the considered risky for the system (depends on the sensitivity of the network and the norms chosen by the designer and/or the manager). After simulating the different configurations, the model selects the participation of each hydrant in that envelope and translate it to the indicator called Hydrant Risk Indicator.

3. The Case Study

The Sinistra Ofanto irrigation scheme, in the province of Foggia (Italy), is divided into seven irrigation districts (Figure 4). District four, in turn, is subdivided into sectors. The study was performed for the sector 25. The irrigation network is an on-demand pressurized irrigation system. It consists of 19 hydrants with a nominal discharge of 10 (L s^{-1}) and an upstream end constant piezometric elevation of 128 m a.s.l. The NP was equal to 10 bars for all installed pipes, as per the information obtained from the managers. The layout of sector 25 is reported in Figure 5.

With an irrigation network of 19 hydrants and four possible operating modes for each of them (open, closed, opening and closing), an enormous number of configurations could be generated, assuming that the network has been designed as having five hydrants simultaneously open. The selection of a rather small system fulfills the objective of clarity and simplification of the comparison between the results, and makes it possible to analyze different configurations of hydrants. Nonetheless, the code supports large-scale networks and the desired combination of hydrants.

The transient flow is defined as the transition from one steady flow to another steady flow. To keep the assumption of having five hydrants simultaneously open, the perturbation in the present study was created by closing two hydrants and substituting them by opening two new ones.

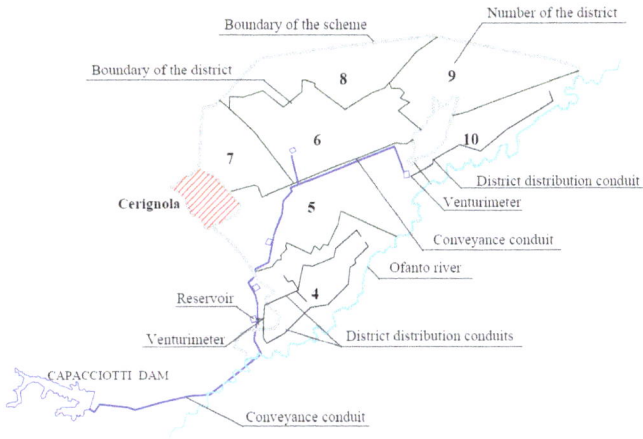

Figure 4. The Sinistra Ofanto irrigation scheme.

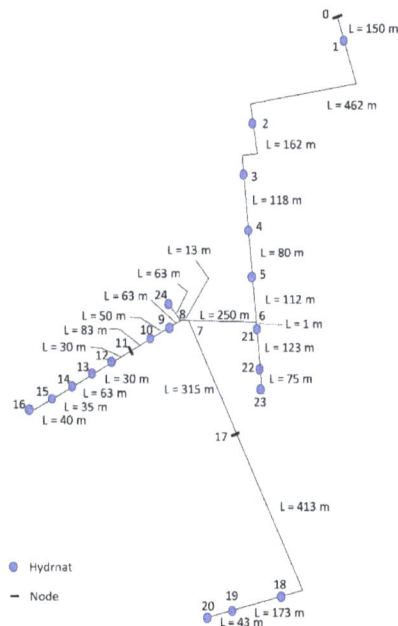

Figure 5. Layout of the network (sector 25).

4. Results and Discussion

4.1. The Uniformity of Random Generated Configurations

The strong variation of the discharges flowing into the network due to the variation in the demand is the first provoker of perturbation in pressurized irrigation systems. That variation was presented through different configurations. With a view to having a good representation, the software used a random number generator to run different configurations following a uniform distribution function (having the same possibility of getting one operating mode for each hydrant), as shown in Figure 6. The reported results refer to the opening/closing of two hydrants, as this situation occurs with a higher probability compared to the simultaneous opening/closing of three, four or five hydrants, and stronger waves with respect to the opening/closing of one hydrant do occur.

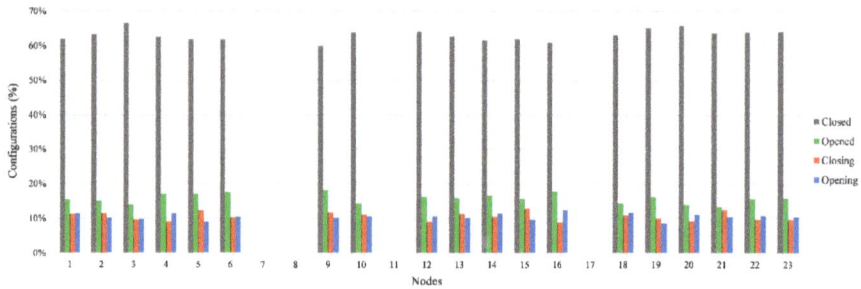

Figure 6. Uniformity distribution of the random number generator (500 hydrant configurations).

4.2. Computation of the Unsteady Flow Pressure Envelopes

In this study, the hydrant closing time Tc = 0 was considered, as it represented the riskiest case that may happen.

The irrigations system consisted of four layout profiles; only the calculated pressures for the profile Res-Node 16 are presented in Figure 7. After the perturbation, the maximum and the minimum pressure waves were recorded along the pipes and presented as 10% equiprobability curves.

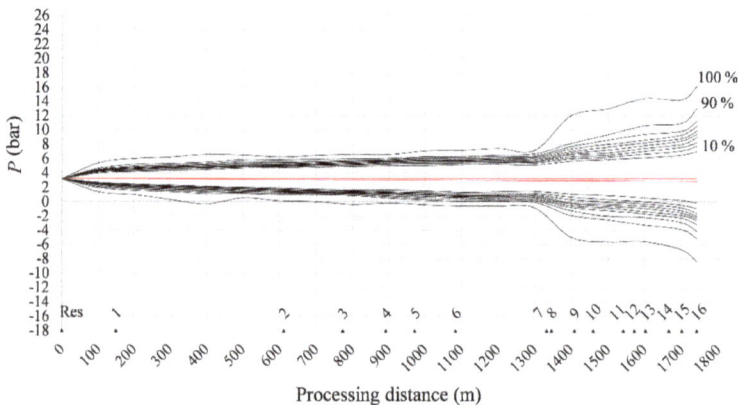

Figure 7. Pressure profile (Res-node 16) for 500 configurations, considering the simultaneous opening and closing of two different hydrants out of five.

Figure 7 shows that the maximum pressure variation for the different steady-state conditions. At node 16, this was around 0.35 bar (red lines) and increased when moving from the upstream end of the studied profile to downstream.

It is worth mentioning that the code imposed a constraint of not having the water column separation, even with the occurrence of low pressure, in line with the assumption mentioned above, that the pipes were assumed to be full and remained full during the transient flow occurrence, which enabled the application of the differential equations.

4.3. Calculation of the HRI

To assess the network response to the different perturbations of configurations, 500 different configurations were generated and analyzed. The number of configurations was run not only to satisfy the uniformity test, but also for the stability of the new indicator. In fact, the HRI graph started to take its final shape at around 350 configurations, where only the real risky hydrants appeared. Once the values were stabilized, the increase in the number of configurations did not significantly affect the results (Figures 8–10).

By running 500 configurations for the allocated system, where all nodes were represented, the contribution of the riskiest hydrants could be easily identified by noticing the extreme irregularities of the generated graphs (Figure 10). In the presented case, hydrants 14, 15, 16 and 20 could be identified as risky hydrants, which could generate positive pressure waves.

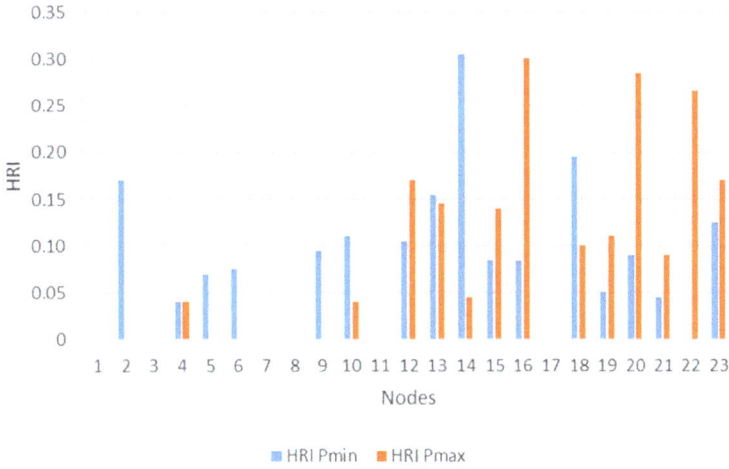

Figure 8. Hydrant risk indicator (HRI) for 50 configurations.

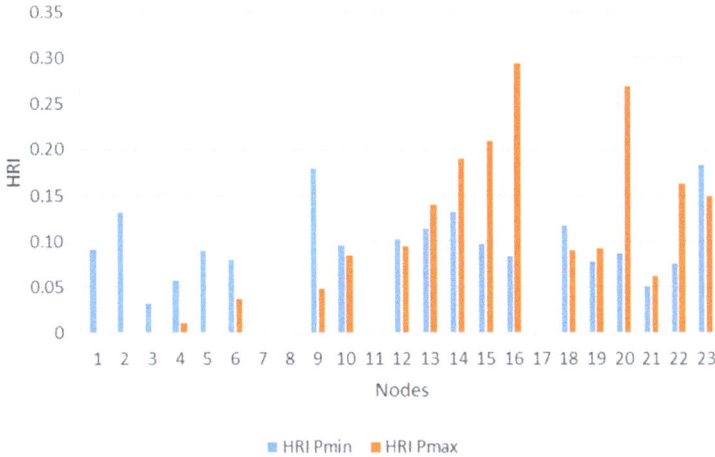

Figure 9. Hydrant risk indicator (HRI) for 350 configurations.

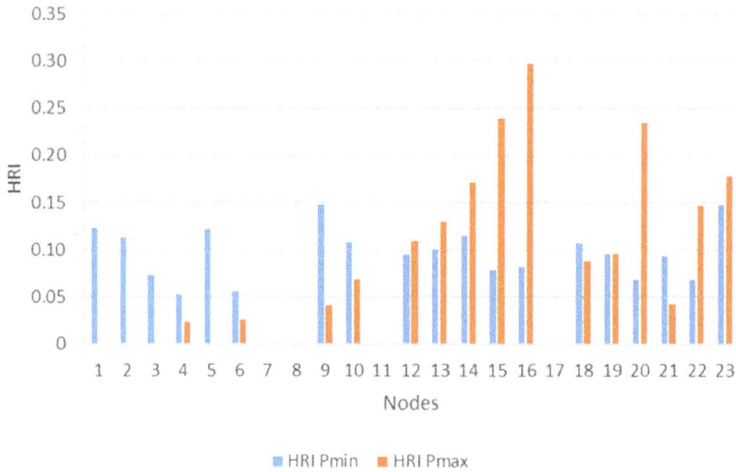

Figure 10. Hydrant risk indicator (HRI) for 500 configurations.

The term "risky configuration" could be precisely defined, in this case, as the configuration, which caused an exceedance of the allowed domain of the variation in pressure set by the manager according to certain criteria (mainly the system infrastructure).

In these extreme cases, risky hydrants had a high probability to cause either positive or negative waves. The impacts of such cases could cause serious problems.

4.4. Calculation of the RPE

The RPE provides a very clear idea about the pipes under risk (P_{max} influence). The pipes are considered to be safe when the RPE values are negative, which means that the maximum occurred pressure does not exceed the nominal pressure. As a value of zero means that the transient pressure is equal to the nominal pressure, from that value onwards the pipes start being under risk.

In Figure 11, the pipes for the main line (Res-Node 8) were in the safe range (RPE < 0). At the level of the node 8, which is the entrance of the branch 8–16, the RPE started to take positive values with less than a 10% probability of occurrence. The zone corresponding to the hydrants from 11 to 13 were potentially subject to failure with a 10% probability of occurrence. The more distant the section from the upstream end, the greater the risk of pipe failure. The failure reached its maximum occurrence probability of 40% (100 − 60%) at the downstream end of the layout profiles (hydrant 16).

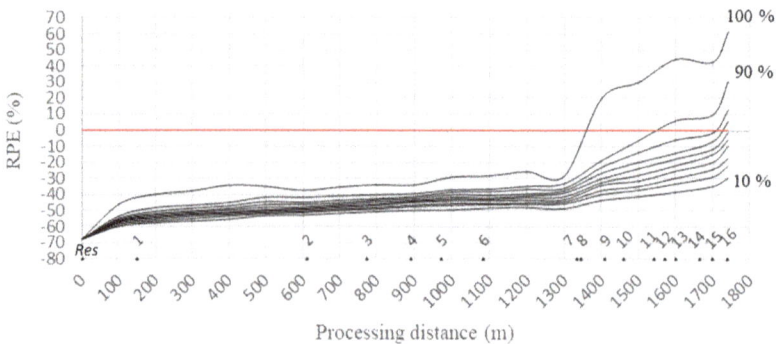

Figure 11. Relative pressure exceedance (RPE) for 500 configurations.

In parallel with the probability of occurrence and the corresponding zones, it is important to mention the role of the RPE that provides an overview of the exceedance severity.

5. Conclusions

The performance of on-demand pressurized irrigation systems is highly affected by unsteady flow, which highlights the importance of identifying appropriate indicators to be integrated into the designing and managing of a pressurized system.

To that end, user-friendly computer code capable of simulating the real operating conditions of a pressurized irrigation system, and consequently, the unsteady flow through the random opening and closing of hydrants was developed. The code made it possible to investigate and quantify the generated effects through two simple indicators developed in the framework of this study (the hydrant risk indicator, or HRI, and the relative pressure exceedance, or RPE).

Such informative indicators could significantly contribute to more efficient operation management of on-demand pressurized systems by avoiding highly risky probabilistic configurations. Moreover, they could be embedded in the designing phase, allowing for a better interpretation of the impacts of different design alternatives.

Author Contributions: The software was developed by B.D. under the supervision of N.L. and U.F.; the methodology, U.F.; data curation, N.L.; writing—original draft preparation, B.L.; writing—review and editing, N.L. and U.F.

Funding: This research received no external funding.

Conflicts of Interest: The authors declare no conflict of interest.

References

1. Lamaddalena, N.; Sagardoy, J.A. *Performance Analysis of On-Demand Pressurized Irrigation Systems*; Food and Agriculture Organization: Roma, Italy, 2000; Volume 59.
2. Calejo, M.; Lamaddalena, N.; Teixeira, J.; Pereira, L. Performance analysis of pressurized irrigation systems operating on-demand using flow-driven simulation models. *Agric. Water Manag.* **2008**, *95*, 154–162. [CrossRef]
3. Lamaddalena, N.; Khadra, R.; Derardja, B.; Fratino, U. A New Indicator for Unsteady Flow Analysis in Pressurized Irrigation Systems. *Water Resour. Manag.* **2018**, *32*, 3219–3232. [CrossRef]
4. Clemmens, A.J. Improving irrigated agriculture performance through an understanding of the water delivery process. *J. Irrig. Drain.* **2006**, *55*, 223–234. [CrossRef]
5. Playán, E.; Mateos, L. Modernization and optimization of irrigation systems to increase water productivity. *Agric. Water Manag.* **2006**, *80*, 100–116. [CrossRef]
6. Renault, D.; Facon, T.; Wahaj, R. *Modernizing Irrigation Management: The MASSCOTE Approach–Mapping System and Services for Canal Operation Techniques*; Food and Agriculture Organization: Roma, Italy, 2007; Volume 63.
7. Abuiziah, I.; Oulhaj, A.; Sebari, K.; Ouazar, D.; Saber, A. Simulating Flow Transients in Conveying Pipeline Systems by Rigid Column and Full Elastic Methods: Pump Combined with Air Chamber. *Eng. Technol. Int. J. Mech. Aerosp. Ind. Mechatron. Manuf. Eng.* **2013**, *7*, 2391–2397.
8. Wichowski, R. Hydraulic transients analysis in pipe networks by the method of characteristics (MOC). *Arch. Hydro-Eng. Environ. Mech.* **2006**, *53*, 267–291.
9. Bergant, A.; Tijsseling, A.S.; Vítkovský, J.P.; Covas, D.I.; Simpson, A.R.; Lambert, M.F. Parameters affecting water-hammer wave attenuation, shape and timing—Part 1: Mathematical tools. *J. Hydraul. Res.* **2008**, *46*, 373–381. [CrossRef]
10. Wylie, E.B.; Streeter, V.L.; Suo, L. *Fluid Transients in Systems*; Prentice Hall: Englewood Cliffs, NJ, USA, 1993; Volume 1.
11. Chaudhry, M.H. *Applied Hydraulic Transients*; Springer: Berlin/Heidelberg, Germany, 1979.
12. Larock, B.E.; Jeppson, R.W.; Watters, G.Z. *Hydraulics of Pipeline Systems*; CRC Press: Boca Raton, FL, USA, 1999.

13. Chaudhry, M.H. Transient-flow equations. In *Applied Hydraulic Transients*; Springer: Berlin/Heidelberg, Germany, 2014; pp. 35–64.
14. Lamaddalena, N.; Pereira, L.S. Pressure-driven modeling for performance analysis of irrigation systems operating on demand. *Agric. Water Manag.* **2007**, *90*, 36–44. [CrossRef]

water

MDPI

Article

A New Water Governance Model Aimed at Supply–Demand Management for Irrigation and Land Development in the Mendoza River Basin, Argentina

Mario Salomón-Sirolesi [1],* and Joaquín Farinós-Dasí [2]

[1] Facultad de Ingeniería, Universidad Nacional de Cuyo, Centro Universitario, Ciudad de Mendoza M5502JMA, Argentina

[2] Departamento de Geografía e Instituto Interuniversitario de Desarrollo Local, Universidad de Valencia.Av. Blasco Ibáñez, 28 (3ª planta), 46010 Valencia, Spain; Joaquin.Farinos@uv.es

*   Correspondence: salomonmario@yahoo.com.ar; Tel.: +54-261-4685181

Received: 26 January 2019; Accepted: 27 February 2019; Published: 5 March 2019

Abstract: This study aimed at achieving an organizational solution for improving the governance of water and land use and, consequently, improving the supply–demand water balance. Related modeling applied to diverse scenarios focus on water and land use development in the Mendoza River basin. A strategic analysis of water organization was performed using causal analysis, producing a Strategic Map (SM) and designing a Balanced Scorecard (BS). To assess the basin's water resources supply and demand, the Water Evaluation and Planning (WEAP) model was applied to the Administrative Management Units existing in the basin, taking into consideration the water availability and the granted water rights. The application of the organizational and governance model to various scenarios referring to 2030 show that by reordering allocations and water use criteria, implementing a better farm irrigation water management, improving capacity building of existing human resources, and adopting more adequate hard- and software for dams and canal management, it will be possible to accommodate demand in 2030 better than at present despite climate change impacts on demand and supply. In addition, users' participation will be enhanced.

Keywords: water and land management; water users' organization; water balance; supply–demand balance model; organizational analysis; participatory management

---

1. Introduction

The water governance model currently used in the Mendoza River basin, Argentina (Figure 1), brings about imbalance and unevenness of management inadequate for an area in full transformation. Spatio-temporal effects of great intensity and magnitude refer to the limited autonomy and self-sufficiency of local management organizations, the inadequate distribution and use of water resources, the separate management of water and land, the impact of territorial, economic, and productive competitiveness, the degradation of soils, all affecting crops production and causing poor service to users [1]. The water management model emerged in the nineteenth century, derived progressively very top–down and technocratic, but functional to the corporate external as well as internal interests. As a result, hierarchical, highly centralized power of decisions does not comply with the principle of accountability. Improved control would be feasible through an administrative decentralization process by watersheds and subareas. The primal social contract for the water administration in Mendoza implemented by the traditional and conservative elite in 1884, in a time of great and rapid agricultural expansion, does not fit well in current conditions; besides, it has caused very negative consequences, difficult to correct, to the territorial–hydrological system. As emblematic cases, demonstrating such trends: a) despite having passed a century since then, it still lacks the implementation of water balances as well as legislative changes facilitating reallocation of water rights; b) loss of water rights in productive areas because use changes from agricultural to recreational ones, affecting the functioning of the system as a whole; c) groundwater overexploitation and speculative use of water resources for more than 40 years, which has caused the loss of the aquifer productivity and quality; and d) failure to consolidate autonomous water organizations for productive local units, according to appropriate monitoring processes.

The consequences of this model's 19th century view, extemporaneous today, affect social, economic, and productive dimensions for not ensuring efficient administration and satisfaction of the demand [2]. Appropriate practices for the sustainable management of water in the basin include real-time water measurement, efficient application of irrigation modules and plans, uniformity in the application of water, clean production associated with water quality, improved land leveling for better performance of irrigation, soil mulching to reduce evapotranspiration, use of adequate irrigation flow rates, monitoring of soil moisture and plant water status, proper design of irrigation units, conservation of drains, and training of water system operators and users [3,4]. Meanwhile, the main issue is to implement a governance model that provides for improved participatory management through local water user organizations.

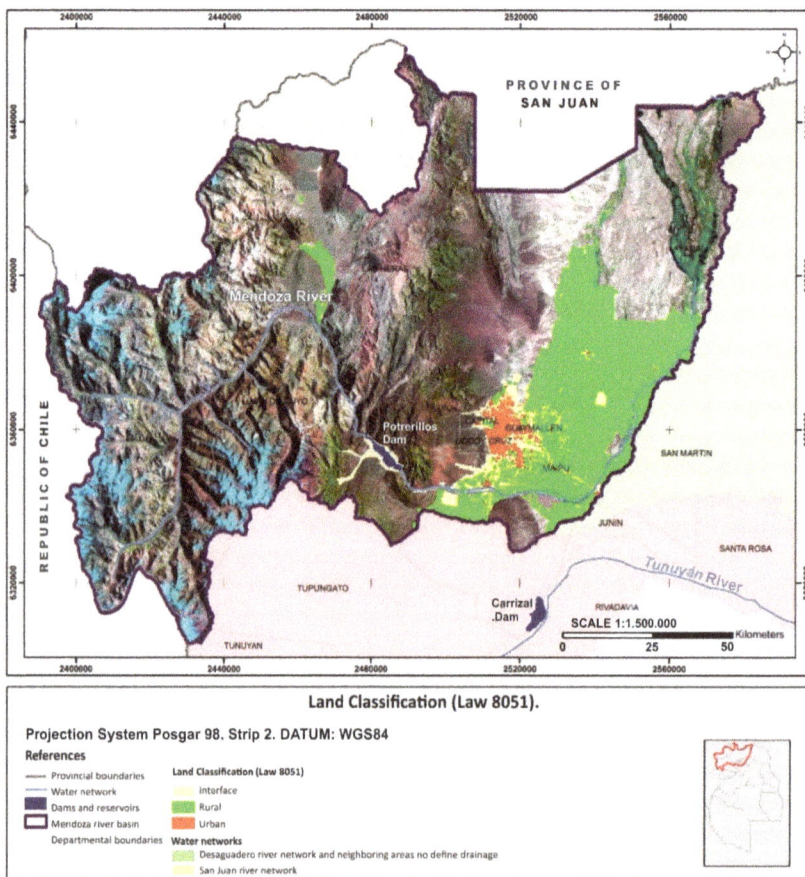

Figure 1. Mendoza River basin [5].

It is worthy of notice that, as part of the decentralization process implemented by the end of the 20th century [6], the Mendoza River Water User Organizations (WUOs) contributed not only to establishing a new governance, but also to the integrated process of regional development, namely evidencing that water management is a critical factor in a semiarid region [7]. WUOs had highly influenced the elaboration of the Land Use Law and the Strategic Development Plan (SDP), the execution of which required an effective supply–demand water balance implementation, restructuring of granted rights, respect for basin autonomy, effective representation and participation of the different water users, and consensual reformulation of the instruments necessary to achieve integrated water management [8]. In this process, there appears that, in the absence of the State, community management and associative modes have been efficient in supplying water at a local scale [9]. Territorial governance is thus a nonexclusive governmental dimension, where public participation influences decision-making and the social and spatial structure of processes involved [10]. Thus, under high social and public control, land and water governance strategies and integrated management services are expected to ensure strategic and productive territorial development [11].

The main objective of this paper was the formulation of a participatory organizational model consistent with current requirements of users, as well as with the water supply–demand balance modeling, in order to reach an equitable and sustainable water resources availability and use in the Mendoza River basin. It is foreseen that the organizational governance proposal will contribute to achieving greater efficiency and effectiveness in the water and land use together with the implementation of the administrative act when considering different prospective modalities relative to various context scenarios. Accordingly, objectives include the identification of strategies to promote the decentralized management of water, namely those referring to the ways in which WUOs may implement water demand management actions based on present supply–demand water balance issues.

The strategic formulation of the governance model was carried out by defining the organizational identity through the determination of the mission, vision, values, and strategies of WUOs. The organizational analysis was developed through the preparation of a strategic map [12] and the linking of the WUOs through the design of the Integral Scorecard (IS) [13]. Secondly, the available water resources of the Mendoza River basin [14] were assessed through updated water balance (WB) studies [15] and the application of the Water Evaluation and Planning model (WEAP) [16]. The latter tool allows contrasting water supply and demand considering a distribution system marked by spatial and temporal variability. An adjusted modeling approach [17] was used for different scenarios according to the peculiarities of the basin relative to each of the basin's Administrative Management Units (AMUs) and considering the availability of water and the granted uses [18,19]. Based on the constitutional law of Mendoza, which previses the water balance and assesses for the reallocation of rights, three scenarios have been considered in water modeling: i) trend, which is to continue with the allocation of water without changing the category of agricultural rights in use and with the current efficiencies; ii) possible, which is to equate the agricultural rights in use and delivery of 100% of the endowment improving the efficiencies, and iii) contrasted, which is to distribute the water with 100% to all the rights registered whatever the use and improving the efficiencies. For this purpose, different criteria for irrigation planning were analyzed [20], irrigation requirements and scheduling were considered [21], and strategies for water management were identified [22].

2. The Study Area

The Mendoza River basin is located in the Andean Central West Region of Argentina, covering an area of 19,553 km^2 with a population of 1,170,000 inhabitants [23]. It includes a densely populated sector corresponding to the urbanized oasis, an intensive irrigated area, and areas of great natural value as well as areas with more extensive uses such as units of mountain range, premountain range, and non irrigated alluvial plains (Figure 1). River flows, namely those aimed at the supply of the urban and irrigated areas, are regulated by the Potrerillos Dam [24].

With the aim to achieve a systematized knowledge of the basin with snow-glacial regime, it must be considered that the main contribution, represented by snowfall, its accumulation–compaction, and freezing–melting phases, generates direct flows complemented, to a lesser extent, with rainfall contributions. These flows cause surface runoff, as well as subsurface and groundwater flows corresponding to the various components of the basin water balance (Figure 2), which must be assessed in order to obtain a comprehensive knowledge of water supply [14].

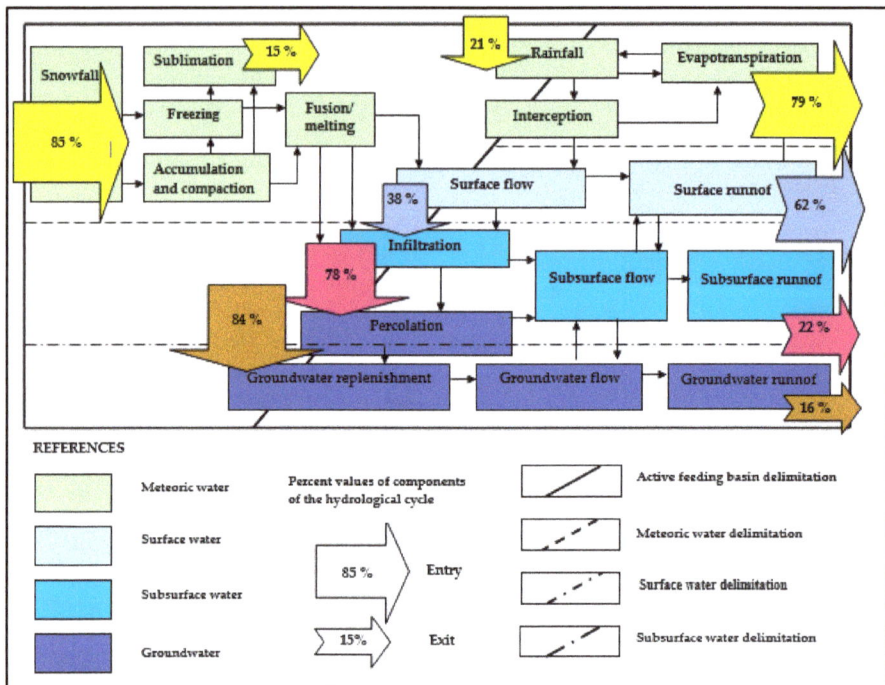

Figure 2. Main water balance components and physical processes in the Mendoza River basin.

3. Formulation of the Organizational Model

The strategic analysis of water organization was performed using the causal analysis method, thus considering that such organization is a constituent part of the development of a continuous, dynamic process. The sequential chart proposed includes several stages and steps corresponding to the main timings, which have been adapted from the strategic map method [12], to identify, organize, and describe strategies within the context of the water management model (Table 1).

Table 1. Sequential description of organizational analysis [12].

Stages	Steps	Description
	Mission	Why do we exist?
Organizational identity	Vision	What do we want to do?
	Values	What is important to us?
	Strategy	Our game plan
Organizational analysis	Strategic map	Translate the strategy
Organizational linkage	Balanced scorecard	To act, measure, and focus

At this stage, we proceeded to describe stakeholders involved in the proposed water organization, based on the present situation and envisaging the future one [25]. The required water management objectives have also been considered in order to understand the complexities and problems of managing the Mendoza River basin mainly in terms of making compatible the balance between supply and demand. At this stage:

- *Mission* is highlighted as the base of the water governance, the reason for its existence and its purpose, which is reflected in its activities.
- *Vision*, in turn, presents an image of the future, the course desired to be adopted and that enables knowing what is to be accomplished.
- *Values* allow knowing which aspects are important to water governance and constitute the reference framework for its image within the community. Values define the set of bases and principles that regulate water management of the organization and allow the building of the institutional philosophy [26].
- *Strategy* is the set of ordered actions that are developed in a dynamic way, according to the context and the capabilities available to implement them. It describes how a water organization intends to create values in the organization in relation to the services it offers, as well as to their implementation. The strategy covers various topics in a simultaneous and complementary manner, depending on their implementation timings, with operational processes usually being quicker than those that include the application of innovation processes. It also requires a specific link between the users and the values proposed to meet their needs [27].

A Strategic Map (SM) presents a causality structure enabling identification of components and interrelations of the organizational model's strategy with the aimed processes and results. It also allows assessing, measuring, and improving the most critical processes leading to their successful implementation. It makes it possible to conduct a strategic analysis, interpret the development stage of the strategy, and visualize the connection between tangible and intangible assets. Furthermore, an SM eases the assessment and selection of strategic options based on quantitative and qualitative criteria. The strategy's most critical factor is its efficient implementation, in order to ensure a sustained creation of value. In turn, this depends on the management of key internal processes, namely financial, operational, relations with costumers and innovation, and social and regulatory processes. Strategic maps, therefore, become visualization tools that facilitate the organizational description and drive the valuation process [13]. SM is thus a tool to measure organizational performance and to analyze the strategy used. It allows creating value from four different perspectives: a) the financial perspective, relative to the strategy for growth, profitability, and risk, viewed from the shareholder's perspective; b) the customer's perspective, referring to the strategy to create value and differentiation from the customer's viewpoint; c) the process perspective, relative to the strategic priorities of the different business processes that create satisfaction for clients and shareholders; and d) the learning and growth perspective, which refers to those intangible assets that are more important to develop strategies, such as human assets, information capital, and organizational culture.

The resulting SM makes use of all four perspectives described and is an important tool for strategy control through continuous monitoring. In addition, it allows explaining the strategy hypotheses in a coherent, integrated, and systemic way [12]. For the purpose of applying the SM method in this research, we adopted it by including an additional perspective related to water management, similar to that adopted in the Water Strategic Plan 2020 [28].

The Balanced Scorecard (BS) is a procedure corresponding to a management model of strategic initiatives. One of its main attributes is controlling financial variables jointly with those related to intangible assets. Using BS requires that objectives and associated indicators, both financial and nonfinancial, derive from the water organization's vision and strategy. For this reason, it is a method to align trends, business units, human resources, and technological means with the water organization's strategy [29]. The BS is generally proposed as an organized process involving different perspectives. Goals to reach are proposed for each perspective, which are causally related to one another. The model explaining these relationships is the above-mentioned Strategic Map, which describes the strategy hypothesis and raises the connection of the desired results of the strategy with the inducers and their linkage that will make them possible through relationships in different perspectives [13].

Social conflicts in the irrigated areas are manifested in: territorial transformations without planning and regulation, poor service to the irrigator, imbalances and inequities in water distribution, soil, water, and plant degradation, centralization in organizations, affecting territorial, economics, and productive competitiveness, with lack of profitability and investments in rural areas under irrigation and exodus of the peasant population to urban areas.

4. Modeling the Supply–Demand Water Balance

4.1. Water Balance Formulation

It is understood as the result of an adjusted model that contrasts water supply and demand at the level of the Administrative Unit of Management (AMU), considering agro-climatic conditions influencing irrigation demand and also nonagricultural uses [30], the following:

$$CWB = GS - GD \qquad (1)$$

where CWB represents the Mendoza River current water balance, GS is the mean Gross Supply from the river and contributing streams, and GD is the total Gross Demand per AMU, with the total GD per AMU defined as the sum of water demand per crop type:

$$GD = \Sigma\ (RA \times RC \times ETc \times Ef) \qquad (2)$$

where RA is the registered area having water rights per type of use in the AMU, RC is the reduction coefficient of water allocation depending upon the category of the granted water rights, ETc is crop evapotranspiration per type of land use, and Ef is the current global irrigation efficiency.

The AMUs were defined according to the available surface and groundwater sources of supply, as well as the existent channel network. Complementarily, we considered the catchment and channeling infrastructure, irrigation performance [31], edaphic conditions [32], and predominant land and water uses. The distribution system in the Mendoza River basin was analyzed, and existing WUOs were grouped into different AMUs according to their sources of water supply, the modality of operation of the system, and homogeneity criteria of management (Figure 3).

The WEAP model (Water Evaluation and Planning) [16] was adopted for calculating the supply–demand water balance. For the estimation of irrigation efficiency [4], two components were analyzed: the transport and distribution efficiencies through the canal system, and the irrigation application efficiencies relative to the farm fields [33]. Field efficiency studies were analyzed and compared with observed ones, particularly referring to the Mendoza River basin [34,35]. The following modeling parameters were considered: registered areas, reduction coefficients according to category of water rights, and average irrigation efficiencies (Table 2).

Table 2. Water rights to irrigation water use and efficiencies. Mendoza River basin.

Categories of Use	Permanent Water Rights	Temporary Water Rights	Public Town Irrigation	Urban Uses
Surface area (ha)	42,147	40,195	3,087	6,498
Allocation reduction coefficient	1	0.8	0.8	1
Farm application efficiency			51.4%	
Canal transport efficiency			81.8%	
Global efficiency			42.0%	

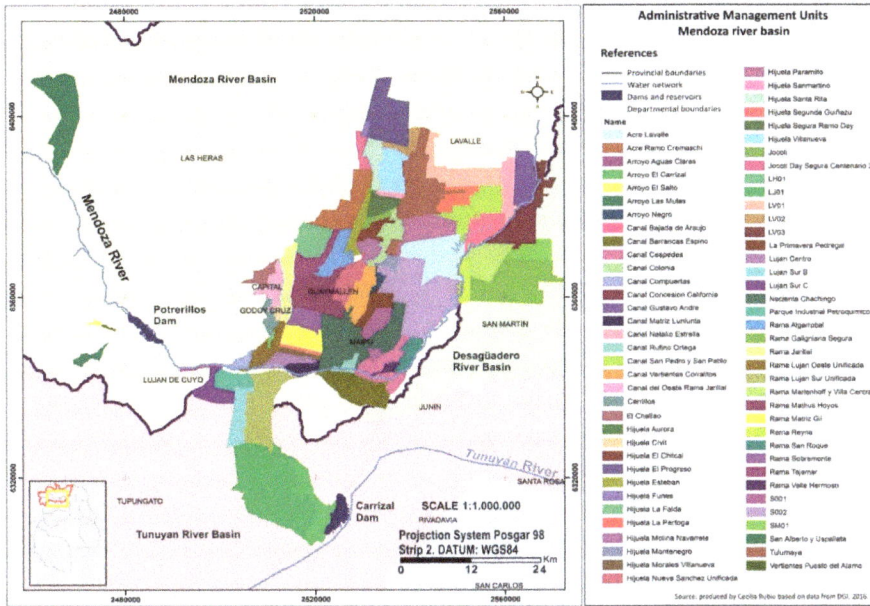

Figure 3. Administrative Management Units in the Mendoza River basin.

4.2. Water Supply

In assessing surface water resources, we considered the mean supply from river and streams, in accordance with the provisions of Provincial Laws 386 and 430, which mention it as a reference value for average flows [36]. The mean water supply from the Mendoza River was obtained from the study of water volumes recorded at the Guido gauging station, located upstream of the Potrerillos Reservoir before the exit to the fluvial valley in the pre-mountain massif (Figure 4).

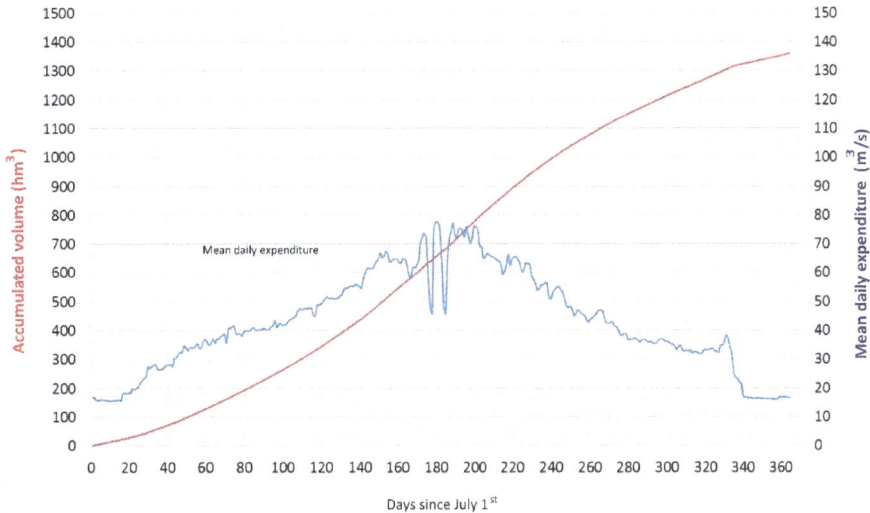

Figure 4. Mean daily water use curve and accumulated volume at Potrerillos Dam. (average values for 1 July 2006 to 30 June 2015).

The series of hydrologic years considered for modeling corresponds to the 2006–2015 period, which has been considered highly representative, and includes data on precipitation, operation, and spills. Yearly data of that period varies from rich to poor in terms of flow. In addition, for the same period, real-time nival and meteorological data from the Horcones and Toscas stations in the High Cordillera are made available to increase the integral hydrological knowledge of the basin. There is also information for this period about the operation of the Potrerillos Dam and Reservoir, the stabilization of management, and the calibration of hydro-mechanical equipment. Average historical volumes were computed from daily records, the average historical spills were assessed, and the average historical discharges were estimated with monthly frequencies for the Cipolletti Dam [5]. The operation of dams and reservoirs is dependent upon weather conditions, service provided by some branch canals, occurrence of rainfall, or compensatory measures that make the rules of operation be dynamic and varying from one cycle to other.

An analysis was made of the operations carried out in the modeling period. The recorded and modeled volumes discharged by the Potrerillos Dam were compared (Figure 5) and the quality of simulations performed was assessed with the Nash–Sutcliffe modeling efficiency indicator [37], resulting NSE = 0.82. This coefficient expresses the relative magnitude of the mean square error relative to the observed data variance. The maximum value is NSE = 1.0, which can only be achieved if there is a perfect match between all observed and simulated values. The closer the values of NSE are to 1.0, the better are the estimates of discharges. The obtained NSE value is quite high and therefore indicates that the simulation model provides confident results. Computations took into consideration the spill losses (Table 3).

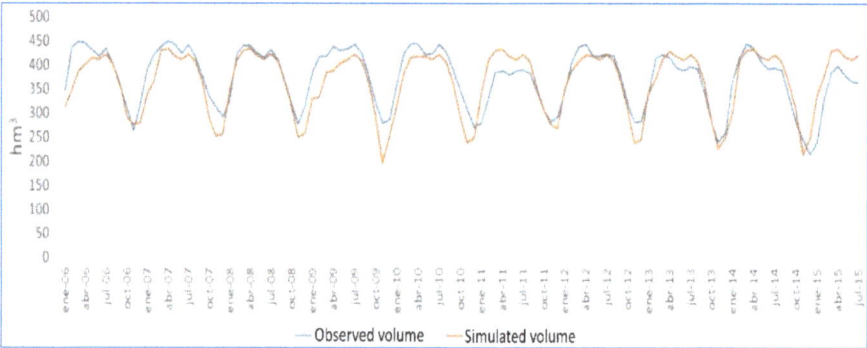

Figure 5. Validation of recorded and modeled volumes discharged by the Potrerillos Dam, 2006–2015.

Table 3. Averages spills at Potrerillos Dam, 2006–2015.

Spills (hm³)												
Aug	Sep	Oct	Nov	Dec	Jan	Feb	Mar	Apr	May	Jun	Jul	Annual
65.0	94.6	124.9	161.4	182.8	163.3	128.4	96.7	73.6	70.8	57.8	41.4	1260.7

4.3. Water Demand

For estimation of water demand, various indicators of the water use performance were used [38]. Interactions between surface water and groundwater could not be omitted [3]. Demand and groundwater supply were estimated in a complementary manner for all AMUs considering groundwater use or a conjunctive use of surface and groundwater. In the case of conjunctive use, it was taken into account that groundwater pumping was only carried out in those fields equipped with wells and it was assumed that this pumped water was used to satisfy seasonal deficits due to scarce supply by the canal system [5].

Land use in agricultural areas has been characterized adopting representative crops for each AMU. Irrigation requirements were determined through the Kc-ETo approach that combines the reference evapotranspiration ETo with a crop coefficient characteristic of each crop [3,39]; net irrigation demands were computed by considering the local agricultural calendars for each type of crop as previously tested [40]. The referred WEAP software [16] accepts that the Kc-ETo approach is calculated externally by the user [41]. Thus, ETo was calculated using the FAO-PM method [39]; crop coefficients (Kc) and other parameters (e.g., crop phonological dates, soil characteristics) were obtained from literature [3,39,40] and adjusted to the AMU areas based upon existing field data [34]. For those computations, available climate data were used and spatially distributed using the method of Thiessen polygons [42]. These data allowed computing crop evapotranspiration and estimating effective precipitation, which directly contribute to meet crop demands [43], as well as other useful precipitation that may influence the efficiency of water use [44]. Spatial information was then obtained with a GIS tool. Data were used as input to the model built in the WEAP software [35]. It was therefore possible to characterize and map the main land uses (Figure 6) and water and irrigation requirements for the Mendoza River basin after aggregating results relative to all AMUs. Thus, the volumetric demand distribution was assessed for all selected series and AMUs. The procedure was replicated for diverse hydrologic years in the 2006–2015 period, with inclusion of demand variability depending on each year's weather conditions.

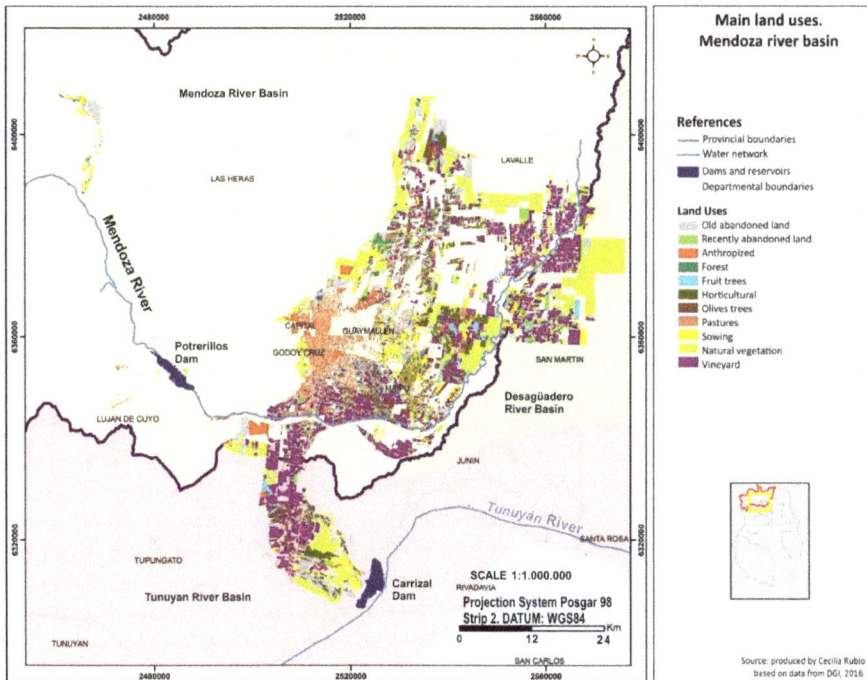

Figure 6. Main land uses. Mendoza River basin.

4.4. Supply–Demand Relationships

Supply and demand relationships have been determined for the current situation considering all granted water rights, the current irrigation efficiency, and the rules of operation of reservoirs, canals, and diversion dams. To this end, we considered two key indicators for achieving an appropriate supply–demand balance in the Mendoza River basin—*demand dissatisfaction* and *demand coverage* (percent of demand covered by the supply)—to estimate the percent guarantee of irrigation water [45], for which the reference value for the region is 81% [46]. Demand dissatisfaction corresponds to the difference between the water volume required to meet gross demand and the amount of supply available to satisfy such demand. This factor also allows determining the missing water volume (i.e., that cannot be covered by the available supply) [47]. There is also a link to the failure total, which is the summation of the monthly failures expressed in volume, understanding that a failure occurs when, in a certain month, supply does not suffice to meet the gross demand. The demand coverage corresponds to the percentage of gross demand effectively satisfied by the available supply, considering the monthly coverage for each AMU. It is computed as the difference to 100% of the ratio between failure and gross demand expressed in percentage. This value is lower than the global annual coverage because, although annual supply could have been enough to meet annual demand, when performing a monthly analysis, it became evident that the unevenness of such supply throughout the year results in months with deficit and others with surplus [22].

The global annual balance is calculated from the summing to the year of the volumes of supplies and demands. It indicates whether the annual supply–demand balance is deficient, excessive, or balanced. In turn, the global annual coverage is the percent value of the ratio between annual supply and annual gross demand. That coverage is limited to 100 percent to take into consideration that monthly surpluses or deficits are unevenly distributed throughout the year. Deficits for each AMU are computed monthly and expressed in hm3. The demand dissatisfaction is expressed in terms of ratios between water volume and area, the latter referring to where water is lacking, and is expressed in mm during the whole cycle. Another indicator is the percentage of gross demand met by the available supply, which corresponds to the inverse of the demand dissatisfaction when expressed in percentage [30].

5. Governance Organization and Forecasted Supply–Demand Balance

5.1. Governance Organizational Model

Mission, vision, values, and strategy are defined for the organizational water management model in an irrigated area of the Mendoza River basin with high social and environmental dynamics:

Mission: Building up an organizational model for local water management that responds to the socio-economic requirements of the land and water users of the irrigated area in the Mendoza River basin, particularly considering the users' participation through WUOs and AMUs.

Vision: Ensuring sustainable water use and productivity by achieving appropriate water demand management in a context of climate variability and territorial transformation in the Mendoza River basin.

Values: Responsibility with the community and the environment; effective participation of users; technical administrative efficiency in management; transparency in actions and communication; integrity and equity; commitment to local and territorial development.

Strategy: Achieving proper water administration performance in irrigated management units in the Mendoza River basin, using an autarchic, technically based organizational model that enables progressive implementation of training and innovation actions for sustainable use of water resources in the territory and that contributes to the local development process.

The proposed organizational model refers to an institution of public nature but not state-run, and with significant alliance with the private sector. Therefore, it must be taken into account that while for public sector organizations, it is relevant to analyze the system water use and management performance in order to achieve the defined mission with the highest level of success, for the private sector organizations, more importance is given to water and land productivity. Considering the above-mentioned aspects, objectives were grouped in agreement with key organizational dimensions, and to determine causal connections between objectives and perspectives, that is, cause–effect relationships, through building a strategic map of the organizational and governance model for the Mendoza River basin (Figure 7).

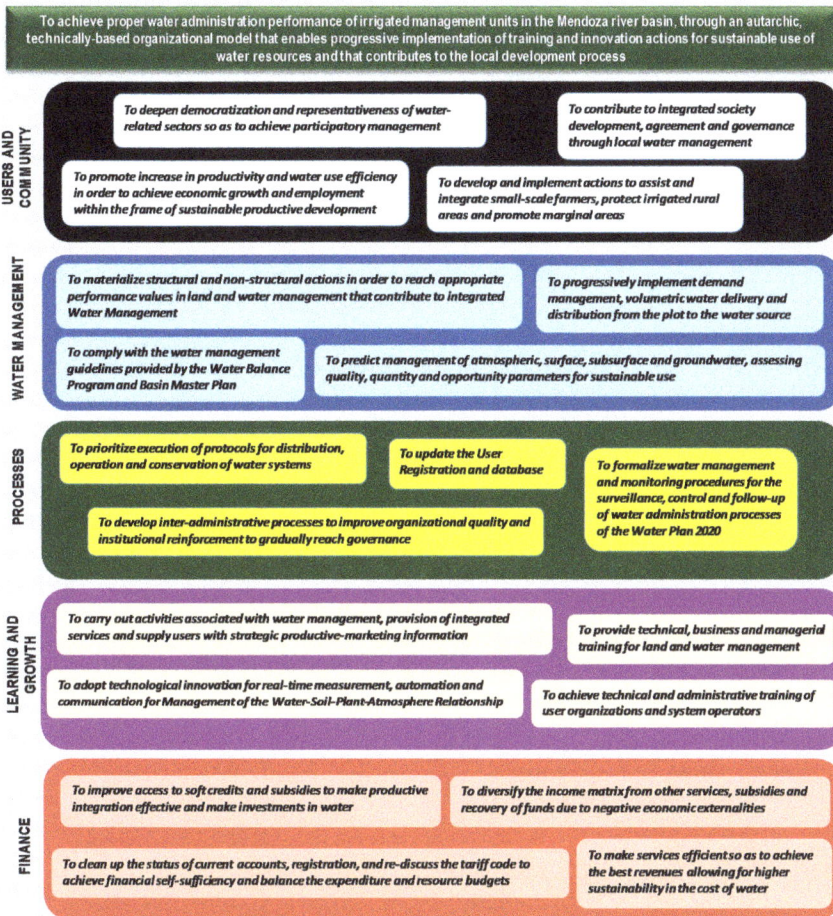

Figure 7. Summary of the strategic map of the organizational model for water governance in the irrigated area of the Mendoza River basin.

Among the most important aspects of the methodological adaptation of the Strategic Map (SM) and of a Balanced Scorecard (BS), the financial perspective was taken as the basis of the organizational model because financial and socio-economic variables have a great relevance in institutional functioning. The following indicators were therefore adopted: (i) financial self-sufficiency, ratio between the income resulting from water rights fees and the costs of operation, maintenance, and management of the water system [45]; (ii) performance of channel water allocation, ratio between effective and predicted recovery of water fees [45]; and (iii) monetary efficiency in water use, computed as the ratio of costs per volume of water allocated [48]. These indicators are used per every AMU. The learning and growth perspective is next in importance relating to the implementation of modern technical management, thus improving the capacity building of human resources and modernizing the hard- and software of the water system.

These water management issues definitely support technological innovation and are paramount to implement and develop both the learning process and the quality of the organizational processes. In addition, BS was methodically executed around different perspectives whose achievable aims were proposed. The organizational model that explains those relations in the strategic map helps describe the strategy's hypothesis aimed at fully achieving the water management goals.

5.2. Scenarios for Simulation

Forecasted supply–demand water balance alternatives were analyzed for the 2030 horizon. The *Demand Dissatisfaction* and *Demand Coverage* indicators were used. Three future scenarios (Table 4) were proposed in addition to the current one (Sc0):

- *Sc1 – when temporary water rights allow the use of up to 80% of the available water (i.e., RC = 0.80)*
- *Sc2 – when temporary water rights allow the use of up to 100% of the available water (i.e., RC = 1.0)*
- *Sc3 – when irrigation aims at satisfaction of the total registered area, including land where irrigation was previously abandoned*

The forecasts on the availability of surface water resources as influenced by climate change consider a reduction of snowfall of 20% by 2080, and an increase in air temperature of 4 °C for the next century [49]. For modeling purposes, an average reduction rate of snowfall close to 0.31% per year was adopted [50]. This decrease, together with the increase in temperature, induces a change in the basin hydrology that, on average and for the projected period of analysis, shall modify the flow regime and requires adaptability in the use of the water [51,52] because there would be higher flows in winter and lower ones in summer compared to present [53]. The induced variation of climate and hydrologic behavior in the context of global climate change [54,55] requires assessing water balance changes and related impacts on different categories of water rights in terms of area and/or water allocation. The cultivation of land presently abandoned but that could have water rights was not considered except for scenario Sc3.

Table 4. Main characteristics of scenarios used for supply–demand water balance modeling, Mendoza River basin.

Scenarios for the 2030 Horizon	Climate-Change-Affected Variables	Efficiency of Water Use	RC of Temporary Water Rights
Sc0 – Current water use and governance	Present condition	51%	0.80
Sc1 – Temporary water rights with RC = 80%	Reduced snowfall, higher temperature in lowlands, increased evapotranspiration	59%	0.80
Sc2 – Temporary water rights with RC = 100%		59%	1.00
Sc3 – Water service to the total registered area		59%	1.00

Factors commonly taken into account for developing and simulating all scenarios (Table 4) consist of: (i) an improved irrigation water management providing for a reasonable efficiency of water use of 59% without requiring structural changes in the canal system; (ii) a decrease in mountain snowfall that would lead to the variation of the river flow regime, thus to adaptation changes in water supply; (iii) an increase in temperature in the lowlands that likely will cause variable increases in evapotranspiration and demand for water, depending on the type of water use; (iv) variation in land use, namely referring to the distribution of areas cropped and urbanized; and (v) changes in management and operation of the basin water system, mainly relative to the governance issues, improvement of the AMUs, and more effective participation of users through the respective WUOs.

5.3. Forecasts for Supply–Demand Water Balance for the 2030 Horizon

The previously defined gross demand, the supply and failure total, the percent coverage of demand by supply (including total failures), the global supply–demand balance, and global annual coverage of demand by supply were estimated using WEAP for both agricultural and other consumptive uses. Results relative to the considered scenarios are presented in Table 5.

Table 5. Comparing results for supply–demand water balance scenarios for the 2030 horizon, Mendoza River basin.

Scenarios for the 2030 Horizon	Registered Cropped Area (ha)	Gross Demand (hm³)	Supply Total (hm³)	Failure Total (hm³)	Demand Coverage (%)	Global Annual Balance (hm³)	Global Annual Coverage (%)
Sc0 – Current water use and governance	54,720	955.49	1131.05	−34.26	96.4%	175.57	100%
Sc1 – Intermittent water rights, RC = 0.8	54,720	793.09	1040.47	−4.07	99.5%	247.38	100%
Sc2 – Intermittent water rights, RC = 1.0	59,342	866.30	1042.79	−5.58	99.4%	176.49	100%
Sc3 – Service to the total registered area	76,534	1170.44	1049.65	−170.74	85.4%	−120.79	86%

The current balance (Sc0) does not include the abandoned old lands, but does take into account the factors of the current climate and the existing efficiency in the farm, which reaches an average value of 51.4%. Analyzing the indicators produced when modeling with scenario Sc1 (temporary water rights and RC = 0.8), it was observed that indicators are generally better than those obtained for the current condition (Table 5) despite global change-influenced climate and hydrologic variables for the 2030 horizon, resulting in increased demand and more varied water supply. That improvement is likely due to higher farm application efficiency, 59% vs 51%, and to an optimized operation of reservoirs and diversion dams, which are expected to contribute to decreasing failures and improving the water supply distribution service, thus resulting in an increased guarantee of irrigation water availability.

For Sc2 (temporary water rights and RC = 1.0), there is a larger registered cropped area due to transformation of temporary into permanent water rights. An increase in annual gross demand of 73.21 hm³ occurs and there is a negligible increase of 1.51 hm³ in the sum of failures, while the base water supply remains the same. An improved supply is considered due to improvements in water distribution and global efficiency, which are expected to keep failures, demand coverage, and global annual coverage at levels similar to Sc1. However, the global annual balance diminishes. Notwithstanding, analyzed in general, this difference continues to be positive for average years.

The modality Sc3 is for a scenario where the whole of the granted area in the basin is predicted to have irrigation, thus including all abandoned lands, recently or not. The sum of failures reaches then a high value of 170.74 hm³, particularly important due to high demand when river runoff is low. Demand coverage falls to 85% while the gross demand increases about 215 hm³ and the global annual coverage reaches 86%. But even so, these values are above the regional reference value of the irrigation water guarantee percentage. Thus, scenario Sc3 requires adopting additional measures not considered in this study.

6. Conclusive Remarks

A strategic formulation of the organizational model was defined after performing a strategic analysis of the organization. Designing the strategic map of the organization from the perspective of users and the community and relating to new perspectives on water management, processes, learning and knowledge, and finance responded well to objectives. Moreover, in the proposed organizational and governance model, the effective participation of users was considered as the main basis for formulating institutional cross-sectoral (horizontal/integral) policies. That participation is aimed at concrete decision-making by all stakeholders involved in every circumstance concerning water management. The result is that the designed SM for the Mendoza River basin is aimed at supporting sustainable productive development and satisfaction of socio-economic requirements through an appropriate supply–demand management and adopting volumetric water delivery to users. The Administrative Units of Management (AMUs) were adopted to manage water at local level, and the areas of influence of the WUOs were reformulated accordingly.

The evaluation of the water resources of the Mendoza River basin was performed and an analysis was made of the operations carried out in the modeling period of 2006–2015. The recorded and modeled volumes at the Potrerillos dam were compared. A high value of 0.82 for the Nash–Sutcliffe modelling efficiency indicator of goodness of fit was obtained, which allowed considering that the available supply was confidently assessed. The demand was estimated on basis of previous studies and used the updated methodology of FAO56 guidelines. The balance supply-demand was computed with model WEAP. For the modeling period of 2006–2015 three scenarios were tested and compared with the base one referring to present conditions. It was observed that more requiring scenarios than present were evaluated positively but not a scenario where irrigated areas would increase much to include presently abandoned ones. Positive results were obtained due to the new organization and governance model, which assumed decentralized water management through the AMUs, users' participation with the Water User Organizations (WUOs), and increased global water efficiency due to improved water management and canal transport and deliveries through adopting innovative approaches related to hard- and software, as well as capacity building.

Author Contributions: M.S.-S. made the characterization of the study basin and the coordination in modeling the water balance and J.F.-D. formulated the model of water governance. Both authors wrote the article.

Acknowledgments: Luis S. Pereira, University of Lisbon for advising on irrigation modernization and management and revising the last version of the article, Joan Mateu and Joan Romero, Universitat de València, and Antonio Rico Amorós, Universidad de Alicante, for contribution of theoretical framework. José Álvarez; Carlos Sánchez; Juan Pina, and Bernardo Herrera for their strategic contributions of the Water 2020 Plan in the water management of Mendoza that were considered in the article. Jorge Chambouleyron, Cecilia Rubio, Elena Abraham, Juan Satlari, Guillermo Cuneo, Gonzalo Muñiz, and Nora Mustoni for the collaboration and scientific review of the materials and methods used; Elma Montaña, Christopher Scott, and Facundo Martin for the conceptual vision of the science and policy dialogue considered in the formulation of the model.

Conflicts of Interest: The authors declare to have no conflicts of interest.

References

1. Salomón, M.; Álvarez, J. *Preservación y Promoción de Zonas Irrigadas en el Marco del Plan Agua 2020 y Plan de Ordenamiento Territorial de Mendoza. I Jornadas Nacionales de Investigación en Ciencias Sociales. Cambio Climático y Desarrollo Sostenible en las Ciencias Sociales*; Facultad de Ciencias Políticas, Universidad Nacional de Cuyo: Mendoza, Argentina, 2016.

2. Salomón, M.; Rosell, M.; Thomé, R.; Albrieu, H.; López, J.; Ruiz Freites, S.; Yapura, J. *El Aporte de la GIRH a Través de las Organizaciones de Usuarios en el Ordenamiento Territorial y Desarrollo Estratégico de Mendoza. XXI Congreso Nacional del Agua*; CONAGUA: Resistencia, Chaco, 2011; Chapter 2; pp. 42–58.

3. Pereira, L.S.; de Juan, J.A.; Picornell, M.A.; Tarjuelo, J.M. *El Riego y sus Tecnologías*; Centro Regional de Estudios del Agua, Universidad de Castilla-La Mancha: Albacete, Spain, 2010; 296p, Available online: http://crea.uclm.es/crea2/sp/index.php (accessed on 16 September 2018).

4. Chambouleyron, J. *Riego y Drenaje. Técnicas Para el Desarrollo de una Agricultura Regadía Sustentable Mendoza*; Editorial Universidad Nacional de Cuyo, EDIUNC: Mendoza, Argentina, 2005; 1026p.

5. Departamento General de Irrigación (DGI). *Datos y estadísticas Cuencas de Mendoza. Dirección de Gestión Hídrica*; Dirección de Recaudación y Finanzas: Mendoza, Argentina, 2016.

6. Salomón, M.; Ruiz Freites, S. La Descentralización en la Administración del Agua. In Proceedings of the XX Congreso Nacional del Agua, Mendoza, Argentina, 9–13 May 2005.

7. Salomón, M. El desafío del riego frente a la escasez hídrica en Mendoza. In Proceedings of the VI Jornadas de Riego y Fertirriego, Luján de Cuyo, Mendoza, Argentina, 7–9 November 2012; Facultad de Ciencias Agrarias, Universidad Nacional de Cuyo: Mendoza, Argentina, 2012.

8. Domínguez, J. *Hacia una Buena Gobernanza Para la Gestión Integrada de los Recursos Hídricos. VI Foro Mundial del Agua. Documento Temático de las Américas: Los Mensajes Clave de las Américas. Los Marcos Legales e Institucionales del agua dan Certidumbre a los Agentes Sociales y Favorecen el Financiamiento para el Sector Hídrico, Ordenamiento Territorial y Desarrollo Estratégico de la Cuenca del Rio Mendoza, Argentina. Adopción de la Ley de Ordenamiento Territorial, Integrando los Principios de la GIRH World Water Fórum*; Colegio de México: Mexico City, Mexico, 2011; Volume 1, pp. 22–23.

9. Farinós, J. *Gobernanza Territorial Para el Desarrollo Sostenible: Estado de la Cuestión y Agenda*; Boletín de la A.G.E. Nº 46; Asociación de Geógrafos Españoles: Madrid, Spain, 2008; Volume 46, pp. 11–32.

10. Pedregal Mateos, B.; Laconi, C.; Mancilla García, M.; Orozco Frutos, G.; Figueroa, A.; Cabello Villarejo, V.; Moral Ituarte, L. Collaborative mapping of wáter conflicts in Andalusia, Spain. In Proceedings of the Open Science Conference Pecs II, Oaxaca, Mexico, 7–10 November 2017.

11. Romero, J.; Farinós, J.; Presentación en Romero, J.; Farinós, J. (Eds.) *Gobernanza Territorial en España. Claroscuros de un Proceso a Partir del Estudio de Casos*; Publicaciones de la Universidad de Valencia/Instituto Interuniversitario de Desarrollo Local, Colección Desarrollo Territorial: Valencia, Spain, 2006; pp. 15–19.

12. Kaplan, R.; Norton, D. *Mapas Estratégicos. Convirtiendo los Activos Intangibles en Resultados Tangibles*; Harvard Business Review Press: Barcelona, Spain, 2004; ISBN 9781591391340.

13. Bustos, G. *Mapa de Objetivos Organizacionales: Una Generalización del Mapa Estratégico*; Pontificia Universidad Católica de Valparaíso: Valparaíso, Chile, 1997.

14. Abraham, E.; Salomón, M.; Rubio, C.; Soria, D. *Aportes Metodológicos para Evaluación Hidrológica de Cuencas Andinas. Estudio Cuenca Río Mendoza*; Revista Zonas Áridas; Centro de Investigaciones de Zonas Áridas, Universidad Nacional Agraria La Molina: Lima, Perú, 2010; Chapter 1; pp. 9–33.

15. FAO-PNUD. *Planes Directores de Ordenamiento de los Recursos Hídricos de la Provincia de Mendoza. Informe Principal Plan Cuenca Río Mendoza*; Proyecto PNUD/FAO/ARG/00/008; FAO-PNUD: Mendoza, Argentina, 2004.

16. Stockholm Environment Institute (SEI). WEAP (Water Evaluation and Planning). Available online: http://www.weap21.org/ (accessed on 8 February 2017).

17. Morábito, J.; Hernández, J.; Martinis, N. Fornero, L. *Modelación Hidrológica de la Cuenca Norte de Mendoza*; IT Nº 146; Centro Regional Andino (CRA): Mendoza, Argentina, 2012.

18. Piccone, L.; Salomón, M. *El manejo del Agua de Riego en la Provincia de Mendoza. Ejemplo de usos y Aprovechamiento de Aguas en el Gran Mendoza*; Mendoza en el 2000; Facultad de Ciencias Agrarias, Universidad Nacional de Cuyo, Centro Coordinador de Ediciones Académicas: Mendoza, Argentina, 1994; Chapter 12; pp. 97–103.

19. GWP. *Manual para la Gestión Integrada de Recursos Hídricos en Cuencas". Network of Basin Organizations. Global Water Partnership International*, 1st ed.; Empresa Gráfica Mosca: Paris, Republiqué Française, 2009; 111p.

20. Estrada, F.; Luján García, L. *Criterios de Garantía en la Planificación de Regadíos*; CEDEX (Centro de Estudios y Experimentación de Obras Públicas): Madrid, Spain, 1993.

21. Pereira, L. Necessidades de agua e programação da rega: Modelação, avanços e tendências. In *Modernización de Riegos y Uso de Tecnologías de Información (Taller Internacional, La Paz, Bolivia, September 2007)*; Ruz, E., Pereira, L.S., Eds.; CYTED and PROCISUR/IICA: Montevideo, Uruguay, 2008; pp. 14–16.

22. Balairón Pérez, L. *Gestión de Recursos Hídricos*; Politext: Barcelona, Spain, 2002; Volume 89.

23. INDEC. *Censo 2010. Año del Bicentenario*; Instituto Nacional de Estadísticas y Censos (INDEC): Buenos Aires, Argentina, 2010.

24. Salomón, M.; Abraham, E.; Sánchez, C.; Rosell, M.; Thomé, R.; López, J.; Albrieu, H. Análisis de los Impactos Ambientales Generados por las Presas Sobre los Sistemas de Riego. In *Tecnologías para o uso Sustentável da Água em Regadió*; Cuenca Río Mendoza, P.L., Brea Victoria, F., Paredes, P., García, M., Palacios, E., Torrecillas, A., Eds.; Edições Colibri: Lisboa, Portugal, 2008; Chapter 5.7; pp. 122–128.

25. Godet, M. *Prospectiva Estratégica: Problemas y Métodos*, 2nd ed.; Cuaderno Nº 20; Ediciones: Donostia-San Sebastián, Spain, 2007.

26. Matus, C. *Política, Planificación y Gobierno. Organización Panamericana de la Salud*; Organización Mundial de la Salud, Fundación Altadir, ILPES: Buenos Aires, Argentina, 1987; pp. 242–359.

27. Miklos, T.; Arroyo, M. *Prospectiva y Escenarios para el Cambio Social*; Serie Working Papers; FCPS-UNAM: Mexico City, Mexico, 2008.

28. Departamento General de Irrigación (DGI). Plan Agua 2020. Mendoza. 2014. Available online: http://www.agua.gob.ar/2020/sobre-el-plan (accessed on 20 January 2017).

29. Martínez Pedros, M.; Milla Gutiérrez, A. *La Elaboración del Plan Estratégico a Través del Cuadro de Mando Integral*; Ediciones Díaz de Santos SA: Madrid, Spain, 2005.

30. Álvarez, J.; Pina, J.; Sánchez, C.; Salomón, M. *Hydrological Balance Implementation in Mendoza's Province. Decision support and modeling tool for integrated management of water resources*, 3rd ed.; Inter-Regional Conference on Land and Water Challenges: Tools for developing, Colonia, Uruguay, 27–30 September 2015; Available online: http://www.ainfo.inia.uy/digital/bitstream/item/5026/1/Agrociencia-Congreso-CIGR2015.pdf (accessed on 21 December 2016).

31. USBR. *Bureau of Reclamation Manual. Volume V. Irrigated Land Use, Part. 2 Land Classification*; US Dept. Interior: Washington, DC, USA, 1953.

32. Vallone, R.; Maffei, J.; Olmedo, G.; Morábito, J.; Mastrantonio, L.; Lipinski, V.; Filippini, M.F. *Mapa de Aptitud de Suelos con Fines de Riego y de Riesgo de Contaminación Edáfica de los Oasis Irrigados de la Provincia de Mendoza*; Informe Técnico Final, Componente de Calidad de Agua y Suelo. Programa de Riego y Drenaje; Departamento General de Irrigación-Programa de Servicios Agrícolas Provinciales–Organización de Estados Iberoamericanos: Mendoza, Argentina, 2007.

33. ASAE—American Society of Agricultural Engineering. *Evaluation of Irrigations Furrows*; EP419.1.USA; ASAE: Washington, DC, USA, 2000; pp. 893–898.

34. Morábito, J. *Desempeño del Riego por Superficie en el Área de Riego del Río Mendoza: Eficiencia Actual y Potencial: Parámetros de Riego y Recomendaciones para un Mejor Aprovechamiento Agrícola en un Marco Sustentable*; Universidad Nacional de Cuyo, Facultad de Ciencias Agrarias: Mendoza, Argentina, 2003; 97p.

35. Satlari, G.; Cúneo, G.; Mustoni, N. *El Balance Hídrico como Herramienta de Planificación*; Congreso Nacional del Agua; Argentina CONAGUA: Paraná, Argentina, 2015.

36. Ruiz Freites, S. *Legislación y administración de Aguas en Mendoza, Capítulo XI Legislación y Administración del Agua en Mendoza, en Mathus Escorihuela, M. (Dir.). Derecho y Administración de Aguas*, 1st ed.; Zeta Editores: Mendoza, Argentina, 2007; pp. 409–465.

37. Nash, J.E.; Sutcliffe, J.V. River flow forecasting through conceptual models: Part 1. A discussion of principles. *J. Hydrol.* **1970**, *10*, 282–290. [CrossRef]

38. Pereira, L.S.; Cordery, I.; Iacovides, I. Improved indicators of water use performance and productivity for sustainable water conservation and saving. *Agric. Water Manag.* **2012**, *108*, 39–51. [CrossRef]

39. Allen, R.; Pereira, L.; Raes, D.; Smith, M. *Evapotranspiración del Cultivo. Guías Para la Determinación de los Requerimientos de Agua de los Cultivos*; Estudio para Riego y Drenaje FAO: Rome, Italy, 2006; Nro. 56; 298p.

40. Salomón, M.A.; Sánchez, C.M.; Pereira, L.S. Estimación del balance hídrico mediante aplicación del modelo ISAREG en el Canal Segundo Vistalba, Lujan de Cuyo, Mendoza (Argentina). In *Modernización de Riegos y Uso de Tecnologías de Información*; Ruz, E., Pereira, L.S., Eds.; CYTED and PROCISUR/IICA: Montevideo, Uruguay, 2008; pp. 115–117.

41. Satlari, G.; Cúneo, G. *La Asignación del agua, a Definir en el Balance Hídrico, Requiere una Mejor Eficiencia de Riego*; CONAGUA 5.230; Congreso Nacional del Agua: Paraná, Argentina, 2015.

42. Chow, V.; Maidment, D.; Mays, L. *Hidrología Aplicada*; Editorial Mc Graw-Hill: Bogotá, Colombia, 1994; 583p.

43. González Loyarte, M.; Menenti, M.; Diblasi, A. Mapa bioclimático para las Travesías de Mendoza (Argentina) basado en la fenología foliar. *Revista FCA UNCuyo* **2009**, *41*, 105–122.

44. Casas, R.; Albarracín, C. (Eds.) *El Deterioro de los Suelos y del Ambiente en la Argentina*; Editado por Fundación para la Educación, la Ciencia y la Cultura Buenos Aires (FECIC-PROSA): Buenos Aires, Argentina, 2015; Tomo I; 604p, Tomo II; 452p, ISBN 978-950-9149-40-3.

45. Bos, G.; Chambouleyron, J. *Parámetros de desempeño de la agricultura de riego de Mendoza, Argentina*; Instituto Internacional del Manejo del Agua: Mexico City, Mexico, 1999.

46. De Mendoza, G. *Aprovechamiento Integral del Río Mendoza. Proyecto Potrerillos. Manifestación General de Impacto Ambiental (Ley N° 5961)*; Ministerio de Ambiente y Obras Públicas, Subsecretaría de Medio Ambiente (MAyOP): Mendoza, Argentina, 1998.

47. Sánchez, C.; Salomón, M.; Pereira, L. Evaluación del desempeño de los sistemas de distribución de riego tradicionales mediante uso del modelo ISAREG en Mendoza (Argentina). In *Tecnologías Para o uso Sustentável da Água em Regadío*; Pereira, L., Brea Victoria, F., Paredes, P., García, M., Palacios, E., Torrecillas, A., Eds.; Edições Colibri: Lisboa, Portugal, 2008; Chapter 5.5; pp. 114–117.

48. Sánchez Cohen, M.; García Vargas, Y. *Índices de Desempeño de Los Distritos de Riego del Noroeste de México*; INIFAP CENID-RASPA Universidad Autónoma Chapingo, Unidad Regional Universitaria de Zonas Áridas: Texcoco, México, 2004; 4p.

49. Martin, F.; Montaña, E.; Mussetta, P.; Salomón, M.; Scott, C. *Science-Policy Dialogue and Adaptation: Fascing short, Medioum and Long Term Scenarios in Mendoza (Argentina)*; Third Interational Cimate Change; I: 147. Ed.; Governo de Brasil, PROVIA-INPE, Fortaleza: Ceará, Brasil, 2014.

50. IPCC. *Fourth Assessment Report: Climate Change 2007 (AR4). Intergovernmental Panel on Climate Change*; Cambridge University Press: Cambridge, UK, 2007.

51. Organización Meteorológica Mundial. *Guía de Prácticas Hidrológicas. Adquisición y Proceso de Datos, Análisis, Predicción y Otras Aplicaciones*; OMM N° 168: Buenos Aires, Argentina, 1994.

52. Salomón, M. Water security and climate adaptation: Bridging science and policy. The case of the Mendoza River Basin. In Proceedings of the XVI World Water Congress, 29 May–3 June 2017; International Water Resources Association (IWRA): Cancún, Mexico, 2017.

53. Bonisegna, J. *Impacto del Cambio Climático en los Oasis del Oeste Argentino. Revista Ciencia e Investigación*; Tomo 64 n°1: Buenos Aires, Argentina, 2014; pp. 45–58.

54. Villalba, R.; Bonisegna, J.; Masiokas, M.; Cara, L.; Salomón, M.; Pozzoli, J. *Cambio Climático y Recursos Hídricos. El Caso de las Tierras Secas del Oeste Argentino*; Revista Ciencia Hoy; Editoriales Asociación Civil Ciencia Hoy AAPC: Buenos Aires, Argentina, May–June 2016.

55. Villalba, R.; Bonisegna, J. *Cambios Climáticos Regionales en el Contexto del Calentamiento Global*; Informe Ambiental 2009; Secretaría de Ambiente, Gobierno de Mendoza: Mendoza, Argentina; Available online: https://es.scribd.com/document/323170636/Informe-Ambiental-2009 (accessed on 17 October 2016).

![water logo] *water*

MDPI

Article

Irrigation Governance in Developing Countries: Current Problems and Solutions

Enrique Playán [1,*], Juan Antonio Sagardoy [2] and Rosendo Castillo [3]

[1] Department of Soil and Water, Estación Experimental de Aula Dei, CSIC, Zaragoza 50059, Spain
[2] Independent Researcher, Rome 00124, Italy; sagardoy22@alice.it
[3] Consultora de Ingeniería Rural (CINGRAL), Santa Cruz 8, Zaragoza 50003, Spain; rcastillo@cingral.com
* Correspondence: enrique.playan@csic.es; Tel.: +34 976 716 087

Received: 14 June 2018; Accepted: 7 August 2018; Published: 23 August 2018

Abstract: The evolution of water governance and societal perception in large, public irrigation systems in developing countries has triggered successive waves of reforms since the 1980s. Among them are Participatory Irrigation Management, Irrigation Management Transfer, Public-Private Partnerships or Market Instruments. Reforms have generalized the implementation of Water Users Associations (WUAs) in continuous interaction with a public Irrigation Agency. This paper set out to review recurrent problems and reported solutions in the governance of irrigated areas in developing countries and to relate solutions to problems in a case study context. The combination of literature review and the experience of the authors permitted identification and characterization of eight problems and eight solutions. A semi-quantitative approach was designed to relate solutions to problems in case study WUAs. The approach is based on the definition of a generic problem-solution matrix and a WUA-specific problem vector. The solution vector indicates the adequacy of each solution to a case study WUA. It can be obtained by multiplying the problem vector with the problem-solution matrix. Application of this approach to seven case study WUAs demonstrated its potential. Local fine-tuning of the coefficients defining the problem-solution matrix seems required to draw conclusions effectively guiding decision-making.

Keywords: reform; Participatory Irrigation Management; Transfer; water users association

1. Introduction

World irrigation development accelerated in the twentieth century, following intense progress in civil engineering machinery and the push of governments and international development agencies. Many developing countries—often former colonies—changed their view on irrigation: from support to colonial agriculture to a national policy for employment and poverty eradication [1]. Between 1970 and 1990, the world area equipped for irrigation increased from 184 to 258 M ha. Irrigation development continued, and by 1992 the area equipped for irrigation reached 324 M ha [2]. Today the typical irrigation project in the world is located in Asia (70%), draws from surface water resources (62%), uses surface irrigation (86%), attains a cropping intensity of 130%, and grows cereals (61%) [3].

Most of these irrigation projects were developed by governmental Irrigation Agencies, which were in charge of construction and delivering irrigation water. Farmers were rarely consulted during the phases of irrigation design and maintenance. By the 1980s, the performance of these new irrigation projects was dissatisfactory [4]. State-owned projects showed financial problems related to the low irrigation fees and to the difficulties in collecting them [5]. Low operation and management funding led to unreliable water delivery [6], which in turn precluded fee collection. Vermillion and Sagardoy [5] identified the lack of managerial accountability of governmental Irrigation Agencies as a critical issue. Deferred maintenance required cyclical rehabilitation interventions, giving birth to the well-known "reform-neglect-rebuild" cycle [1]. These perceptions led to the development and application of several reform procedures.

By the 1980s, irrigation organizations started seeking new irrigation management models. Vermillion and Sagardoy [5] reported that Irrigation Management Transfer (IMT) processes were implemented to involve water users in the governance of irrigated areas. These authors defined IMT as "the relocation of responsibility and authority for irrigation management from government agencies to non-governmental organizations". Garcés-Restrepo et al. [7] reviewed the motivations leading to IMT processes, and singled out some relevant factors: reduction of costs for the public sector, improvement of irrigation profitability, recovery of irrigation fees, efficiency and equitability of water allocation, infrastructure maintenance and settlement of disputes. During IMT processes, Water Users Associations (WUAs) were identified as a key elements for the new management models, although other organizations (such as cooperatives or municipalities) have adopted responsibilities in irrigation operation and management [4]. By 2007, more than 57 countries had embarked on some type of irrigation sector reform based on IMT [7].

Participatory Irrigation Management (PIM) was introduced as an attitudinal change, often accompanying the reforms introduced by IMT. According to Poddar et al. [6], PIM refers to users' involvement in all aspects of irrigation management, and consequently is a subset of the broader concept of IMT. Kloezen et al. [8] analyzed the successful results of IMT at the *Alto Río Lerma* irrigation district in Mexico, and identified the following key traits: a strong organizational base, cooperation between the public Irrigation Agency and the users, training, and political commitment.

By the end of the 20th century, water markets were promoted as a tool to improve irrigation governance [9]. Water markets make the opportunity cost of water use explicit, and therefore feed economic efficiency. According to Meinzen-Dick [10], the implementation of water markets requires physical infrastructure to mobilize water among users, effective, specific government organizations, and effective user groups. These requirements were met in Chile, where markets were developed in the 1990s.

A decade later, Public-Private Partnerships (PPP) were advocated by Prefol et al. [11] as a logical sequence of the preceding reform efforts. According to these authors, PPP can apply to investment, regulation, operation and management, and agricultural production. PPP can be formalized as public contracts or as public service delegations. Trier [12] reviewed PPP experiences in developing countries. According to this author, private financial contributions can be appropriate for modern systems with commercial farmers. In the context of subsistence agriculture, management contracts can be appropriate.

In the last forty years, countries and international developing agencies have promoted different waves of reforms, counting on a variety of funding arrangements. In general, reforms have focused more on operation and management than on construction and rehabilitation [1]. Reforms have followed the four models discussed in the preceding paragraphs (IMT, PIM, water markets and PPP). The combination of PIM/IMT has so far been the most applied model by far. Meinzen-Dick [10] reviewed reform organizational models, which she qualified as alleged "panaceas" to denote that each of them had been proposed to solve all existing governance problems. In her view, solutions are needed that "fit the local biophysical, social and economic conditions". This diagnose suggests that despite all the reported efforts, the problems found in the governance of irrigated areas have not been solved.

After decades of irrigation reform practice, several hot issues remain in irrigation governance: Garcés-Restrepo et al. [7] outlined the need for more efficient and equitable water delivery, and for an endogenous capacity to solve disputes. Kloezen et al. [8] reported on the complex issue of redeploying the large public staff in the Irrigation Agency following IMT. Yildirim and Çakmak [13] stated that for sustainable PIM, users need to invest in the construction of the irrigation systems. Teamsuwan and Satoh [14] recognized that it was easy to create and strengthen WUAs in Thailand, but anticipated that their sustainability would be problematic. Bhatt [15] reported on several farmers', organizational, systemic and operational issues preventing the success of PIM in India. Ul Hassan [16], analyzing IMT governance in Asia, reported shortcomings stemming from poor accountability and transparency in the legal and policy environment. Veldwisch and Mollinga [17] observed that the local IMT policy in Uzbekistan established WUAs under strong state tutelage. Analyzing IMT implementation in Indonesia, Suhardiman [18] concluded that the Irrigation Agency and international donors had common interests in implementing IMT using a business-as-usual approach. As a corollary of all these hot issues, after reviewing 230 case studies in the literature, Senanayake et al. [19] found it impossible to conclude about the success or failure of IMT processes, owing to the weaknesses of the methods used to infer impact and to problems inherent to the available information.

These findings call for a structured approach to the actual problems of irrigation governance in developing countries and to their solutions. MASSCOTTE [20] is a contribution to the problem-solution analysis, focusing on large canal systems and on a comprehensive approach to operation modernization. It is based on a Rapid Appraisal Process systematically covering all aspects of engineering, governance, maintenance, and operation.

While many developed countries have found a way to reduce the public burden of irrigation governance, ensuring adequate system maintenance and satisfaction of users' needs, most developing countries continue to be trapped in a cycle of successive reforms yielding moderate progress. Asthana [21] reported disenchantment with irrigation projects in developing countries, quoting reasons such as environmental issues, displacement, time and cost overruns, inequities and corruption. According to this author, projects are creating less irrigation potential than planned.

The objectives of this paper are to: (1) critically review the problems and reported solutions in the governance of irrigated areas in developing countries; and (2) relate solutions to problems in a case study context, contributing to the creation and utilization of value in irrigated areas. The paper draws from analyses and examples in the literature and from the experience of the authors.

2. Irrigation Infrastructure: Models for Ownership and Management

For this paper, an irrigation system can be divided in five elements:

- General-purpose, large water works (serving all types of water uses);
- Rural, large water works (delivering water to irrigated areas);
- Rural, small water works (delivering water to farms);
- The hydrant or the WUA canal turnout (separating the public or collective part from a farm); and
- The on-farm irrigation structures.

In this section, different models are discussed for the ownership and the management of these five elements (see Figure 1 for a chart and acronyms representing different models).

Regarding ownership, there are three basic models: (O1) public delivery networks with private farms; (O2) public, collective and private ownership; and (O3) collective and private ownership. Farmers' organizations (such as a WUA, a cooperative or a municipality) own the collective part of the system.

Regarding management, the situation is much more complex, with many variants. Most irrigated areas have started with a model based on a very strong public intervention (M1). IMT processes led to models M2 and M3, in which users' associations manage (and often own) rural water works and the delivery point to farms (the canal turnout or the hydrant). In the most advanced models, a users' organization in a developed area can manage private farmers' irrigation and large (even general-purpose) water infrastructure (M4 and M5).

Public management is performed by the Irrigation Agency (IA). This generic name corresponds to a variety of organizations in different countries and irrigated areas. IAs can take the form of Ministries, River Basin Authorities, Irrigation project authorities or even municipalities. These organizations operate in the name of the Government. If PPP schemes have been adopted, management can be performed by companies operating in the name of the Government.

In this paper, problems and solutions are analyzed for irrigation systems having public and/or collective infrastructure. This is not the case, for instance, of most groundwater systems and of the systems composed of individual pumps along a riverbank.

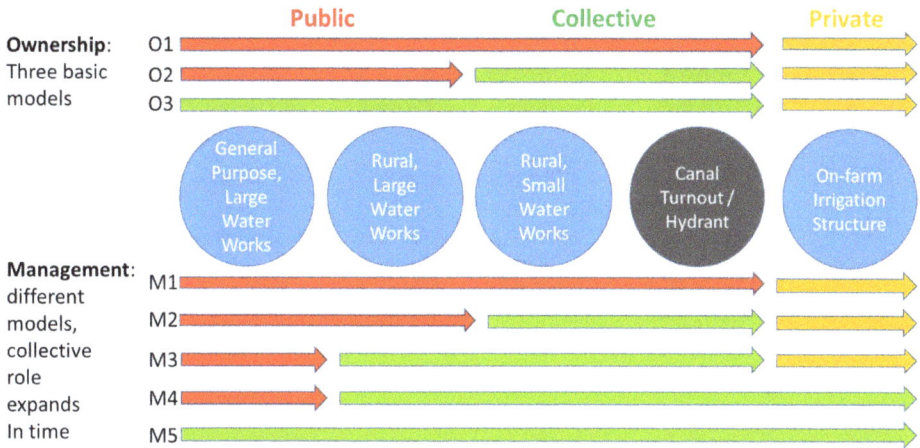

Figure 1. Models for ownership and management of irrigation infrastructure: combination of public, collective and private intervention.

3. The Time Line of an Irrigation System

Table 1 presents the idealized time line of the development of an irrigation system. The process has been divided in four phases: planning, implementation, consolidation, and maturity. In a development context, the duration of the first three phases has been estimated in 3–5 years, 2–4 years and a decade, respectively. The last phase, maturity, has an undetermined duration. For every phase, three focus areas have been considered: (1) technical; (2) legal and regulatory; and (3) socioeconomic. Technical issues have been attributed to the Government, the IA, the WUAs and the farmers. Thirty-two actions have been identified, leading to the successful completion of the irrigation system timeline.

The need for reform can appear at any point within the time line. However, reform needs are a common outcome of monitoring and evaluation. Market instruments and PPPs are often planned for and even introduced in the planning phase of the irrigation system. PIM/IMT can be introduced at the planning phase but will typically be implemented during the consolidation phase. This may be planned for (as in Table 1) or may arise with the need for reforms.

The consolidation phase is therefore a critical point in the time line. Problems may appear in this phase that compromise the maturity of the project and seriously threat its sustainability. In many projects worldwide, reform has been identified as a critical requirement to consolidate the irrigation system or to pass from the consolidation phase to the maturity phase. As previously discussed, the implementation of reforms does not necessarily lead to the consolidation or maturity of an irrigation system.

Table 1. Idealized time line of an irrigation system, covering the phases of planning, implementation, consolidation, and maturity from the technical, legal and regulatory, and socioeconomic perspectives. A set of 25 actions has been identified and associated to phases and perspectives.

Actions / Types of issues	Planning						Implementation						Consolidation						Maturity					
	Government	Agency	WUAs	Farmers	Legal regulatory	Socio-economic	Government	Agency	WUAs	Farmers	Legal regulatory	Socio-economic	Government	Agency	WUAs	Farmers	Legal regulatory	Socio-economic	Government	Agency	WUAs	Farmers	Legal regulatory	Socio-economic
1. Structuring the project development team	X																							
2. Setting up a demonstration farm near the project area: experimental fields, irrigation facilties and cooperating farmers	X	X		X																				
3. Visiting similar areas with the project team and future WUA personnel	X	X	X																					
4. Developing solutions adapted to different farming models: subsistence, family, entrepreneurs and agribusiness	X	X																						
5. Defining the legal framework of the project					X																			
6. Avoiding overlapping organisms, defining governance roles and responsibilities					X																			
7. Developing financial systems adapted to solutions above	X	X				X																		
8. Developing procedures for the procurement, contracting and control of the construction works					X																			
9. Building irrigation infrastructure counting on local companies. Iniciating future WUA personnel in the use of machinery							X																	
10. Analyzing staple and agribusiness crops. Testing in real farms procedures developed in the demonstration farm								X	X	X														
11. Providing technical support: from Government and/or Agency to WUAs and farmers							X	X																
12. Developing geodatabases for plots, users, crops, meteorology, infrastructures							X	X																
13. Developing on-farm irrigation systems fit to the local constraints and the farming models							X	X	X															
14. Initiating WUA personnel and farmers in IMT/PIM: a) O&M for canals, ditches and open drains; b) WUA management and participation							X	X	X	X														
15. WUA by-laws, porcedures for conflict resolution											X													
16. Seeking actors' commitments on WUA O&M							X	X	X	X														
17. Fostering market orientation of farmers. Selection of models (cooperatives, agribusiness…)							X					X						X						
18. Considering gender issues in WUAs and farming models: governance and participation												X						X						
19. Professionalizing WUAs: hiring technical staff, controlling the water cycle and mastering O&M														X										
20. Setting up Irrigation Service Fees covering O&M cost														X	X									
21. Materializing PIM: electing of WUA representatives, consolidating WUA authority on the part of the users														X	X	X								
22. Materializing IMT/PIN: allocating responsibilities														X	X	X								
23. Technifying cooperatives to compete in the market																		X						
24. Fostering interaction between WUAs and Cooperatives: Fertilizers, plant protection, Irrigation techniques for market penetration														X	X			X						
25. Managing for all types of water users: hydropower, irrigation and cities													X	X	X			X						
26. Continuous monitoring and evaluation activities													X	X	X	X	X	X	X	X	X	X	X	X
27. Seeking WUA independence from the Government and the Agency																				X				X
28. Considering modernization actions in WUA infrastructure																			X	X				X
29. Providing legal security to users and investors																								X
30. Gaining access to new technologies													X	X	X									
31. Controlling irrigation return flows and non-point source pollution													X	X	X									
32. Integrating WUAs in national and international scene. Sharing experiences, improving self-steem																				X				

4. Current Problems

In the following sections, problems precluding the success of the abovementioned reform models are presented. The list of problems is loosely based on the analysis performed by Garcés-Restrepo et al. [7] in their Table 14. These authors presented an analysis of world implementation of IMT, identifying and ranking a set of 19 problems and issues. Ranking was performed according to the global relevance of each item, although the regional relevance of each problem was also provided. In this paper, discussion has been extended to all reform models (not just IMT), the number of problems has been reduced, and additional sources of evidence have been considered. Table 2 presents the eight current problems, and their correspondence to those identified by Garcés-Restrepo et al. [7]. Instead of following a relevance ranking, our list of problems follows a functional approach: resistance to reform, reform framework, construction project, infrastructure, organization, capacities, training, and cost recovery. Seven of the eight problems correspond to some of the items in the list of Garcés-Restrepo et al. [7]. The poor conception of the irrigation system project does not correspond to any of the items in that list.

For each problem below, the first paragraph contains a description based on scientific evidence. The second paragraph derives from the experience of the authors in irrigation research, engineering, and international consulting.

4.1. Agency and Government Officials Resist Reform

This is the most frequently identified problem, according to Garcés-Restrepo et al. [7]. Agency and government resistance to reform ranked first in the list of problems in all regions of the world. Suhardiman and Giordano [1] realized that policy intervention in government-managed irrigation systems had not significantly improved performance, and set out to analyze the nature of "irrigation bureaucracies" and their effect on the outcome of reform. According to these authors, bureaucracies are actors rather than instruments in reforms, and have their own missions and interests: the "hydraulic mission". In application of this mission, the Agency and its officials see their budget and power legitimized by the role of irrigation infrastructure and water security in national economic development. Suhardiman and Giordano [1] stated that irrigation bureaucracies created the "reform-neglect-rebuild" cycle to sustain their autonomy, attracting a continuous flow of development funds. Resistance reduces the scope of reform to cost recovery, recentralizes rather than decentralizes, and can ultimately neutralize reforms.

We have identified three additional reasons why Irrigation Agencies resist irrigation system reform. The first one is that political support to the program is insufficient. Agencies will generally not be enthusiastic about reform if they fear that the continuity or depth of the reform is not granted. The second reason is that reforms such as PIM/IMT require new expertise that Agency officials most likely will not have. The third reason is that reform may be designed to downscale the role, the size, and the responsibilities of the agency. This may be a threat to the employment and the careers of Agency officials.

Table 2. Correspondence between the ranked list of problems and issues in implementing IMT (Table 14 in Garcés-Restrepo et al., 2007 [7]), and the list of current problems in irrigation governance discussed in this paper.

Problems and Issues in Implementing IMT, Table 14 in Garcés-Restrepo et al. (2007) [7]	Current Problems in Irrigation Governance							
	Agency and Government Officials Resist Reform	Farmers Resist Reform	Weak legal and Financial Reform Framework	Poor Conception of the Irrigation System Project	Irrigation Systems Deteriorated	Unsuitable WUA Organization Structure	Weak Capacities of WUAs	Inadequate WUA Cost Recovery
Resistance to IMT by agency	X							
Inadequate training of WUA								
Difficult for govt. to finance IMT			X					
Irrigation systems heavily deteriorated					X			
Weak capacity to train WUA								
Weak legal framework for IMT			X			X		
Inadequate farmer payment for O&M								X
Weak techn. and mngt. capacity of WUA							X	
Inadequate training for govt. staff	X							
Agency reform and staff disposition								
Farmers resist IMT		X						
No clear/single IMT policy or Programme			X					
Resistance to IMT by local government	X							
Democratic elect. of WUA officers difficult						X		
Conflicts between farmers/villages		X				X		
Politicians resist IMT								
Inadequate support services							X	
WUA cannot apply sanctions								
Farmers lack access to credit						X		X

4.2. Farmers Resist Reform

Bhatt [15] presented a rather negative profile of Indian farmers when confronting PIM. In her perspective, farmers were suspicious, reluctant to adopt responsibilities and had little knowledge of technical issues. Farmers did not trust that the WUAs created under PIM reforms would adequately satisfy their demand for irrigation water. Meinzen-Dick et al. [22] analyzed the factors affecting farmers' collective action in the management of water resources in India. Several the key factors were intangible in nature, such as the leadership and social capital. Hu et al. [23] reported that the creation of WUAs in China in the context of PIM reforms did not necessarily lead to increased farmers' participation. Surveys indicated that farmers did not understand the nature, function, and participatory approach of the WUAs. Yami [24] claimed that the little attention paid to community inputs when building WUAs in Ethiopia seriously affected their acceptability and effectiveness. Aydogdu et al. [25] analyzed farmers' satisfaction with WUAs in the Turkish GAP project. These authors reported on the general opinion that WUAs failed to perform their duties completely. According to Garcés-Restrepo et al. [7], Farmers' resistance to IMT was in the middle of the ranked list of problems. Another item on the list, the conflicts between farmers and villages, can contribute to the resistance of farmers to IMT.

It is somehow natural that farmers resist: reform always represents a profound change in the way water is distributed and managed. More importantly, reforms such as PIM/IMT represent a sociological change that intends to involve users in water management decision-making in countries or local situations where such tradition does not exit. It is also logical that farmers do not trust a new management system that is generally imposed by the same IA that failed in creating an atmosphere of mutual confidence. Knowing that this is a potential problem, the conclusion may be advanced that not enough was done in many cases to disseminate the reform messages. Most of the references in the previous paragraph refer to the initial phase of WUA establishment. Not enough energy was used to let farmers understand the nature of their new responsibilities. On the other hand, expectations of farmers' participation are often placed too high. This is particularly true in large irrigation systems, where day-to-day operations are carried out by technical staff. In such systems, as long as water services are satisfactory, farmers limit their participation to raise complaints and to participate in formal institutional meetings.

4.3. Weak Legal and Financial Reform Framework

This heading directly relates to three elements in the list of problems by Garcés-Restrepo et al. [7] (Table 2): the difficulty for governments to finance IMT, the weak legal framework for IMT, and the fact that there is no clear/single IMT policy or program. The first step to a sound legal reform framework is the explicit registration of water rights. When discussing the IMT reform in Mexico, Kloezen et al. [8] pointed out that the new Water Act established in 1992 did not provide water rights to individual users, and did not establish allocation rules in the case of water scarcity. However, the Act permitted the signing of tradable water concession agreements and the utilization of districts' infrastructure by private parties. Analyzing water management in India, Poddar et al. [6] stated that since water rights were not attributed to individuals, reforms based on water markets could not be introduced. However, as recognized by these authors, explicit, nominal water rights are needed for farmers to invest and maintain irrigation infrastructure. Regarding the institutional framework, Meinzen-Dick [10] observed that donors and policy makers would like to apply similar reforms in all contexts, applying the panacea paradigm. To overcome the failures resulting from this ill approach, she advocated for a polycentric governance model in which state, collective, and market institutions played a role.

During IMT processes, WUAs have often been established without a proper legislation that recognizes them as legal entities. In Jordan, for instance, the lack of specific legislation for WUAs led to cooperatives being used for collective water management. This induced specific problems in water management. In other cases, the legal framework establishes that WUAs are associations of private nature, which is contradictory with an association that manages a public good: water. One of the most successful PIM/IMT programs has been carried in Turkey. From the legal point of view, this program was based on an internal regulation of the General Directorate of State Hydraulic Works (DSI) that allowed the institution to transfer the management of irrigation systems to water users' organizations when appropriate. Only recently, a law has been published regulating the whole process.

4.4. Poor Conception of the Construction/Rehabilitation Projects of the Irrigation System

Critical design deficiencies limit irrigation system performance and cannot be exclusively solved by governance measures. Examples of design shortcomings include selection of incorrect solutions, errors in the sizing of system structures and use of inadequate technologies. Orstrom [26] pointed out frequent engineering misconceptions in irrigation projects. The benefit-cost analysis, the area to be irrigated and the agricultural yields were often overestimated. On the contrary, the recurrent costs of operation and maintenance were systematically underestimated. This perception is in agreement with the profile of the irrigation bureaucracies described by Suhardiman and Giordano [1]. While reviewing the effectiveness of PIM reforms in Cambodia, Asthana [21] stated that competition to get a project approved can lead to exaggerate the expected benefits and to understate the project costs. Orstrom [26] and Vos [27] stated that government-managed projects were often too large to be efficient. These large projects focused on the engineering works, and often left water distribution, maintenance, and farmers' organization unattended. In the best of cases, as reported by Orstrom [26], government officials would only describe the main structure of the farmers' organization. This author described the process of institution crafting, that she anticipated would become more intense than engineering works in the decades to come. This approach was later on adopted by the World Bank in their Comprehensive Development Framework [28]. Ortstrom [26] concluded that farmers must be involved in the physical and financial design of the irrigation systems, and that they must have decision-making capacity. These statements were adopted by Vermillion and Sagardoy [5], when declaring that WUAs must be in the driver's seat of IMT reforms. This enables users to identify and prioritize options, and to make financial decisions. This participative approach to infrastructure development was widely accepted in the 20th century [4,13].

The ex-ante cost-benefit assessment of public projects continues to be optimistic in all types of projects (not just in irrigation) and in all regions of the world. Large and complex irrigation systems have proven successful in developed regions or for commercial irrigated agriculture. However, small, community-targeted projects are currently being designed for developing regions or in contexts of subsistence farming. In such environments, users' institutions develop with the project (actions 3, 9, 11, 14 in Table 1), and often do not have enough strength to guide project decisions, particularly at the time of the initial construction. Rehabilitation projects making part of reform policies often count on more organized users. Participation in project physical and financial design will greatly facilitate users' implication in O&M and cost recovery.

4.5. Irrigation Systems Deteriorated

This item was considered relevant in the list by Garcés-Restrepo et al. [7] (Table 2). Vermillion and Sagardoy [5] described the processes of rapid irrigation system deterioration. The area actually irrigated decreased in time, as water distribution became undependable. Water misuse induced waterlogging and salinity problems. System deterioration widened the gap between actual and potential irrigation performance. This gap is one of the factors triggering the reform of government-managed irrigation systems [22]. According to Vermillion and Sagardoy, an assessment of the performance gaps is needed to evaluate the actions required at the irrigation system, taking into account two options: enhancement or reform [5]. These authors stated that a history of enhancement actions with a continuously widening gap is a clear sign of the need for reform. They also warned that political feasibility (enhancement) may override the real needs of the system (reform). Suhardiman and Giordano [1] stated that the vicious cycle of poor financial and institutional arrangements, deferred maintenance, infrastructure deterioration and poor performance explains the typical low performance of government-managed irrigation systems.

Irrigation system deterioration remains a fundamental problem and is one of the major reasons for implementing PIM/IMT programs. Insufficient government budget to finance rehabilitation/modernization projects and the difficulties to borrow from national or international banks may lead to the transfer of degraded irrigation systems. This creates a challenge for the new WUAs, which may confront serious operating problems. If irrigation cannot proceed normally, farmers will have limited financial capacity to contribute to the repayment of investment costs, and therefore the full cost of the rehabilitation/modernization will have to be covered by the Government. In some cases, rehabilitation projects have been implemented without establishing WUAs. These are examples of irrigation system enhancement, in which the need for successive projects will most likely arise again in the following years.

4.6. Unsuitable WUA Organization Structure

Vermillion and Sagardoy [5] recognized that one of the key drivers for IMT was fostering managerial accountability in the irrigation sector. According to these authors, these user-driven organizations needed to incorporate incentives, sanctions, and transparency. After forty years of practice, it can be stated that PIM/IMT has improved collective water management and empowered users. In general, the process counted on a transition period characterized by Agency-WUA cooperation [8]. Veldwisch and Mollinga [17] reported that government control of Uzbek WUAs continued to be strong even after the transition period, with WUAs implementing state control over water distribution and agricultural production. According to these authors, WUAs were held responsible for state production targets, and WUA leaders were appointed by the Government. In fact, Garcés-Restrepo et al. [7] included the difficulty in holding democratic elections for WUA leaders in their list of IMT problems. Despite these difficulties, PIM/IMT has resulted in significant independence of WUAs from their governments after long development periods [15]. Several authors have analyzed WUA organizational problems hindering PIM implementation in different areas of the world [15,23–25,29–31]. These authors identified the following additional issues: (1) interference between WUAs, village governments and agricultural cooperatives; (2) unclear duties and authority of WUA managers; (3) strong government policies, hindering WUA functionality; (4) marginalization of rural women in WUA decision-making; and 4) lack of supportive policy and legal WUA regulations.

The case of Turkey demonstrates that several types of organizations can be used in the same PIM/IMT program if each type is applied taking into consideration the local conditions and the possibilities to provide adequate supporting services to farmers. Often the problem is that serious limitations are imposed to WUAs. In some cases, not all farmers benefiting from the irrigation system are WUA members because association is not mandatory. In other cases, key management functions are still carried by government agencies. This can imply the operation of key irrigation structures (pumping stations, canals, dams, etc.) essential for efficient WUA management. As a result, WUAs have limited responsibility in managing the irrigation system. In some irrigation systems, WUAs are not authorized to act as contractors for certain rehabilitation works. Finally, WUAs may lack a section in charge of operation and maintenance or established procedures for internal conflict solving.

4.7. Weak Capacities of WUAs (Technical, Managerial)

Reforms analyzed in this paper have permitted more progress in developed than in developing environments [6,19]. While this can be attributed to several economic, societal and policy factors, WUA capacities stem as a major cause. In fact, Garcés and Restrepo [7] listed several WUA training requirements following IMT. The early success of the Mexican IMT program has been linked to efforts in increasing the capacities of the emerging WUAs [8]. Part of the problems related to WUA capacities derive from the continuity of Agency personnel in the newly created WUAs and from the absence of WUA staff selection procedures [7,8]. Hamada and Samad [4] identified some of the unfair factors that can appear in PIM areas when WUA capacities are far from optimum: dominance of upstream and large-scale farmers, low participation of female farmers, failure to collect fees, and lack of financial transparency. Bhatt [15] compiled operational problems for PIM implementation in India, and underlined the importance of financial provisions as drivers for capacities, and lack of knowledge about tasks such as record-keeping and accounting. Omid et al. [32] reported that Iranian WUA success relied on WUA capacities for administrative procedures, operation of irrigation facilities and water fee collection.

At the root of these problems is the absence of professionals to carry out the daily WUA responsibilities. This is often caused by the insufficiency of the collected irrigation service fees (ISF) to hire external personnel. Ultimately, this depends on the size and ambition of the WUAs. In some areas of the world, WUAs are large and can mobilize funds and staff to maximize performance; in other areas, WUAs are very small and focus on maintaining watercourses. Since small WUAs cannot afford to hire technical staff, high levels of performance are rarely achievable in small WUAs, increasing the risk of system deterioration. Limitations in WUA capacities difficult the implementation of monitoring and evaluation procedures. Most of these problems could be solved with the training of WUA personnel. However, traditional (classroom) training has proven ineffective in many instances. On-the-job training in communities of practice stands chances of being effective in this environment.

4.8. Inadequate WUA Cost Recovery

The collection of ISF was identified as a cause for irrigation reform since the first applications of IMT in Mexico [8]. Several reasons were provided to explain weak fee collection: distance to the payment offices, need to wait in line, bribing government officials and ditch tenders, and the lack of consequences. In fact, ISF recovery was one of the major strengths of the Mexican IMT: collection rates approached 100%, the cost to farmers did not increase, and WUA self-sufficiency increased from 50 to 120%. Even after PIM/IMT, many WUAs remain trapped in a vicious circle of "insufficient maintenance—downgrading infrastructure—poor performance—low-cost recovery" [11]. Cases of this vicious circle have been documented in Tunisia and Pakistan [31,33]. WUAs have been reported to approve deficit budgets to match farmers' willingness to pay. Governmental subsidies have been documented after IMT, in cash or in kind [34]. In many countries, WUAs are not allowed by the Government to apply sanctions to farmers not paying their bills. In some cases, WUAs are exposed to high inflation rates or to unexpected repairs or investments, and they find it difficult to face the related expenses [8]. In other cases, problems are created within the WUA, with corruption, fraud and bribing [5]. According to Hamada and Samad [4], one of the PIM objectives was to stimulate a more productive and self-reliant irrigated agriculture, which could support reliable, professional water management. Ounvichit et al. [35], when analyzing an irrigation system in southern Thailand, reported that explicit economic returns from farming would increase all types of farmers' participation in WUAs. Unfortunately, donors and governments often have not focused on economic sustainability when selecting project proposals. In this sense, the criteria for investment in irrigation systems published by Orstrom [26] continue to be relevant.

In our view, this is the most serious problem that WUAs face, since it is not possible to operate permanently under economic deficit. In many cases, it remains unclear if farmers cannot pay or if they do not want to pay. The subsidies applied to WUAs in many countries are a source of controversy. In general, subsidies distort the market and create perverse economic effects. However, in some cases, subsidies can be a solution rather than a problem. Subsidies should be seen in the context of the public interest in the irrigation project and its infrastructures. In fact, in some countries and development contexts, subsidies applied to part of the operation and management costs can be a cost-effective alternative to the public rehabilitation of the whole irrigation system. Of course, this is not a generalizable solution, but it can be useful in specific cases. Regarding WUA cost recovery, problems arise when the WUA General Assembly approves lower ISFs than recommended by the management, when—as in the case of Jordan—WUAs are not entitled to recover water fees from farmers, when tariff setting is not transparent or when the ISF collection system is not adequate. The final point in WUA cost recovery is the economic profit from irrigated farming. When farmers make relevant benefits WUAs tend to operate satisfactorily. Therefore, securing farmers' wellbeing is a way to ensure irrigation system maintenance and smooth operation. Unfortunately, the economic result of irrigated farming is the responsibility of the Ministry of Agriculture, while PIM/IMT is the responsibility of the IA. Coordination of efforts between these two institutions is infrequent.

5. Current Solutions

The following sections present eight solutions to the abovementioned problems of water governance in irrigated areas. These solutions that been proposed, described, and/or experimented in different contexts. Table 3 presents a classification of the contribution of the identified solutions to the problems. Each solution addresses several problems with different degrees of contribution. In the following description of each problem, two paragraphs are also used: a description according to the scientific literature (readers should note that there is much less literature on solutions than there is on problems), and the professional experience of the authors of this paper. The order in which solutions are presented is loosely based on the order in which the related problems were described.

5.1. High-Level Political Commitment

One of the causes for the early case of IMT in Mexico was the high-level political commitment [8]. This was very important in redefining the role of the IA after IMT, and in fostering cooperation between all actors. Garcés-Restrepo et al. [7] in their assessment of worldwide IMT efforts, identified high-level political commitment as a key recommendation for on-going programs. According to these authors, commitment needs to be strong at the highest possible level for a sustained period. They recommended study tours to irrigation systems with successful reform applications to encourage political support.

The authors agree on the importance of political commitment, and on the need for commitment at the highest political level. However, experience indicates that this is very difficult to obtain and sustain. Therefore, reforms are commonly performed under partial commitment conditions.

5.2. Additional Legal Reforms Concerning the WUAs and the Agency

The relation between the Agency and the WUAs is guided by the principle of subsidiarity: decisions are made at the lowest possible level [7]. The application of PIM/IMT has led to a reorganization of the Agency focusing on decentralization and downsizing. This implies withdrawing from the downstream-end levels of the irrigation network (Figure 1), focusing on WUA capacity-building, regulating the irrigation sector, river hydrology, canal management, and developing monitoring and evaluation capacities [7]. On the side of the WUAs, reform must ensure that the WUA roles are clear and sufficient [4], and that farmers enter PIM processes with sufficient clarity and well-defined responsibilities [32]. These changes must be framed in a supportive policy, regulatory and legal environment [16], which needs to address all aspects of irrigation governance: financial transparency, equity in water allocation [4], dispute resolution [15], collective decision-making, equity in cost allocation, and a mutually beneficial relation with the agency [32]. The establishment of WUAs requires clarifying responsibilities with parallel rural organizations, such as cooperatives or villages [23]. These reforms must be complemented by specific provisions to achieve financial self-sufficiency through planning and effective cost recovery [15]. In some cases reforms also include the establishment of tradable water rights [8]. In recent years, the need to reform the irrigation bureaucracies as part of IMT processes has gained importance [18]. This issue transcends irrigation governance, to address the way donors and government agencies approach development assistance.

Table 3. Semi-quantitative contribution of the identified solutions to the problems identified in this paper. Contribution has been evaluated by the authors in a scale from 0 to 5, with 5 expressing the highest contribution. The table also presents the Difficulty of the Problems and the Power of the Solutions.

		High-Level Political Commitment	Additional Legal Reforms Concerning the WUAs and the Agency	Flexible, Adaptive Reform Programs	Rehabilitation/ Modernization Programs	Capacity-Building (WUAs, Agency and Government Officials)	Monitoring and Evaluation at the WUAs	External Support	Benefit from Farming and Participation	Difficulty of the Problem, DP
Problems	Agency and government officials resist reform	5	3	3	1	2	2	1	1	2.8
	Farmers resist reform	2	2	3	5	2	2	2	3	2.4
	Weak legal and financial reform framework	5	5	3	1	1	1	1	2	2.6
	Poor conception of the construction/rehabilitation projects of the irrigation system	1	1	1	3	2	4	3	2	2.9
	Irrigation systems deteriorated	3	3	2	5	3	3	3	4	1.8
	Unsuitable WUA organization structure	2	5	3	1	2	2	3	3	2.4
	Weak capacities of WUAs (technical, managerial)	1	1	2	2	5	4	4	4	2.1
	Inadequate WUA cost recovery	1	3	3	2	2	5	3	5	2.0
Power of the Solution, PS		2.5	2.9	2.5	2.5	2.4	2.9	2.5	3.0	

Reforms must explicitly declare WUA or farmer water rights and establish a clear WUA legal profile (public, private or public-private). The legal scope of WUA activities must also be addressed, specifying if activities other than water management are allowed for WUAs. On the other hand, as water governance becomes more hydrologic, explicit, and participative, it is important that WUAs are represented at the River Basin Organizations in charge of water allocation, emergency response and environmental management. To do so, WUAs need to be structured in associations or federations representing their interests at higher organizational levels. It is equally important that the technical bodies of the WUAs engage in communities of practice at the basin or national levels. This will ensure the propagation of good practices and innovations in water management. The transition of low-level water management from the Agency to the WUAs does not automatically generate the required authority at the WUAs to address conflicts in water management and to ensure ISF recovery. Legal reforms must address this problematic issue to prevent WUA discredit.

5.3. Flexible, Adaptive Reform Programmes

Effective collective arrangements are one of the keys to success in the self-governance of irrigation systems [16]. Garcés-Restrepo et al. [7] introduced the need for flexible and adaptive reforms, claiming for effective awareness campaigns and stakeholder consultations. Local boundaries need to be found for the responsibilities of both the Agency and the WUA, which need to complement each other in the new institutional arrangements [18]. Although the PIM/IMT model is well-established, Veldwisch and Mollinga [17] reported that its application in different areas of the world led to very different results. These authors addressed the term "institutional bricolage", coined by Sehring [36] as the "partly purposeful and partly unintentional process of the combination and transformation of institutional elements that results in a qualitatively new type of institution". The need for participation of WUAs in reform processes is firmly established today. However, the relative importance of the public, collective or private actors in irrigation system ownership and management is currently open to discussion. Meinzen-Dick [10] opted for a polycentric combination of each type of institutions after taking into consideration the strengths of the local institutions and after evaluating the possible synergies. The success of the early Mexican IMT case has been partially explained by its flexibility and adaptation to the local scene [8].

The main limitation to the application of this solution is the time required to assess the nature and characteristics of the local institutions, to consult the stakeholders and to arrive to conclusions about the institutional, technical, and financial aspects of the required reform. The requirements of the donor institutions and/or the local governments often put pressure on this phase of the process, resulting in incomplete assessments and fragile agreements with local stakeholders.

5.4. Rehabilitation/Modernization Programmes

According to Orstrom [26], users need to participate in the design of an infrastructure producing localized benefits, Conversely, they also have to contribute with their own resources, so that the public decision on investment should only be taken if users contribute both to the project design and funding. There is a general agreement on the need for rehabilitation/modernization, as related to the reform process. However, there have been different experiences on the relative timing of the reform and the rehabilitation/modernization. Kloezen et al. [8] reported how, in the successful Mexican IMT case, farmers believed that rehabilitation/modernization should have been performed prior to transference. However, the Agency did not engage in construction projects before transfer. Shortly afterwards, WUAs bought heavy construction machinery using ISFs. Farmers often express fear that IMT will imply full WUA responsibility and financial liability for rehabilitation/modernization. The commitment of the Agency to get technically and financially involved in these projects after IMT has facilitated the transfer process [5]. In Turkey, for instance, modernization/rehabilitation projects in transferred WUAs are assessed, guided and funded by the Agency (DSI) [37]. Incremental infrastructure improvement has been advocated as a continuous process of rehabilitation/modernization, opposed to the complete reconstruction of the irrigation system following a few decades of operation [5]. At the on-farm level, traditional surface irrigation systems have been replaced in some projects by sprinkler and drip irrigation systems, following high expectations on the applicability of these new systems [38]. The recent recognition of irrigation management as a key factor in improving irrigation performance has often reduced attention on the importance of updating the irrigation structures [39]. In fact, the combination of management and structural improvements constitutes a complete rehabilitation/modernization project.

The experience of the authors indicates that rehabilitation/modernization projects in developing regions often focus on the construction aspects and fail to explore the relation of the infrastructure with the natural resources and the farming systems. A wide margin for improvement exists in the adaptation of construction projects to local conditions. Aspects such as hydraulics and structural analysis receive much more attention than agricultural systems, on-farm irrigation systems, drainage, salinity, irrigation organization and operation and management.

5.5. Capacity-Building (WUAs, Agency and Government Officials)

Reform constitutes a new game in irrigation management, for which none of the actors have the required capacities. The change in the roles of the government officials, the Agency and the WUAs requires new capacities. Moreover, at the onset of reform, the public sector does not have the capacities required to train the WUAs [7]. The largest deficit in training appears at the WUAs, which need to build technical, financial, administrative, extension, agribusiness and marketing capacities *ex-novo* and in a short time [7,24]. A common solution is to organize training from the public sector for the leaders and the personnel of some of the new WUAs. In some cases, capacity-building programs have been extended to farmers.

Public budgets for capacity-building are always shorter than required. Infrastructure continues to find a more effective access to public funding than training does. Capacity-building programs have to beat the additional problem of focusing on small groups distributed throughout the complete irrigated area. Therefore, many projects resort to training trainers or WUA managers. An additional problem is that rural communities are not fit for standard training, and require long "on-the-job" capacity-building programs. Such programs could be developed in cooperation with nearby projects, promoting peer-to-peer, informal training. An evolution of the currently used capacity-building tools is required, since all actors seem to be dissatisfied with the current costs and results.

5.6. Monitoring and Evaluation at the WUAs

The importance of monitoring and evaluation in irrigated agriculture was first outlined in the 1980s with the aims of assessing the impact on the project beneficiaries (income, employment, quality of life) [40]. Vermillion and Sagardoy [5] structured the concept in the context of IMT. These authors issued a set of recommendations that – two decades later – continue to be critical for success. Among them, the importance of selecting a reduced number of information-efficient key indicators, the need to combine top- and bottom-directed monitoring, and the benefits of periodic reporting of monitoring and evaluation information to WUA governance bodies. Several monitoring and evaluation tools have been proposed in the last decades for irrigated areas, focusing on database technologies [41,42], remote sensing applications [43] and other information technologies.

An efficient, sufficiently funded monitoring and evaluation system should be able to provide real-time detection of the operational, financial and management problems of the WUA. The abundance of monitoring and evaluation tools for irrigated areas is in contrast with the low level of implementation of these procedures. In fact, examples of systematic monitoring and evaluation efforts in WUAs are not numerous, particularly in development contexts.

5.7. External Support

External support is understood as personnel other than water users that contribute to perform the activities attributed to the WUA. These personnel can be hired by the WUA, can belong to an irrigation services company hired by the WUA or can belong to a firm participating in a PPP process. The importance of counting on sufficient ISF collection to hire WUA personnel was recognized by Orstrom [26], and later stated by Vermillion and Sagardoy [5]. These authors introduced the concept when describing IMT: farmers are not supposed to implement the service previously lent by the Agency. External services can cover a range of professional capacities, such as technical, legal, accounting, engineering, or communication. The goal of external support is to implement WUA policies. In the context of PPP, operation and management have been recognized as crucial tasks for external support [11].

WUAs established in developed regions have already gone a long way in securing external support. The last decade has seen the hiring of engineers (and other professionals) by WUAs to assist them during the process of rehabilitation and modernization. Later, these professionals have led the operation and management of modern infrastructure. In developing regions, the process is only incipient, and irrigation professionals are more frequent in the Public Administration than they are in the WUAs. The lack of professionals in the WUAs makes communication with the Government, the Agency, and service providers difficult. A key aspect in the hiring of professionals is the relation between the WUA budget and the cost of a professional. The minimum WUA size (in hectares) required to hire a professional changes with the local development level. The Federation of several local WUAs can help attain the minimum required WUA size.

5.8. Benefit from Farming and Participation

When analyzing the literature, the reform of irrigation systems is often perceived by authors as dissociated from the agricultural systems that these supply. Some authors have recognized the importance of linking irrigation and agriculture. For instance, when analyzing the sustainability of WUAs in Thailand, Ounvichit et al. [35] signaled that explicit economic returns from irrigated agriculture would have farmers participate in WUAs and to support their activities. Similar opinions were cast by Hamada and Samad [4] when listing the requisites for sustainable PIM processes: farmers must receive tangible benefits from using water. A similar principle was expressed by Ul Hassan [16] when addressing water users: the benefits of governance and management must exceed the cost.

While this is a common-sense principle, it can explain several experiences lived by the authors when analyzing irrigation projects in developing regions. Inactivity by farmers is often associated to non-profitable crops, governmental or non-governmental subsidies in the form of food or commodities, or completely deteriorated irrigation/drainage structures or soil fertility. Conversely, the existence of a profitable economic model – leading families beyond subsistence—often constitutes a sufficient incentive to establish a vigorous WUA. This issue is linked to the conception of the rehabilitation/modernization programs: the new WUA infrastructure needs to be designed for agricultural success in family farming, company farming or both. Unspecific irrigation system designs, not paying attention to the final purpose of irrigation water delivery (creating wealth through farming) are bound to meet relevant sustainability problems. WUAs can also contribute to farmers' benefits by promoting the creation of cooperatives, the cultivation of market-oriented crops and the establishment of companies providing irrigation services.

6. A Case Study Approach to Problems and Solutions in Irrigation Governance

Seven case studies are described in this section. These involve different types of WUAs and different degrees of success in performance and service to the farmers and to the public interest. The case studies are in seven countries located in four continents. Two of these countries are included in the list of Low-Income Developing Countries produced by the International Monetary Fund [44]

(Cameroon and the Kyrgyz Republic). Three countries are included in the list of Emerging Market Economies (Chile, Georgia, and Jordan). Finally, two advanced economies are included (Portugal and Spain). Even in advanced economies, irrigation systems are deeply related to rural development. Therefore, the seven case studies provide insights of different rural development contexts and different types of reforms. In the following sections, case studies are presented in country alphabetical order. The name of the specific WUAs has been omitted.

6.1. Cameroon

The WUA makes part of an irrigated area supplied by a 35-year old canal with a capacity of 14 m^3/s. Water is abstracted from a reservoir essentially used for hydropower. The original project included irrigation in both riverbanks, covering eleven thousand hectares of public land. However, the currently irrigated area only extends to six hundred hectares. The project focuses on rice production, with family farms of about 0.5 ha. Water governance is complex, with several Ministries and public bodies implied in the project at different levels. Government officials do not have experience in planning or managing irrigated agriculture. An additional level of complexity is introduced by the ethnic profile of the farmers, which leads to different approaches to farming. Very few locals have capacities in irrigated agriculture. The project area has received intense migrations from Cameroon and from neighboring countries. This has exponentially increased the population of the local capital city in the last decade. The current population exceeds three hundred thousand inhabitants, which are exposed to relevant food security problems. This is the most problematic case study in this paper, with serious difficulties in the approach of the Agency and the Government, in the reform framework, infrastructure and in the capacities and organization of the WUA, which is far from recovering the operation and management costs. Discussions are currently being held with international funding agencies to identify the most adequate solutions to this complex case.

6.2. Chile

This WUA, located in the arid coastal north of Chile, is reasonably organized, and has relevant capacities to maintain and improve its irrigation infrastructure. It integrates mining companies (the largest water right holders in the area) in addition to farmers. The WUA organizational model is well developed, participative and offers guarantees to members, although its small size constitutes a relevant challenge. During the recent liberalization of water in Chile (1998–2005), perpetual water rights were assigned. These rights are not coupled to land tenure, leading to an active water market. Mining companies have recently bought water rights to profit from the high price of copper ore. Farmers selling water rights go out of business because local aridity requires irrigation for any agribusiness operation. In general, farmers have very good knowledge of their production system, and grow fruits and fresh produce for international markets. A few negative aspects can be highlighted. For instance, several legal and organizational aspects could be strengthened. The absence of an IA guiding and organizing local WUAs, and managing large hydrologic systems is resulting in poor coordination and poor hydrological planning. In an area where water rights exceed renewable water resources by far, this is a critical issue, one that could eventually lead to water bankruptcy.

6.3. Georgia

The case study makes part of a thirty thousand hectare irrigation system is managed by LLC Georgian Amelioration, a public agency acting as the IA. The irrigated area is currently experiencing a rehabilitation process. After decades of abandonment and migration to the nearby capital city, not much social capital remains: an aged society is in place, with little interest in irrigation. In fact, irrigation is currently performed in less than 10% of the system area. Past attempts to constitute WUAs have failed. LLC has a bureaucratic approach, characterized by slow administrative progress. There is a strong need for success of the WUA model, and complete remodeling of the public governance. The sustainability of the area is seriously threatened by the lack of a profitable agricultural model, compatible with the local social traits. This may be a typical case of Public Administration failure, leading to PIM/IMT. However, the process is not adequately grounded, and success seems very unlikely in the current circumstances. The combination of a bottom-up, credible project for local agricultural production and a concessional model for infrastructure development could reverse the current trends, accelerating the process and counting with local support.

6.4. Jordan

The cooperative performing the WUA role is in the Jordan Valley. It covers an irrigable area of 1026 ha, with 254 farm units frequently using localized irrigation. The cooperative is part of a government PIM/IMT program whereby several irrigation systems were transferred to farmers between 2001 and 2010. By 2012, 18 water cooperatives were established, covering some 20,000 ha (60% of the irrigated area of the country). The irrigation transfer process started with a technical report by Jordan Valley Authority (JVA, the IA) regarding the condition of the irrigation system to be transferred and following the training of leaders and staff. WUAs in Jordan are registered as independent financial and administrative cooperatives (under cooperative law and by laws) and are under the umbrella of the Jordan Cooperative Corporation (JCC). JCC provides a variety of supporting services: training in accounting matters, auditing, and some scattered monitoring. JVA intends to move WUAs under its oversight, but so far this has not taken place. Farmers in the JVA are mostly producers of fruit trees and vegetables, exporting to Europe and Gulf countries. Most of the farmers are middle to old age and have performed relevant investments in greenhouses and farm reservoirs. Lack of network rehabilitation and water scarcity limit farmers' satisfaction. The cooperative has promoted some training activities for the farmers that were widely appreciated. Fruitful cooperation exists between JVA and the cooperative. Neither the Cooperative nor its members have water rights, and this the general case in Jordan. Members pay very low fees to JVA, covering only a fraction of the actual operation and management costs. The cooperative is heavily subsidized by JVA, resulting in low independence. Less than 50% of the farmers covered by the irrigation network are members of the cooperative, since this is not mandatory to have access to water. The irrigation infrastructure is more than 60 years old, and maintenance is poor. The Government is aware of the unsustainable financial situation, but a strong increase of water fees would not be socially acceptable. Cooperatives are demoralized by their scarce level of autonomy. Unless urgent actions are taken, the survival of this system will remain linked to the continuity of the subsidies to the cooperatives.

6.5. Kyrgyz Republic

The WUA is in Fergana, in a traditional irrigation scheme characterized by large irrigation water supply and very small WUAs. The irrigation system was the target of a USAID project on WUA development. Therefore, farmers and the IA officers are aware of the theory and have moderate interest on the practice: farmers do not believe in the WUA model. Infrastructure maintenance is delayed by organizational problems, and by factors such as the lack of construction materials. The combination of poor water management and poor infrastructure maintenance results in poor water service. When this is combined with an aged farmer's community, the result is abandonment of irrigation practice and emigration to nearby cities. There is currently no perspective for a modern, profitable farming system based on irrigation. Under these conditions, the sustainability of the area is seriously threatened. The Government could eventually attract an international donor for irrigation rehabilitation / modernization and PIM/IMT. The abundance of irrigation water brings no guarantee to the sustainability of the irrigation system. Local change should start by the farming systems and the institutional arrangements for water management.

6.6. Portugal

The irrigation system involves a hundred and ten thousand hectares of new irrigated land in the *Alentejo* region. The main canal supplies water to WUAs and large farms. The case study WUA involves farmers of all sizes. The project is managed by a public company, EDIA, acting as the IA. EDIA has well-trained directive and technical personnel, and performs design, tendering, supervision, and management of all infrastructure: from source to farm. The project is based on the irrigation of Mediterranean crops: 50% olive, 25% cereals, 25% horticulture. Under this project, WUAs were quickly formed and federated. In general, the case study WUA is a success story. Success is only partial, for there are a few partial negative aspects, related to the WUA organization structure and capacities. Despite the public nature of EDIA, it effectively responds to users' needs, and promotes efficiency in water management and profitability in farming.

6.7. Spain

The case study WUA is in a PPP irrigation development in the north of the country. The large presence in Spain of multinational urban water utility companies and the control of national debt by the European Union have facilitated the development of this model in Spain. In fact, two PPP irrigation projects are on-going, exceeding fifty thousand hectares each. In both cases, the PPP is developing new irrigated land. These two projects show large organizational differences. One of the key positive aspects of these projects is that the WUA remains in the hands of farmers, as required by the Spanish water law. The technical bodies of the WUA and the IA are provided by the company (or group of companies) running the PPP for the local authorities. These companies build and operate the project under a concessional model, similar to well-known examples (highways). On the negative side, the PPP model can generate large costs that the public system and/or the farmers cannot afford. In fact, companies carry more overheads and taxes, as compared to direct action by public authorities. The expansion of the PPP model is limited by the low cost of WUA management operations, since WUAs benefit from their deep knowledge of the local environment and from their non-profit nature. Political interference can always be a key negative factor in irrigation PPPs. The two on-going projects in Spain show relevant differences in economic and managerial performance. In the long run, convergence between irrigation and urban water seems to be the goal for multinational water utility corporations. This perspective could count on the support of some political leaders.

7. From Problems to Solutions: A Semi-Quantitative Approach

A semi-quantitative approach was developed to assess the potential contribution of each solution to the documented problems (Table 3). Quantitative marks were used for this purpose, ranging from 0 to 5 (where 0 implies that a particular solution does not contribute to the solution of a problem, and 5 implies that the solution is very relevant to this problem). Table 3 constitutes the problem-solution, or M matrix. The numerical values of the coefficients of M presented in Table 3 are not aimed to be universal. They correspond to the appreciation of the authors, and we believe that different persons would end up with different coefficients. The same can be said of the region of the world where the matrix is to be applied: different matrices would probably apply to different regions. The application of the proposed semi-quantitative approach to an irrigated area should start with the stakeholders' discussion of the matrix coefficients in the local context.

Table 3 presents two additional indicators:

$$\text{Difficulty of the Problem}: DP_i = 5 - \frac{\sum_j M_{ij}}{ns} \tag{1}$$

$$\text{Power of the Solution}: PS_j = \frac{\sum_i M_{ij}}{np} \tag{2}$$

where i is the sub index for problems, j is the sub index for solutions, and np and ns are the number of problems and solutions respectively. In this case, $np = ns = 8$. DP_i and PS_j can range between 0 and 5.

According to Table 3, the most difficult problems were: the poor conception of the construction/rehabilitation projects of the irrigation system, and the resistance to reform by the

Agency and Government officials. The reason for this classification is that a few solutions can address these problems with moderate or high contributions. Conversely, the most powerful solutions were: farmers benefiting from participation and farming, additional reforms, and monitoring and evaluation. These solutions intensely contribute to a wide number of problems. The authors did not assign a value of zero to any matrix element. They considered that each of the identified solutions could have a positive influence of each problem, even if this influence was very small.

The next step is to assess the Intensity of the Problems of a given WUA using a scale of integer numbers also ranging from 0 to 5, where 0 indicates that the problem has the lowest local importance, and 5 indicates maximum local importance. These scores constitute the problem vector (*P*):

$$P = (p_1, p_2, p_3, p_4, p_5, p_6, p_7, p_8) \tag{3}$$

Table 4 presents the elements of the problem vectors for the case study WUAs. For convenience, problem vectors are presented for each WUA as table columns. Again, the authors attributed non-zero values to each problem of the case study WUAs. The intensity of the problems in each area can be estimated by adding the elements of the problem vectors. Case study areas in Cameroon, Georgia, Kyrgyz Republic, and Jordan (in this order), ranked high in the list. On the other hand, case study areas in the most developed countries (Chile, Spain, and Portugal) presented relatively low problems. The most intense problems in the study case areas were the unsuitable organization structure and the capacities of the WUAs.

Table 4. Intensity of the problems in each of the case study WUAs. Problems were assessed by the authors in a scale from 0 to 5, with 5 expressing the highest problem intensity.

Problems \ Case Study WUAs	Cameroon	Chile	Georgia	Jordan	Kyrgyzstan	Portugal	Spain
Agency and government officials resist reform	5	1	2	4	5	1	1
Farmers resist reform	2	2	5	1	1	1	3
Weak legal and financial reform framework	5	4	2	5	5	1	1
Poor conception of the construction/rehabilitation projects of the irrigation system	3	1	4	3	3	1	2
Irrigation systems deteriorated	4	1	4	4	4	1	1
Unsuitable WUA organization structure	5	3	5	5	5	2	2
Weak capacities of WUAs (technical, managerial)	5	3	5	2	5	2	2
Inadequate WUA cost recovery	5	2	5	5	3	1	2

The pre-solution vector (*S'*) can be obtained as:

$$S' = P \times M \tag{4}$$

where operator * is matrix multiplication. The pre-solution vector can be expressed as:

$$S' = (s'_1, s'_2, s'_3, s'_4, s'_5, s'_6, s'_7, s'_8) \tag{5}$$

Elements of the pre-solution vector represent the contribution of the solutions to the problems of each case study WUA. Elements take integer numbers ranging between 0 and 8×5^2. The solution vector (*S*) is obtained by standardizing *S'* so that its elements are real numbers ranging from $\min_i(p_i)$ to $\max_i(p_i)$. This ensures that for a given case study area the elements of the solution vector use the same range of values as the problem vector. Following this approach, element j of the solution vector can be obtained as:

$$s_j = \min_i(p_i) + \frac{\left(s'_j - \min_j(s'_j)\right)}{\left(\max_j(s'_j) - \min_j(s'_j)\right)}\left(\max_i(p_i) - \min_i(p_i)\right) \tag{6}$$

The higher the value of s_j, the more this solution can contribute to address the problems of the case study WUA.

The results of the application of the semi-quantitative approach to the case study WUAs are presented in Table 5. For convenience, solution vectors are presented for each WUA as table columns. Because of the procedure described in Equation (6), the intensity of the solutions in each area follows the same order as the problems. The most intensely recommended solutions were: users benefiting from farming and participation, additional reforms concerning the WUA and the Agency, and Monitoring and Evaluation of the WUAs.

Table 5. Contribution of the solutions to the problems of each case study WUA, as estimated using the proposed semi-quantitative approach. Contributions are expressed in a scale from 0 to 5, with 5 expressing the highest contribution intensity.

Solutions \ Case Study WUAs	Cameroon	Chile	Georgia	Jordan	Kyrgyzstan	Portugal	Spain
High-level political commitment	4	2	2	3	4	1	1
Additional legal reforms concerning the WUAs and the Agency	5	4	4	5	5	2	2
Flexible, adaptive reform programs	3	2	3	3	3	1	2
Rehabilitation/modernization programs	2	1	4	1	1	1	2
Capacity-building (WUAs, Agency and Government officials)	3	2	3	1	2	1	2
Monitoring and Evaluation at the WUAs	4	2	5	3	4	2	3
External support	3	2	4	2	3	2	2
Benefit from farming and participation	5	4	5	4	4	2	3

A detailed discussion of the solutions is presented for the case study in Jordan, which is in the middle of the ranking list for the intensity of its problems. According to the proposed methodology, the solution contributing most to the identified problems is to perform additional reforms in the WUAs and the Agency. Additional reforms should address the cooperative nature of the users association, the close relation to the JVA, and the low-cost recovery. A new legislative framework is required to create a specific WUA legal entity, to separate the WUA from the Agency, and to attain WUA financial security. The next identified solution by intensity is to increase the benefit of water users from farming and participation. While farmers export most of their products, participation in the cooperative is not sufficiently beneficial for them (only half of the farmers participate and pay fees). The next priorities include political commitment, flexible, adaptive reforms, and monitoring and evaluation. Solutions to the Jordan case study did not prioritize rehabilitation and modernization, capacity-building or external support.

The proposed methodology has provided insight on the governance problems and solutions of the seven case study areas. However, shortcomings can be expected when it is applied to specific areas. A key shortcoming is related to the relative timing of the problems and the solutions. Discussions among the authors of this paper evidenced that the contribution of the solutions to the problems depends on when problems appear, and solutions are planned in a given irrigation system. For instance, we attributed a contribution of 3 points to rehabilitation/modernization programs when addressing the problem of poor conception of the construction/rehabilitation projects. If this problem was detected during the first years of project implementation, rehabilitation/modernization could have a small contribution. If—on the contrary—at the time of the analysis the poorly conceived project had been in operation for decades, the contribution of rehabilitation/modernization would be at the maximum level. Consequently, results would be more accurate if the exercise was applied to WUAs sharing the same timing of problems and solutions. The problems with the timing of problems and solutions can be illustrated by the case study in Spain, where the solution "rehabilitation/modernization programs" obtained 2 points. This is a low mark, but the solution makes no sense in a system largely under construction. This solution received the same mark as other, clearly more suited solutions, such as capacity-building of the WUAs or flexible and adaptive PPP processes.

Finally, the results of the proposed semi-quantitative approach need to be carefully interpreted and screened for unexpected results, which can appear in all proposed solutions. This methodology must be seen as a source of insight and a way to integrate and structure stakeholders' discussions. In fact, a participative approach should be used to elaborate the problem-solution matrix and the problem vectors, as well as to discuss and interpret the solution vectors. The assessment of the solutions for each case study area can be used as a starting point to discuss the prioritization of public reforms and WUA developments.

8. Conclusions

The critical review the problems and reported solutions in the governance of irrigated areas in developing countries permitted elaboration of a list of eight problems and eight solutions. The elements of the list were produced from the evidence in the literature and the experience of the authors. Dozens of scientific and technical papers were analyzed from the 1980s onwards, and their conclusions were analyzed from the scientific, technological, managerial, and engineering experience of the authors. We recognize that a different set of authors could have produced a different structure of problems and solutions. However, we believe that the elements of our lists would constitute the pillars of the findings of different authors.

We have presented a semi-quantitative approach to relate solutions to problems in a case study context. Seven case study WUAs were analyzed from seven countries in four continents. The approach is based on the calculation of a solution vector for each case study area by multiplying a problem vector with a problem-solution matrix. The method rests upon the quantification of problems, solutions, and their interactions using a 0–5-point scale.

Our analysis has shown the usefulness of this concept, which seems particularly adequate in the context of stakeholders' discussions. The solution vectors provide insight on the adequacy of policy, infrastructure and organizational solutions contributing to improved irrigation governance. A local fine-tuning of the coefficients defining the problem vector and the problem-solution matrix seems required to draw conclusions effectively guiding decision-making. The approach seems particularly suited for WUAs having similar timing between the identified problems and solutions. Finally, variability in the problems of the analyzed WUAs is required to obtain variability in the solutions, facilitating policy development or prioritization.

Author Contributions: Enrique Playán, Juan Antonio Sagardoy and Rosendo Castillo contributed to all efforts leading to the production of this paper.

Funding: This research received no external funding.

Conflicts of Interest: The authors declare no conflict of interest.

References

1. Suhardiman, D.; Giordano, M. Is there an alternative for irrigation reform? *World Dev.* **2014**, *57*, 91–100. [CrossRef]
2. AQUASTAT. *Area Equiped for Irrigation*; Prepared by AQUASTAT, FAO's Global Information System; Food and Agriculture Organization of the United Nations: Roma, Italy, 2014.

3. AQUASTAT. *Irrigated Crops*; Prepared by AQUASTAT, FAO's Global Information System; Food and Agriculture Organization of the United Nations: Roma, Italy, 2014.
4. Hamada, H.; Samad, M. Basic principles for sustainable participatory irrigation management. *JARQ Jpn. Agric. Res. Q.* **2011**, *45*, 371–376. [CrossRef]
5. Vermillion, D.L.; Sagardoy, J.A. *Transfer of Irrigation Management Services*; Food and Agriculture Organization of the United Nations: Rome, Italy, 1999; p. 99.
6. Poddar, R.; Qureshi, M.E.; Syme, G. Comparing irrigation management reforms in Australia and India—A special reference to participatory irrigation management. *Irrig. Drain.* **2011**, *60*, 139–150. [CrossRef]
7. Garcés-Restrepo, C.; Vermillion, D.L.; Muñoz, G. *Irrigation Management Transfer: Worldwide Efforts and Results*; Food and Agriculture Organization of the United Nations: Rome, Italy, 2007; p. 63.
8. Kloezen, W.H.; Garcés-Restrepo, C.; Johnson, S.H. *Impact Assessment of Irrigation Management Transfer in the Alto Rio Lerma Irrigation District, Mexico*; International Irrigation Management Institute: Colombo, Sri Lanka, 1997; p. 34.
9. Rosegrant, M.W.; Binswanger, H.P. Markets in tradable water rights—Potential for efficiency gains in developing-country water-resource allocation. *World Dev.* **1994**, *22*, 1613–1625. [CrossRef]
10. Meinzen-Dick, R. Beyond panaceas in water institutions. *Proc. Natl. Acad. Sci. USA* **2007**, *104*, 15200–15205. [CrossRef] [PubMed]
11. Prefol, B.; Tardieu, H.; Vidal, A.; Fernandez, S.; Plantey, J.; Darghouth, S. Public-private partnership in irrigation and drainage: Need for a professional third party between farmers and government. *Irrig. Drain.* **2006**, *55*, 253–263. [CrossRef]
12. Trier, R. Review of international experience with public-private partnership in the irrigation subsector. *Irrig. Drain.* **2014**, *63*, 212–220. [CrossRef]
13. Yildirim, Y.E.; Cakmak, B. Participatory irrigation management in turkey. *Int. J. Water Resour. Dev.* **2004**, *20*, 219–228. [CrossRef]
14. Teamsuwan, V.; Satoh, M. Comparative analysis of management of three water users' organizations: Successful cases in the Chao Phraya delta, Thailand. *Paddy Water Environ.* **2009**, *7*, 227–237. [CrossRef]
15. Bhatt, S. How does participatory irrigation management work? A study of selected water users' associations in Anand district of Gujarat, Western India. *Water Policy* **2013**, *15*, 223–242. [CrossRef]
16. Ul Hassan, M.M. Analyzing governance reforms in irrigation: Central, south and West Asian experience. *Irrig. Drain.* **2011**, *60*, 151–162. [CrossRef]
17. Veldwisch, G.J.A.; Mollinga, P.P. Lost in transition? The introduction of water users associations in Uzbekistan. *Water Int.* **2013**, *38*, 758–773. [CrossRef]
18. Suhardiman, D. The power to resist: Irrigation management transfer in Indonesia. *Water Altern.* **2013**, *6*, 25–41.
19. Senanayake, N.; Mukherji, A.; Giordano, M. Re-visiting what we know about irrigation management transfer: A review of the evidence. *Agric. Water Manag.* **2015**, *149*, 175–186. [CrossRef]
20. Renault, D.; Facon, T.; Wahaj, R. *Modernizing Irrigation Management—The Masscote Approach: Mapping System and Services for Canal Operation Techniques*; Food and Agriculture Organization of the United Nations: Rome, Italy, 2007; volume 63, p. 207.
21. Asthana, A.N. Is participatory water management effective? Evidence from Cambodia. *Water Policy* **2010**, *12*, 149–164. [CrossRef]
22. Meinzen-Dick, R.; Raju, K.V.; Gulati, A. What affects organization and collective action for managing resources? Evidence from canal irrigation systems in India. *World Dev.* **2002**, *30*, 649–666. [CrossRef]
23. Hu, X.J.; Xiong, Y.C.; Li, Y.J.; Wang, J.X.; Li, F.M.; Wang, H.Y.; Li, L.L. Integrated water resources management and water users' associations in the arid region of northwest china: A case study of farmers' perceptions. *J. Environ. Manag.* **2014**, *145*, 162–169. [CrossRef] [PubMed]
24. Yami, M. Sustaining participation in irrigation systems of Ethiopia: What have we learned about water user associations? *Water Policy* **2013**, *15*, 961–984. [CrossRef]
25. Aydogdu, M.H.; Yenigun, K.; Aydogdu, M. Factors affecting farmers' satisfaction from water users association in the Harran plain-gap region, turkey. *J. Agric. Sci. Technol.* **2015**, *17*, 1669–1684.
26. Orstrom, E. *Crsfting Institutions for Self-Governing Irrigations Systems*; Center for Self-Governance: San Francisco, CA, USA, 1992.

27. Vos, J. Understanding water delivery performance in a large-scale irrigation system in Peru. *Irrig. Drain.* **2005**, *54*, 67–78. [CrossRef]
28. Cleaver, F.; Toner, A. The evolution of community water governance in Uchira, Tanzania: The implications for equality of access, sustainability and effectiveness. *Nat. Resour. Forum* **2006**, *30*, 207–218. [CrossRef]
29. Kiymaz, S. Problems and Solutions for Water User Associations in the Gediz Basin Example. Ph.D. Thesis, University of Çukurova, Çukurova, Turkey, 2006.
30. Caizhen, L. Gender issues in water user associations in china: A case study in Gansu province. *Rural Sociol.* **2008**, *18*, 150–160. [CrossRef]
31. Mahdhi, N.; Sghaier, M.; Smida, Z. Efficiency of the irrigation water user association in the Zeuss-Koutine region, South-Eastern Tunisia. *New Medit* **2014**, *13*, 47–55.
32. Omid, M.H.; Akbari, M.; Zarafshani, K.; Eskandari, G.H.; Fami, H.S. Factors influencing the success of water user associations in Iran: A case of moqan, tajan, and varamin. *J. Agric. Sci. Technol.* **2012**, *14*, 27–36.
33. Mekonnen, D.K.; Channa, H.; Ringler, C. The impact of water users' associations on the productivity of irrigated agriculture in Pakistani Punjab. *Water Int.* **2015**, *40*, 733–747. [CrossRef]
34. Johnson, S.H. *Management Transfer in Mexico: A Strategy to Achieve Irrigation District Sustainability*; Resarch Report 16; International Irrigation Management Institute: Colombo, Sri Lanka, 1997; p. 39.
35. Ounvichit, T.; Ishii, A.; Kono, S.; Thampratankul, K.; Satoh, M. An alternative approach to sustainable water users' organization in national irrigation systems: The case of the Khlong Thadi weir system, Southern Thailand. *Irrig. Drain.* **2008**, *57*, 23–39. [CrossRef]
36. Sehring, J. Path dependencies and institutional bricolage in post-soviet water governance. *Water Altern.* **2009**, *2*, 61–81.
37. Koc, C. Sustainability of irrigation schemes transferred in Turkey. *Irrig. Drain.* **2018**, *67*, 242–250. [CrossRef]
38. Postel, S.; Polak, P.; Gonzales, F.; Keller, J. Drip irrigation for small farmers—A new initiative to alleviate hunger and poverty. *Water Int.* **2001**, *26*, 3–13. [CrossRef]
39. Plusquellec, H. Is the daunting challenge of irrigation achievable? *Irrig. Drain.* **2002**, *51*, 185–198. [CrossRef]
40. Biswas, A.K. Monitoring and evaluation of irrigated agriculture—A case-study of Bhima project, India. *Food Policy* **1987**, *12*, 47–61. [CrossRef]
41. Mateos, L.; López-Cortijo, I.; Sagardoy, J.A. Simis: The fao decision support system for irrigation scheme management. *Agric. Water Manag.* **2002**, *56*, 193–206. [CrossRef]
42. Playán, E.; Cavero, J.; Mantero, I.; Salvador, R.; Lecina, S.; Faci, J.M.; Andrés, J.; Salvador, V.; Cardeña, G.; Ramón, S.; et al. A database program for enhancing irrigation district management in the Ebro valley (Spain). *Agric. Water Manag.* **2007**, *87*, 209–216. [CrossRef]
43. Ambast, S.K.; Keshari, A.K.; Gosain, A.K. Satellite remote sensing to support management of irrigation systems: Concepts and approaches. *Irrig. Drain.* **2002**, *51*, 25–39. [CrossRef]
44. International Monetary Fund. *Seeking Sustainable Growth: Short-Term Recovery, Long-Term Challenges*; International Monetary Fund: Washington, DC, USA, 2017.

water

MDPI

Article

The Application of a Modified Version of the SWAT Model at the Daily Temporal Scale and the Hydrological Response unit Spatial Scale: A Case Study Covering an Irrigation District in the Hei River Basin

Zheng Wei [1,2,*], Baozhong Zhang [1,2,*], Yu Liu [1,2] and Di Xu [1,2]

1   State Key Laboratory of Simulation and Regulation of Water Cycle in River Basin, China Institute of Water Resources and Hydropower Research, Beijing 100038, China; liuyu@iwhr.com (Y.L.); xudi@iwhr.com (D.X.)
2   National Center of Efficient Irrigation Engineering and Technology Research, Beijing 100048, China
*   Correspondence: weiz1983@126.com (Z.W.); zhangbz@iwhr.com (B.Z.); Tel.: +86-10-6878-5226 (Z.W.)

Received: 1 May 2018; Accepted: 31 July 2018; Published: 10 August 2018

Abstract: As a well-built, distributed hydrological model, the Soil and Water Assessment Tool (SWAT) has rarely been evaluated at small spatial and short temporal scales. This study evaluated crop growth (specifically, the leaf area index and shoot dry matter) and daily evapotranspiration at the hydrological response unit (HRU) scale, and SWAT2009 was modified to accurately simulate crop growth processes and major hydrological processes. The parameters of the modified SWAT2009 model were calibrated using data on maize for seed from 5 HRUs and validated using data from 7 HRUs. The results show that daily evapotranspiration, shoot dry matter and leaf area index estimates from the modified SWAT2009 model were satisfactory at the HRU level, and the *RMSE* values associated with daily evapotranspiration, shoot dry matter, and leaf area index were reduced by 17.0%, 1.6%, and 71.2%, compared with SWAT2009. Thus, the influences of various optimal management practices on the hydrology of agricultural watersheds can be effectively assessed using the modified model.

Keywords: calibration; irrigation district; evapotranspiration; crop growth; validation

1. Introduction

An irrigation district is a composite ecosystem with anthropogenic and natural elements [1]. In an irrigation district, the processes of artificial irrigation and evapotranspiration are critical for water resource management and are important in the hydrological cycle. To understand and analyse watershed processes and interactions, assess management scenarios, test research hypotheses, and evaluate the influence of changing irrigation [2], a coupled hydro-agronomic model is needed.

Numerous agricultural watershed models, such as the Agricultural Non-Point Source (AGNPS) model [3] and Soil and Water Assessment Tool (SWAT) [4,5], have been used to support water quality management, water resource analysis, and soil erosion assessment in agricultural watersheds. Among these models, SWAT is one of the best for the long-term simulation of watersheds dominated by agricultural land use. For example, SWAT has been used to evaluate the influence of irrigation diversion on river flow [4,6–9], to simulate climate change and the associated effects under various scenarios [10,11], to calculate nutrient and sediment yields [12], and to estimate the water balance [13]. SWAT can be effectively calibrated, by comparing the sediment yield and/or simulated surface runoff and nutrient concentrations in runoff to observations at the outlets of watersheds at the subbasin level or at large spatial scales [7,14–18]. The modified SWAT model was developed to improve the simulation of particular watershed processes. Additionally, the modified version includes enhanced flow predictions (including interflow and percolation, as well as hydraulic conductivity), the evaluation of phosphorus derived from bank erosion in the upper soil layers, organic nitrogen losses, fast percolation, a groundwater dynamics sub-model, amended dynamic functions for crop growth, and channel and drainage losses [12,19–27].

Many studies of runoff hydrological processes have been performed using the SWAT model, but crop growth dynamics (represented by metrics such as shoot dry matter and the leaf area index) and evapotranspiration (ET_c) are more significant than rainfall–runoff processes in irrigation districts. Notably, additional studies must be performed in irrigation districts. First, ET_c simulations have yielded satisfactory results at the field spatial scale and the daily temporal scale. Many studies have described the dual crop coefficients of many crops, such as cotton, winter wheat, maize, sorghum and soybean [28–33]. In SWAT, studies have used satellite-estimated monthly evapotranspiration values to calibrate the ET_c parameters at the subbasin scale [34–37], daily lysimeter evapotranspiration values to test the ET_c parameters at the subbasin scale [38]. The weekly or daily temporal scales and hydrological response unit (HRU) spatial scales of such studies are coarse. Existing research has shown that SWAT generally underestimates daily and monthly ET_c. In addition, automatic irrigation in SWAT can be improved to accommodate limited irrigation scheduling strategies by adding parameters that allow irrigation levels to be set as a percentage of ET_c. Some studies have mentioned potential flaws in SWAT's automatic irrigation capabilities to simulate real irrigation conditions. At present, the irrigation trigger factor of the soil moisture content method is not reported as the percentage of plant water demand option, but as soil water, which is easily ignored [38–40]. Data availability at the daily scale is often limited in hydrological modelling. Notably, more attention should focus on evapotranspiration in various irrigation domains and the validation and calibration of model parameters in combined distributed hydrological models (e.g., SWAT) and dual crop coefficient models (e.g., SIMDualKc). Second, the irrigation quota is limited to the field capacity, and the excess water above the field capacity is returned to the source and is not taken into account in the calculation of the daily soil water balance. In fact, the irrigation quota is always larger than the field capacity, and such a constraint is not suitable for surface irrigation and flooding irrigation. Additionally, various irrigation schedules should be considered in different irrigation domains, and SWAT should be used for various applications, instead of creating a "unified optimum irrigation schedule", developing methods for "auto-irrigation", etc., to accurately describe the cyclic processes involved in regional irrigation. Third, while many studies have assessed the crop water productivity index on the basin scale [41,42] and yield predictions on the HRU scale [43], crop growth dynamics have rarely been evaluated at small scales. Four, the leaf area index (*LAI*) curve is determined via linear regression after leaf senescence, but this method is not suitable for irrigated crops, *LAI* follow and have logistic relationships with climatic and soil variables [44].

The objectives of this paper are as follows: (1) to modify SWAT and improve its ability to model crop growth parameters (shoot dry matter and *LAI*) and daily evapotranspiration; (2) to conduct a sensitivity analysis of the modified model and identify the parameters that strongly influence estimates of crop growth and ET_c; and (3) to calibrate and validate the parameters of *LAI*, shoot dry matter, and evapotranspiration in the modified model at small spatial (the HRU level) and temporal (daily) scales.

2. Materials and Methods

2.1. Study Area

The Yinke experimental site is located in an artificial oasis in the central Hei River Basin of Northwest China (Figure 1). The elevation of the basin ranges from 1456 to 1600 m, and the basin encompasses a total irrigation area of 18.65 km^2 (Figure 1). The irrigation water supply is conveyed by the Yinke Canal [45], which is located adjacent to the Hei River and forms part of the irrigation system that connects the southwest region to the northeast region. In 2012, the main cultivated crop was maize (84.13%). Other land cover types included forestry (0.18%), pasture (0.82%), water (0.69%), and residential land (14.18%). The climate is cold, and the arid region receives a mean annual precipitation of 125 mm. The reference evapotranspiration in the area is 1972 mm. additionally, the mean annual temperature is 6.7 °C, and the temperature difference between winter and summer is large. The soil texture is mainly homogeneous sandy loam with a gravelly bottom layer.

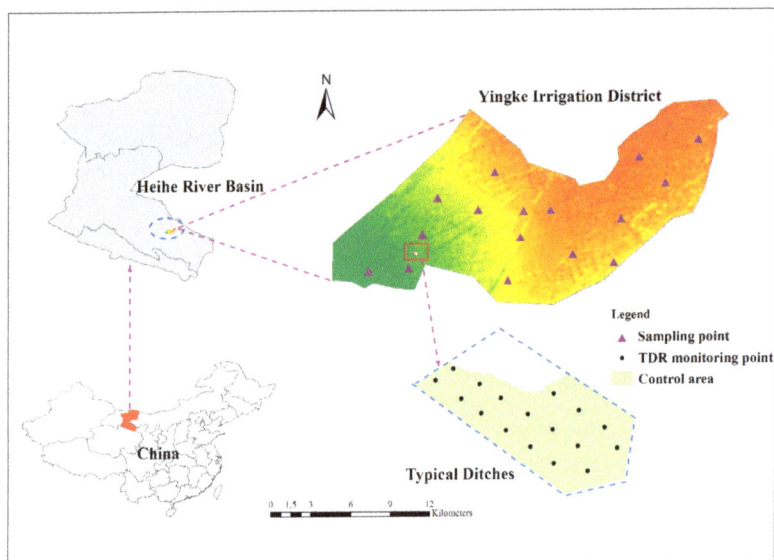

Figure 1. Location of the Yinke irrigation district.

2.2. Field Experiment

Sixteen points that represent different branch canals and soil types were arranged in the irrigation district, and 17 points that represent typical ditches were arranged in the second rural canal (Figure 1). Maize for seed was the crop chosen for the experiment (local variety name: series 13).

At each monitoring point of the second rural canal, a field soil moisture measurement instrument (TRIME-PICO IPH T3/44) was installed to monitor the soil water profile within each 0.2–0.4 m layer at 15-day intervals. At each monitoring point of different branch, oven drying method was used to monitor the soil water profile at 15-day intervals. The *LAI* and shoot dry matter were included in the crop growth observations. The assessments of *LAI* and biomass were conducted at 15-day intervals. The application ratios of fertilizer were 180 kg/ha for N and 150 kg/ha for P, which were implemented based on local fertilizer management practices [46]. An eddy covariance (EC) instrument was established to measure the latent heat flux at 100°24′37″ E 38°51′25″ N and an elevation of 1519 m. The raw EC data were collected at a sampling frequency of 10 Hz, and they were processed using the post-processing software EdiRe [47].

2.3. Model Description

2.3.1. SIMDualKc Model

The SIMDualKc model calculates daily crop evapotranspiration by considering soil evaporation and crop transpiration components according to the water balance and the dual coefficient method. The SIMDualKc software application was developed over a range of cultural practices and to provide

ET_c information for use in irrigation scheduling and hydrologic water balances [48]. The actual crop evapotranspiration in the model can be computed as follows:

$$ET_c = (K_s K_{cb} + K_e)ET_0 \tag{1}$$

where K_s is the water stress coefficient, K_e is the soil evaporation coefficient, and K_{cb} is the basal crop coefficient. Additionally, ET_0 is the reference evapotranspiration (mm/d) [49].

The SIMDualKc model, described in the companion paper, was developed to compute crop ET_c using many recent refinements [50]. The SIMDualKc model calibration procedure involves the adjustment of standard soil parameters (e.g., readily evaporable water *REW*, total evaporable water *TEW*, and the depth of the surface soil layer Z_e) and crop parameters (e.g., K_{cb} and the depletion fraction p) to minimize the differences between observed and estimated ET_c values [29].

2.3.2. SWAT Model

SWAT is a temporally continuous, physically-based, and spatially semi-distributed model [51]. A watershed can be categorized into multiple sub-watersheds that can be further divided into particular land/soil utilization characteristic units or HRUs. The water balance of each HRU is based on four volumes of storage, including storage in the soil profile (0–2 m), snow, deep aquifers (>20 m), and shallow aquifers (typically 2–20 m). Chemical loadings, flow generation and the sediment yield were calculated in each HRU, and the resulting loads were routed via ponds, channels, and/or reservoirs to the watershed outlet. The soil profile was further divided into multiple layers with different soil water processes, such as evaporation, infiltration, percolation, lateral flow and plant uptake.

A storage routing method is used in the soil percolation module of SWAT to simulate flow in every soil layer in the root zone. Crop evapotranspiration is simulated as a function of root depth, *LAI*, and potential evapotranspiration [52].

Crop evapotranspiration (ET_c) can be determined using the Penman-Monteith method, and the surface runoff from daily rainfall was estimated using the modified Soil Conservation Service (SCS) curve number [51].

The distributed modelling was carried out through a coupling of modified SWAT2009 and SIMDualKc model (Figure 2). Measured data (Soil water, daily crop evapotranspiration by EC) in the second rural canal were used to calibrate and validate the SIMDualKc model, measured data (Soil water, crop data) in different branch canals and daily crop evapotranspiration calculated by the SIMDualKc model were used for analysis performance of the modified SWAT2009.

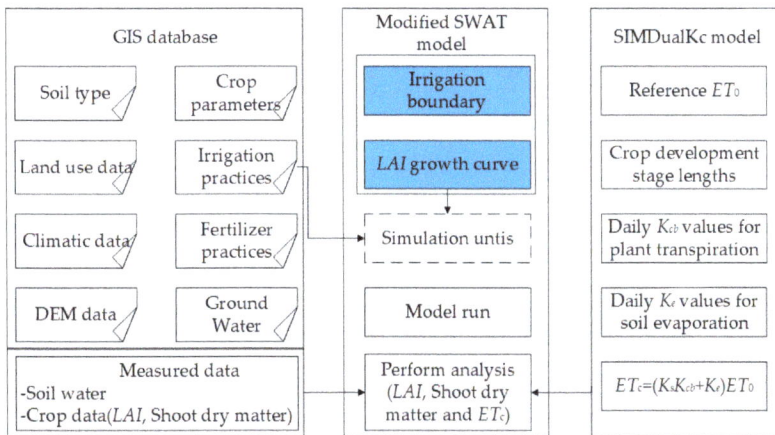

Figure 2. Schematic diagram of the coupling modified SWAT2009 and SIMDualKc model.

2.3.3. Sensitivity Analysis

The key of SWAT model is the selection of hydrological model parameters. At present, the optimization of SWAT model parameters can be divided into manual adjustment and automatic adjustment. Manual calibration parameters require certain calibration experience of hydrologists, which requires a long working time. Due to the large influence of human interference factors on the model, there is no good evaluation standard. SWAT-CUP can conduct sensitivity analysis, uncertainty analysis and parameter automatic rate determination for the output results of SWAT2009. To assess the parameters that affect crop growth and ET_c based on daily values, the LH-OAT (Latin Hypercube-One-factor-At-a-Time) method was used in the study, through the LH-OAT, the dominant parameters were determined and a reduction of the number of model parameters was performed [53,54]. The parameters which have influences on crop growth and ET_c were used for sensitivity analysis are presented in Section 3.3.1.

2.4. SWAT Input Data

Three basic files are required to divide the basin into HRUs and subbasins: a digital elevation model (DEM), a soil map and a land use/land cover map. The topographic characteristics (the drainage network, slope length, slope, the number of subbasins and the delimited watersheds) were generated from the DEM of the Hei River Basin (30 × 30 m grid size). The soil and land use data were based on soil distribution maps and crop data obtained from the online portal of the Ecological and Environmental Science Data Centre of Western China. In total, 137 HRUs and 13 subbasins were delineated in the study area (Figure 3). The weather input data, which included minimum and maximum daily air temperatures, relative humidity, wind speed and solar radiation, were obtained from Zhangye meteorological station. This station which is located at 100°25′48″ E 38°55′48″ N and an elevation of 1470 m. Irrigation scheduling was shown in Table 1 for different branch canals, it is reflected in the SWAT input file by setting the *.mgt files of 13 subbasins separately. The results of the specific settings and process were shown in Figure 3. Current farm management strategies, such as planting and tillage scheduling and harvesting and fertilization practices, were used as inputs to the model.

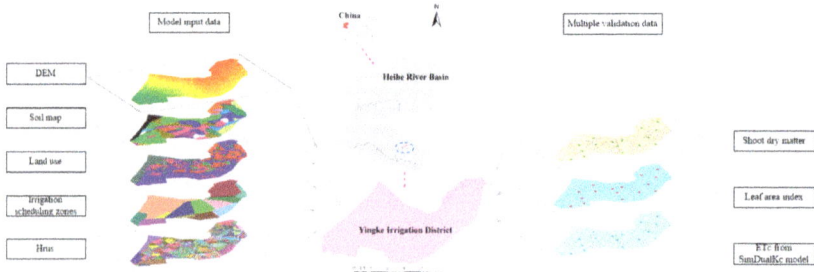

Figure 3. Model input data and multiple validation data.

Table 1. Irrigation quota/irrigation scheduling in the Yingke irrigation district.

Brach Canals		The Second Rural Canal			
Sub Basin	Irrigation Quota (mm)	Sub Basin	Irrigation Quota (mm)	Irrigation Time (Month/Day)	Irrigation Water (mm)
1	786	8	652	5/26	120
2	786	9	885	6/22	180
3	748	10	861	7/21	160
4	858	11	634	8/13	150
5	1375	12	861		
6	1415	13	1176		
7	652				

2.5. Modification of the Model

Since excess irrigation is returned to the source of irrigation instead of being considered in the surface runoff calculations and daily soil water balance calculations, the original version of SWAT2009 could not be used to simulate agricultural irrigation practices [55]. The linear decreasing *LAI* curve after senescence could underestimate ET_c. Thus, the following modifications were included in the source code of the modified version of SWAT to include the excess water, as previously noted in the soil water balance and *LAI* growth calculations:

1. The maximum amount of water used was in accordance with the depth of irrigation water used in every HRU, as specified by the irrigation operation scheme, rather than the amount of water in the soil profile based on the field capacity by Farida et al. [56]. This modification can be expressed as follows:

Original version: vmm = sol_sumfc

Modified version: vmm = irr_amt

where vmm refers to the maximum amount of water used (mm); irr_amt refers to the depth of irrigation water used in each HRU (mm), as specified by the user; and sol_sumfc refers to the amount of water in the soil profile at field capacity (mm).

2. *LAI* values had subsequent effects on ET_c estimation by Gary Marek et al. (2016) [38]. The *LAI* curve after senescence was originally linear, but it should be represented using a logistic growth curve for irrigated crop. This modification can be expressed as follows:

Original version:

$$\text{LAI} = \frac{LAI_{mx}}{\left(1 - fr_{phu,sen}\right)^2} \cdot \left(1 - fr_{phu}\right)^2, fr_{phu} \geq fr_{phu,sen} \tag{2}$$

Modified version:

$$\text{LAI} = \frac{LAI_{mx}}{\left(1 + a \cdot \exp\left(1 + b \cdot fr_{phu}\right)\right)}, fr_{phu} \geq fr_{phu,sen} \tag{3}$$

where a and b are empirical parameters, *LAI* is the leaf area index on a given day, LAI_{mx} is the maximum leaf area index, fr_{phu} is the fraction of potential heat units accumulated by a plant on a given day during the growing season, and $fr_{phu,sen}$ is the fraction of the growing season (PHU) in which senescence is the dominant growth process.

2.6. Model Performance

Three statistical methods and time series plots were used to assess the performance of SWAT and SIMDualKc according to the data. The five statistical criteria used to assess the effectiveness of the validation and calibration results were as follows: (i) the Nash-Sutcliffe efficiency (*NSE*); (ii) the root mean squared error (*RMSE*); and (iii) the coefficient of determination (R^2). The calibration objectives for *LAI*, shoot dry matter, and ET_c were to minimize the *RMSE* and maximize the R^2 and *NSE* values.

3. Results

3.1. Calibration and Validation of the SIMDualKc Model

3.1.1. Calibration of the SIMDualKc Model

The model simulations were initiated using a table of K_{cb} values; p and soil evaporation parameter values were those recommended by Allen et al. [49]; and the initial Z_e, *TEW* and *REW* values of 0.1 m, 28 mm and 8 mm, respectively, were used for silt-loam soils. The suggested initial values of a_p and b_p, which are deep percolation parameters, were 370 and −0.0173, respectively [57]. The soil parameters was provided in Table 2, and the irrigation schedule was shown in Table 1 for the second rural canal.

The results of comparing the observed and simulated values of available soil water (*ASW*) based on the calibration data sets of maize for seed are presented in Figure 4. The figure shows that the *ASW* dynamics were well simulated, and there was no apparent bias in the estimation. The calibrated values of K_{cbini} and p exhibited good agreement with those proposed by Allen et al. [49], and the calibrated values of K_{cbmid} were slightly smaller than those of Allen et al. [49] and Duan et al. [58]. These differences were likely caused by crop and application differences, as this crop was planted for seed and was sent to local companies who buy from the farmers. The reference values discussed above are presented in Table 3.

Table 2. Selected soil parameters and values used in SIMDualKc model.

Depth (mm)	Bulk Density (g/cm^3)	Clay Content (% Soil Mass)	Silt Content (% Soil Mass)	Sand Content (% Soil Mass)
0–200	1.46	13.88	51.17	34.96
200–800	1.48	15.19	50.45	34.36
800–1400	1.57	16.59	50.26	33.16

Table 3. Initial and calibrated values of the crop and soil parameters appropriate for maize for seed: crop coefficients, depletion fractions under conditions of no stress, soil evaporation and deep percolation parameters.

	Initial Values	Calibrated Values
Crop coefficients		
K_{cbini}	0.15	0.15
K_{cbdev}	0.15–1.2	0.15–0.95
K_{cbmid}	1.2	0.95
K_{cbend}	0.35	0.35
Depletion fractions		
P_{ini}	0.5	0.5
P_{dev}	0.5	0.5
P_{mid}	0.5	0.5
P_{end}	0.5	0.5
Soil evaporation		
REW (mm)	8	10
TEW (mm)	28	34
Z_e (m)	0.1	0.15
Deep percolation		
a_p	370	366
b_p	−0.0173	−0.065

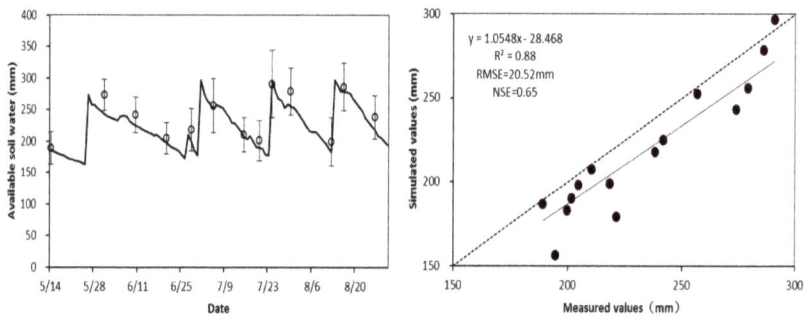

Figure 4. Comparison between simulated and observed available soil water of maize for seed.

The results show that the regression coefficient was less than 1.0, indicating that the data plotted slightly below the 1:1 line of the observed data. The coefficient of determination was 0.88, indicating that most of the variance could be explained by the model. The *RMSE* reached 20.52 mm, representing approximately 6.7% of the total available water (*TAW*). Additionally, the *NSE* value was 0.65. The statistical test values of determining coefficient, regression coefficient and intercept are 9.98, 0.52, and 1.14, respectively, with *t*-test ($t_{0.05}$ = 2.131, *n* = 15). The results suggest that the SIMDualKc model effectively accounted for the variation in the observed ASW and accurately predicted the ASW value of maize for seed.

3.1.2. Validation of the SIMDualKc Model

The results of comparing the observed and simulated ET_c values are presented in Figure 5. The coefficient of determination was 0.79, indicating that most of the variance could be explained by the model. The *RMSE* reached 1.01 mm/d, and the *NSE* value was 0.56. The statistical test values of determining coefficient, regression coefficient and intercept are 15.55, 11.53, and 8.09, respectively, with *t*-test ($t_{0.05}$ = 1.996, *n* = 67). The results show that there are some schematic errors in the model that can reducing ET_c predictive power. The observed value ET_c is the water consumption calculated by the energy balance of the eddy-related system, while the SIMDualKc model calculates the ET_c based on the water balance formula. Results of previous research indicate that the ET_c calculated by the vorticity correlation system is lower than that calculated by the water balance formula [59], the comparison of ET_c at different scales may be the main reason of schematic errors.

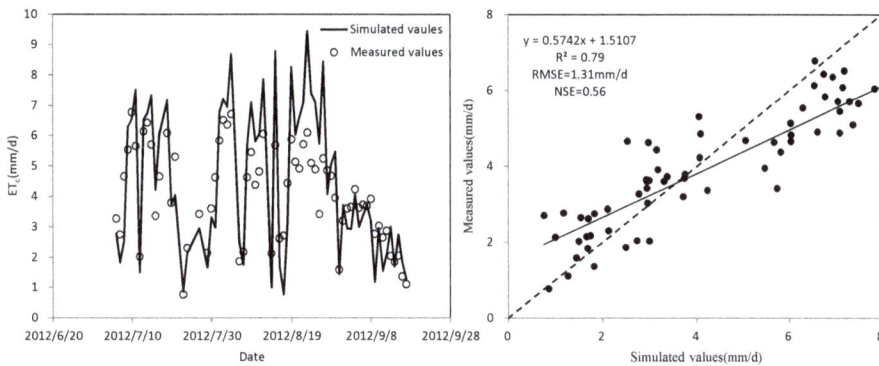

Figure 5. Comparison between the simulated and observed ET_c values of maize for seed.

3.2. SWAT2009 vs. Modified SWAT

3.2.1. Water Balance

After calibrating and validating SWAT2009 and modified SWAT2009 independently, the irrigation quota of different SWAT versions is shown in Figure 6a, and there were larger differences between SWAT 2009 and modified SWAT2009 versions, the differences ranging from 0–430 mm. the modification of Equation (1) could increase the irrigation water in the soil, and more percolation in the water cycle. In the Figure 6b, there were some differences between modified SWAT2009 and original SWAT2009 versions, and the differences ranged from 46–68 mm. The total evapotranspiration ranged from 540–580 mm of modified SWAT2009, the other is 492–519 mm. Yongyong Zhang et al. (2016) showed that evapotranspiration of seed maize is 545 mm during the growing season [60]. The results is similar with the results of modified SWAT2009 versions. The modification of Equation (2) could increase evapotranspiration, especially after peak *LAI* was reached. In the Figure 6c, there were larger difference between modified SWAT2009 and original SWAT2009 versions, and the differences ranged from −42 to

401 mm. the variation of percolation is similar with the variation of irrigation quota. In the Figure 6d, there were some differences in the senescence stage, the *ASW* decline rate of modified SWAT2009 was faster than that of original SWAT2009. Therefore, there was a different water balance between modified SWAT2009 and original SWAT2009 versions.

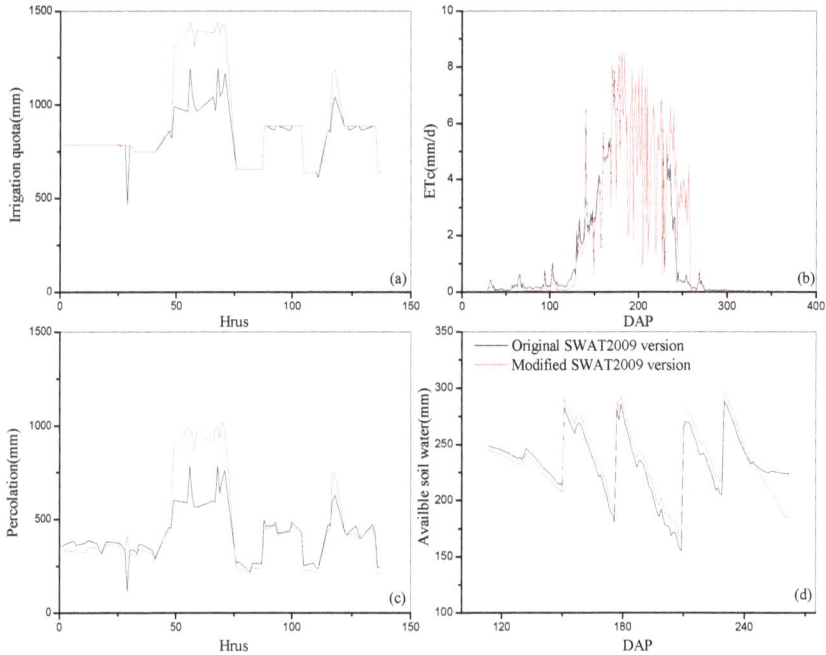

Figure 6. Water balance element of different SWAT versions. ((**a**) irrigation quota; (**b**) evapotranspiration; (**c**) perolation; and (**d**) available soil water).

3.2.2. Difference of Model Performance

Crop growth (specifically, the leaf area index and shoot dry matter) and daily evapotranspiration were used to test the SWAT2009 model at the hydrological response unit (HRU) scale, the comparisons between the estimated and measured values of ASW, shoot dry matter, *LAI* and estimated daily ET_c values using SIMDualKc, SWAT2009, and the modified SWAT model exhibited large differences.

The daily ET_c results exhibited some differences between the observed data and values simulated using SIMDualKc, SWAT2009, and the modified SWAT model. The less differences were observed for the shoot dry matter and ASW based on SWAT2009 and the modified SWAT2009 model (Table 4). These observations suggest that the parameters values of shoot dry matter and *ASW* used in the calculation process are nearly equal.

Large differences were observed between SWAT2009 and the modified SWAT model (Figure 7). The largest difference occurred during the leaf senescence stage, when the statistical parameters were much lower than those in the modified SWAT model (Table 4). In SWAT2009, the *RMSE* is 1.84 and the *NSE* is negative. These differences are because of the linear *LAI* functions used in the leaf senescence stage.

Based on this comparison between SWAT2009 and the modified SWAT model, the largest improvement was associated with *LAI*, followed by the ET_c and available soil water, the values of shoot dry matter were nearly equal in both models.

Table 4. Comparison between SWAT2009 and the modified SWAT2009 model.

Parameter	Version	R^2	*RMSE*	*NSE*
Available soil water	SWAT2009	0.71	18.30	0.70
	Modified SWAT2009	0.80	15.98	0.77
Evapotranspiration	SWAT2009	0.35	2.41	0.31
	Modified SWAT2009	0.53	2.00	0.52
Shoot dry matter	SWAT2009	0.89	2.64	0.85
	Modified SWAT2009	0.89	2.60	0.86
LAI	SWAT2009	0.15	1.84	−1.12
	Modified SWAT2009	0.84	0.53	0.82

The units of available water content, evapotranspiration, shoot dry matter, and *LAI* are mm, mm/d, t/ha, and none, respectively.

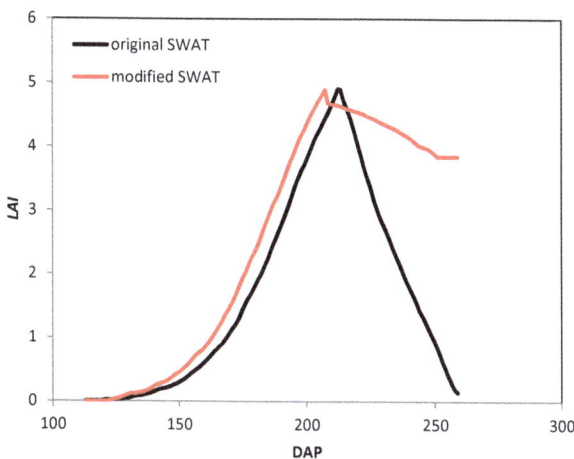

Figure 7. Variations in the *LAI* based on the original and modified SWAT models.

3.3. Modified SWAT Model Calibration and Validation

3.3.1. Sensitivity Analysis

During the process of simulating *LAI* using the SWAT model, a sensitivity analysis was performed with the observed data. The results indicated that *BLAI*, *LAI*, LAI_{MX2}, and LAI_{MX1} were the primary parameters in the mgt, crop, sol, gw, and rte input files used in the SWAT model (Table 5). Additionally, *LAI* development depended on the accumulation of plant heat units and the environmental stress indexes. Thus, the shape coefficients are very important for obtaining accurate *LAI* development curves [55]. In the process of simulating shoot dry matter, *BIO_E*, *T_OPT*, and *T_BASE* were the primary parameters (Table 6). Biomass production was coupled with the radiation-use efficiency (*BIO_E*) and intercepted, photosynthetically active radiation [55]. The higher the *BIO_E* value is, the more biomass can be produced [26]. In the process of simulating daily ET_c, *SOL_K*, *ESCO* and *GSI* were the primary parameters (Table 7). *ESCO* is an important parameter used to estimate unsaturated soil evaporation [61], and *GSI* is a significant parameter used to calculate r_c in the Penman-Monteith equation [49]. Additionally, *ESCO* is the soil evaporation compensation factor, which affects all the water balance components [62].

Table 5. Global sensitivities of the parameters that affect the leaf area index.

Parameter [a]	*t*-Value [b]	*p*-Value [c]	Initial Value	Calibrated Value
FRGRW2	1.67	0.09	0.5	0.71
FRGEW1	4.27	<0.01	0.15	0.05
BLAI	9.79	<0.01	6	5.14
DLAI	12.42	<0.01	0.7	0.67
LAI_{MX2}	18.17	<0.01	0.95	0.77
LAI_{MX1}	20.11	<0.01	0.05	0.01

[a] Parameter definitions can be found in the theoretical documentation of SWAT [63]. [b] The *t*-value indicates parameter sensitivity; the larger the *t*-value is, the more sensitive the model output is to the parameter. [c] The *p*-value indicates the significance of the *t*-value; the smaller the *p*-value is, the less chance that a parameter has of being falsely identified as sensitive.

Table 6. Global sensitivities of the parameters that affect shoot dry matter.

Parameter [a]	*t*-Value [b]	*p*-Value [c]	Initial Value	Calibrated Value
HVSTI	0.88	0.38	—	—
USLE_C	1.61	0.11	0.5	0.26
EXT_COEF	1.68	0.09	0.5	0.78
BIO_E	7.15	<0.01	39	28.85
T_OPT	10.58	<0.01	25	23.68
T_BASE	12.36	<0.01	8	12.26

[a] Parameter definitions can be found in the theoretical documentation of SWAT [63]. [b] The *t*-value indicates parameter sensitivity; the larger the *t*-value is, the more sensitive the model output is to the parameter. [c] The *p*-value indicates the significance of the *t*-value; the smaller the *p*-value is, the less chance that a parameter has of being falsely identified as sensitive.

Table 7. Global sensitivities of the parameters that affect daily evapotranspiration.

Parameter [a]	*t*-Value [b]	*p*-Value [c]	Initial Values	Calibrated Values
ALPHA_BF	0.14	0.89	—	—
GWQMN	0.14	0.89	—	—
SOL_BD	0.22	0.83	—	—
SOL_ZMX	0.31	0.76	—	—
CH_N2	0.33	0.74	—	—
GW_REVAP	0.38	0.70	—	—
CO2HI	0.41	0.68	—	—
CH_K2	0.57	0.57	—	—
GW_DELAY	0.67	0.50	—	—
SOL_AWC_B	0.82	0.41	—	—
CN2	0.82	0.41	—	—
CANMX	0.86	0.39	—	—
SOL_AWC_C	1.39	0.17	0.25	0.20
ALPHA_BNK	1.58	0.11	0.5	0.42
EPCO	1.8	0.07	0.1	0.69
SOL_AWC_D	2.11	0.04	0.18	0.21
SOL_K	5.63	<0.01	20	30
ESCO	3.37	<0.01	0.1	0.57
GSI	23.93	<0.01	0.007	0.01

[a] Parameter definitions can be found in the theoretical documentation of SWAT [63]. [b] The *t*-value indicates parameter sensitivity; the larger the *t*-value is, the more sensitive the model output is to the parameter. [c] The *p*-value indicates the significance of the *t*-value; the smaller the *p*-value is, the less chance that a parameter has of being falsely identified as sensitive.

3.3.2. *LAI* Calibration and Validation

Only those parameters with the highest sensitivities were considered in the calibration process. Table 5 shows the initial values and calibrated values of each parameter considered in the calibration process.

Regional monitoring stations were distributed over 16 HRUs, of which four HRUs were sparsely planted with vegetables (greenhouse crops). Hence, they were not considered in this study. Calibration and validation were performed sequentially [54]. The *LAI* data from HRU-1, HRU-18, HRU-24, HRU-30, and HRU-58 were adopted for calibration, and the *LAI* data from HRU-68, HRU-82, HRU-99, HRU-100, HRU-104, HRU-115, and HRU-118 were used for validation.

The calibration and validation curves and statistical parameters are shown in Figure 8 and Table 8, respectively. In the calibration phase, the coefficient of determination ranged from 0.90 to 0.98, the *NSE* ranged from 0.72 to 0.93, and the *RMSE* ranged from 0.32 to 0.98. In the validation phase, the coefficient of determination ranged from 0.90 to 0.97, the *NSE* ranged from 0.22 to 0.89, and the *RMSE* ranged from 0.41 to 0.95. The simulated and measured *LAI* values in the calibration and validation phases exhibited good agreement, and the simulated results effectively described the growth process represented by the *LAI* of maize for seed.

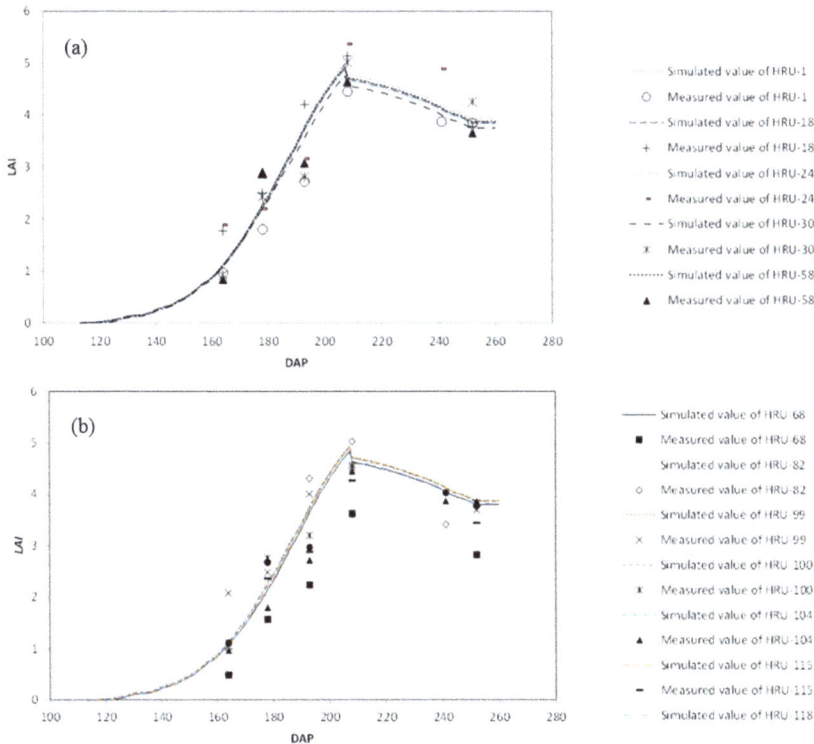

Figure 8. Typical variations in simulated and measured *LAI* values ((**a**) values for calibration; and (**b**) values for validation; DAP is day after planting).

Table 8. Statistical parameters related to the simulation of *LAI* in the calibration and validation phases.

	HRU	R^2	RMSE	NSE
	1	0.94	0.32	0.93
	18	0.96	0.54	0.80
Calibration	24	0.85	0.71	0.75
	30	0.81	0.98	0.72
	58	0.89	0.42	0.89
	68	0.96	0.95	0.22
	82	0.88	0.51	0.84
	99	0.81	0.71	0.42
Validation	100	0.84	0.53	0.77
	104	0.94	0.41	0.89
	115	0.94	0.45	0.88
	118	0.85	0.54	0.69

3.3.3. Calibration and Validation of Shoot Dry Matter

The default values and adjusted values of each parameter considered in the calibration process are presented in Table 6. The calibrated and validated HRUs were the same as those used in the process of *LAI* calibration and validation. The calibration and validation curves and statistical parameters are shown in Figure 9 and Table 9.

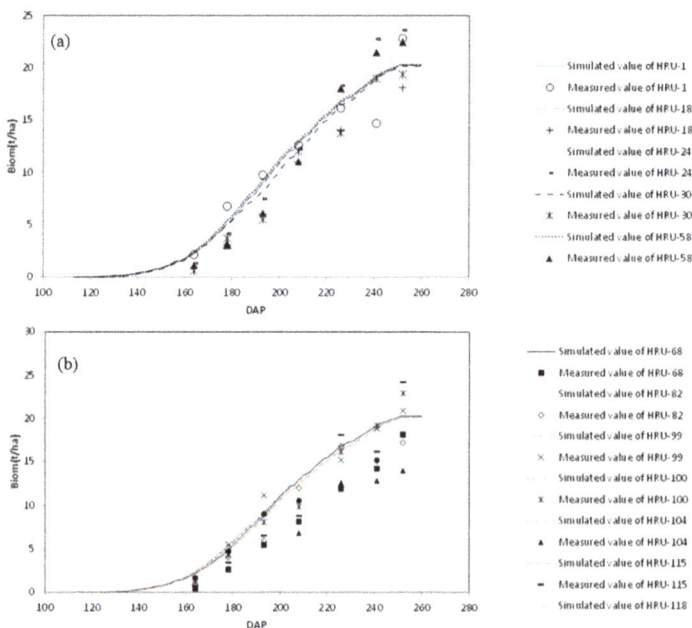

Figure 9. Typical variations in simulated and measured values of shoot dry matter ((**a**) values for calibration; and (**b**) values for validation; DAP is day after planting).

In the calibration phase, the coefficient of determination between the simulated and measured values of shoot dry matter ranged from 0.92 to 1.0, the *NSE* ranged from 0.66 to 0.97, and the *RMSE* ranged from 1.64 to 5.97 t/ha. In the validation phase, the coefficient of determination ranged from 0.92 to 0.99, the *NSE* ranged from 0.75 to 0.94, and the *RMSE* ranged from 1.14 to 2.91 t/ha. The measured values were consistently less than simulated values for the validation period, the inaccuracy of shoot dry matter estimation might also be caused by the errors in response to soil water stress in the cumulative temperature model.

Table 9. Statistical parameters related to shoot dry matter in the calibration and validation phases.

	HRU	R^2	RMSE (t/ha)	NSE
	1	0.90	2.89	0.78
	18	0.85	3.58	0.77
Calibration	24	0.93	5.07	0.68
	30	0.98	1.64	0.94
	58	0.97	5.97	0.66
	68	0.91	2.55	0.81
	82	0.85	2.91	0.78
	99	0.93	2.17	0.88
Validation	100	0.95	1.14	0.94
	104	0.98	2.65	0.87
	115	0.98	2.45	0.75
	118	0.93	3.19	0.84

3.3.4. ET_c Calibration and Validation

The default values and adjusted values of each parameter considered in the calibration process are presented in Table 7. The HRUs used for calibration and validation were the same as those used in the process of *LAI* calibration and validation. The calibration and validation curves and statistical parameters are presented in Figures 10 and 11 and Table 10.

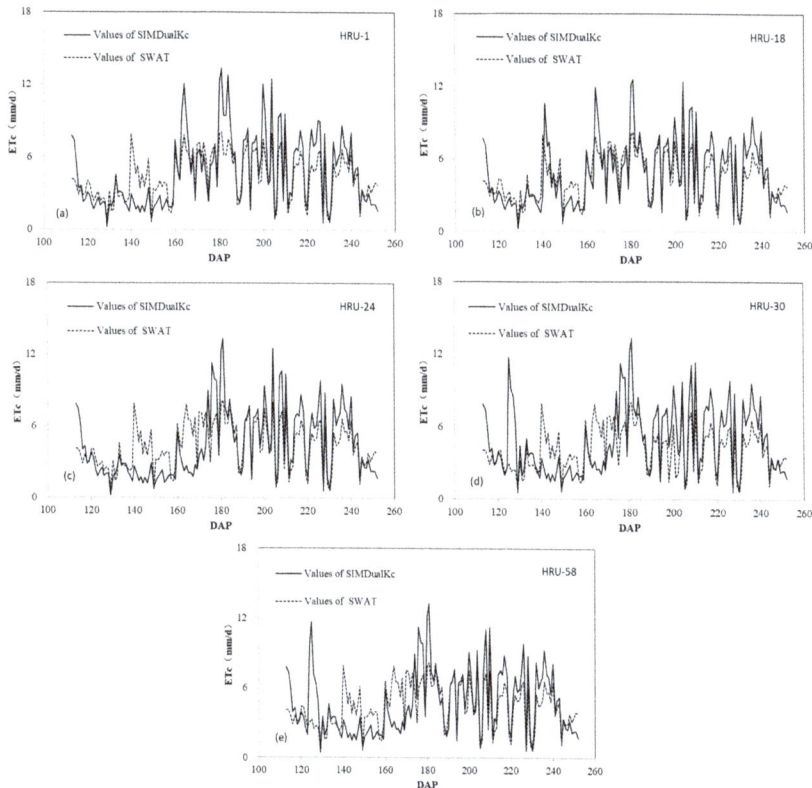

Figure 10. Variations in ET_c simulated using SIMDualKc and SWAT ((**a**) HRU-1; (**b**) HRU-18; (**c**) HRU-24; (**d**) HRU-30; and (**e**) HRU-58, which are used for calibration; DAP is day after planting).

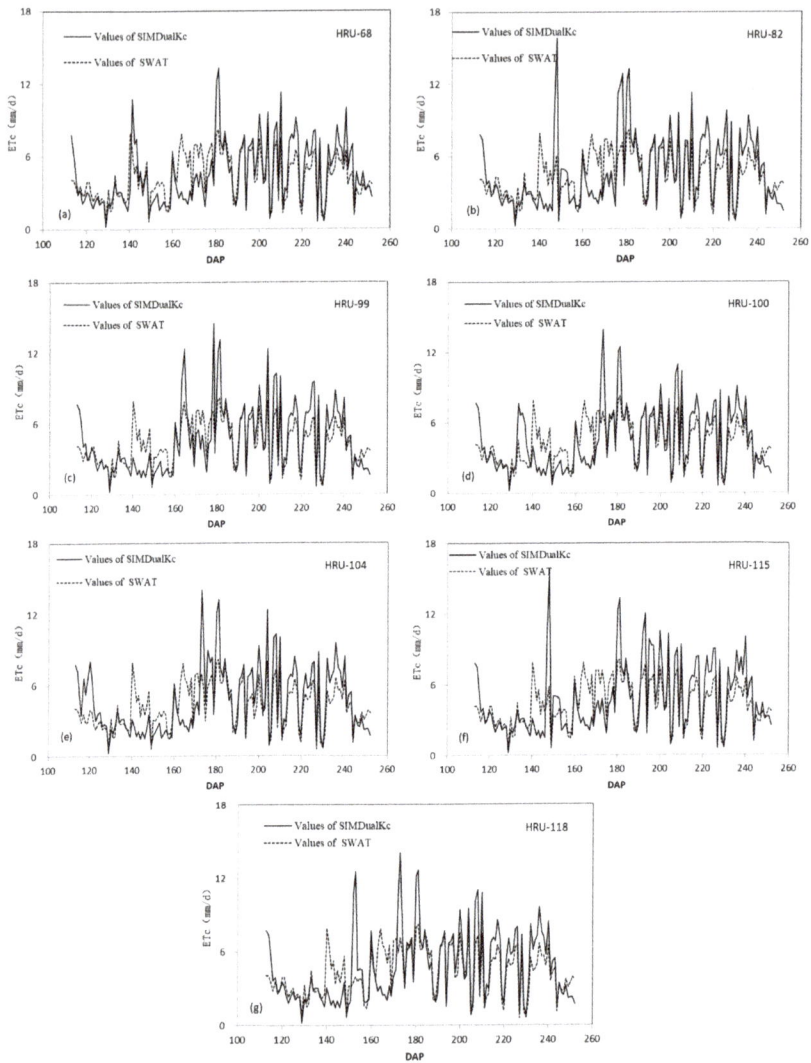

Figure 11. Variations in ET_c simulated using SIMDualKc and SWAT ((**a**) HRU-68; (**b**) HRU-82; (**c**) HRU-99; (**d**) HRU-100; (**e**) HRU-104; (**f**) HRU-115; and (**g**) HRU-118, which are used for validation; DAP is day after planting).

During the calibration phase, the coefficient of determination between the SIMDualKc and SWAT simulations ranged from 0.62 to 0.86, the *NSE* ranged from 0.33 to 0.70, and the *RMSE* ranged from 1.52 to 2.40 mm/d. Additionally, in the validation phase, the coefficient of determination between the SIMDualKc and SWAT simulations ranged from 0.70 to 0.80, the *NSE* ranged from 0.48 to 0.62, and the *RMSE* ranged from 1.69 to 2.19 mm/d. The ET_c values simulated by SWAT and SIMDualKc exhibited good agreement in the calibration and validation phases. Marek et al. (2016) showed that The *NSE* value could reach 0.7 in the validation period [38], the inaccuracy of ET_c estimation might also stem from the errors in the soil water stress function and pre-set GSI.

Table 10. Statistical parameters related to ET_c in the calibration and validation phases.

Phase	HRU	R^2	RMSE (mm/d)	NSE
Calibration	1	0.71	1.79	0.65
	18	0.74	1.52	0.70
	24	0.51	2.07	0.51
	30	0.39	2.40	0.33
	58	0.42	2.20	0.41
Validation	68	0.59	1.69	0.58
	82	0.52	1.91	0.52
	99	0.49	2.19	0.48
	100	0.64	1.80	0.62
	104	0.52	2.03	0.50
	115	0.51	2.07	0.50
	118	0.49	2.16	0.48

3.3.5. Available Soil Water Test

Results showed that the simulated and observed available soil water are in agreement at all observation points (Figure 12). The determination coefficients ranged from 0.56 to 0.96, the regression coefficient ranged from 0.76 to 1.02, *RMSE* ranged from 7.36 to 25.46 mm, and the average *NSE* was 0.59, thus indicating that the model explained only a relatively small fraction of observed variance.

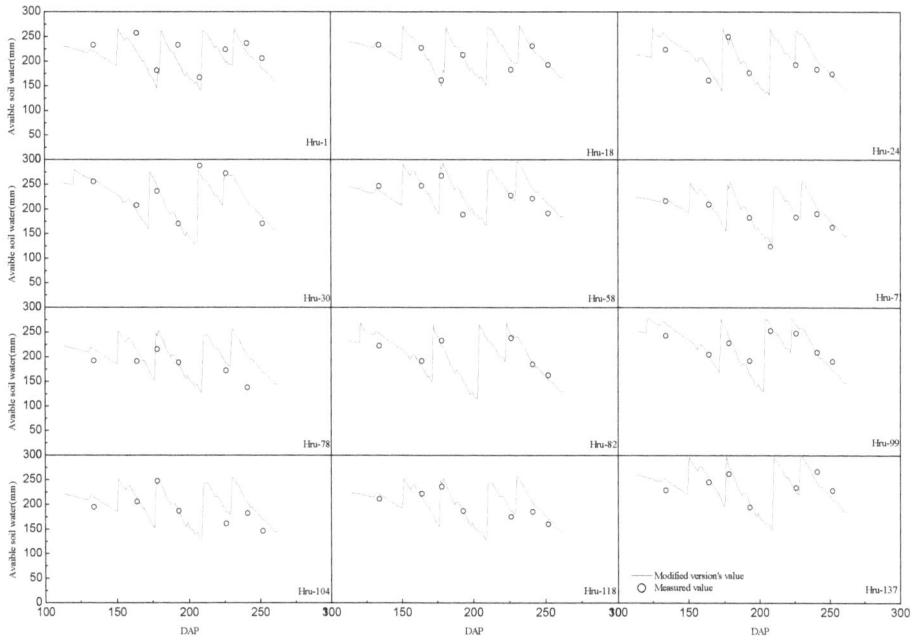

Figure 12. Simulated and measured available soil water.

4. Discussion

4.1. Crop Growth Dynamics

The typical *LAI* growth patterns (HRU-1) used in the original and modified SWAT models are shown in Figure 7, which illustrates considerable differences in the senescence stage. Since the original SWAT model uses a linear attenuation formula to simulate the *LAI*, it reflects the dynamics of plant growth under natural conditions but not for agricultural cultivated crops. The parameters were established using empirically fitted functions, although the *LAI* growth-related modules were modified for the senescence phase. Therefore, the growth mechanism of the leaf area was not properly represented. The *LAI* can influence the radiation-use efficiency, which influences the production of biomass because the shape coefficients are flexible in the optimal *LAI* curves [55]. When senescence is the major growth process, other shape coefficients can be used to modify the *LAI* growth sub-module [26]. Notably, the modified sub-module did not consider the effects of nitrogen stresses on leaf senescence. Thus, this process should be added in subsequent studies [14].

The calculation of shoot dry matter in SWAT is based on empirical formulas. Specifically, biomass production is calculated using Beer's law [64], which can assess the proportion of shoots to the total crop biomass. The fraction of roots in the total biomass varies from 0.40 in the emergence stage to 0.20 at maturity. Biomass production can be affected by the light extinction coefficient (K), the biomass fraction, and the radiation-use efficiency.

Cabelguenne et al. divided the entire period of crop development into various physiological stages to improve the simulation results and assess differences in water stress between phenological stages [14]. Similarly, future versions of SWAT should consider additional processes related to crop growth to enhance the simulation of crop yields and the yield response to irrigation management.

4.2. Evapotranspiration

According to Figure 11, the ET_c values simulated using SWAT and SIMDualKc are not identical. ET_0 is the foundation for the calculations of ET_c. Slight deviations between ET_0 in SWAT and SIMDualKc occur because SWAT uses alfalfa as a reference crop, whereas SIMDualKc ET_0 uses grass. For grass, r_s equals 70 m s^{-1}, and the reference crop height is 0.12 m. For alfalfa, r_s is 100 m s^{-1}, and the reference height is 0.40 m. The input radiation in SWAT was iteratively increased in each subbasin until the value of ET_0 in SWAT matched the value of ET_0 in SIMDualKc to correct these differences. Generally, ET_c in SIMDualKc is different from ET_c in SWAT for three potential reasons.

First, one hypothetical conditions are used in SWAT. One condition states that evaporation will decrease firstly if the soil water availability is insufficient (Appendix A, Equation (A1)). The water stress unit (*wstrs*) in SWAT2009 uses the ratio of the actual and maximum plant transpiration (Appendix A, Equation (A2)), an overestimated value of actual plant transpiration will lead to no or little water stress, and the associated material and water cycles are affected.

Second, the *LAI* can affect evapotranspiration when the Penman-Monteith method is used (Appendix A, Equations (A3) and (A4)). In this case, *LAI* decreases linearly in the senescence stage, potentially leading to underestimated values of ET_c. Gary Marek et al. (2016) also showed that SWAT generally underestimated ET_c at both the daily and monthly levels [38].

Third, the module that represents agricultural irrigation management in the SWAT model is relatively simple. The effects of irrigation and precipitation on crop transpiration are not effectively reflected in SWAT. Nonetheless, this issue can be resolved in the SIMDualKc model by the daily water balance of the topsoil layer [49]. Although there is inherent uncertainty involved in using this approach, the SWAT model eliminates some of the uncertainty by establishing an upper irrigation limit based on the soil water content.

4.3. Soil Water

In SWAT2009, soil moisture is simulated using the method of layered infiltration, and the model calculates actual crop evapotranspiration based on the available water in different soil layers. The water stress coefficient is then calculated based on the ratio of actual to potential crop evapotranspiration, and this coefficient restricts crop production.

SWAT2009 assumes that water stress occurs when the content of available soil water decreases to ASW/4 (Appendix A, Equation (A5)). Allen et al. and Luo et al. used 0.55 as the threshold value of soil water stress instead of 0.25 [26,49]. Water extraction decreases below this threshold, despite the differences between crop and soil types. Nonetheless, the soil water holding capacity is varied, based on the different crop and soil types. The water stress coefficient calculated by Equation (A5) uses an exponential formula. There are many applications of the water stress coefficient from FAO56 using the ratio of total and readily available water [49], and from Feddes et al. (1978) considering the effects of water potential on the uptake rate to be multiplicative [65]. The water stress coefficient formula of SWAT should accept more tests for irrigation crops.

5. Conclusions

The highest sensitivity parameters are *FRGRW2*, *FRGEW1*, *BLAI*, *DLAI*, *LAImx2*, and *LAImx1* for the leaf area index, *EXT_COEF*, *BIO_E*, *T_OPT*, and *T_BASE* for the shoot dry matter, *EPCO*, *SOL_AWC_D*, *SOL_K*, *ESCO*, and *GSI* for daily evapotranspiration.

The modified version of the SWAT model exhibits better performance on the daily evapotranspiration, shoot dry matter, and leaf area index at the daily temporal scale and the HRU spatial scale.

Based on the performance statistics of modified SWAT model, there may be errors related to water stress functions in shoot dry matter and ET_c estimation, and the most important thing is that the stress factor cannot be well reflected into the crop growth. Therefore, users should be aware of the response of crop growth to soil moisture movement.

Author Contributions: Z.W. and B.Z. conceived and designed the study. Z.W. wrote the initial draft. Y.L. and D.X. reviewed and revised the manuscript.

Funding: This research was supported by the National Key R&D Program of China (2017YFC0403202), the Chinese National Natural Science Fund (91425302, 51479210, and 51379217), IWHR Research & Development Support Program (ID0145B082017 and ID0145B742017), and the Special Fund of State Key Laboratory of Simulation and Regulation of Water Cycle in River Basin, China Institute of Water Resources and Hydropower Research (SKL2018CG).

Conflicts of Interest: The authors declare no conflict of interest.

Appendix A

See Equations (A1)–(A5):

$$\begin{aligned} &if(pet < es\_max + ep\_max)then \\ es\_max &= pet \times es\_max/(es\_max + ep\_max) \\ &if(pet\_day < es\_max + ep\_max)then \\ es\_max &= pet_{day} - ep\_max \end{aligned} \quad (A1)$$

where *ep_max* is the maximum transpiration, *es_max* is the maximum evaporation, *pet_day* is the potential evapotranspiration, and *pet* is the amount of *pet_day* remaining after the water stored in the canopy evaporates.

The water stress unit (*wstrs*) in SWAT2009 is as follows:

$$wstrs = 1 - E_{tact}/E_t \quad (A2)$$

where E_t is the maximum plant transpiration on a given day and E_{tact} is the actual plant evapotranspiration. The plant transpiration rate and *wstrs* range from 0.0 to 1.0 when *wstrs* equals 0.0.

$$r_c = \frac{r_1}{0.5 \cdot LAI} \quad (A3)$$

$$\lambda ET_c = \frac{\Delta \cdot R_{net} + \rho \cdot \gamma \cdot \frac{VPD}{r_a}}{\Delta + \gamma \cdot \left(1 + \frac{r_c}{r_a}\right)} \quad (A4)$$

where ET_c is the daily evapotranspiration (mm d^{-1}), Δ is the slope of the saturation vapour pressure-temperature curve (kPa $^{\circ}$C^{-1}), ρ is the air density (kg m^{-3}), γ is the psychrometric constant (kPa $^{\circ}$C^{-1}), VPD is the vapour pressure deficit (kPa), r_a is the atmospheric resistance (s m^{-1}), r_c is the canopy resistance (s m^{-1}), and r_l is the minimum effective stomatal resistance of a single leaf (s m^{-1}).

SWAT2009 assumes that water stress occurs when the available soil water content decreases to ASW/4:

$$reduc = exp\left[5\left(\frac{1}{4}\frac{SW_l}{ASW_l} - 1\right)\right] when\ SW_l \leq \frac{1}{4}ASW_l \tag{A5}$$

where SW_l refers to the volumetric soil water content above the wilting point, *reduc* is the water uptake reduction factor, *l* is the number of soil layers, and *ASW* is the available soil water.

References

1. Wang, W.-Y.; Luo, W.; Wang, Z.-R. Surge flow irrigation with sediment-laden water in northwestern China. *Agric. Water Manag.* **2005**, *75*, 1–9. [CrossRef]
2. He, C. Integration of geographic information systems and simulation model for watershed management. *Environ. Model. Softw.* **2003**, *18*, 809–813. [CrossRef]
3. Young, R.A.; Onstad, C.; Bosch, D.; Anderson, W. AGNPS: A nonpoint-source pollution model for evaluating agricultural watersheds. *J. Soil Water Conserv.* **1989**, *44*, 168–173.
4. Arnold, J.G.; Allen, P.M.; Bernhardt, G. A comprehensive surface-groundwater flow model. *J. Hydrol.* **1993**, *142*, 47–69. [CrossRef]
5. Gassman, P.W.; Reyes, M.R.; Green, C.H.; Arnold, J.G. *The Soil and Water Assessment Tool: Historical Development, Applications, and Future Research Directions*; Center for Agricultural and Rural Development, Iowa State University: Heady Hall, AI, USA, 2007.
6. Behera, S.; Panda, R. Evaluation of management alternatives for an agricultural watershed in a sub-humid subtropical region using a physical process based model. *Agric. Ecosyst. Environ.* **2006**, *113*, 62–72. [CrossRef]
7. Bosch, D.; Sheridan, J.; Batten, H.; Arnold, J. Evaluation of the SWAT model on a coastal plain agricultural watershed. *Trans. ASAE* **2004**, *47*, 1493–1506. [CrossRef]
8. Santhi, C.; Muttiah, R.; Arnold, J.; Srinivasan, R. A GIS-based regional planning tool for irrigation demand assessment and savings using SWAT. *Trans. ASAE* **2005**, *48*, 137–147. [CrossRef]
9. Sophocleous, M.; Perkins, S.P. Methodology and application of combined watershed and ground-water models in Kansas. *J. Hydrol.* **2000**, *236*, 185–201. [CrossRef]
10. Edmonds, J.A.; Rosenberg, N.J. Climate change impacts for the conterminous USA: An integrated assessment summary. In *Climate Change Impacts for the Conterminous USA*; Springer: Berlin/Heidelberg, Germany, 2005; pp. 151–162.
11. Thomson, A.M.; Brown, R.A.; Rosenberg, N.J.; Izaurralde, R.C.; Benson, V. Climate change impacts for the conterminous USA: An integrated assessment. In *Climate Change Impacts for the Conterminous USA*; Springer: Berlin/Heidelberg, Germany, 2005; pp. 43–65.
12. Lenhart, T.; Van Rompaey, A.; Steegen, A.; Fohrer, N.; Frede, H.G.; Govers, G. Considering spatial distribution and deposition of sediment in lumped and semi-distributed models. *Hydrol. Process.* **2005**, *19*, 785–794. [CrossRef]
13. Cau, P.; Cadeddu, A.; Lecca, G.; Gallo, C.; Marrocu, M. Calcolo del bilancio idrico della regione Sardegna con il modello idrologico SWAT. *L'acqua* **2005**, *5*, 29–38.
14. Cabelguenne, M.; Debaeke, P. Experimental determination and modelling of the soil water extraction capacities of crops of maize, sunflower, soya bean, sorghum and wheat. *Plant Soil.* **1998**, *202*, 175–192. [CrossRef]
15. Cabelguenne, M.; Jones, C.; Marty, J.; Dyke, P.; Williams, J. Calibration and validation of EPIC for crop rotations in southern France. *Agric. Syst.* **1990**, *33*, 153–171. [CrossRef]
16. Chanasyk, D.; Mapfumo, E.; Willms, W. Quantification and simulation of surface runoff from fescue grassland watersheds. *Agric. Water Manag.* **2003**, *59*, 137–153. [CrossRef]
17. Du, B.; Saleh, A.; Jaynes, D.; Arnold, J. Evaluation of SWAT in simulating nitrate nitrogen and atrazine fates in a watershed with tiles and potholes. *Trans. ASAE* **2006**, *49*, 949–959. [CrossRef]

18. Shrestha, M.K.; Recknagel, F.; Frizenschaf, J.; Meyer, W. Assessing SWAT models based on single and multi-site calibration for the simulation of flow and nutrient loads in the semi-arid Onkaparinga catchment in South Australia. *Agric. Water Manag.* **2016**, *175*, 61–71. [CrossRef]

19. Van Griensven, A.; Bauwens, W. Multiobjective autocalibration for semidistributed water quality models. *Water Resour. Res.* **2003**, *39*, 1348–1356. [CrossRef]

20. Van Griensven, A.; Bauwens, W. Application and evaluation of ESWAT on the Dender basin and the Wister Lake basin. *Hydrol. Process.* **2005**, *19*, 827–838. [CrossRef]

21. Hattermann, F.; Krysanova, V.; Wechsung, F.; Wattenbach, M. Integrating groundwater dynamics in regional hydrological modelling. *Environ. Model. Softw.* **2004**, *19*, 1039–1051. [CrossRef]

22. Baffaut, C.; Benson, V. Modeling flow and pollutant transport in a karst watershed with SWAT. *Trans. ASABE* **2009**, *52*, 469–479. [CrossRef]

23. Cau, P.; Paniconi, C. Assessment of alternative land management practices using hydrological simulation and a decision support tool: Arborea agricultural region, Sardinia. *Hydrol. Earth Syst. Sci.* **2007**, *11*, 1811–1823. [CrossRef]

24. Zheng, J.; Li, G.-Y.; Han, Z.-Z.; Meng, G.-X. Hydrological cycle simulation of an irrigation district based on a SWAT model. *Math. And Comput. Model.* **2010**, *51*, 1312–1318. [CrossRef]

25. Kannan, N.; Jeong, J.; Srinivasan, R. Hydrologic modeling of a canal-irrigated agricultural watershed with irrigation best management practices: Case study. *J. Hydrol. Eng.* **2010**, *16*, 746–757. [CrossRef]

26. Luo, Y.; He, C.; Sophocleous, M.; Yin, Z.; Hongrui, R.; Ouyang, Z. Assessment of crop growth and soil water modules in SWAT2000 using extensive field experiment data in an irrigation district of the Yellow River Basin. *J. Hydrol.* **2008**, *352*, 139–156. [CrossRef]

27. Wang, J.; Cui, Y. Modified SWAT for rice-based irrigation system and its assessment. *Trans. CSAE* **2011**, *27*, 22–28. (In Chinese)

28. Liu, Y.; Pereira, L. Validation of FAO methods for estimating crop coefficients. *Trans. Chin. Soc. Agric. Eng.* **2000**, *16*, 26–30. (In Chinese)

29. Pereira, L.; Cai, L.; Hann, M. Farm water and soil management for improved water use in the North China plain. *Irrig. Drain.* **2003**, *52*, 299–317. [CrossRef]

30. Liu, Y.; Luo, Y. A consolidated evaluation of the FAO-56 dual crop coefficient approach using the lysimeter data in the North China Plain. *Agric. Water Manag.* **2010**, *97*, 31–40. [CrossRef]

31. Tolk, J.; Howell, T. Measured and simulated evapotranspiration of grain sorghum grown with full and limited irrigation in three high plains soils. *Trans. ASAE* **2001**, *44*, 1553–1558.

32. Zhao, N.; Liu, Y.; Cai, J.; Paredes, P.; Rosa, R.D.; Pereira, L.S. Dual crop coefficient modelling applied to the winter wheat–summer maize crop sequence in North China Plain: Basal crop coefficients and soil evaporation component. *Agric. Water Manag.* **2013**, *117*, 93–105. [CrossRef]

33. Odhiambo, L.O.; Irmak, S. Evaluation of the impact of surface residue cover on single and dual crop coefficient for estimating soybean actual evapotranspiration. *Agric. Water Manag.* **2012**, *104*, 221–234. [CrossRef]

34. Immerzeel, W.; Droogers, P. Calibration of a distributed hydrological model based on satellite evapotranspiration. *J. Hydrol.* **2008**, *349*, 411–424. [CrossRef]

35. Cai, X.; Xu, Z.; Su, B.; Yu, W. Distributed simulation for regional evapotranspiration and verification by using remote sensing. *Trans. CSAE* **2009**, *25*, 411–424.

36. Qian, K.; Ye, S.; Zhu, Q. Evapotranspiration simulation with different scenarios analysises of Fangshan District by SWAT model. *Trans. CSAE* **2011**, *27*, 99–105.

37. Liu, X.; Wang, S.; Xue, H.; Singh, V.P. Simulating Crop Evapotranspiration Response under Different Planting Scenarios by Modified SWAT Model in an Irrigation District, Northwest China. *PLoS ONE* **2015**, *10*, e0139839. [CrossRef] [PubMed]

38. Marek, G.W.; Gowda, P.H.; Evett, S.R.; Baumhardt, R.L.; Brauer, D.K.; Howell, T.A.; Marek, T.H.; Srinivasan, R. Estimating Evapotranspiration for Dryland Cropping Systems in the Semiarid Texas High Plains Using SWAT. *J. Am. Water Resour. Assoc.* **2016**, *52*, 298–314. [CrossRef]

39. Marek, G.W.; Gowda, P.H.; Marek, T.H.; Porter, D.O.; Baumhardt, R.L.; Brauer, D.K. Modeling long-term water use of irrigated cropping rotations in the Texas High Plains using SWAT. *Irrig. Sci.* **2017**, *35*, 111–123. [CrossRef]

40. Chen, Y.; Marek, G.W.; Marek, T.H.; Brauer, D.K.; Srinivasan, R. Assessing the Efficacy of the SWAT Auto-Irrigation Function to Simulate Irrigation, Evapotranspiration, and Crop Response to Management Strategies of the Texas High Plains. *Water* **2017**, *9*, 509. [CrossRef]

41. Ahmadzadeh, H.; Morid, S.; Delavar, M.; Srinivasan, R. Using the SWAT model to assess the impacts of changing irrigation from surface to pressurized systems on water productivity and water saving in the Zarrineh Rud catchment. *Agric. Water Manag.* **2015**, *175*, 15–28. [CrossRef]

42. Huang, F.; Li, B. Assessing grain crop water productivity of China using a hydro-model-coupled-statistics approach: Part I: Method development and validation. *Agric. Water Manag.* **2010**, *97*, 1077–1092. [CrossRef]

43. Sinnathamby, S.; Douglas-Mankin, K.R.; Craige, C. Field-scale calibration of crop-yield parameters in the Soil and Water Assessment Tool (SWAT). *Agric. Water Manag.* **2017**, *180*, 61–69. [CrossRef]

44. Luo, T.; Pan, Y.; Ouyang, H.; Shi, P.; Luo, J.; Yu, Z.; Lu, Q. Leaf area index and net primary productivity along subtropical to alpine gradients in the Tibetan Plateau. *Glob. Ecol. Biogeogr.* **2004**, *13*, 345–358. [CrossRef]

45. Ge, Y.; Xu, F.; Zhuang, J. HiWATER: Dataset of investigation on channel flow and socio-economy in the middle reaches of the Heihe River Basin. *Heihe Plan Sci. Data Cent.* **2013**. [CrossRef]

46. Ge, Y.; Zhuang, J.; Ma, C.; Xu, F. HiWATER: Dataset of investigation on crop phrenology and field management in the middle reaches of the Heihe River Basin. *Heihe Plan Sci. Data Cent.* **2013**. [CrossRef]

47. Xu, T.; Liu, S.; Xu, L.; Chen, Y.; Jia, Z.; Xu, Z.; Nielson, J. Temporal upscaling and reconstruction of thermal remotely sensed instantaneous evapotranspiration. *Remote. Sens.* **2015**, *7*, 3400–3425. [CrossRef]

48. Rosa, R.D.; Paredes, P.; Rodrigues, G.C.; Alves, I.; Rui, M.F.; Pereira, L.S.; Allen, R.G. Implementing the dual crop coefficient approach in interactive software. 1. Background and computational strategy. *Agric. Water Manag.* **2012**, *103*, 8–24. [CrossRef]

49. Allen, R.G.; Pereira, L.S.; Raes, D.; Smith, M. *Crop Evapotranspiration-Guidelines for Computing Crop Water Requirements—FAO Irrigation and Drainage Paper 56*; FAO—Food and Agriculture Organization of the United Nations: Rome, Italy, 1998; p. 6541.

50. Rosa, R.D.; Paredes, P.; Rodrigues, G.C.; Rui, M.F.; Alves, I.; Pereira, L.S.; Allen, R.G. Implementing the dual crop coefficient approach in interactive software: 2. Model testing. *Agric. Water Manag.* **2012**, *103*, 62–77. [CrossRef]

51. Arnold, J.G.; Srinivasan, R.; Muttiah, R.S.; Williams, J.R. Large area hydrologic modeling and assessment part I: Model development1. *JAWRA J. Am. Water Resour. Assoc.* **1998**, *34*, 73–89. [CrossRef]

52. Williams, J. The erosion-productivity impact calculator (EPIC) model: A case history. *Philos. Trans. R. Soc. London. Ser. B Boil. Sci.* **1990**, *329*, 421–428. [CrossRef]

53. Abbaspour, K.C.; Vejdani, M.; Haghighat, S. SWAT-CUP calibration and uncertainty programs for SWAT. In *MODSIM 2007 International Congress on Modelling and Simulation*; Modelling and Simulation Society of Australia and New Zealand Inc.: Canberra, Australia, 2007; pp. 1603–1609.

54. Izady, A.; Davary, K.; Alizadeh, A.; Ziaei, A.; Akhavan, S.; Alipoor, A.; Joodavi, A.; Brusseau, M. Groundwater conceptualization and modeling using distributed SWAT-based recharge for the semi-arid agricultural Neishaboor plain, Iran. *Hydrogeol. J.* **2015**, *23*, 47–68. [CrossRef]

55. Neitsch, S.; Arnold, J.; Kiniry, J.; Williams, J.; King, K. *Soil and Water Assessment Tool: Theoretical Documentation Version 2005*; Texas Water Resources Institute: College Station, TX, USA, 2005.

56. Dechmi, F.; Burguete, J.; Skhiri, A. SWAT application in intensive irrigation systems: Model modification, calibration and validation. *J. Hydrol.* **2012**, *470*, 227–238. [CrossRef]

57. Liu, Y.; Pereira, L.; Fernando, R. Fluxes through the bottom boundary of the root zone in silty soils: Parametric approaches to estimate groundwater contribution and percolation. *Agric. Water Manag.* **2006**, *84*, 27–40. [CrossRef]

58. Duan, A.; Sun, J.; Liu, Y.; Xiao, J. *The Main Crop Irrigation Water Quota in Northern Area*; China Agricultural Science and Technology Press: Beijing, China, 2004. (In Chinese)

59. Zhang, B.; Xu, D.; Liu, Y.; Li, F.; Cai, J.; Du, L. Multi-scale evapotranspiration of summer maize and the controlling meteorological factors in north China. *Agric. For. Meteorol.* **2016**, *216*, 1–12. [CrossRef]

60. Zhang, Y.; Zhao, W.; He, J.; Zhang, K. Energy exchange and evapotranspiration over irrigated seed maize agroecosystems in a desert-oasis region, northwest China. *Agric. For. Meteorol.* **2016**, *223*, 48–59. [CrossRef]

61. Vazquez-Amabile, G.; Engel, B. Use of SWAT to compute groundwater table depth and streamflow in the Muscatatuck River watershed. *Trans. ASAE* **2005**, *48*, 991–1003. [CrossRef]

62. Kannan, N.; White, S.; Worrall, F.; Whelan, M. Sensitivity analysis and identification of the best evapotranspiration and runoff options for hydrological modelling in SWAT-2000. *J. Hydrol.* **2007**, *332*, 456–466. [CrossRef]

63. Neitsch, S.L.; Arnold, J.G.; Kiniry, J.R.; Williams, J.R. *Soil and Water Assessment Tool: Theoretical Documentation. Version 2009*; Texas Water Resources Institute: College Station, TX, USA, 2011.

64. Monsi, M.; Saeki, T. Uber den Lictfaktor in den Pflanzengesellschaften und sein Bcdeutung fur die Stoffproduktion. *Jpn. J. Bot.* **1953**, *14*, 22–52.

65. Simunek, J.; Saito, H.; Sakai, M. *The HYDRUS-1D Software Package for Simulating the One-Dimensional Movement of Water, Heat, and Multiple Solutes in Variably-Saturated Media*, version 3.0; Department of Environmental Sciences, University of California: Riverside, CA, USA, 1998.

water MDPI

Article

Five-Year Experimental Study on Effectiveness and Sustainability of a Dry Drainage System for Controlling Soil Salinity

Changshu Wang, Jingwei Wu *, Wenzhi Zeng *, Yan Zhu and Jiesheng Huang

State Key Laboratory of Water Resources and Hydropower Engineering Science, Wuhan University, Wuhan 430072, China; wcswhu@whu.edu.cn (C.W.); zyan@whu.edu.cn (Y.Z.); huangjiesheng1962@gmail.com (J.H.)
* Correspondence: jingwei.wu@whu.edu.cn (J.W.); zengwenzhi1989@whu.edu.cn (W.Z.);
  Tel.: +86-27-6877-5466 (J.W.)

Received: 5 December 2018; Accepted: 4 January 2019; Published: 10 January 2019

Abstract: The dry drainage system (DDS) is an alternative technique for controlling salinization. To quantify its role in soil salinity control, a five-year field observation from 2007 to 2011 was completed in a 2900 ha experimental plot in Yonglian Experimental Station, Hetao Irrigation District, China. Results showed that the groundwater table depth in the fallow areas quickly responded to the lateral recharge from the surrounding croplands during irrigation events. The groundwater electrical conductivity (GEC) of fallow areas increased from 5 mS·cm^{-1} to 15 mS·cm^{-1}, whereas the GEC below croplands produced small fluctuations. The analysis of water and salt balance showed that the excess water that moved to fallow was roughly four times that moved by an artificial drainage system and with 7.7 times the corresponding salt. The fallow areas act as a drainage repository to receive excess water and salt from surrounding irrigated croplands. Slight salt accumulation occurred in irrigated croplands and salts accumulated, with an accelerating trend over the final two years. The evaporation capability weakened, partly due to the salt crust in the topsoil, and the decrease in soil permeability in the soil column, which was almost impermeable to water. Using halophytes may be an effective method to remove salts that have accumulated in fallow areas, having great economic and ecological value. A DDS may be effective and sustainable in situations where the fallow areas can sustain an upward capillary flux from planted halophytes.

Keywords: dry drainage system; water and salt balance; groundwater; evaporation; salinity

1. Introduction

Salinization is a long-standing problem that has threatened crop production and food security in many arid and semi-arid regions, and has impeded sustainable irrigation over the past 2000 years. Salt accumulation has played a significant role in the abandonment of settlements and in the breakdown of ancient civilizations throughout history [1–5]. The overall loss from groundwater-associated and irrigation-induced salinity in the world was estimated to be $27.3 USD billion/year [6,7]. To alleviate the impact of soil salinity, various technical measures and strategies have been developed since the early 1900s. Irrigation systems can be sustainable if excess salt and drainage water are adequately removed from the subsoil [8]. Many soil-related problems could be minimized by installing various types of drainage ditches. An artificial drainage system (ADS) is a popular method used to drain salt away from the root zone by deliberate flood irrigation events [9–13]. However, conventional drainage methods produce seemingly intractable economic [14] and environmental problems [9–12,15–17].

The dry drainage system (DDS) is a technique that was first proposed in 1992 for areas with a shallow groundwater table depth (GTD), high evaporation, and intensive irrigation [18]. In the DDS operating process, a part of the land that is generally low-lying is permanently or seasonally fallowed to act as a sink for excess water and salts from neighboring irrigated croplands. The GTD in irrigated croplands rises along with the irrigation events, whereas that in the fallow area falls with the aid of evaporation. This produces a hydraulic gradient that induces lateral groundwater migration flow from the croplands to the fallow area. Thus, the excess salt is eventually transported to the fallow areas. As the fallow area is not irrigated and the evaporation in the salinity-hazard area is generally strong, the DDS continuously functions. The salt balance in the irrigated areas can be maintained providing the fallow areas are large enough to evaporate the excess water. The DDS does not need extra investment in the operation and maintenance, and can mitigate various potential environmental concerns compared with ADS, including degrading water quality, destroying wetlands, and increasing drain outfall erosion [9,15,18–20].

Previous studies have discussed and tested the effects of DDSs on soil salinity control, for example, the key design technology of the DDS and studies based on field-scale numerical simulations [9,11]. These technologies have contributed to the sustainability of the agro-ecosystem by helping to maintain the water and salt balance in the root zone of irrigated croplands, controlling soil salinity and minimizing environmental threats [12,15,20–22]. The long-term effectiveness of the DDS in the Hetao Irrigation District (HID), China, has been studied by various methods (e.g., remote sensing, field experiments, and conceptual/numerical models) [15,19,21–24]. However, previous studies mainly investigated surface salinity dynamics and did not measure the salt accumulation in the soil profile, especially below the root zone [9,11]. The effects of salt accumulation in fallow areas on the evaporation rate and the sustainability of the DDS are still unclear [9,11,15]. Therefore, the main objective of this study was to quantify the capability of the DDS via the salt balance in specific monitoring wells and to investigate and discuss the sustainability of DDSs on soil salinity control with a five-year field observation. Based on the five-year archived data, the soil salt content (SSC) was measured in the vertical profiles of irrigated and fallow areas, and we analyzed the dynamic change in GTD, groundwater electrical conductivity (GEC), and SSC. We evaluated the effectiveness and sustainability of DDS in the HID, and propose some suggestions for improving the DDS.

2. Materials and Methods

2.1. Brief Description of the HID and YES

The observations were performed at the Yonglian Experimental Station (YES) (Figure 1a) in the mid-HID, upper Yellow River basin ($40°19'$–$41°20'$ N, $106°10'$–$109°30'$ E; Figure 1b), located in the Inner Mongolia Autonomous Region, China (Figure 1c). The area totaled 2900 ha, which included 71.4% irrigated cropland, 17.9% fallowed areas, 9.7% villages, and 1.0% water bodies (detected by Landsat 5 images, date: 1 August 2007 and 30 July 2009). There were four blocks of fallow areas that were randomly distributed, 10 blocks of inhabited areas, and three water bodies. Sunflower and various melon seeds accounted for half of the cropland areas, as the HID is the largest sunflower-growing region in China [25,26]. The predominant soil was silt loam, and the soil texture was comprised of sand (8.91–16.69%), silt (52.32–70.57%), and clay (13.72–35.66%). The bulk density ranged from 1.35 to 1.51 $g \cdot cm^{-3}$, differing with the soil profile [25].

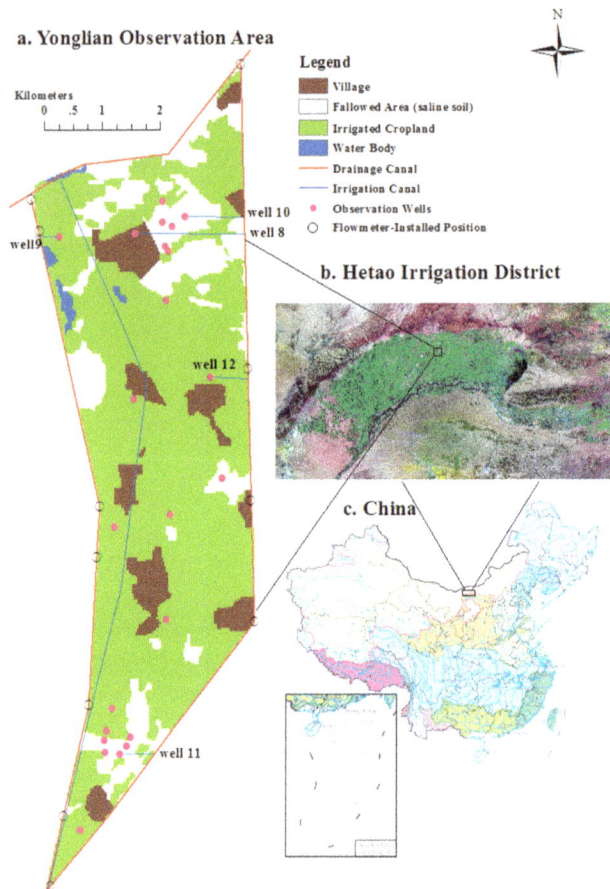

Figure 1. Location of the Hetao Irrigation District (HID) and the observation field: (**a**) observation area; (**b**) HID; (**c**) China.

The mean annual potential evapotranspiration was as high as 2200 mm (Φ20 cm evaporation pan), while the average precipitation was only 170 mm annually, mostly (63–70%) during the monsoon season (June to September) (Figure 2). The hydrometeorological data (2007–2011, hourly meteorological data, e.g., P (precipitation), air temperature, the wind speed at a 2 m height above the ground surface, radiation, and relative humidity) were automatically recorded by the meteorological station, and the mean monthly air temperature together with the relative humidity are shown in Figure 3.

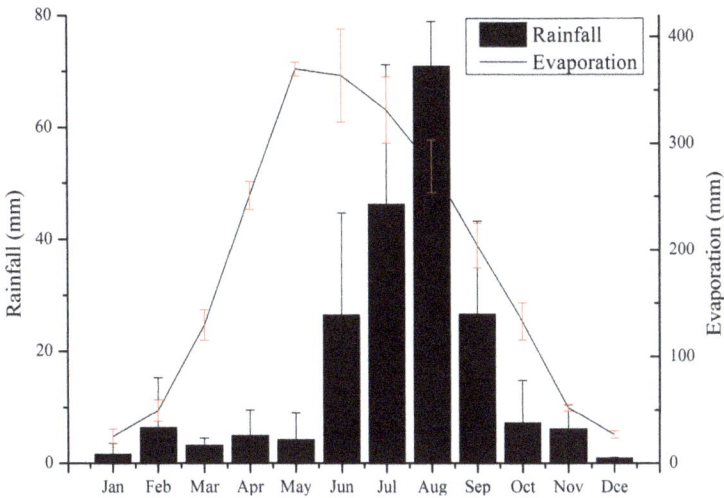

Figure 2. Distribution of the monthly average rainfall and evaporation from 2007 to 2011.

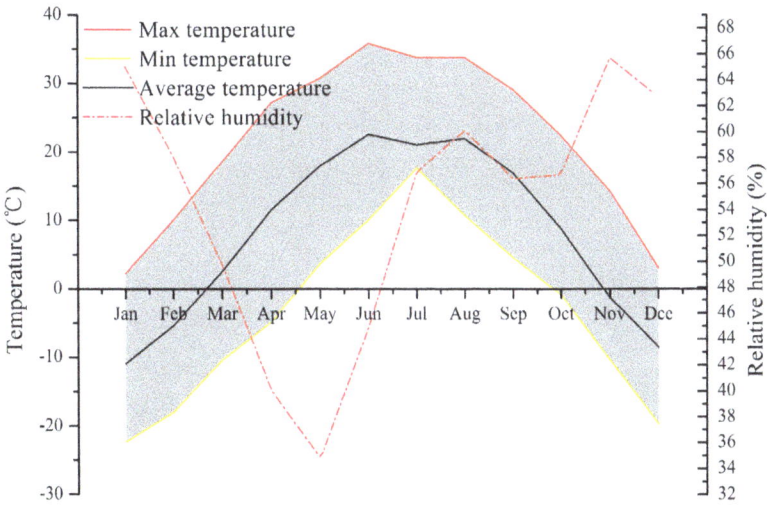

Figure 3. Distribution of the monthly maximum and minimum temperature and average values of temperature and relative humidity from 2007 to 2011.

2.2. Observations and Data Collection Design

The irrigation and artificial drainage water amounts were measured daily at the inlet and several control points with flowmeters. Eleven flowmeter-installed positions (FIP) were set up along the inlet irrigation and drainage canals (Figure 1, left). The local farmer irrigated the crop with water from the Yellow River by flood (surface) irrigation. The drainage water amount for the observation field was calculated by sectioning according to the controlled areas for each flowmeter with the percentage of the irrigation areas on both sides of the drainage ditches. Observation wells were installed to monitor the groundwater variation, and the wells were placed perpendicular to the water flow. The four wells were chosen to calculate the water/salt balance and to analyze the amount of dry drainage water of the DDS with the distributed locations and cover areas of the fallow area (well 8 and well 10 located in the north fallow, well 12 in middle, and well 11 in the south fallow, the areas of which were 90, 110, 135, and 105 ha, respectively). Well 9 represented irrigated cropland in the north for the comparison of dynamics with fallow areas in the analysis of water and salt transport and salt accumulation in the DDS.

GTD and the corresponding electrical conductivity (EC at 25 °C) were monitored every five days and once per day during an irrigation event. The water samples (e.g., irrigation water, drainage water, and groundwater) were collected in 550 mL clean polyethylene bottles. Soil samples were taken by hand auguring near the observation well at depths of 0–10 cm, 10–30 cm, 30–50 cm, 50–70 cm, 70–100 cm, and 100–140 cm. Soil moisture was determined using a regular gravimetric method (oven-dry) and measured every 10 days. The $EC_{1:5}$ of the soil water extract and the EC of the water samples were measured using a digital conductivity meter (DDSJ-308A, Leici, Yidian Co., Ltd., Shanghai, China), and the SSC was measured using a drying method.

The water EC (mS·cm^{-1}) was converted to total dissolved solids (TDS, g·L^{-1}) using the empirical formula developed from laboratory testing as follows [21,27]:

$$TDS \text{ (g·L}^{-1}) = 0.69EC \text{ (mS·cm}^{-1}) \tag{1}$$

where 0.69 is an empirical coefficient calibrated by water samples from local groundwater and surface water.

2.3. Conceptual Model Description

The amount of lateral water migration from irrigated to fallow areas is represented by the dry drainage depth, which was estimated from the water and salt balance in the fallow areas. The water and salt balance was then calculated from the amount of water measured in the inlet and the outlet, and their corresponding salinities. For each fallow block (*i*), regarded as a balanced unit of soil water and groundwater, the water balance was calculated using the following [15,28]:

$$D_{di} = E_i + \Delta S + \mu\Delta H - P, \tag{2}$$

where D_{di} (mm) is defined as the dry drainage depth for the fallow area (block *i*, dimensionless), using the wells that were located in the fallow area for presenting the blocks of wells 8, 10, 11, and 12 for calculation in each fallow; ΔS (mm) and $\mu\Delta H$ (mm) are the storage capacity change in the soil water and in the groundwater, respectively, with the corresponding time interval from the first irrigation event in May to the autumn irrigation in October every year, which are estimated from the dynamics of soil water content and GTD; μ is the specific yield, measured by the local hydrological bureau with a dimensionless value of 0.033 [15,28]; *P* is the precipitation of the corresponding time interval (mm), where the daily *P* is collected from the field meteorological station (self-recording); and E_i (mm) is the total evapotranspiration of the fallow area, which was calculated using the Penman-Monteith equation and the corresponding crop coefficient. As there was no irrigation event in the fallow area, E_i (mm) was mainly due to the groundwater evaporation. We used Block *i* to verify the result of E_i. The groundwater evaporation was calculated for each block using the empirical equation in Equation (5) as follows [29]:

$$C = E_g/\varepsilon_0, \tag{3}$$

$$C = f(H) = 0.3356 - 0.2929 \ln H \tag{4}$$

$$E_g = (0.3356 - 0.2929 \ln H_i)\varepsilon_0 \tag{5}$$

where *C* is the groundwater evaporation coefficient defined by the ratio of the evaporation from the free water surface to the evaporative intensity of the groundwater, as shown in Equation (3); E_g is the evaporative intensity of the groundwater (mm); and ε_0 is the evaporation from the surface of free water (mm). *C* is a function of the GTD for the same soil, which was silt loam in the study area, as shown in Equation (4), where H_i is the average GTD for the calculation block (*i*) with the corresponding period (m).

3. Results and Discussion

3.1. Dynamics of GTD and GEC in the DDS

The fluctuations in GTD from May 2007 to December 2011 in the four fallow blocks are shown in Figure 4a, represented by wells 8, 10, 11, and 12, respectively. The average GTD was about 1.6 m during the growing season in the observation area, varying from 2.8 m in February up to the soil surface (less than 0.5 m) in November. The dynamics of the GTD for the four fallow wells fluctuated periodically with the irrigation events from the surrounding cropland. The lateral migration process of groundwater between irrigated and fallow areas was obvious and intense, particularly during the autumn irrigation period, where the duration usually ranged from mid-October to early December. Autumn irrigation was a flood (surface) irrigation event for all cropland in the fallow period, and the irrigation amount was almost two to four times the other irrigation events during the crop growth stages, which were not full irrigations and depended on the water demand of different crop types. Irrigation in May was mainly for sunflower seeding, winter wheat, and melon pre-irrigation, whereas irrigation from June to September was for maize and various pastures.

For a closer examination of the dynamics of the GEC between the irrigated and fallow areas, the GEC from 2007 to 2011 in four fallow blocks is shown in Figure 4b, which indicates that the GEC of the fallow areas always experienced considerable fluctuations following the irrigation events of the surrounding irrigated cropland. The greatest variations in EC ranged from 3.5 mS·cm^{-1} to more than 10 mS·cm^{-1}. The maximum and minimum values occurred before and after the autumn irrigation, respectively. The dynamics of the GEC were relatively stable during the crop growing stage.

We chose the dynamics of well 9, which was located in irrigated cropland (Figure 5b), and the abovementioned four fallow wells during the autumn irrigation period for a specific description for one month from 14 October to 14 November 2007. Autumn irrigation commenced on October 14 for well 9 and the GTD rose steeply from 2.3 m to the soil surface about two days later, before decreasing to 1.5 m with redistribution and drainage (Figure 5b). The GEC below the irrigated croplands fluctuated with an initial slight decrease, reaching the lowest value of 0.9 mS·cm^{-1}, and finally exceeding the initial value, with little change around the value of 1.5 mS·cm^{-1}. The whole process of GEC below the irrigated croplands varied minimally, showing a trend of slight increase due to the deep percolation and the salt in the root zone leaching out continuously with the end of redistribution. Notably, the GTD in the fallow areas quickly responded to the irrigation from the surrounding cropland without being directly recharged by irrigation (Figure 5a,c–e). The GTD increased in a straight line more slowly compared with the cropland (well 9), and varied from 2.5 m to the soil surface along with the progressive groundwater lateral migration recharge. The increasing process was sustained until the freezing process began, which was different from the croplands. The GEC below the fallow area experienced different variation trends with different initial values, irrigation times, and fallow locations. The GEC below well 8, which was located in the village and adjacent to a road, was not like the other fallow wells located in the middle fallow areas. Well 8 had a decreasing trend and large variations during the final autumn irrigation period (Figure 5a). The other fallow wells continuously increased between the autumn irrigation period, especially for well 10, which increased from 5 mS·cm^{-1} to 15 mS·cm^{-1} (Figure 5c). Well 12 demonstrated the same increasing trend to 2.8 mS·cm^{-1}. Well 11 fluctuated with an initial slight increase, reaching a value of 8 mS·cm^{-1}, and finally showing a decreasing trend with redistribution and drainage (Figure 5d).

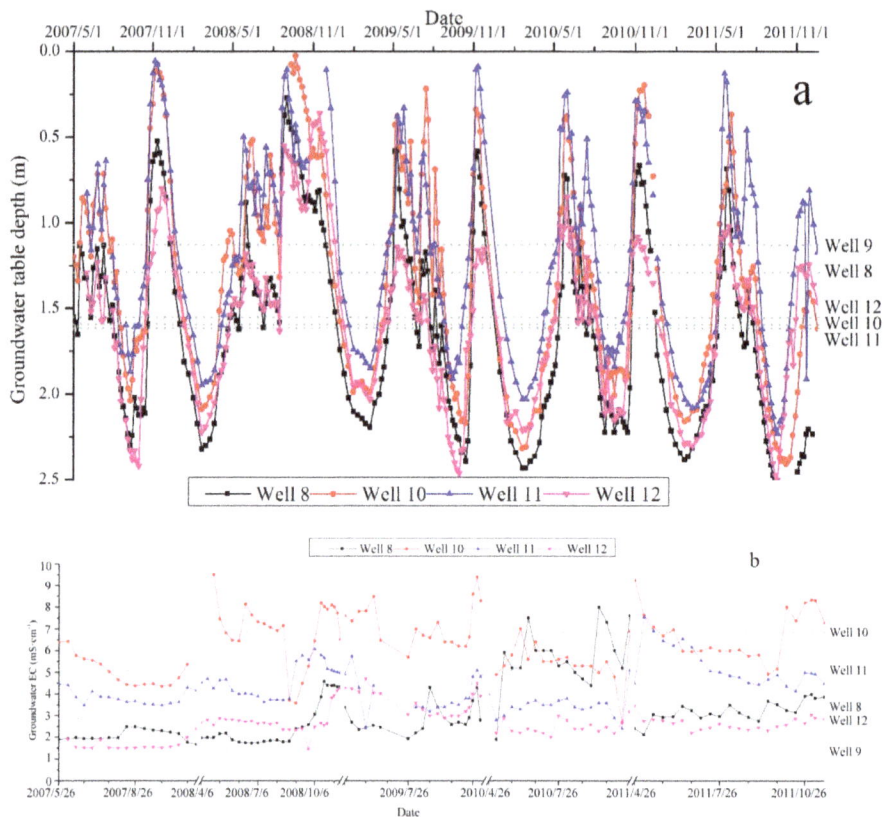

Figure 4. The fluctuation of the groundwater table depth (GTD) and groundwater electrical conductivity (GEC) in the fallowed area between 2007 and 2011: (**a**) GTD dynamics and (**b**) GEC dynamics. The dashed line represents the average value of four fallow wells (well 8, well 10, well 11, and well 12) and one irrigated well (well 9).

Another interesting finding was the process of increase in GTD, which presented some differences between irrigated and fallow areas. The increase in GTD below the fallow areas decreased for 10 days after the autumn irrigation event, and then displayed a slightly decreasing trend toward the soil surface, whereas the GEC sharply increased with the end of the water redistribution. The GTD reached the soil surface after two days and 15 days of autumn irrigation in the irrigated croplands and the fallow areas, respectively. During the increase in the GTD below the fallow areas, the salts that had accumulated in the fallow soil profile began to dissolve until the lateral groundwater recharge reached equilibrium. This means that a steeply increasing process for the GEC below the fallow areas occurred. However, the GEC finally decreased if the initial value was larger than the lateral migration flowing water. So, as shown in Figure 5a,d, the dynamic trend depends on the initial value, the location, and distance, and the irrigation management. The salt also drained away from the soil profile of the irrigated and fallow areas during the irrigation and drainage processes. This means that the salt in the soil profile redistributed under the two processes, making room for its storage and accumulation in the next crop growth stage.

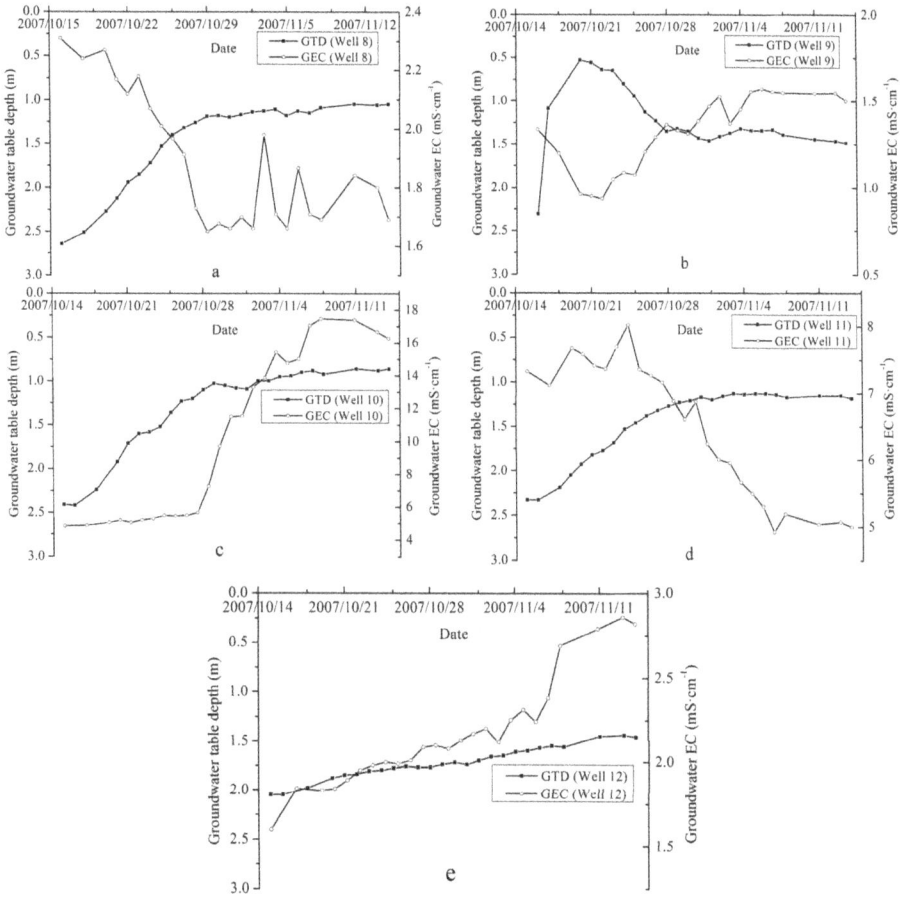

Figure 5. The GTD and GEC variation in the fallow and irrigated area during autumn irrigation: (**a**) well 8, (**b**) well 9, (**c**) well 10, (**d**) well 11, and (**e**) well 12.

As we compared the processes of water diversion and the dynamics of GTD, the fallow areas produced good responses to the irrigation events of the surrounding croplands, and the GTD rose near the ground surface with the recharge of the groundwater transverse flow from cropland irrigation. However, this process had little salt leaching function; the salt accumulation in the observed areas is shown in detail in Table 1. As Table 1 shows, the salt accumulation in the soil profile had an increasing trend from 2007 to 2011. The croplands (we use superscript 1 to represent the Location line of Table 1) have different increasing trends during five years. The dynamic change in GTD in fallow areas indicates that the process of lateral groundwater migration from the irrigation croplands recharges the fallow area continuously and increases the salt content simultaneously.

Table 1. Different average values in salt profile (%; 2011 minus 2007).

Location	Soil Profile (cm)						Average
	0–10	10–30	30–50	50–70	70–100	100–140	
Well 1 [1]	0.06	−0.01	−0.02	0.03	0.32	0.32	0.07
Well 2 [1]	0.06	0.07	0.08	0.11	0.18	0.04	0.09
Well 3 [1]	0.13	0.05	0.02	−0.05	0.02	−0.04	0.03
Well 4 [1]	0.05	0.09	0.02	0.07	0.15	−0.05	0.06
Well 5 [1]	0.19	0.01	0.16	0.13	0.24	0.20	0.15
Well 6 [1]	0.04	0.03	0.02	0.07	0.09	0.03	0.04
Well 7 [1]	0.02	0.02	0.05	0.11	0.12	0.07	0.06
Well 8 [2]	−0.44	−0.28	−0.06	0.10	0.06	0.06	−0.09
Well 9 [1]	0.01	0.05	0.10	0.08	0.05	0.08	0.06
Well 10 [2]	0.85	0.10	0.13	0.12	0.12	0.17	0.25
Well 11 [2]	0.01	0.17	0.03	0.08	0.12	0.13	0.09
Well 12 [2]	0.04	−0.03	−0.03	−0.03	−0.05	−0.04	−0.02

Note: [1] in the table represents croplands and [2] represents fallow lands.

3.2. Inter-Exchanges of Water Transport of the DDS

The computed water transport of the DDS for each fallow block from May to October over the five-year analysis is shown in Table 2. The computed items in the conceptual model included the changes in soil and groundwater storage, the evapotranspiration in the fallow areas, and the precipitation. The computed period was between May (the first irrigation event) and October (the autumn irrigation) each year. We found that the storage changes in the soil water content and the groundwater during every irrigation event presented considerably variations between each block and time. The former and latter varied from −128 mm to 2 mm and −75 mm to 165 mm, respectively. The comprehensive results together ranged from −109 mm for well 11 in 2011 to 128 mm for well 12 in 2008. Evapotranspiration from the fallow areas of each block was as high as 751 mm during the five-year ongoing data collection, and the lowest degree of evaporation, occurring in well 8 in 2011, was 216 mm. The precipitation was also different: precipitation in 2008 reached twice the annual average. The dry drainage depth was unstable, varying from 60 mm to 557 mm. The average water transport by the DDS for each block varied between 170 mm and 422 mm, demonstrating strong spatio-temporal variability.

The relations between storage change, evaporation, and rainfall and dry drainage depth are displayed in Table 2. All the computed items had positive correlations with the dry drainage depth during the various processes, especially the evapotranspiration in fallow areas, which determined the capability of the DDS in a given block or district. The rainfall and storage changes had weak correlations due to the different soil properties, GTD, soil quality, and the crop type. The synthesis of these numerous factors exerted a considerable influence on the capability of the DDS. The precipitation was discontinuous and heavy intensity rainfall was uncommon. Thus, it may not percolate to the deep soil profile or recharge the groundwater considering the high evaporation rate, so its effect on the DDS was not obvious. In general, the capability of the DDS was relative to a variety of influential factors, such as water table depth and size, the ratio of fallow to irrigated area, the crop species, and the distribution of the fallow lands in the landscape. It was difficult to regularly obtain generalized results and uniform reference criteria for the DDS, so the evaluation should depend on a typical and specific block or district with distinct management strategies.

Table 2. Dry drainage depth from the four fallow blocks (2007–2011).

Location	Year	ΔS (mm)	$\mu\Delta H$ (mm)	E_i (mm)	P (mm)	D_{di} (mm)	Average (mm)
Well 8	2007	−38	119	339	140	280	
	2008	−7	80	431	314	190	
	2009	−93	75	261	92	152	170
	2010	−95	120	319	174	170	
	2011	−101	19	216	74	60	
Well 10	2007	−59	122	473	140	396	
	2008	1	23	752	314	462	
	2009	−95	109	634	92	557	388
	2010	−30	29	433	174	258	
	2011	−55	−22	419	74	268	
Well 11	2007	−11	43	493	140	385	
	2008	−12	30	751	314	455	
	2009	−128	139	625	92	545	422
	2010	−16	−7	545	174	348	
	2011	−38	−71	563	74	380	
Well 12	2007	−28	25	327	140	184	
	2008	2	126	459	314	273	
	2009	−103	63	349	92	218	248
	2010	−5	99	340	174	260	
	2011	−117	165	330	74	304	

Note: ΔS (mm) and $\mu\Delta H$ (mm) are the storage capacity changes in the soil water and in the groundwater, respectively; E_i is the total evapotranspiration of the fallow area, block i (mm); and P is the precipitation within the calculation period (mm).

The distribution type and location of the fallow areas exerted strong influences on the dry drainage depth. The fallow area where well 8 was located is situated in the margin of a village and adjacent to a road; it had the lowest drainage depth among the four blocks. The evaporation rate was mostly constricted by the hard road surface and the varieties of buildings. Observation wells 10 and 11 were located in the center of fallow areas, and the capability of the DDS here was greater than in the other blocks. This indicated that the distribution pattern of the fallow areas was a critical aspect when designing the DDS. However, the DDS is more likely to need a larger drainage area than artificial drainage, especially for subsurface drainage. Thus, whether land is a limiting factor for agricultural applications and for the long-term economic profit of the DDS should be evaluated. The dry drainage depth was one of the main factors for designing the ratio of irrigated to fallow lands for creating a sink area that also depended on other influencing factors mentioned above.

3.3. Water and Salt Balance

The five-year observations of water and salt balance are shown in Table 3. The application of irrigation water had little difference over the five years, and reached over 1000×10^4 m^3, except in 2008, when it only reached 856×10^4 m^3. The main reason for this was that a substantial precipitation event of 314 mm occurred, which was roughly twice the multi-year average precipitation. The dry drainage water amount was obtained from the multiplication of the DDS depth and block areas, ranging from 199×10^4 m^3 (2011) to 296×10^4 m^3 (2009). The amount of water transported from irrigated croplands to fallow areas was 1258×10^4 m^3 over the five years, which was 4.3 times that which was moved by ADS. The TDS values of the irrigation and drainage water were 0.5 g·L^{-1} and 1.2 g·L^{-1}, respectively, and the groundwater TDS ranged from 1.6 g·L^{-1} to 2.43 g·L^{-1}. The salt imported with the irrigation ranged from 4366 t to 7657 t over the five years, and the sum of the imported salt was 33,013 t. The effects of the DDS and the artificial drainage were multiplied by the corresponding salinities of the groundwater and the drainage water, respectively. The corresponding salt movement by the DDS was, on average, 7.7 times that of the artificial drainage, with values of

27,472 t and 3575 t, respectively, and it was as high as 10 times in 2008 and 2010. The residual salt in the irrigated croplands varied over the five-year period according to the balance analysis. In the first three years, salt balance was well maintained. However, over the final two years, more than 30% of the imported salt was left inside the irrigated areas. Considering the 4.9% salt being desalinated from the irrigated croplands according to the five-year balance results, the DDS had an obvious function in maintaining the salt balance of the croplands. The salt balance for quantifying the effectiveness of the DDS was 27,472 t over the five-year timespan.

Table 3. Five-year water and salt balance of the observation field.

Balance Item	2007	2008	2009	2010	2011	2007–2011
Irrigation (10^4 m^3)	1320	856	1382	1151	1418	6127
Dry drainage water (10^4 m^3)	266	282	296	215	199	1258
Artificial drainage water (10^4 m^3)	73	43	69	30	78	293
Irrigation water salinity (g/L)	0.54	0.51	0.47	0.64	0.54	
Groundwater salinity (g/L)	2.30	2.43	2.24	2.18	1.60	
Drainage water salinity (g/L)	1.20	1.20	1.25	1.28	1.20	
Imported salt with irrigation (ton)	7128	4366	6495	7367	7657	33,013
Dry drainage salt (ton)	6118	6853	6630	4687	3184	27,472
Artificial drainage salt (ton)	876	516	863	384	936	3575
Residual salt in the irrigated area (ton)	134	−3003	−998	2296	3537	1966
Residual salt in the irrigated area (kg/ha)	65	−1450	−482	1109	1708	950
Residual ratio	1.9%	−68.8%	−15.4%	31.2%	46.2%	−4.9%

Figures 6 and 7 show the SSC of the soil profile between 0 and 140 cm in the fallow areas and the irrigated croplands, respectively. The SSC increased over the five years, both in the irrigated and fallow lands, and the dynamic change in salt in the fallow areas had similar trends in the 140 cm soil profiles, as all the soil layers had almost equivalent increases, except in the topsoil. The soil salinity of the fallow areas increased four times and the deep soil profile had different degrees of increase (Figure 6). Well 8 was located in the margin of a road, and it had a thick and compact soil profile, so that upward salt movement was difficult, and salt mainly accumulated in the deep soil. Well 10 was located in the middle of the fallowed region; the salt accumulated in the topsoil more obviously. However, a variation in the increasing degree of SSC occurred in the irrigated croplands; this may have been decided by the different crop types and the irrigation plans (especially autumn irrigation). Figure 7 shows the dynamic change in two different seasons' salt profiles. The irrigated cropland had slight salinity accumulations in the seeding and harvest periods. We concluded, by referring to Table 3, that the residual salt concentration in the irrigated areas was about 950 kg·ha^{-1}. The SSC in the irrigated area reached a balanced state under the drainage effects of the DDS; with other drainage methods, the salinity in the soil profile did not reach this threshold salinity for crops [17]. Table 1 also shows that, although the distribution of SSC was maintained at an appropriate degree, the salinization areas did not increase in the irrigated croplands, but the areas of moderate salinization increased, mostly in the fallow regions. Sunflower, the main economic crop species, is salt tolerant and exhibited a moderate sensitivity to salt. Residual salt in the irrigated areas gradually increased, which was partly due to the intensification of agriculture and massive land reclamation. Farmers irrigated fallow lands with the same scheme as used with irrigated croplands for salt leaching a few years after land reclamation, even though no crops were planted. The quantity of the fallow areas declined steeply with the government management strategy encouraging farmer activities in the reclamation of low-lying natural ponds or natural patches. With a lack of effective and timely management irrigation and drainage measures, more sediment is deposited in the drainage canal bed. Without an adequate outlet to remove leaching salts from irrigation districts, the excess water and salt remain and accumulate in the soil profile and the groundwater.

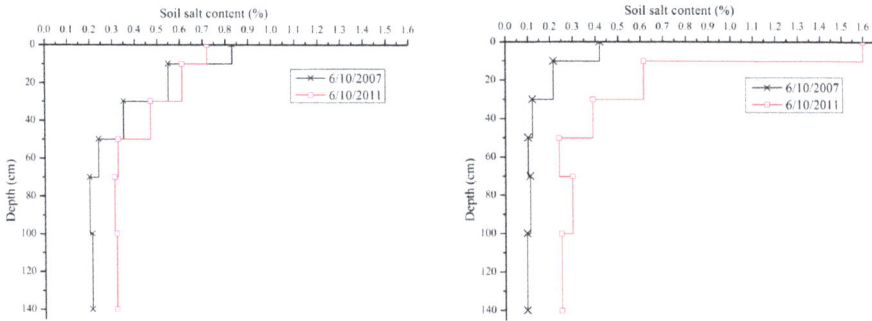

Figure 6. Soil salt content (SSC) profile in the fallowed areas (**Left**: Well 8; **Right**: Well 10).

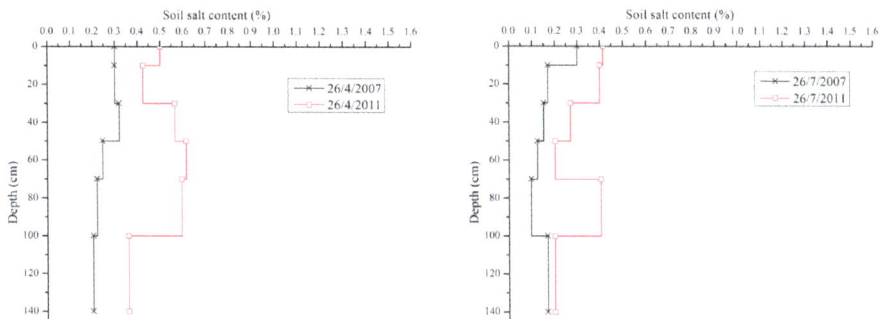

Figure 7. Soil salt content (SSC) profile in the irrigated cropland, well 9 (**Left**: 26 April; **Right**: 26 July).

3.4. Effectiveness and Sustainability of a Dry Drainage System

One of the most important aspects for operating a DDS is the frequent and intense transverse flow of water from irrigated croplands to fallow regions [22]. The main concern with the management of fallow areas in a DDS is how to increase, or at least maintain, the evaporation rate from the bare soil surface and how to improve the efficiency of salt removal to accommodate large-scale intensive agriculture in the future [11]. A DDS would satisfy the water and salt balance demands, which also depend on the acreage of the irrigation district, the hydrogeological and climatic conditions, the ratio of fallow to irrigated croplands, and their distribution pattern [9,11,12,15,21,30]. The ratio of irrigated to fallow areas is a critical aspect for public policy makers when designing a DDS. In previous studies, it varied from 1 to 15, with many variations in influencing factors [9,11,18]. In this study, the ratio is about four. We deem that DDS is effective if the evaporation rates in the fallow areas exceed the lateral migration of water from the croplands. The average volume of transported water from the four blocks was 307 mm during the five-year observation, which was less than the evaporation of 453 mm in the fallow areas, which means that the current ratio could be higher.

To improve the capability of a DDS, a shallow GTD is often adopted. In the HID, natural patches maintain the DDS function, with its distribution regions usually being 30 cm lower than the adjacent croplands [21]. A study in San Joaquin Valley, California, U.S.A., proposed a 30-cm-deep excavation of a depression for achieving only the necessary evaporation flux from the groundwater in fallow areas [11]. However, the appropriate GTD should balance the ecological water demand of crops and vegetation, bare soil evaporation, and salt accumulation in the soil profile. This means that an optimum GTD is critical for leaching practices [9,12] and for agro-ecosystem sustainability [17]. An optimum depth of 1.4–2 m for tamarisk in natural patches obtained a desalination rate of 45–100% compared with the initial salinity. However, 26–48% of bare soil evaporation was reduced [21,22]. Therefore, a suitable GTD helps balance crop growth, regional ecology, and the DDS function.

The fallow areas not only act as an evaporation sink, but also as a salt repository. Water and soluble salts transport upward through capillary action that causes salt capping on the topsoil, deterioration of soil hydraulic properties, and ecological degradation in natural patches [9,21,31,32]. The driving force for capillary rise that is caused by a gradient of vapor pressure near the soil surface decreases since the soil profile is plugged by imported salts, which significantly impacts evaporation in the fallow areas [32,33]. As the salts continuously accumulated in the fallow areas (Table 1), the effectiveness of DDS may be questionable. Therefore, suitable management measures, such as excavation, tillage, or others, are required for maintaining a DDS's capability. This needs additional costs (i.e., earthwork and machinery expenses) [11]. Recent studies showed that planting halophytes and salt-tolerant crops help mitigate the adverse effects of salt accumulation on soil permeability and porosity in fallow soil [34]. In the HID, natural patches distributed around the croplands, which grow varieties of halophytes, also promote the hydraulic gradient through root water uptake [19,21–23].

To summarize, we propose the following for the effective and sustainable use of a DDS. Firstly, intensive irrigation in accordance with leaching requirements (LR), about 0.2 in the local area, and sufficient fallow with an optimal distribution pattern for the DDS, are required to achieve a uniform distribution of salt accumulation. Secondly, a relatively shallow GTD must be carefully selected that satisfies both the LR and the balance as a valuable resource for plant root uptake of water in irrigation croplands with high evaporation rates at fallow [9,11,34]. Thirdly, suitable management measures, such as tillage, excavation, and cultivation with halophytes, are required to sustain the operation of the DDS.

4. Conclusions

The DDS is a potential approach for controlling soil salinity induced by irrigation, and can contribute to attaining and sustaining the salt balance in irrigated lands with insufficient ADS. The GTD in the fallow areas quickly responded to the lateral recharge from the surrounding croplands. The GEC of the fallow areas increased from 5 mS·cm^{-1} to 15 mS·cm^{-1}, the excess water moving to the fallow lands was roughly four times greater than that moved by an ADS, and it contained 7.7 times more salts. The DDS played a more important role compared to ADS in draining excess water and salt in terms of many impact factors, including environmental and economic factors, and long-term interests. However, a slight salt accumulation occurred in the irrigated croplands and the SSC greatly increased in the fallow land. It is necessary to pay attention to the effectiveness and sustainability of the DDS in the long-term operation, and management practices must be applied to fallow areas, including groundwater table management and land use planning. From an aggregated perspective, more large-scale experiments, observations, and simulations should be performed on the DDS, including adequate demonstrations of both theoretical and practical concepts in multi-scale field tests in future research. The effectiveness and sustainability of DDSs need to be substantially evaluated by a co-operative effort with different expert disciplines in the fields of hydrology, agronomy, physiology, soil, and genomic science before engineering applications are introduced. Then, this alternative method for controlling soil salinity will be economical and more practical.

Author Contributions: The research article presented here was carried out in collaboration with several authors. C.W. and J.W. conceived the article idea and designed this study. C.W. and W.Z. performed the field experiment. C.W. and J.W. analyzed the data and wrote the first draft of the manuscript. W.Z. and Y.Z. modified and improved the manuscript. J.W., J.H., W.Z., and Y.Z. made significant suggestions for the methodology, data analysis, and manuscript writing.

Acknowledgments: This research was jointly supported by the 13th Five-year National Key Research and Development Program of the Chinese Ministry of Science and Technology (Grant Nos. 2017YFC0403304 and 2016YFC0501304), and the National Science Foundation of China (Grant Nos.51790532, 51709175, and 51790533). Authors appreciate Liya Zhao's help for performing the field experiment.

Conflicts of Interest: The authors declare no conflict of interest.

Abbreviations

EC   electrical conductivity ($mS \cdot cm^{-1}$)
GEC   groundwater EC ($mS \cdot cm^{-1}$)
TDS   total dissolved solids ($g \cdot L^{-1}$)
P   precipitation (mm)
DDS   dry drainage system
YES   Yonglian Experimental Station
HID   Hetao Irrigation District
ADS   artificial drainage system
FIP   flowmeter-installed position
GTD   groundwater table depth
SSC   soil salt content
LR   leaching requirement

References

1. Jacobsen, T.; Adams, R.M. Salt and silt in ancient Mesopotamian agriculture. *Science* **1958**, *128*, 1251–1258. [CrossRef] [PubMed]
2. Wichelns, D.; Qadir, M. Achieving sustainable irrigation requires effective management of salts, soil salinity, and shallow groundwater. *Agric. Water Manag.* **2015**, *157*, 31–38. [CrossRef]
3. Liu, Y.; Tian, F.; Hu, H.; Sivapalan, M. Socio-hydrologic perspectives of the co-evolution of humans and water in the Tarim River basin, Western China: The Taiji–Tire model. *Hydrol. Earth Syst. Sci.* **2014**, *18*, 1289–1303. [CrossRef]
4. van Schilfgaarde, J. Irrigation—A blessing or a curse. *Agric. Water Manag.* **1994**, *25*, 203–219. [CrossRef]
5. Rengasamy, P. World salinization with emphasis on Australia. *J. Exp. Bot.* **2006**, *57*, 1017–1023. [CrossRef] [PubMed]
6. Vries, F.; Acquay, H.; Molden, D.; Scherr, S.; Valentin, C.; Cofie, O. *Integrated Land and Water Management for Food and Environmental Security*; IWMI: Colombo, Sri Lanka, 2003.
7. Qadir, M.; Quillérou, E.; Nangia, V.; Murtaza, G.; Singh, M.; Thomas, R.J.; Drechsel, P.; Noble, A.D. Economics of salt-induced land degradation and restoration. *Nat. Resour. Forum* **2014**, *38*, 282–295. [CrossRef]
8. FAO. *The State of the World's Land and Water Resources for Food and Agriculture (SOLAW)—Managing Systems at Risk*; Food and Agricultural Organization of the United Nations and Earth Scan: Abingdon, UK, 2011.
9. Konukcu, F.; Gowing, J.W.; Rose, D.A. Dry drainage: A sustainable solution to waterlogging and salinity problems in irrigation areas? *Agric. Water Manag.* **2006**, *83*, 1–12. [CrossRef]
10. Nadeem, A.M. Computer Simulation of Salinity Control by Means of an Evaporative Sink. Ph.D. Thesis, University of Newcastle upon Tyne, Newcastle upon Tyne, UK, 1996.
11. Khouri, N. Potential of dry drainage for controlling soil salinity. *Can. J. Civ. Eng.* **1998**, *25*, 195–205. [CrossRef]
12. Konukcu, F. Potential of dry drainage as a sustainable solution to waterlogging and salinisation. In *Biosaline Agriculture and Salinity Tolerance in Plants*; Öztürk, M., Waisel, Y., Khan, M.A., Görk, G., Eds.; Birkhäuser Basel: Basel, Switzerland, 2006; pp. 129–135.
13. Thayalakumaran, T.; Bethune, M.G.; McMahon, T.A. Achieving a salt balance—Should it be a management objective? *Agric. Water Manag.* **2007**, *92*, 1–12. [CrossRef]
14. Datta, K.K.; Jong, C.D. Adverse effect of waterlogging and soil salinity on crop and land productivity in northwest region of Haryana, India. *Agric. Water Manag.* **2002**, *57*, 223–238. [CrossRef]
15. Wu, J.; Zhao, L.; Huang, J.; Yang, J.; Vincent, B.; Bouarfa, S.; Vidal, A. On the effectiveness of dry drainage in soil salinity control. *Sci. China Ser. E Technol. Sci.* **2009**, *52*, 3328–3334. [CrossRef]
16. Wichelns, D.; Oster, J.D. Sustainable irrigation is necessary and achievable, but direct costs and environmental impacts can be substantial. *Agric. Water Manag.* **2006**, *86*, 114–127. [CrossRef]
17. Ayars, J.E.; Shouse, P.; Lesch, S.M. In situ use of groundwater by alfalfa. *Agric. Water Manag.* **2009**, *96*, 1579–1586. [CrossRef]
18. Gowing, J.W.; Wyseure, G.C.L. Dry-drainage a sustainable and cost-effective solution to waterlogging and salinisation. In Proceedings of the 5th International Drainage Workshop, Vol. 3, ICID-CIID, Lahore Pakistan, 8–15 February 1992; pp. 6–26.

19. Yu, B.; Jiang, L.; Shang, S. Dry drainage effect of Hetao irrigation district based on remote sensing evapotranspiration. *Trans. CSAE* **2016**, *32*, 1–8.

20. Lei, Z.D.; Shang, S.H.; Yang, S.X.; Qu, J.L.; He, C.D.; Zhu, W.D. Preliminary analysis on the dry drainage effect of low-lying lands in Yerqiang Oasis in Xinjiang. *J. Irrig. Drain.* **1998**, *17*, 1–4.

21. Ren, D.; Xu, X.; Ramos, T.B.; Huang, Q.; Huo, Z.; Huang, G. Modeling and assessing the function and sustainability of natural patches in salt-affected agro-ecosystems: Application to tamarisk (*Tamarix chinensis* Lour.) in Hetao, upper Yellow River basin. *J. Hydrol.* **2017**, *552*, 490–504. [CrossRef]

22. Ren, D.; Xu, X.; Engel, B.; Huang, G. Growth responses of crops and natural vegetation to irrigation and water table changes in an agro-ecosystem of Hetao, upper Yellow River basin: Scenario analysis on maize, sunflower, watermelon and tamarisk. *Agric. Water Manag.* **2018**, *199*, 93–104. [CrossRef]

23. Yue, W.; Yang, J.; Tong, J.; Gao, H. Transfer and balance of water and salt in irrigation district of arid region. *J. Hydraul. Eng.* **2008**, *36*, 623–626.

24. Wu, J.; Vincent, B.; Yang, J.; Bouarfa, S.; Vidal, A. Remote Sensing Monitoring of Changes in Soil Salinity: A Case Study in Inner Mongolia, China. *Sensors* **2008**, *8*, 7035–7049. [CrossRef]

25. Hong, M.; Zeng, W.; Ma, T.; Lei, G.; Zha, Y.; Fang, Y.; Wu, J.; Huang, J. Determination of growth stage-specific crop coefficients (Kc) of sunflowers (*Helianthus annuus* L.) under salt stress. *Water* **2017**, *9*, 215. [CrossRef]

26. Zeng, W.; Wu, J.; Hoffmann, M.P.; Xu, C.; Ma, T.; Huang, J. Testing the APSIM sunflower model on saline soils of Inner Mongolia, China. *Field Crops Res.* **2016**, *192*, 42–54. [CrossRef]

27. Burkhalter, J.P.; Gates, T.K. Agroecological impacts from salinization and waterlogging in an irrigated River Valley. *J. Irrig. Drain. Eng.* **2005**, *131*, 197–209. [CrossRef]

28. Yue, W.F. Study on the Mechanism of Consumption in Yichang Irrigation Sub-District of Irrigation District of Inner Mongolia along the Yellow River. Master's Thesis, Wuhan University, Wuhan, China, 2004.

29. Wang, L.P.; Chen, Y.X.; Zeng, G.F. *Irrigation, Drainage and Salinity Control in Hetao Irrigation District, Inner Mongolia*; China Waterpower Press: Beijing, China, 1993; pp. 52–53.

30. Ren, D.; Xu, X.; Huang, Q.; Huo, Z.; Xiong, Y.; Huang, G. Analyzing the role of shallow groundwater systems in the water use of different land-use types in arid irrigated regions. *Water* **2018**, *10*, 634. [CrossRef]

31. Letey, J. Soil salinity poses challenges for sustainable agriculture and wildlife. *Calif. Agric.* **2000**, *54*, 43–48. [CrossRef]

32. Shi, W.; Shen, B.; Wang, Z. Water and salt transport in sand-layered soil underevaporation with the shallow underground water table. *Trans. CSAE* **2005**, *9*, 23–26.

33. Shimojima, E.; Yoshioka, R.; Tamagawa, I. Salinization owing to evaporation from bare-soil surfaces and its influences on the evaporation. *J. Hydrol.* **1996**, *178*, 109–136. [CrossRef]

34. Singh, A. Soil salinization and waterlogging: A threat to environment and agricultural sustainability. *Ecol. Indic.* **2015**, *57*, 128–130. [CrossRef]

Article

Evapotranspiration of the Brazilian Pampa Biome: Seasonality and Influential Factors

Gisele Cristina Rubert [1,*]**, Débora Regina Roberti** [1]**, Luis Santos Pereira** [2]**,**
Fernando L. F. Quadros [3]**, Haroldo Fraga de Campos Velho** [4] **and Osvaldo Luiz Leal de Moraes** [5]

[1] Departamento de Física, Universidade Federal de Santa Maria (UFSM), Santa Maria 97105-900 RS, Brazil; debora@ufsm.br
[2] Centro de Investigação em Agronomia, Alimentos, Ambiente e Paisagem (LEAF), Instituto Superior de Agronomia, Universidade de Lisboa, Tapada da Ajuda 1349-017 Lisboa, Portugal; luis.santospereira@gmail.com
[3] Departamento de Zootecnia, Universidade Federal de Santa Maria (UFSM), Santa Maria 97105-900 RS, Brazil; flfquadros@yahoo.com.br
[4] Laboratório de Computação Aplicada, Instituto Nacional de Pesquisas Espaciais, São José dos Campos 12227-010 SP, Brazil; haroldo.camposvelho@inpe.br
[5] Centro Nacional de Monitoramento e Alertas de Desastres Naturais, São José dos Campos 12227-010 SP, Brazil; osvaldo.moraes@gmail.com
* Correspondence: girubert@gmail.com; Tel.: +55-55-98124-0290

Received: 8 November 2018; Accepted: 11 December 2018; Published: 15 December 2018

Abstract: Experimentally characterizing evapotranspiration (ET) in different biomes around the world is an issue of interest for different areas of science. ET in natural areas of the Brazilian Pampa biome has still not been assessed. In this study, the actual ET (ET$_{act}$) obtained from eddy covariance measurements over two sites of the Pampa biome was analyzed. The objective was to evaluate the energy partition and seasonal variability of the actual ET of the Pampa biome. Results showed that the latent heat flux was the dominant component in available energy in both the autumn–winter (AW) and spring–summer (SS) periods. Evapotranspiration of the Pampa biome showed strong seasonality, with highest ET rates in the SS period. During the study period, approximately 65% of the net radiation was used for the evapotranspiration process in the Pampa biome. The annual mean ET rate was 2.45 mm d^{-1}. ET did not show to vary significantly between sites, with daily values very similar in both sites. The water availability in the Pampa biome was not a limiting factor for ET, which resulted in a small difference between the reference ET and the actual ET. These results are helpful in achieving a better understanding of the temporal pattern of ET in relation to the landscape of the Pampa biome and its meteorological, soil, and vegetation characteristics.

Keywords: actual evapotranspiration; Pampa biome; eddy covariance; evaporative fraction; hysteresis loops

---

1. Introduction

Many experimental studies have been carried out to quantify the evapotranspiration (ET) of different ecosystems and biomes around the world using the eddy covariance (EC) methodology, which is main methodology employed to estimate actual ET from site micrometeorological measurements [1,2]. However, the Pampa biome, characteristic of southern Brazil and part of Argentina and Uruguay, is still not well characterized using that methodology or the quantification of ET and its relationship with environmental variables. In southern Brazil, the Pampa biome presents mostly grassland vegetation interspersed with gallery forests [3]. It is a complex biome, which has different vegetation, among which the most representative is fields dominated by grasses. Related studies are lacking and the only recent one refers to characterizing the actual ET of Tifton 85 Bermudagrass as affected by the frequency of cuttings [4].

Although the flora of the Pampa biome totals more than 3000 species, it is dominated by about 450 species of forage grasses and more than 150 species of legumes. The development of this flora is due to the different effects associated with latitude, altitude, and soil fertility. Thus, the Pampa biome presents unique characteristics in terms of vegetation/grass cover [5]. This plant diversity among the species characteristic of the biome's vegetation is what determines the vegetation growth capacity in the seasons of the year, thus defining the balance of the annual production of forage.

Human activities have converted or degraded many areas of this biome [6]. The main use of this region is cattle raising, with a variation of the animal load between the seasons [7]. As a biome that faces fragility of soil, flora, and fauna, sustainable land use in the Brazilian Pampa is only possible if economic activities are adequately conditioned by the soil and environment capacities and adaptations of its plant and animal communities, as well as the related dynamic conditions throughout the biome. It is therefore essential to improve knowledge of vegetation changes due to anthropogenic and/or climatic causes.

In this study, estimates of actual evapotranspiration (ET_{act}) of natural vegetation of the Pampa biome using the eddy covariance methodology [1,2] were analyzed for two sites in southern Brazil—Santa Maria (SMA) and Pedras Altas (PAS)—located about 300 km from each other. Both sites are within the Pampa and have quite similar vegetation and climate. Their selection owed mainly to differences in soil hydraulic characteristics, with SMA having deep soil with a high water holding capacity and PAS having shallow sandy and stony soil with a small water holding capacity. These differences allowed us to capture the influences of soil characteristics on evapotranspiration fluxes and, thus, on the behavior of actual ET throughout the year.

Only a few eddy covariance studies exist for comparison of evapotranspiration in natural pastures. In Arizona, United States, Krishnan et al. [8] obtained daily values varying from 2.8 to 3.6 mm d^{-1}. Hu et al. [9] showed annual variations of ET of up to 4 mm d^{-1} in a pasture in Inner Mongolia. Trepekli et al. [10], for a coastal grassland in the Mediterranean climate of Greece, reported maximum ET of 8.2 mm d^{-1} in the period of the highest air temperature (summer), while ET values were negligible in winter. Rajan et al. [11] reported that ET was strongly affected by on-site water availability in a managed pasture in the Southern Great Plains, Texas, United States.

In Brazil, evapotranspiration in some different ecosystems has been characterized using eddy covariance data, for example, in the Amazon rainforest [12–14], in the Pantanal biome [15], and in the Cerrado biome [16,17]. However, these ecosystems are in the tropical region, where the seasonality is dominated by precipitation rates. Differently, the Pampa biome is in the subtropical climate, where the seasonality is dominated by solar radiation and presents regular precipitation rates. Therefore, in this study, we intend to answer the following scientific questions about ET in the Pampa biome: (i) How is the partition of energy in the surface? (ii) Which are the physical processes determining the evaporative fraction? (iii) What is the seasonal variation of ET? (iv) What is the relation between actual and reference ET? (v) What is the form of the hysteresis between ET and meteorological variables on a daily timescale?

2. Materials and Methods

The study period was from 1 September 2014 to 1 September 2016 for two sites in the Pampa biome in the State of Rio Grande do Sul (RS), Brazil: Santa Maria (SMA) and Pedras Altas (PAS). This period was divided into subperiods that comprised autumn–winter (AW) and spring–summer (SS). The AW period refers to the months of April–September and SS to the months of October–March.

2.1. Site Description

The experimental site of SMA—29°43′27.502″ S; 53°45′36.097″ W; 88-m elevation—is located in the experimental area of the Federal University of Santa Maria, covering 24 ha of natural vegetation characteristic of the Pampa biome, near the city of Santa Maria (Figure 1). The experimental area is part of the International Long Term Ecological Research (ILTER) network. The vegetation in the study area, used as pasture for beef cattle, mainly consists of native grasses with a predominance of *Andropogon lateralis*, *Axonopus affinis*, *Paspalum notatum*, and *Aristida laevis* [18]. This composition is uniformly distributed in the study area [19]. Some studies have already been carried out in this experimental site with objectives focused on the study of the morphogenesis of native species [20,21]. The regional soil type is locally known as a "Planossolo Háplico Eutrofico" according to the exploratory map of soils of RS state [22]. These soils have high fertility and high water retention capacity. The physical properties of the soil in SMA are described in Table 1, representing a deep silt–clay–loam soil. According to the classification of Köppen [23], the climate belongs to the Cfa group, defined as temperate humid with a hot summer.

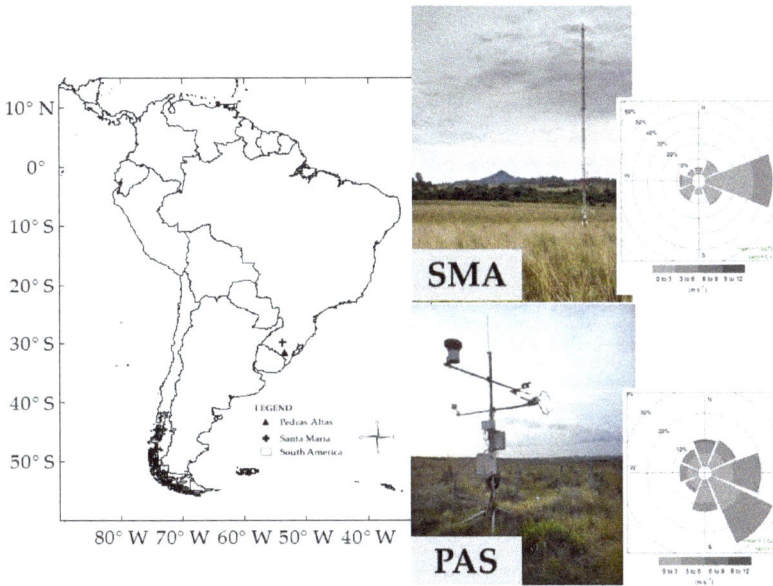

Figure 1. Location of study areas in the Pampa biome—Santa Maria (SMA) and Pedras Altas (PAS)—with the respective wind rose for the study period.

The second site, PAS—31°43.556′ S; 53°32.036′ W; 395-m elevation—is located on a private farm near the city of Pedras Altas (Figure 1). As with the Santa Maria site, there is no evidence that the area was used for another type of management than livestock. Both sites are situated within similar vegetation physiognomies in the Pampa, with a great diversity of grass species [24]. The phytophysiognomy of the PAS has double extract with predominance of creeping species, mainly stoloniferous and rhizomatous species, and also includes *Axonopus affinis*, *Paspalum notatun*, *Aeristida laevis*, and *Iriantus angustifolium*. The soil consists of Neosols and Cambisols [22], with rocky outcrops. The soil textural characteristics described in Table 1 correspond to a shallow sandy-loam soil. The climate also belongs to the Cfa classification.

The footprint of the EC was calculated using the software EddyPro® according to the methodology proposed by Kljun et al. [25]. The fetch analysis indicated that about 90% of the flux originated within a circle with a radius of 115 m for SMA and 90 m for PAS. At both sites, there were no obstacles within the footprint area. Wind in the SMA was predominantly eastern (46%), with its maximum intensity reaching 9 m s^{-1}. The wind direction at the PAS was southeast (approximately 27%), with maximum intensity for the period reaching 11 m s^{-1}.

Table 1. Physical properties of the soil at the experimental sites in the SMA and PAS sites in southern Brazil.

Depth * (cm)	Sand (%)	Clay (%)	Silt (%)	Field Capacity θ_{FC} ($m^3\ m^{-3}$)	Permanent Wilting Point θ_{WP} ($m^3\ m^{-3}$)	Macroporosity ($m^3\ m^{-3}$)	Microporosity ($m^3\ m^{-3}$)
				Santa Maria			
05	47.12	16.90	35.97	0.34	0.12	0.10	0.36
10	42.85	20.09	37.05	0.31	0.11	0.12	0.33
25	40.03	22.61	37.36	0.24	0.12	0.13	0.35
45	34.36	32.70	32.94	0.35	0.13	0.13	0.37
75	26.47	43.71	29.82	0.41	0.21	0.12	0.42
100	28.81	40.06	31.13	0.41	0.16	0.08	0.42
110	17.45	34.28	48.27	0.40	0.26	0.05	0.40
125	17.74	27.95	54.31	0.37	0.25	0.08	0.41
145	14.94	30.65	54.41	0.42	0.17	0.05	0.43
				Pedras Altas			
05	59.30	0.81	39.89	0.31	0.03	0.12	0.32
15	58.11	0.90	40.98	0.26	0.03	0.13	0.28
30	56.33	1.01	42.66	0.23	0.04	0.16	0.24
50	54.35	1.18	44.45	0.23	0.05	0.16	0.24

* Soil depths refer to the horizons not affected by saturation water.

2.2. Meteorological and Flux Measurements

The experimental data for both sites were obtained with flux towers. In Santa Maria, the sensor set included a 3D sonic anemometer (Wind Master Pro; Gill Instruments, Hampshire, UK), measuring wind and air temperature components, and a gas analyzer (LI7500, LI-COR Inc., Lincoln, NE, USA), measuring the H_2O/CO_2 concentration at 3-m height sampled at a 10-Hz frequency from 1 September 2014 to 15 June 2016. After this period, the gas analyzer and the anemometer were replaced by the sensor Integrated CO_2 and H_2O Open-Path Gas Analyzer and a 3D Sonic Anemometer (IRGASON, Campbell Scientific Inc., Logan, UT, USA). At the Pedras Altas site, the same variables were obtained using a sonic anemometer (CSAT3, Campbell Scientific Inc., Logan, UT, USA) and an open path infrared gas analyzer (LI7500, LI-COR Inc., Lincoln, NE, USA), set at the height of 2.5 m for the entire period.

Atmospheric variables were measured at both sites with the following sensors placed at 3-m height: air temperature and relative humidity (RH) with a thermo-hygrometer (HMP155, Vaisala, Finland), and precipitation with a rainfall sensor (TR525USW, Texas Electronics, Dallas, TX, USA). At the Santa Maria site, net radiation and short-wave incident radiation sensors (CNR4, Kipp & Zonen, Delft, The Netherlands) were used. At Pedras Altas, measurements of short-wave incident radiation (Li 200S Pyranometer—LI-COR, Lincoln, NE, USA) and net radiation (CNR2—Campbell Scientific Inc., Logan, UT, USA) were performed. At both sites, soil heat flux was measured with soil heat plates (HFP01, Hukseflux Thermal Sensors B.V., Delft, The Netherlands) placed at 0.10-m depth and soil water content was measured using water content reflectometers (CS 616, Campbell Scientific Inc., Logan, UT, USA) at a depth of 0.10 m.

2.3. Flux Data Processing and Gap Filling

The eddy covariance method was used in the high-frequency data (10 Hz) for the determination of the sensible heat flux (H) and latent heat flux (LE) over 30-min block average using EddyPro® software version 6.1 (Li-Cor, Lincoln, NE, USA). The configurations used were double rotation and correction for the effects of density [26], and the high frequency spectral correction was based on the mathematical formulations to model the attenuations due to the instrumental configuration [27]. High and low-pass filter corrections followed the methodology of Moncrieff et al. [28] and Moncrieff et al. [29], respectively. Flux quality tests followed the Mauder and Foken [30] methodology. Angle of attack correction for wind components was used according to the Nakai and Shimoyama [31] methodology. Finally, for statistical analysis, spikes removal followed the method of Vickers and Mahrt [32].

In postprocessing, LE and H data for periods that showed physically inconsistent and during precipitation events (and up to 60 min after the event) were discarded. The inconsistency referred to LE values < -40 W m^{-2} or > 650 W m^{-2} at SMA and LE < -200 W m^{-2} or > 650 W m^{-2} at PAS; H values < -60 W m^{-2} or > 300 W m^{-2} at SMA and H < -100 W m^{-2} or > 300 W m^{-2} at PAS. After data filtering, which also included malfunctioning periods, the total LE data gap was 30% at both sites and the total H data gap was 19% at SMA and 30% at PAS.

The energy balance closely followed the Foken et al. [33] methodology. With this methodology, the residual between the available energy ($R_n - G$) and the energy used for turbulent processes (H + LE) was partitioned between LE and H using the experimental Bowen ratio ($\beta =$ H/LE) for each site.

Gaps in LE and H data were filled using the method proposed by Reichstein et al. [34] with the REddyProc package. In order to complete the missing data of the meteorological variables used in the gap-fill method, data of air temperature, relative humidity, precipitation, and solar radiation collected by the automatic stations of the National Institute of Meteorology (INMET) were used. For the SMA site, the nearest INMET station is approximately 4 km from the flux tower (29.72° W, 53.72° S, 103-m elevation). The nearest INMET station to the PAS site is located in the city of Bagé, 67 km away (at the position: 31.34° W, 54.01° S, 226-m elevation).

2.4. Evapotranspiration

The actual evapotranspiration of the Pampa biome (ET_{act}) was estimated from the mean LE in W m^{-2}, measured with the EC method, thus resulting in

$$ET_{act} = \frac{LE}{L_v \rho_w}(1000)(\Delta t) \tag{1}$$

where L_v is the latent heat of vaporization (2.45×10^6 J kg^{-1}), ρ_w is the water density (998 kg m^{-3}), and Δt is the time scale used in the analysis (the same time scale of LE average). To obtain daily ET (mm d^{-1}), the Δt was 86400 s, and to obtain hourly ET (mm h^{-1}), Δt was 3600 s.

The daily reference evapotranspiration, ET_o (mm d^{-1}), was computed with the FAO-PM equation [35], defined as

$$ET_0 = \left(\frac{0.408\Delta(R_n - G) + \gamma\frac{900}{T+273}u_2(e_s - e_a)}{(\Delta + \gamma(1 + 0.34u_2))} \right) \tag{2}$$

where R_n is the net radiation flux density at the crop surface (MJ m^{-2} d^{-1}), G is the soil heat flux density (MJ m^{-2} d^{-1}), T is the air temperature at 2-m height (°C), u_2 is the wind speed at 2-m height (m s^{-1}), e_s is the saturation vapor pressure (kPa), e_a is the actual vapor pressure (kPa), Δ is the slope of the vapor pressure curve (kPa °C^{-1}), and γ is the psychometric constant (kPa °C^{-1}). The wind speed at 2-m velocity was estimated using the Allen et al. [35] methodology applied to the measured data.

3. Results and Discussion

3.1. Atmosphere and Soils

The daily variation of the meteorological variables is shown in Figure 2. The incident short-wave or global radiation, R_g, presented strong seasonality. On average, R_g in the SS period was 51% and 56% higher than in the AW period in the SMA and PAS sites, respectively. For the entire study period, the mean value for R_g was 173.8 and 182.1 W m^{-2} for SMA and PAS, respectively. Mean values of air temperature were 19.3 and 17.1 °C, which were close to the climatological normal values (18.8 °C for SMA and 17.9 °C for PAS) [36]. Comparing AW and SS periods, the air temperature presents a well-defined seasonal variation, with daily minimum close to 4 °C in the AW season and a daily maximum of 30 °C in the SS period for both sites.

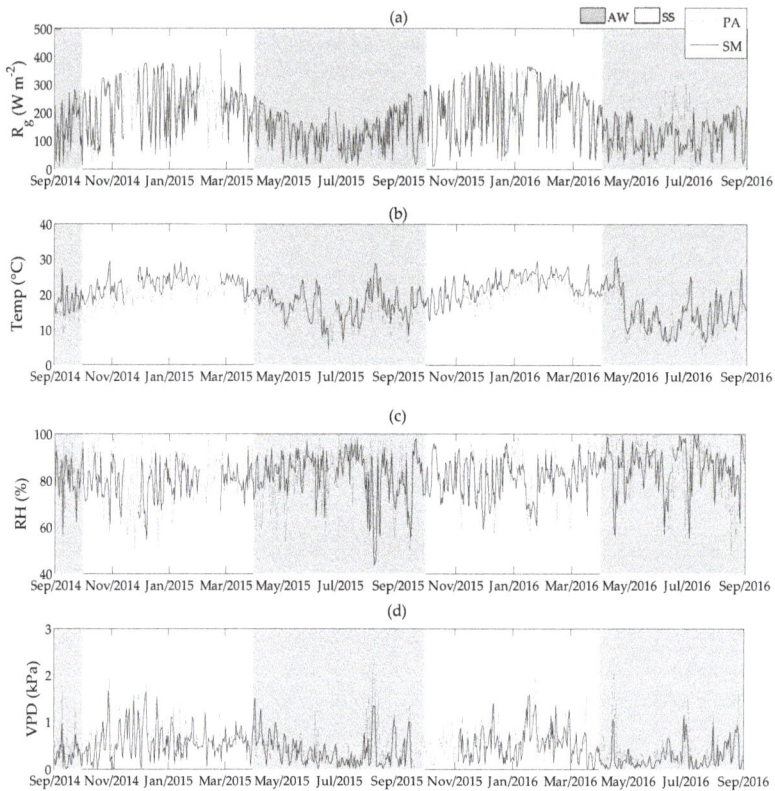

Figure 2. Daily values of meteorological variables: (**a**) Global Radiation—R_g (W m^{-2}), (**b**) Air Temperature—Temp (°C), (**c**) Relative Humidity—RH (%), and (**d**) Vapor Pressure Deficit—VPD (kPa). The hatched areas indicate the autumn–winter (AW) periods.

The RH of the air presented great variability throughout the year, without characterizing a seasonal pattern, and with daily mean values ranging from 43% to 100% in SMA and from 48% to 98% for PAS. The mean values of RH were 82% for both sites. The vapor pressure deficit (VPD), defined by the difference between the saturation and actual vapor pressure, ranged from 0 to 2.58 kPa per day. Its dynamics showed a pattern similar to that of air temperature in both sites. On average, for the entire period, the VPD value for the SMA site was greater than for PAS.

Figure 3 shows daily accumulated precipitation, Prec (mm), and soil water content, θ (m^3 m^{-3}). The precipitation was generally well distributed throughout the year and the total values were similar for each site in different years. For the 2014/2015 year (September 2014 to September 2015), the precipitation was 1922 mm at SMA and 1753 mm at PAS, and in 2015/2016, it was 2050 mm at SMA and 1788 mm at PAS. These values were greater than the climatological average for both sites, 1300 mm for PAS and 1617 mm for SMA. The high precipitation values in the study period, especially 2015/2016, were due to the El Niño phenomenon [37], which tends to increase precipitation in southern Brazil. For the AW and SS periods, the accumulated precipitation was 1085.6–1020 mm and 489–1377 mm for SMA and 1105–901 mm and 433.2–1102 mm for PAS, respectively, in 2014/2015–2015/2016. Although the highest cumulative precipitation occurred in the SS period for both sites, the maximum daily precipitation occurred in winter for PAS, 122 mm on 19 July 2015, and in spring for SMA, 126 mm on 7 October 2015.

The mean for soil water content was approximately 60% higher in the SMA site than at the PAS site because the soil at SMA is a clay loam soil and thus has a higher water holding capacity than at PAS, where the soil is a shallow sandy loam soil with rocky outcrops, thus having a small water holding capacity. On days when precipitation occurred, the mean values of θ reached 0.35 m^3 m^{-3} for both sites. The PAS site showed greater variability of θ throughout the year.

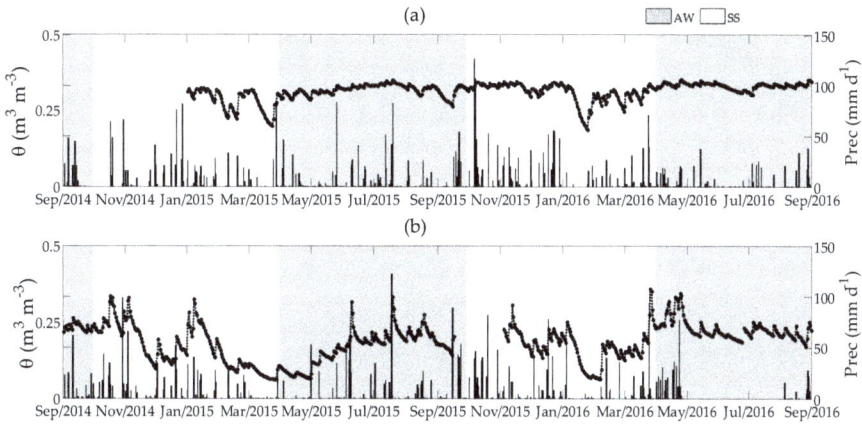

Figure 3. Daily accumulated precipitation, Prec (mm d^{-1}), and volumetric soil moisture at 0.1-m depth, θ (m^3 m^{-3}), for (**a**) SMA and (**b**) PAS. The bar plot represents accumulated daily precipitation and the lines represents θ. The hatched areas indicate the AW periods. Observations were not performed from September 2014 to January 2015 at SMA.

3.2. Energy Balance Closure

The relationship between the available energy (R_n − G) and the turbulent heat fluxes (H + LE) was used as an indicator of the accuracy of the energy fluxes, H and LE [38]. This relationship (Figure 4) was obtained with the experimental data but excluded periods when turbulent fluxes were gap filled. The slope of the linear regression was 0.75 for SMA and 0.72 for PAS, representing an underestimation of approximately 25% of the available energy by the turbulent fluxes.

Most studies in the literature have shown a nonclosure of the energy balance of around 10%–30% [39–43], which is typically related to the underestimation of the turbulent fluxes measured by the eddy covariance method. Aubinet et al. [39] reported that an energy imbalance is expected since, in the accounting of the results, not all the exchanges and processes involved are considered. According to Foken et al. [44], the phenomenon of nonclosure of the energy balance on the surface is not an eddy covariance method problem but a problem related to the heterogeneity of the terrain and its influence on turbulent exchanges. They suggest that sensible and latent heat fluxes can be corrected by using the Bowen's ratio under the assumption that the scalar similarity is fulfilled. This technique was followed in this work and the turbulent fluxes, H and LE, were corrected after partitioning the residuals of the energy balance using the Bowen's ratio.

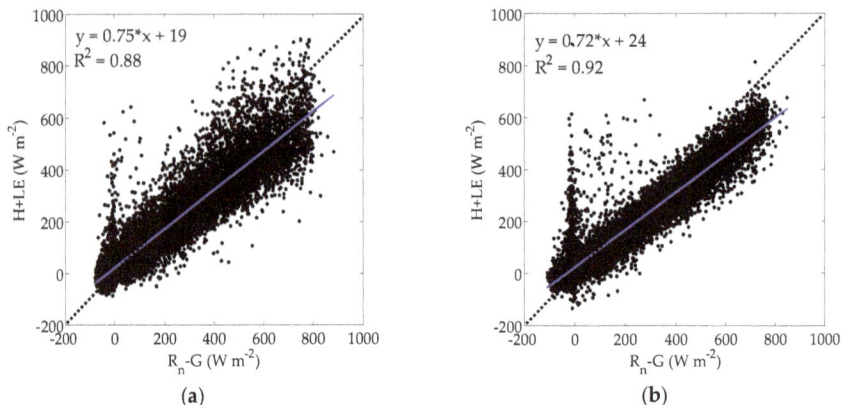

Figure 4. Scatterplots of the half-hour energy balance closure: the available energy ($R_n - G$) vs. turbulent heat fluxes (H + LE) at (**a**) SMA and (**b**) PAS. The solid blue line represents the linear fit of the data. For reference, the 1:1 line is also shown (dashed black line).

3.3. Energy and Water Availability in the Pampa Biome

The partition of the net radiation, R_n, between H (heat fraction, HF = H/R_n) and LE (evaporative fraction, EF = LE/R_n) is shown in Figure 5. Both sites presented a significant difference in the daily values of fractions HF and EF. In general, EF values were greater than HF. EF presented a more significant seasonal variation and EF decreased during the AW period, with minimums between May and August and higher values between September and October, the period of plant growth (increase in biomass production) in the Pampa [45]. Moreover, in the AW season, HF and EF presented greater daily variability, while in the SS period, that variability was small. The magnitude of the difference between HF and EF was higher during SS periods, as influenced by the water and energy availability, with increasing EF values.

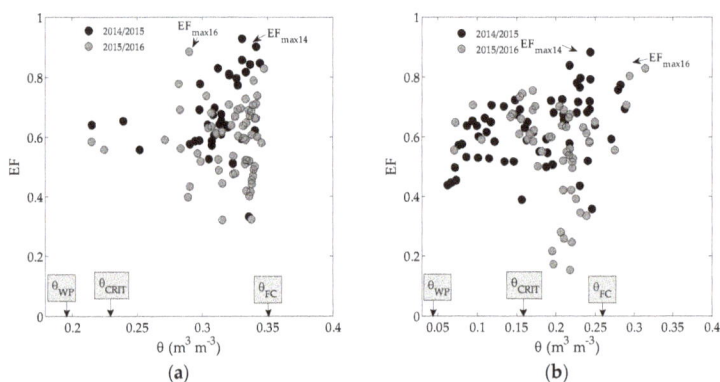

Figure 5. Six-day average of the evaporative fraction (EF = LE/R_n) and heat fraction (HF = H/R_n) for (**a**) SMA and (**b**) PAS from September 2014 to September 2016. The hatched areas refer to the AW periods.

The evaporative fraction in the year 2014/2015 was 0.69 for the PAS site and 0.66 for the SMA site. In the year 2015/2016, these values decreased by 8% and 11% in the PAS and SMA sites, respectively. The annual mean evaporative fraction for the total study period was 0.64 and 0.66 for the SMA and PAS sites, respectively (Table 2). Therefore, approximately 65% of the R_n was used for the evapotranspiration process in the Pampa biome. In pasture vegetation near the Mediterranean coast, Trepekli et al. [10] found that 67% of the available energy was transformed into LE during the growing season and 20% during the vegetation senescence period.

Table 2 presents the values of the evaporative fraction in other Brazilian biomes. Da Rocha et al. [13], in the Amazon rainforest, and Sanches et al. [15], in Pantanal, obtained average annual values for the evaporative fraction of about 25% and 30%, respectively, higher when compared to Pampa. Cabral et al. [17] found for Cerrado an annual mean EF similar to the results obtained in this study. Diferently, Giambelluca et al. [16] obtained EF values of 27% for Cerrado Denso and 39% for the Campo Cerrado lower than Pampa.

Table 2. Annual mean of evaporative fraction (EF) of various Brazilian biomes.

Biome	EF = LE/R$_n$	Reference
Pampa	0.66	This study, PAS site
Pampa	0.64	This study, SMA site
Amazon rainforest	0.86	Da Rocha et al. [13]
Pantanal	0.80	Sanches et al. [15]
Cerrado	0.68	Cabral et al. [17]
Cerrado Denso (CD)	0.48	Giambelluca et al. [16]
Campo Cerrado (CC)	0.40	Giambelluca et al. [16]

The evaporative fraction can be characterized into two main evapotranspiration regimes as proposed by Seneviratne et al. [46]: soil water limited or energy limited. The reduction of soil water content influences the evaporative fraction because when theta is low, the extraction of water by the roots faces additional resistances and the rate of water extraction results are smaller than the potential (not stressed) ET rate [4,35]. In other words, soil moisture provides a first-order control on land–atmosphere exchanges when it is limiting. Figure 6 shows the relationship between soil water content (θ) at 0.10-m depth and the evaporative fraction for our sites. The results did not show any peculiar relationship that could help identify those two ET regimes, so a specific analysis was required. From the experimentally measured field capacity (θ_{FC}) and permanent wilting point (θ_{WP}), the average values for these parameters for 0.1 m were θ_{FC} = 0.35 and 0.26 m^3 m^{-3}, respectively, for SMA and PAS, and θ_{WP} = 0.17 and 0.04 m^3 m^{-3} for SMA and PAS, respectively. Shuttleworth [47] defined the critical soil water content (θ_{CRIT}) distinguishing those two evapotranspiration regimes (water or energy limited ET) as a typical value of 50%–80% of θ_{FC}. We assumed θ_{CRIT} as 65% of θ_{FC} (a mean value between 50% and 80%), thus resulting in θ_{CRIT} values of 0.23 and 0.17 m^3 m^{-3} for SMA and PAS, respectively.

EF reached a maximum value EF$_{max}$ = 0.93 for θ = 0.33 m^3 m^{-3} in 2014/2015 and of EF$_{max}$ = 0.88 for θ = 0.29 m^3 m^{-3} in 2015/2016 at the SMA site. For the PAS site, the maximum EFs were EF$_{max}$ = 0.88 for θ = 0.24 m^3 m^{-3} in 2014/2015 and of EF$_{max}$ = 0.83 for θ = 0.31 m^3 m^{-3} in 2015/2016. These values of θ for the maximum EF were between θ_{CRIT} and θ_{FC} at both the SMA and PAS sites for 2014/2015. However, for PAS in 2015/2016, the value of θ for EF$_{max}$ exceeded the values of field capacity. At the SMA site, for both years of study, ET regimes were energy limited, corresponding to values of soil water content above the critical value, θ_{CRIT}. In other words, the evaporative fraction was independent of the soil water content and the ET process was controlled by the available energy. At the PAS site, where the soil had a smaller soil water holding capacity and the soil water content remained lower than at SMA, the ET regimes were not well defined. During the two years of analysis, a greater dependence on the soil water content was observed, but values of θ higher than θ_{CRIT} were observed, namely, for EF$_{max}$, due to the large and frequent precipitation observed. These results demonstrated that the energy partitioning response to precipitation was more important at the PAS site because its soil had a low water holding capacity.

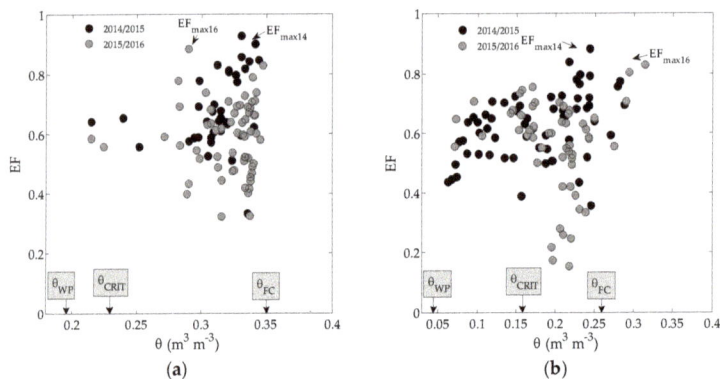

Figure 6. Evaporative fraction, EF, vs. soil water content, θ (m^3 m^{-3}), for (**a**) SMA and (**b**) PAS. Gray rectangles indicate the position of values of θ_{CRIT} (critical soil water content), θ_{FC} (experimentally measured field capacity), and θ_{WP} (permanent wilting point).

The analysis of the limiting factors, as well as the partitioning of the available energy in the turbulent fluxes, is of essential importance for the modeling of these surface processes. Bagley et al. [48] pointed out that the variables influencing the surface flux partition may change seasonally, depending on the local vegetation conditions. It could also be noticed from our results that the soil and climate behavior were influenced by seasonality, which also affected the partitioning of the turbulent flux. Gokmen et al. [49] emphasized that the overestimation of ET usually occurs in the hydrological regime where water availability is the limiting factor. Although the relationship between soil water content and evapotranspiration depends on soil type, vegetation type, and vegetation adaptation to dryness [50], the role of soil water content near the surface is significant.

3.4. Evapotranspiration Variability in the Pampa Biome

The ET_{act} averages for the entire period at the SMA and PAS sites, 2.36 ± 1.4 and 2.56 ± 1.7 mm d^{-1}, respectively, were very similar. Assuming the average of both sites, the Brazilian Pampa biome presented a mean annual ET_{act} of 2.45 mm d^{-1}. The daily ET_{act} values ranged from close to 0 to almost 7.1 mm d^{-1} at both sites, with the lowest values in the AW period and the highest values in the SS period (Figure 7), coinciding with the seasonal behavior of global radiation (Figure 2a).

The daily ET_{act} decreased dramatically on cloudy days and subsequently increased sharply. On average, the ET_{act} in PAS was around 8% greater than in SMA (Table 3). This may be influenced by the lower wind speed and lower R_g in SMA, as described in Section 2.1 [51]. The mean ET_{act} values during the AW period at the SMA and PAS sites were, respectively, 50% and 38% of the mean ET_{act} in the SS period.

Between July and September 2015, higher values for ET_{act} were observed at the SMA site. By separately analyzing the environmental variables for these days, we observed that high VPD, low relative humidity of the air, high wind intensity reaching 4.7 m s^{-1}, and a high air temperature of about 30 °C did occur (Figure 2). These atypical days of high ET_{act} values during AW were characterized by the "north wind phenomenon", corresponding to the occurrence of hot and humid wind coming from Amazonia and which intensifies near Santa Maria due to its particular topography [52]. For pasture land near the Mediterranean coast, Trepekli et al. [10] reported maximum values for ET_{act} of 8.2 mm d^{-1} in summer, the value of which is 15% greater than those found in our study. Those authors also reported that the values obtained for ET_{act} greater than 8 mm d^{-1} were the highest ET rates previously reported in comparable study areas. These values occurred due to advection flows from the sea. However, negligible values for ET_{act} were found in winter.

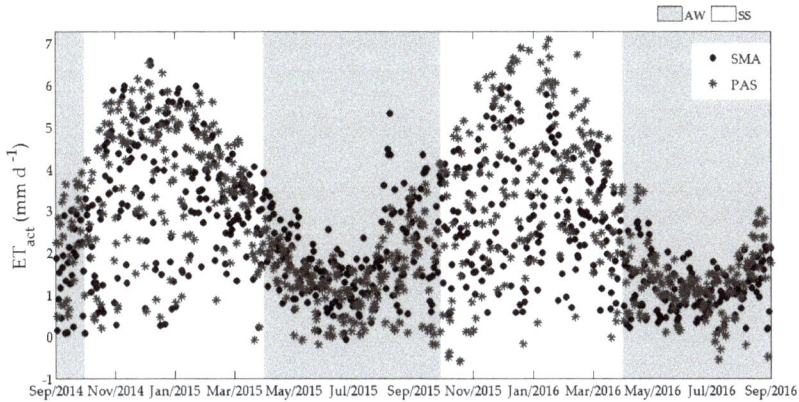

Figure 7. Seasonal variability of actual daily evapotranspiration (ET$_{act}$) at SMA and PAS sites from September 2014 to September 2016. The hatched areas refer to the AW periods.

Table 3. Daily mean of actual evapotranspiration, ET$_{act}$ (mm d^{-1}), and reference evapotranspiration, ET$_{o}$ (mm d^{-1}), for the Pampa biome sites of Santa Maria and Pedras Altas relative to various periods.

Period		Site	ET$_{act}$ (mm d^{-1})	ET$_{o}$ (mm d^{-1})
Autumn and Winter, AW		Santa Maria	1.58	1.93
		Pedras Altas	1.42	1.73
Spring and Summer, SS		Santa Maria	3.15	3.83
		Pedras Altas	3.68	3.74
Annual	2014/2015	Santa Maria	2.56	2.89
		Pedras Altas	2.60	2.72
	2015/2016	Santa Maria	2.16	2.79
		Pedras Altas	2.49	2.82
Two-year period		Santa Maria	2.36	2.84
		Pedras Altas	2.56	2.76

There was a small difference between annual ET$_{act}$ average values for the SMA and PAS sites (Table 3), approximately 2% and 13% in the years 2014/2015 and 2015/2016, respectively. These differences may be associated with the precipitation and the soil characteristics of each site. Figure 3 shows higher soil moisture for the SMA site because the soil water storage was larger there than at PAS. Annual ET$_{act}$ at the PAS site represented 54% and 51% of the annual precipitation for 2014/2015 and 2015/2016, respectively. Differently, at the SMA site, ET$_{act}$ represented 48% of the precipitation for 2014/2015, while for 2015/2016, this percentage was reduced to 39%. On average, the actual ET$_{act}$ represented 44% of the observed precipitation for SMA and 53% for PAS in the two years of study. Therefore, the average annual ET$_{act}$ in the Pampa biome (898 mm year^{-1}) represented 48% of the accumulated precipitation (1878 mm year^{-1}). Krishnan et al. [8] reported a study in a semiarid pasture in North America where precipitation explained more than 80% of the variance in annual evapotranspiration. Paoloni et al. [53] found on the Argentina Pampa an accumulated annual ET$_{act}$ of 1220 mm year^{-1}, around 35% greater than in the Brazilian Pampa. For other Brazilians biomes, Goulart et al. [54] reported a cumulative ET of 1337.5 mm year^{-1} for the Pantanal, Shuttleworth [55] obtained annual ET of 1393 mm year^{-1} for the Amazon forest, and Almeida [56] reported 1350 mm year^{-1} for the Atlantic Forest biome. Da Rocha et al. [13], studying the Amazonia region, reported daily values of 3.96 mm d^{-1} in the dry season and 3.18 mm d^{-1} in the wet season. Therefore, the Brazilian Pampa biome presents a smaller annual ET than most Brazilian biomes (Table 2) but with greater seasonality. Furthermore, Giambelluca et al. [16] reported small seasonal variability for Cerrado Denso

(CD) and Campo Cerrado (CC) vegetation in the central region of Brazil, with a variation of around 1–1.5 mm d^{-1}. The annual mean was 2.25 and 1.91 mm d^{-1} for ET in CD and CC, respectively. Thus, the ET$_{act}$ in the Pampa biome was greater and had higher seasonality than both Cerrado biomes in central Brazil.

Daily mean ET$_{act}$ and ET$_o$ exhibited a similar seasonal pattern for both sites in the Pampa biome in southern Brazil. Different from ET$_{act}$, ET$_o$ was greater (around 3%) in SMA than at PAS (Table 3). On average, ET$_{act}$ represented 83% of the ET$_o$ for the SMA site and about 92% for PAS (Table 3). These high values of the ratio of ET$_{act}$/ET$_o$ indicate that the Pampa has high water availability, since natural grass has an ET rate close to the ideal grass crop assumed as a reference crop. Daily maximum ET$_o$ values were 8 and 7.2 mm d^{-1} for the SMA and PAS sites, respectively. The greatest differences between ET$_{act}$ and ET$_o$ occurred by March 2016 in SMA and by early 2015 in PAS, which were observed in transitional periods between climate seasons, likely due to greater thermal amplitude and higher net radiation that occurred during those periods.

ET$_{act}$ was significantly correlated with ET$_o$ at both sites analyzed (Figure 8). The Pearson correlation coefficient (r) relating ET$_{act}$ and ET$_o$ at the SMA and PAS sites was 0.92 and 0.95, respectively. These results suggest that ET$_o$ was a good indicator of ET$_{act}$ in the Pampa biome because soil water availability did not restrict ET$_{act}$, the values of which were close to ET$_o$. This indicates that the seasonal variability of ET$_{act}$ was determined by atmospheric forcing, mainly, R$_n$ and VPD. The lowest values of ET$_{act}$ and ET$_o$ were observed during the AW period, whereas the opposite was observed in the SS period, then associated with a greater variability of differences between ET$_{act}$ and ET$_o$. The higher correlation between ET$_{act}$ and ET$_o$ was observed at the PAS site, which is coherent with the previous analysis about the energy and water availability in Section 3.4.

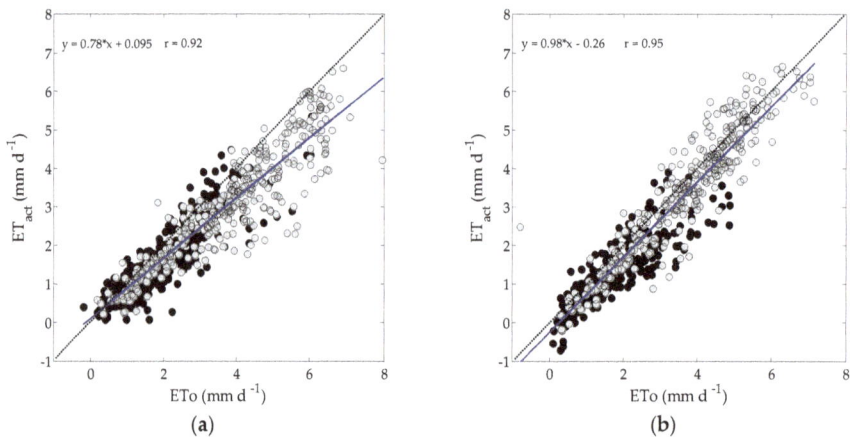

Figure 8. Scatterplots of daily ET$_{act}$ and ET$_o$ for (**a**) SMA and (**b**) PAS. The black dots are for the AW period and the gray dots are for the SS period. The solid blue line represents the linear fit of all data. For reference, the 1:1 line is also shown (dashed black line).

3.5. Hysteretic Relations between ET$_{act}$ and Meteorological Variables

The concept of hysteresis is related to the ability of a system to absorb and recover from disturbances. Hysteresis can therefore be defined as the dependence of a variable response not only on the value of a variable driving but also on its past history [57]. Hysteretic relations were also found between the diurnal variation in evapotranspiration and vapor pressure deficit.

Here, hysteresis loops clockwise in the relationships between ET$_{act}$ and vapor pressure deficit (VPD) and air temperature (Temp) were observed in the average daily cycles of 2014/2015 and 2015/2016 (Figure 9), respectively. No hysteretic loops were observed for both sites in ET$_{act}$

relations with the net radiation; instead, a better ET_{act} correlation with this variable was observed. Zheng et al. [58] also reported hysteresis loops on air temperature and VPD relationships with ET_{act}. Likewise, they did not observe an ET hysteresis response to the net radiation variability.

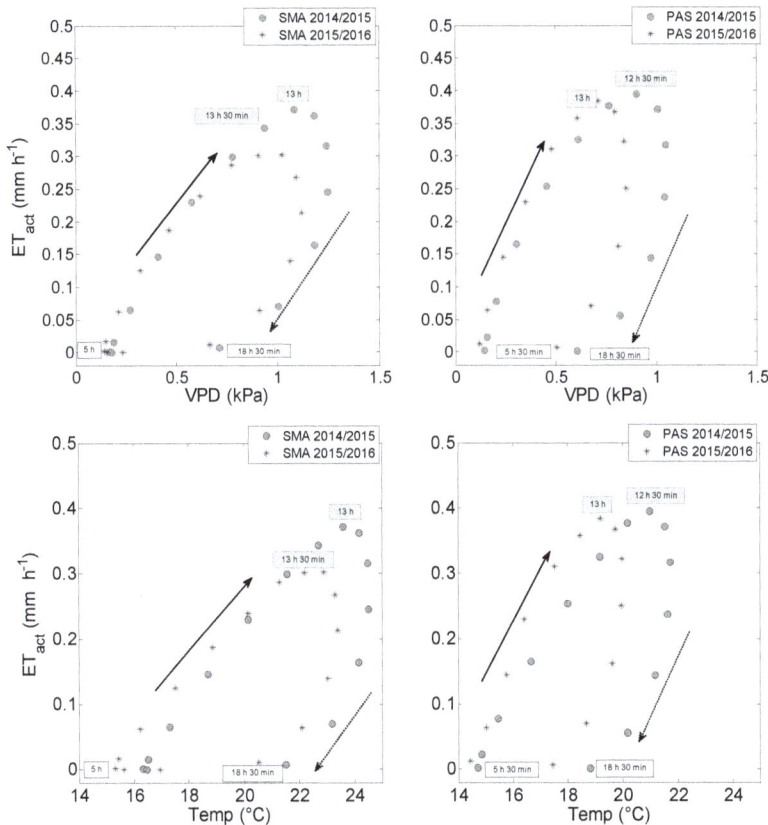

Figure 9. Hysteretic loops relating actual ET_{act} and VPD (**top panels**) and air temperature (**bottom panels**) for SMA (**left panels**) and PAS (**right panels**) for different years. The arrows indicate the direction of the hysteresis effect—morning with a solid arrow and afternoon with a dashed arrow. Gray rectangles indicate the timing (hour) when maximum ET occurs. White rectangles indicate the inversion timing (hour) of the ET signal.

ET_{act} values during the night were around zero and were removed from the hysteresis curves. Frequently, the latent heat flux was negative at night in the sites analyzed, indicating dew formation [40]. At night, ET_{act} values showed slight variations, but at the beginning of the morning, the ET_{act} values increased rapidly and peaked at approximately 12:30 to 1:30 p.m. (local time).

For both sites, ET_{act} increased as soon as VPD increased in the morning. ET_{act} decreased with the reduction of VPD in the afternoon. A similar behavior was observed by Zheng et al. [58] and Ahrends et al. [59]. The increase in ET_{act} in the morning and a subsequent decrease in the afternoon may be directly related to the available energy and the soil water content available to the plants. The opening and/or closing of the stomata may contribute to the appearance of hysteresis. Previous studies have found that stomata responded differently to changes in environmental factors during stomatal control processes, with greater stomatal conductance in the morning [60,61], indicating higher surface conductance and, therefore, a higher evapotranspiration rate compared to the afternoon. Summarizing, stomatal control affects canopy transpiration and avoids excessive water loss on days when the vapor pressure deficit is greater.

The hysteresis loops between ET_{act} and air temperature for SMA and PAS sites were similar to those found between ET_{act} and VPD. The maximum ET_{act} for the year 2014/2015 occurred with the air temperature 1 °C higher than in the year 2015/2016, for both sites. Also, for the 2015/2016, the hysteresis loops shifted to the right side, representing ET_{act} that were small in the morning and greater in the afternoon. In 2015/2016, ET_{act} in the morning was higher for the same air temperature when compared to the year 2014/2015, which agreed with the fact that the average air temperature was then higher. The climatic conditions of the sites controlled the evaporative rate. Thus, in Pedras Altas, the occurrence of stronger winds favored evapotranspiration.

Regarding the intensity of response of ET_{act} to the meteorological variables, we can conclude that the hysteresis was weaker in ET_{act}–Temperature relations when compared to the ET_{act}–VPD relation. The intensity of the hysteresis could be measured through the area formed by the hysteresis loops (normalizing both axes). Thus, the highest areas obtained were in the ET_{act}–VPD ratios for the Pedras Altas site, averaging 7% of the difference between sites. The values obtained for the areas are presented in Table 4. For ET_{act}–Temperature relations, the areas had close values for both sites. The largest difference, approximately 8.3%, occurred in hysteresis loops in the ET_{act}–VPD relationship for the Santa Maria site between the years 2014/2015 and 2015/2016. In this way, we can conclude that for the Pampa biome, the ET_{act} responded more strongly to the VPD than to temperature, with a hysteresis mean area 74% larger than that relative to the ET_{act}–Temperature relationship.

Table 4. Normalized area of hysteresis between ET_{act} and VPD and air temperature for SMA and PAS for different years. The average is represented by mean $\pm$ standard deviation.

	Site	2014/2015	2015/2016
	SMA	0.0764	0.0833
ET_{act}–VPD	PAS	0.0876	0.0842
	Average	0.0829 $\pm$ 0.0047	
	SMA	0.0220	0.0229
ET_{act}–Temperature	PAS	0.0228	0.0197
	Average	0.0218 $\pm$ 0.0015	

4. Conclusions

The evapotranspiration over the Brazilian Pampa biome was assessed at two sites where observations were performed using eddy covariance. The results allowed us to find answers to various scientific questions:

(i) How is the partition of energy in the surface?

During the study period, approximately 65% of the available energy was used for evapotranspiration in the Pampa biome. Although there were significant differences in the soil moisture conditions between the sites analyzed, there was no apparent distinction in the energy partition. The latent heat flux was the main energy balance component in both the AW and SS periods. While the latent heat flux presented strong seasonality, the sensible heat flux presented a low amplitude through the year. In general, the energy balance components were higher in the SS period than during AW.

(ii) Which are the physical processes determining the evaporative fraction?

In the analyzed sites of the Pampa biome, both soil moisture and available energy were not identified as limiting factors for evapotranspiration. Although the vegetation of both sites was composed of the same species and the soil properties were different, the high water availability in both sites did not affect the energy partition.

(iii) What is the seasonal variability of ET_{act}?

The Pampa biome presented strong seasonality of evapotranspiration, with the highest evapotranspiration rates in the SS period, where the vegetation was in active growth and, therefore, had higher biomass production. Even in periods when there was less biomass production of the vegetation in the Pampa, during AW, evapotranspiration corresponded to a high fraction of the available energy. The annual mean of ET_{act} was 2.45 mm d^{-1}. Daily values of ET_{act} were very similar between sites.

(iv) What is the relation of the actual to the reference ET?

The water availability in the Pampa biome was not a limiting factor for ET_{act} since LE was the principal fraction of net radiation, which resulted in few differences between the reference ET and the actual ET. These conditions allowed ET_0 to be used as an indicator of ET in the Pampa biome. ET_{act} was higher in PAS, while ET_0 was higher in SMA. These differences were not fully clarified in this study, but it could be hypothesized that this behavior related to differences in soil water dynamics influencing ET_{act}. It is also assumed that those differences are likely associated with vegetation biophysical control, such as canopy surface conductance and canopy structure, which needs to be further investigated.

(v) What is the relationship between ET_{act} and meteorological variables on a daily timescale?

The ET_{act} responded linearly to the net radiation but showed hysteresis when related to VPD and air temperature, with similar values for both sites, although responding more strongly to VPD. Clockwise hysteresis loops were observed relative to both variables.

This first study on evapotranspiration of the Pampa biome provides some understanding of the processes and driving forces and factors influencing ET variability. Further studies are required for other areas where ecosystems are differently affected by grazing or by agricultural systems, as well as where vegetation may be influenced by elevation, namely, relative to lowlands. In addition, relating biome ET studies with crop ET studies may also aid in understanding the complexity of these processes.

Summarizing, the Pampa biome is a complex ecosystem where the processes of surface–atmosphere interaction are dependent on weather, climate, soil, and vegetation. In this way, the preservation of this biome is essential for the maintenance of climate, animal, and plant species, as well as local culture.

Author Contributions: The study was planned by G.C.R. and D.R.R., lab data analysis was performed and writing was performed by G.C.R. L.S.P. discussed a few features of the experiments and revised the article writing together with D.R.R. F.L.F.Q. conceived the experiment in Santa Maria Site. D.R.R., H.F.d.C.V., and O.L.L.d.M. obtained the instrumented flux towers.

Funding: The authors acknowledge the National Council for Scientific and Technological Development (CNPq—Brazil), the Coordination for the Improvement of Higher Education Personnel (CAPES—Brazil), the Foundation for Research of Rio Grande do Sul State (FAPERGS), and the Financier of Studies and Projects (FINEP—Brazil) for funding the reported study.

Acknowledgments: The authors acknowledge the staff of the Micrometeorology Lab of the Federal University of Santa Maria for the technical support provided, particularly relative to the flux towers and the eddy covariance instrumentation.

Conflicts of Interest: The authors declare no conflict of interest.

References

1. Baldocchi, D.D.; Hincks, B.B.; Meyers, T.P. Measuring biosphere atmosphere exchanges of biologycally related gases with micrometeorological methods. *Ecology* **1988**, *69*, 1331–1340. [CrossRef]
2. Allen, R.G.; Pereira, L.S.; Howell, T.A.; Jensen, M.E. Evapotranspiration information reporting: I. Factors governing measurement accuracy. *Agric. Water Manag.* **2011**, *98*, 899–920. [CrossRef]
3. Overbeck, G.E.; Müller, S.C.; Fidelis, A.; Pfadenhauer, J.; Pillar, V.D.; Blanco, C.C.; Boldrini, I.I.; Both, R.; Forneck, E.D. Brazil's neglected biome: The South Brazilian Campos. *Perspect. Plant Ecol. Syst.* **2007**, *9*, 101–116. [CrossRef]

4. Paredes, P.; Rodrigues, G.J.; Petry, M.T.; Severo, P.O.; Carlesso, R.; Pereira, L.S. Evapotranspiration partition and crop coefficients of Tifton 85 bermudagrass as affected by the frequency of cuttings. Application of the dual Kc approach. *Water* **2018**, *10*, 558. [CrossRef]

5. Boldrini, I.I.; Overbeck, G.E.; Trevisan, R. Biodiversidade de plantas. In *Os Campos do Sul*, 1st ed.; Pillar, V.P., Lange, O., Eds.; UFRGS: Porto Alegre, Brazil, 2015; pp. 53–70. ISBN 978-85-66106-50-3.

6. Pillar, V.D.P.; Müller, S.C.; Castilhos, Z.M.S.; Jacques, A.V.A. *Campos Sulinos—Conservação e uso Sustentável da Biodiversidade*, 3rd ed.; Ministério do Meio Ambiente: Brasília, Brazil, 2009; 443p, ISBN 978-85-7738-117-3.

7. Nabinger, C.; Ferreira, E.T.; Freitas, A.K.; Carvalho, P.C.F.; Sant'Anna, D.M. Produção animal com base no campo nativo: Aplicações de resultados de pesquisa. In *Campos Sulinos—Conservação e uso Sustentável da Biodiversidade*, 3rd ed.; Ministério do Meio Ambiente: Brasília, Brazil, 2009; pp. 175–199, ISBN 978-85-7738-117-3.

8. Krishnan, P.; Meyers, T.P.; Scott, R.L.; Kennedy, L.; Heuer, M. Energy exchange and evapotranspiration over two temperate semi-arid grasslands in North America. *Agric. For. Meteorol.* **2012**, *153*, 31–44. [CrossRef]

9. Hu, Z.; Wen, X.; Sun, X.; Li, L.; Yu, G.; Lee, X.; Li, S. Partitioning of evapotranspiration through oxygen isotopic measurements of water pools and fluxes in a temperate grassland. *J. Geophys. Res. Biogeosci.* **2014**, *119*, 358–371. [CrossRef]

10. Trepekli, A.; Loupa, G.; Rapsomanikis, S. Seasonal evapotranspiration, energy fluxes and turbulence variance characteristics of a Mediterranean coastal grassland. *Agric. For. Meteorol.* **2016**, *226–227*, 13–27. [CrossRef]

11. Rajan, N.; Maas, S.J.; Cui, S. Extreme drought effects on summer evapotranspiration and energy balance of a grassland in the Southern Great Plains. *Ecohydrology* **2015**, *8*, 1194–1204. [CrossRef]

12. Malhi, Y.; Pegoraro, E.; Nobre, A.D.; Pereira, M.G.P.; Grace, J.; Culf, A.D.; Clement, R. Energy and water dynamics of a central Amazonian rain forest. *J. Geophys. Res.* **2002**, *107*, 8061. [CrossRef]

13. Da Rocha, H.R.; Goulden, M.L.; Miller, S.D.; Menton, M.C.; Pinto, L.D.V.O.; de Freitas, H.C.; Silva Figueira, A.M. Seasonality of water and heat fluxes over a tropical forest in eastern Amazonia. *Ecol. Appl.* **2004**, *14*, S22–S32. [CrossRef]

14. Da Rocha, H.R.; Manzi, A.; Cabral, O.M.; Miller, S.D.; Goulden, M.L.; Saleska, S.R.; Coupe, N.R.; Wofsy, S.C.; Borma, L.S.; Artaxo, P.; et al. Patterns of water and heat flux across a biome gradient from tropical forest to savanna in Brazil. *J. Geophys. Res.* **2009**, *114*, 1–8. [CrossRef]

15. Sanches, L.; Vourlitis, G.L.; Alves, M.C.; Pinto-Júnior, O.B.; Nogueira, J.S. Seasonal patterns of evapotranspiration for a Vochysia divergens Forest in the Brazilian Pantanal. *Wetlands* **2011**, *31*, 1215–1225. [CrossRef]

16. Giambelluca, T.W.; Scholz, F.G.; Bucci, S.J.; Meinzer, F.C.; Goldstein, G.; Hoffmann, W.A.; Franco, A.C.; Bucherta, M.P. Evapotranspiration and energy balance of Brazilian savannas with contrasting tree density. *Agric. For. Meteorol.* **2009**, *149*, 1365–1376. [CrossRef]

17. Cabral, O.M.R.; da Rocha, H.R.; Gash, J.H.; Freitas, H.C.; Ligo, M.A.V. Water and energy fluxes from a woodland savanna (Cerrado) in southeast Brazil. *J. Hydrol. Reg. Stud.* **2015**, *4*, 22–40. [CrossRef]

18. Santos, A.B. dos; de Quadros, F.L.F.; Confortin, A.C.C.; Seibert, L.; Ribeiro, B.S.R.; Severo, P.O.; Casanova, P.T.; Machado, G.G. Rio Grande do Sul State's (Brazil) native grasses morphogenesis under rotational grazing during spring and summer. *Ciência Rural* **2014**, *44*, 97–103. [CrossRef]

19. Quadros, F.L.F.; Pillar, V.P. Dinâmica vegetacional em pastagem natural submetida a tratamentos de queima e pastejo. *Ciência Rural* **2001**, *31*, 863–868. [CrossRef]

20. Oliveira, L.B.; Soares, E.M.; Jochims, F.; Tiecher, T.; Marques, A.R.; Kuinchtner, B.C.; Rheinheimer, D.S.; de Quadros, F.L.F. Long-Term Effects of Phosphorus on Dynamics of an Overseeded Natural Grassland in Brazil. *Rangel. Ecol. Manag.* **2015**, *68*, 445–452. [CrossRef]

21. Confortin, A.C.C.; Quadros, F.L.F.; Santos, A.B.; Seibert, L.; Severo, P.O.; Ribeiro, B.S.R. Leaf tissue fluxes of Pampa biome native grasses submitted to two grazing intervals. *Grass Forage Sci.* **2016**, *71*, 1–9. [CrossRef]

22. Instituto Brasileiro de Geografia e Estatística. Available online: http://mapas.ibge.gov.br/tematicos/solos (accessed on 25 May 2017).

23. Kottek, M.; Grieser, J.; Beck, C.; Rudolf, B.; Rubel, F. World Map of the Köppen-Geiger climate classification updated. *Meteorol. Z.* **2006**, *15*, 259–263. [CrossRef]

24. Overbeck, G.E.; Boldrini, I.I.; Carmo, M.R.B.; Garcia, E.N.; Moro, R.S.; Pinto, C.E.; Trevisan, R.; Zannin, A. Fisionomia dos Campos. In *Os Campos do Sul*, 1st ed.; Pillar, V.P., Lange, O., Eds.; UFRGS: Porto Alegre, Brazil, 2015; pp. 33–41. ISBN 978-85-66106-50-3.

25. Kljun, N.; Calanca, P.; Rotach, M.W.; Schmid, H.P. A simple parameterisation for flux footprint predictions. *Bound.-Lay. Meteorol.* **2004**, *112*, 503–523. [CrossRef]

26. Webb, E.K.; Pearman, G.I.; Leuning, R. Correction of flux measurements for density effects due to heat and water vapor transfer. *Q. J. R. Meteorol. Soc.* **1980**, *106*, 85–100. [CrossRef]

27. Gash, J.H.C.; Culf, A.D. Applying linear de-trend to eddy correlation data in real time. *Bound.-Lay. Meteorol.* **1996**, *79*, 301–306. [CrossRef]

28. Moncrieff, J.; Clement, R.; Finnigan, J.; Meyers, T. Averaging, Detrending, and Filtering of Eddy Covariance Time Series. In *Handbook of Micrometeorology*; Lee, X., Massman, W., Law, B., Eds.; Springer: Dordrecht, The Netherlands, 2004; Volume 29, pp. 7–31. ISBN 978-1-4020-2265-4.

29. Moncrieff, J.B.; Massheder, J.M.; Bruin, H.; Elbers, J.; Friborg, T.; Heusinkveld, B.; Kabat, P.; Scott, S.; Soegaard, H.; Verhoef, A. A system to measure surface fluxes of momentum, sensible heat, water vapor and carbon dioxide. *J. Hydrol.* **1997**, *188–189*, 589–611. [CrossRef]

30. Mauder, M.; Foken, T. Impact of post-field data processing on eddy covariance flux estimates and energy balance closure. *Meteorol. Z.* **2006**, *15*, 597–609. [CrossRef]

31. Nakai, T.; Shimoyama, K. Ultrasonic anemometer angle of attack errors under turbulent conditions. *Agric. For. Meteorol.* **2012**, *18*, 162–163. [CrossRef]

32. Vickers, D.; Mahrt, L. Quality control and flux sampling problems for tower and aircraft data. *J. Atmos. Ocean. Technol.* **1997**, *14*, 512–526. [CrossRef]

33. Foken, T. The energy balance closure problem: An overview. *Ecol. Appl.* **2008**, *18*, 1351–1367. [CrossRef]

34. Reichstein, M.; Falge, E.; Baldocchi, D.; Papale, D.; Aubinet, M.; Berbigier, P.; Bernhofer, C.; Buchmann, N.; Gilmanov, T.; Granier, A.; et al. On the separation of net ecosystem exchange into assimilation and ecosystem respiration: Review and improved algorithm. *Glob. Chang. Biol.* **2005**, *11*, 1424–1439. [CrossRef]

35. Allen, R.G.; Pereira, L.S.; Raes, D.; Smith, M. *Crop Evapotranspiration—Guidelines for Computing Crop Water Requirements*; FAO Irrigation and Drainage Paper 56; FAO: Rome, Italy, 1998; p. 300.

36. INMET—Instituto Nacional de Meteorologia. Available online: http://www.inmet.gov.br/portal/ (accessed on 21 March 2018).

37. Climate Prediction Center—NOAA. Available online: http://origin.cpc.ncep.noaa.gov/products/analysis_monitoring/ensostuff/ONI_v5.php (accessed on 19 October 2017).

38. Wilson, K.B.; Hanson, P.J.; Mulholland, P.J.; Baldocchi, D.D.; Wullschleger, S.D. A comparison of methods for determining forest evapotranspiration and its components: Sap-flow, soil water budget, eddy covariance and catchment water balance. *Agric. For. Meteorol.* **2001**, *106*, 153–168. [CrossRef]

39. Aubinet, M.; Grelle, A.; Ibrom, A.; Rannik, U.; Moncrieff, J.; Foken, T.; Kowalski, A.S.; Martin, P.H.; Berbigier, P.; Bernhofer, C.; et al. Estimates of the annual net carbon and water exchange of forests: The EUROFLUX methodology. *Adv. Ecol. Res.* **2000**, *30*, 113–175. [CrossRef]

40. Wilson, K.B.; Allen, G.; Falge, E.; Aubinet, M.; Baldocchi, D.; Berbigier, P.; Bernhofer, C.; Ceulemans, R.; Dolman, H.; Field, C.; et al. Energy balance closure at FLUXNET sites. *Agric. For. Meteorol.* **2002**, *113*, 223–243. [CrossRef]

41. Kanda, M.; Inagaki, A.; Letzel, M.O.; Raasch, S.; Watanabe, T. LES study of the energy imbalance problem with eddy covariance fluxes. *Bound.-Lay. Meteorol.* **2004**, *110*, 381–404. [CrossRef]

42. Sánchez, J.M.; Caselles, V.; Rubio, E.M. Analysis of the energy balance closure over a FLUXNET boreal forest in Finland. *Hydrol. Earth Syst. Sci.* **2010**, *14*, 1487–1497. [CrossRef]

43. Barr, A.G.; Van der Kamp, G.; Black, T.A.; McCaughey, J.H.; Nesic, Z. Energy balance closure at the BERMS flux towers in relation to the water balance of the White Gull Creek watershed 1999–2009. *Agric. For. Meteorol.* **2012**, *153*, 3–13. [CrossRef]

44. Foken, T.; Leuning, R.; Oncley, S.R.; Mauder, M.; Aubinet, M. Corrections and data quality control. In *Eddy Covariance: A Practical Guide to Measurement and Data Analysis*; Aubinet, M., Vesala, T., Papale, D., Eds.; Springer: Dordrecht, The Netherlands, 2012; pp. 85–131. ISBN 978-94-007-2350-4.

45. Kuplich, T.M.; Moreira, A.; Fontana, D.C. Série temporal de índice de vegetação sobre diferentes tipologias vegetais no Rio Grande do Sul. *Rev. Bras. Eng. Agríc. Ambient.* **2013**, *17*, 1116–1123. [CrossRef]

46. Seneviratne, S.I.; Corti, T.; Davin, E.L.; Hirschi, M.; Jaeger, E.B.; Lehner, I.; Orlowsky, B.; Teuling, A.J. Investigating soil moisture–climate interactions in a changing climate: A review. *Earth Sci. Rev.* **2010**, *99*, 125–161. [CrossRef]

47. Shuttleworth, W.J. Evaporation. In *Handbook of Hydrology*; Maidment, D.R., Ed.; McGraw-Hill Inc.: New York, NY, USA, 1993; pp. 4.1–4.53.

48. Bagley, J.E.; Kueppers, L.M.; Billesbach, D.P.; Williams, I.N.; Biraud, S.C.; Torn, M.S. The influence of land cover on surface energy partitioning and evaporative fraction regimes in the U.S. Southern Great Plains. *J. Geophys. Res. Atmos.* **2017**, *122*, 5793–5807. [CrossRef]

49. Gokmen, M.; Vekerdy, Z.; Verhoef, A.; Verhoef, W.; Batelaan, O.; van der Tol, C. Integration of soil moisture in SEBS for improving evapotranspiration estimation under water stress conditions. *Remote Sens.* **2012**, *121*, 261–274. [CrossRef]

50. Teuling, A.J.; Seneviratne, S.I.; Williams, C.; Troch, P.A. Observed timescales of evapotranspiration response to soil moisture. *Geophys. Res. Lett.* **2006**, *33*, L23403. [CrossRef]

51. Mortarini, L.; Stefanello, M.; Degrazia, G.; Roberti, D.; Castelli, S.T.; Anfossi, D. Characterization of wind meandering in low-wind-speed conditions. *Bound.-Lay. Meteorol.* **2016**, *161*, 165–182. [CrossRef]

52. Arbage, M.C.A.; Degrazia, G.A.; Welter, G.S.; Roberti, D.R.; Acevedo, O.C.; Moraes, O.L.L.; Ferraz, S.T.; Timm, A.U.; Moreira, V.S. Turbulent statistical characteristics associated to the north wind phenomenon in Southern Brazil with application to turbulent diffusion. *Phys. A Stat. Mech. Appl.* **2008**, *387*, 4376–4386. [CrossRef]

53. Paoloni, J.D.; Sequeira, M.E.; Fiorentino, C.E.; Amiotti, N.M.; Vazquez, R.J. Waterresources in the semi-arid Pampa–Patagonia transitional region of Argentina. *J. Arid Environ.* **2003**, *53*, 257–270. [CrossRef]

54. Goulart, M.A.; Sanches, L.; Vilani, M.T.; Pinto Júnior, O.B. Análise da evapotranspiração por wavelet de Morlet em área de Vochysia divergens Pohl no Pantanal. *Rev. Bras. Eng. Agríc. Ambient.* **2015**, *19*, 93–98. [CrossRef]

55. Shuttleworth, W.J. Evaporation from Amazonian rain forest. *Proc. R. Soc.* **1988**, *233*, 321–346. [CrossRef]

56. Almeida, A.C.; Soares, J.V. Comparação entre uso de água em plantações de eucalyptus grandis e floresta ombrófila densa (mata atlântica) na costa leste do Brasil. *Rev. Árvore* **2003**, *27*, 159–170. [CrossRef]

57. Zuecco, G.; Penna, D.; Borga, M.; van Meerveld, H.J. A versatile index to characterize hysteresis between hydrological variables at the runoff event timescale. *Hydrol. Process.* **2016**, *30*, 1449–1466. [CrossRef]

58. Zheng, H.; Wang, Q.; Zhu, X.; Li, Y.; Yu, G. Hysteresis responses of evapotranspiration to meteorological factors at a diel timescale: Patterns and causes. *PLoS ONE* **2014**, *9*, e98857. [CrossRef]

59. Ahrends, H.E.; Lind, H.R.; Schween, J.H.; Crewell, S.; Stadler, A.; Rascher, U. Diurnal Dynamics of Wheat Evapotranspiration Derived from Ground-Based Thermal Imagery. *Remote Sens.* **2014**, *6*, 9775–9801. [CrossRef]

60. Takagi, K.; Tsuboya, T.; Takahashi, H. Diurnal hysteresis of stomatal and bulk surface conductances in relation to vapor pressure deficit in a cool temperate wetland. *Agric. For. Meteorol.* **1998**, *91*, 177–191. [CrossRef]

61. Bai, Y.; Zhu, G.; Su, Y.; Zhang, K.; Han, T.; Ma, J.; Wang, W.; Ma, T.; Feng, L. Hysteresis loops between canopy conductance of grapevines and meteorological variables in an oasis ecosystem. *Agric. For. Meteorol.* **2015**, *214–215*, 319–327. [CrossRef]

water

MDPI

Article

The Water–Energy–Food Nexus: A Fuzzy-Cognitive Mapping Approach to Support Nexus-Compliant Policies in Andalusia (Spain)

Pilar Martinez *, Maria Blanco and Bente Castro-Campos

Department of Agricultural Economics, Universidad Politécnica de Madrid, ETSIAAB, Avda. Puerta Hierro 2, 28040 Madrid, Spain; maria.blanco@upm.es (M.B.); bente.castro@upm.es (B.C.-C.)
* Correspondence: mpilar.martinez@upm.es; Tel.: +34-91-336-3684

Received: 6 April 2018; Accepted: 15 May 2018; Published: 19 May 2018

Abstract: Water, energy and food are essential resources for economic development and social well-being. Framing integrated policies that improve their efficient use requires understanding the interdependencies in the water–energy–food (WEF) nexus. Stakeholder involvement in this process is crucial to represent multiple perspectives, ensure political legitimacy and promote dialogue. In this research, we develop and apply a participatory modelling approach to identify the main interlinkages within the WEF nexus in Andalusia, as a starting point to developing a system dynamic model at a later stage. The application of fuzzy cognitive mapping enabled us to gain knowledge on the WEF nexus according to opinions from 14 decision-makers, as well as contributing to raising awareness and building consensus among stakeholders. Results show that climate change and water availability are key drivers in the WEF nexus in Andalusia. Other variables with significant interlinkages within the WEF nexus are food production, irrigated agriculture, energy cost, socio-economic factors, irrigation water use, environmental conservation, and farm performance indicators. The scenario analysis reveals the interdependencies among nexus sectors and the existence of unanticipated effects when changing variables in the system, which need to be considered to design integrated policies.

Keywords: water–energy–food nexus; policy-making; stakeholder engagement; fuzzy cognitive maps; Spain; Andalusia

---

1. Introduction

Sustainable resource management requires a nexus approach in decision-making that considers trade-offs and synergies across sectors. Water availability and use influence the food and energy sectors and, simultaneously, are influenced by them. Hence, these cross-sectoral connections need to be considered to promote efficiency in the use of resources [1]. The international community increasingly perceives the water–energy–food nexus (WEF) as an overarching concept to address these complex and interconnected resource management challenges [1–3]. This WEF nexus is particularly relevant because expected socio-economic development and climate change will increase pressure on resources and drive conflicts between the different sectors over the coming decades [2,4]. Nevertheless, some studies show the existence of policy incoherence in the WEF nexus throughout the world [5]. Understanding the interlinkages in the WEF nexus is crucial for framing integrated policies that improve the natural environment, upon which economic development and social well-being depend. To that end, increasing attention is devoted to developing conceptual and methodological frameworks to describe the interlinkages in the nexus as well as to guide decision-makers in applying the WEF nexus approach. Over recent years, many global studies can be cited [1,2,6]. In addition, since targeted action is required to address local challenges, examples also exist of specific applications of WEF nexus assessments at the river basin level [7–9]. These frameworks offer valuable insights into the nexus interconnections that can help to identify conflicts and explore solutions.

Stakeholder engagement is crucial to obtaining relevant bottom-up information, representing multiple perspectives and overcoming governance obstacles. A multi-stakeholder approach is used in the ongoing EU Horizon 2020 project SIM4NEXUS (sustainable integrated management for the nexus of water–land–food–energy–climate for a resource-efficient Europe), which aims to bridge the knowledge gap and develop innovative tools to support policy decision-making in the nexus under climate change conditions. Building on stakeholder knowledge and complexity science modelling, SIM4NEXUS develops system dynamic models (SDM) and serious games for 12 different case studies (from global to regional scales) in order to evaluate nexus-compliant policies [10].

This article focusses on stakeholder engagement for the case study of Andalusia (Spain) within the SIM4NEXUS project. Policy incoherence across sectors in this region has resulted in increasing conflicts among WEF nexus sectors that have negatively affected resource sustainability [11–13]. Major pressures on water due to farming intensification and urbanization, over-allocation of water resources and inadequate consideration of the linkages between water saving technologies and energy use have led to environmental degradation [14,15].

Most previous studies with reference to the nexus in Andalusia partially examine the WEF nexus considering two-sector interrelationships; in particular, water and energy or water and agriculture [16,17]. Nevertheless, full comprehension of the interdependencies in the WEF nexus in the region is crucial to avoid unexpected effects and to design effective nexus-compliant strategies. This can only be achieved with the involvement of decision-makers in the process, to ensure integration of multiple perspectives, political legitimacy, as well as consensus building [18].

Participatory modelling with stakeholders has been increasingly applied over the last years to share knowledge and improve understanding of a system. Voinov and Bousquet [19] review the state of the art on stakeholder engagement and discuss the main methodologies applied to engage stakeholders in resource and environmental modelling, among them Bayesian networks (BN) [20], Agent-based models (ABM) [21] and fuzzy cognitive mapping (FCM) [22]. In a recent paper, Voinov et al. [23] review the progress made in applying participatory modelling in environmental decision-making and stress the need to develop comprehensive guidelines to select the best-suited methodology—among the wide range of existing tools and techniques—depending on the modelling objectives as well as the participants involved. In this regard, several frameworks have been proposed to support the process and to achieve the successful participation of stakeholders in conceptual modelling by, among others, Argent et al. [24], Basco-Carrera et al. [25], Gray et al. [26] and Halbe et al. [27].

Among the participatory modelling methodologies, in this study we select FCM because it focusses on knowledge share and stakeholder involvement [28] and allows for interactions and feedbacks between variables (while other methodologies such as BN are based on unidirectional connections). FCM has proven to be very useful in a variety of domains, as emphasized by Papageorgiou and Salmeron [29] in a review of recent applications of this approach. In the environmental field, FCMs have been used to support watershed ecosystems management [30–34], landscape modelling [35,36], forest management [37], farming systems analysis [38], fisheries management [39], bioethanol production from biowaste [40], and intrusion of mining into landscape [41]. FCMs have been also applied to assess climate change impacts and effects of adaptation measures [42,43]. Nevertheless, examples of the implementation of FCMs to the WEF nexus are scarce [44].

Therefore, in this research, we apply FCM to identify the main interconnections within the WEF nexus in Andalusia according to stakeholders' perceptions. Likewise, our work seeks to raise awareness and build consensus for the WEF nexus among the different stakeholders. In addition, the outcome of this study enables the validation of a conceptual model on the WEF nexus in Andalusia, which is a crucial step to developing an SDM at a later stage.

The article is structured as follows: Section 2 presents a brief description of the FCM methodology, the case study region, stakeholders' participation, the data collection and analysis process; Section 3 summarizes the main results obtained from the analysis of the maps and the simulation of scenarios; Section 4 presents the discussion of results; Section 5 includes the main conclusions of the research.

2. Materials and Methods

2.1. Fuzzy Cognitive Maps

Fuzzy cognitive maps (FCMs) represent the behaviour of complex systems based on people's perceptions. FCMs are based on cognitive maps that are graphical representations of causal relationships among variables in a system. These variables can be quantifiable (e.g., temperatures) or not (e.g., water policy). The direction of the causal relationship is represented with signed arrowheads. A positive sign represents a direct causal relationship (a rise in variable A will increase variable B), while a negative sign shows an inverse relationship (a rise in variable A will reduce variable B). Axelrod [45] was the first to use binary cognitive maps to elicit people's perceptions of a system. As causality is uncertain, Kosko [22] extended the binary causal relationships (0 or 1) by introducing fuzziness in cognitive maps, allowing for in-between states [−1 to 1]. Each interrelation between two variables v_i and v_j is assigned a weight w_{ij} that takes a value between −1 and 1: $w_{ij} > 0$ indicates a positive causal relationship, meaning that an increase (decrease) in the value of v_i leads to an increase (decrease) in the value of v_j (the higher the weight the stronger the causal relationship); $w_{ij} < 0$ determines a negative causal relationship, where an increase (decrease) in the value of v_i drives to a decrease (increase) on the value of v_j; $w_{ij} = 0$ depicts no causality between v_i and v_j [46].

FCM builds on stakeholder understanding and experience of the system. The knowledge elicitation process to develop the FCMs can be performed from questionnaires, by literature review, from data or from interviews where participants individually or in groups develop a map [47]. The number of stakeholders engaged in the process varies widely among FCM works from, for example, seven considered by Solana-Gutierrez et al. [33] to 188 stakeholders considered by Reckien [42].

FCMs are represented in the matrix form $E = [w_{ij}]$, where variable v_i is placed on the vertical axis and variable v_j on the horizontal axis. The causal relationship between two variables (w_{ij}) is indicated in the matrix (range from −1 to 1 as mentioned before). Aggregation of individual maps to create the group or social map enables the combination of multiple knowledge sources and helps in understanding complex systems [47,48]. To that end, variables from individual maps are subjectively grouped into broader concepts. Based on these concepts, individual matrices are augmented and condensed through matrix aggregation to create the group matrix [47].

The structure of FCMs can be analysed through matrix indices from graph theory [49]:

- Density (D) provides information on the connectivity of the maps. This index is estimated as the relationship between the number of connections and the maximum number of possible connections between the map's variables:

$$D = \frac{C}{N(N-1)} \tag{1}$$

where D is the density of the map, C is the total number of connections and N is the number of variables.

- Outdegree (*od*) represents the total strength of connections outflowing from the variable. The index is calculated as the horizontal summation of a variable's absolute values:

$$od(v_i) = \sum_{k=1}^{N} w_{ik} \tag{2}$$

where $od(v_i)$ is the outdegree of variable v_i and w_{ik} represents the weight in rows.

- Indegree (*id*) computes the total strength of connections inflowing to the variable. The index is estimated as the vertical summation of a variable's absolute values:

$$id(v_i) = \sum_{k=1}^{N} w_{ki} \qquad (3)$$

where $id(v_i)$ is the indegree of variable v_i and w_{ki} represents the weight in columns.
- Centrality (*c*) of a variable shows the cumulative strength of connections of this variable to other variables. The index is calculated as the sum of the variable's indegree and outdegree:

$$c_i = td\,(v_i) = od(v_i) + id(v_i) \qquad (4)$$

where c_i is the centrality of the variable v_i, $td\,(v_i)$ is the total degree of the variable v_i, $od(v_i)$ is the outdegree of variable v_i and $id(v_i)$ is the indegree of variable v_i.

The variables within a map can be classified into different types according to their outdegree and indegree: Transmitter variables (positive outdegree and zero indegree) are considered drivers of the system, receiver variables (zero outdegree and positive indegree) are dependent variables and ordinary variables (both zero outdegree and indegree) are means. The relationship between the number of receiver and transmitter variables (R/T) is an index of complexity of the system. A high number of receiver variables means that many results can emerge from the system. In contrast, many transmitter variables reflect numerous forcing functions, suggesting that causal relationships are poorly developed [46,47].

The system's behaviour can be analysed through scenario simulation [46] following a two-step process: First, the initial value for each of the vector's elements, also called activation level, is multiplied with the adjacency matrix to create a new vector. The activation level takes a value ranging between 0 and 1: 0 implies that the variable does not exist in the system in a specific iteration, while 1 means that the variable is present to its full extent. In-between values depict intermediate activation levels. After assigning the activation level (equals 1 for all the variables), the converted vector is iteratively multiplied by the adjacency matrix. The state value of each variable v_i in a subsequent iteration is determined by its value in the previous iteration and the value of all the interconnected variables [46].

$$A_i^{(k+1)} = f \left(A_i^k + \sum_{\substack{j \neq i \\ j=1}}^{N} A_j^k * w_{ji} \right) \qquad (5)$$

where $A_i^{(k+1)}$ is the value of the variable v_i at the simulation step $k + 1$, A_i^k is the value of variable v_i at step k, A_j^k is the value of variable v_j at step k, w_{ji} is the weight of the interrelation between variable v_j and variable v_i and f is the activation function.

To keep results within the interval [0, 1], a logistic function is used:

$$f(x) = \frac{1}{1 + e^{-x}} \qquad (6)$$

where x is the value of $A_i^{(k)}$ in the equilibrium point.

After a number of iterations, the system can achieve steady state equilibrium (the state value of each variable v_i converges over a number of iterations), move towards a limit cycle (only some values are repeated) or show a chaotic pattern (different values in each iteration) [22].

Second, simulation scenarios are run following the same procedure but, in this case, the activation levels of the variables defining the scenario remain the same during the iterations [22]. Results from simulation scenarios are presented as a relative change to the steady state. Alternative simulation scenarios allow for exploring the behaviour of the system and understanding its complex interrelations.

2.2. Study Region

Andalusia is a region located in the south of Spain (Figure 1). It is the most populous Spanish region with 8.4 million inhabitants (18.0% of the national population) and the second largest region with 87,600 km² (17.3% of the total national area). The strong agricultural sector (5.1% of GDP in 2016) has turned the region into one of the major fruit and vegetable suppliers to the European market. The industrial sector is mainly based on agricultural and consumer goods (17.0% of GDP in 2016) and the services sector, which dominates the regional economy (76.9% of GDP in 2016), depending heavily on tourism [50].

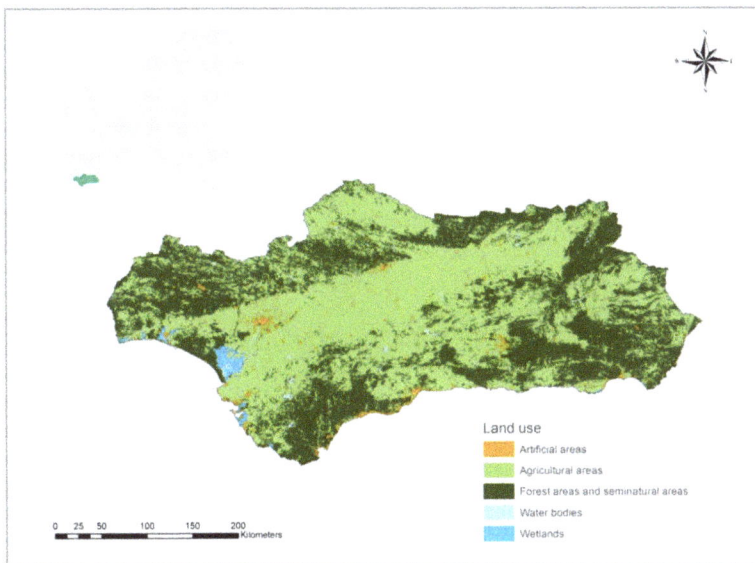

Figure 1. Location of Andalusia and main land uses. Source: Own elaboration.

Andalusia has approximately 4.4 million ha of utilised agricultural area (UAA), accounting for roughly 18.7% of the national total. Of the total UAA, one million ha are irrigated; i.e., 29.3% of the total irrigated land in Spain [51]. Irrigation is mainly used for olives, cereals, industrial crops, fruits and vegetables. This area is concentrated in the Guadalquivir river basin, which amounts to approximately 86% of the total irrigated land in Andalusia. The regional agricultural potential heavily relies on irrigated agriculture, which accounts for more than 80% of total water withdrawals, generates more than 64% of production in Andalusia, represents 67% of farm income and is responsible for 63% of agricultural employment [52].

The variability of available water owing to differences in precipitation, along with high demand, has created a delicate water balance in the different river basins in Andalusia. Over recent years, both the National and the Regional Administrations have developed different actions to save water in the agricultural sector [53,54]. Implementation of different irrigation plans have resulted in short-term water savings but might lead to an increase in water demand in the medium term. This is due to the increased interest in high value crops due to higher water supply guarantees and better quality of irrigation systems, but also due to decoupled payments of the Common Agricultural Policy (CAP) that encourage agricultural production adaptation to market [11,55].

Beyond the impact on water and food production, the irrigation plans led to an increase in energy consumption in the agricultural sector because of the replacement of open channels and surface irrigation systems with pressurised systems. The higher energy needs and system maintenance costs combined with the energy market liberalisation in 2008 drove up irrigation costs [56,57]. Thus, energy has become the key irrigation driver; when energy requirements are high, farmers tend to apply deficit irrigation even having water available [58].

2.3. Stakeholders' Participation

Stakeholders from the water, food and energy sectors in Andalusia were identified through an online investigation and snowball sampling [59]. The selection criteria were to gather experts from the different nexus sectors both at the public and private levels. Therefore, 14 representatives from the regional administration, professional organisations, NGOs and research centres and universities were selected and contacted. The number of stakeholders engaged brought together a broad range of perspectives needed to interpret the interlinkages in the nexus.

As a preliminary step, we conducted phone or face-to-face semi-structured interviews with the selected stakeholders. Seven guiding questions served to get a preliminary understanding of stakeholders' views on major nexus challenges in Andalusia. The guiding questions were sent via email to the stakeholders before the interview. The semi-structured interviews lasted for approximately 45 min to one hour.

A one-day workshop held in the Regional Government of Agriculture, Fishing and Rural Development in Seville in October 2017 brought together the stakeholders. The portion of the workshop dedicated to stakeholders' participation consisted of two sessions: (1) Individual mapping and (2) discussion of policy objectives. After explaining cognitive maps using an example, each participant was asked to build a cognitive map to respond to the question: What are the main interlinkages in the water–energy–food nexus in Andalusia? Participants were free to select the map variables to represent their perspectives without restraint. The magnitudes of the causal relationships between the variables were weighted with a value between −1 and +1. As a result of the first session of the workshop, 11 individual maps built by six representatives from the regional administration, two professional organisations, one NGO and two research centres were obtained. Figure 2 shows one of those individual maps, illustrating the type of information provided by workshop's participants. Three stakeholders either left early or arrived late so that they missed the section that was dedicated to individual map drawing.

After the individual mapping, a round-table discussion during the second half of the workshop offered insights into main nexus challenges in Andalusia. According to these challenges, the group discussion enabled the description of two policy objectives and the corresponding measures to achieve each of them. This valuable information constituted the base for the scenario simulation in this article.

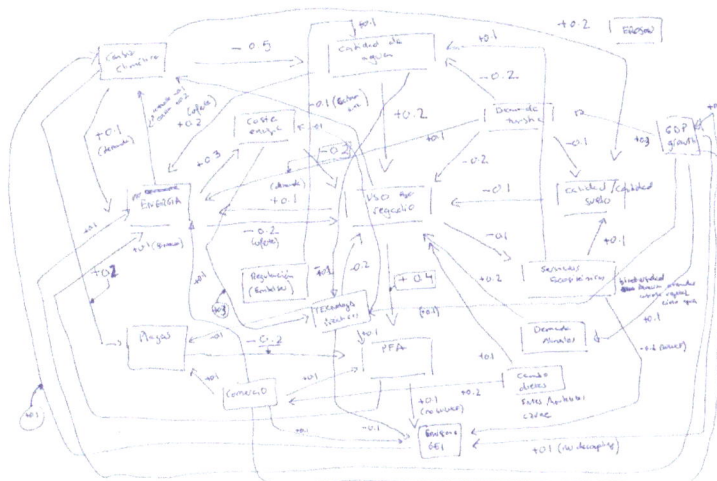

Figure 2. Example of an individual map drawn by stakeholders. Boxes represent the variables identified by the stakeholder. Arrows depict the cause–effect relationship between variables. Number values represent the strength of the causal relationship, where the sign depicts a positive (+) or negative (−) causal relationship and the value between 0 and 1 represents the magnitude.

2.4. Data Processing and Analysis

The eleven individual maps built by stakeholders represented the raw data to develop the FCM. The first step in the data processing was to list all variables included in the individual maps and to relate them to the nexus sectors. As stakeholders used different words to define the same concept, we harmonised variable terminology (e.g., crop intensification, production intensification and agricultural intensification were harmonised as "intensification of agricultural production"). Less-repeated variables were aggregated into broader categories according to similarities (e.g., plagues and exotic species were merged in the variable "plagues") and the names of the variables were standardised across maps. This subjective aggregation condensed 206 variables into 35 standardised variables facilitating the understanding and the analysis of the maps.

Using the standardised variables, we converted each individual map into a matrix. When a causal relationship between two variables was identified by the stakeholder, the same strength of influence provided in the individual map (ranging between −1 and 1) was coded in the matrix (weight w_{ij}). When no causal relationship was given in the individual map, zero was used. Once the individual maps were converted into matrices, all those individual matrices were augmented and merged into a single aggregated matrix to deliver the group map. The connection values between two specific variables were summed and averaged over the total number of individual maps. In the aggregated map, contradictory relationships between two given variables in different individual maps (opposite sign) decrease the causal relationship, while consensus on interlinkages (same sign) strengthens the causal relationship.

2.5. Scenario Analysis

The group matrix is used to run alternative simulation scenarios with the FCMapper software [60]. The definition of scenarios built on the policy objectives identified by stakeholders during the workshop as mentioned before:

- Scenario 1: Climate change mitigation

Stakeholders highlighted climate change as one of the major challenges in the WEF nexus. To address this challenge, the reduction in diffuse emissions is established as a major target. Some of the measures proposed to achieve this objective were to promote efficient agricultural machinery and good agricultural practices (carbon capture, reduction of tillage, precision agriculture, use of sensors, smart agriculture), reduce methane emissions and enhance renewable energies.

Based on this information, we simulated a first scenario considering an improvement in agricultural practices (variable "farming practices" set to 1) and an increase in the use of renewable energies (variable "use of renewable energies" set to 1).

- Scenario 2: Sustainable water resources management

Water availability is another key factor in the WEF nexus, especially for irrigated agriculture, which is by far the largest water user in Andalusia. Stakeholders considered that a reduction in water demand for irrigation is crucial to promote sustainable resource management. To that end, they proposed improvements in water infrastructures and management, as well as measures to reduce diffuse pollution.

According to this information, the second scenario simulated the improvement of water infrastructures (variable "water infrastructures" set to 1) and water quality (variable "water quality" set to 1).

3. Results

3.1. FCM Outcomes

The 11 individual maps drawn by stakeholders comprised 206 variables that, after data processing, were aggregated into 35 variables. Individual maps included an average number of 19 variables, which when grouped resulted in 14 variables on average. Most mentioned variables were climate change (climate change aggregates the variables climate, temperature, precipitation, extreme events, droughts and floods) and water availability, which were presented in all of the 11 maps, followed by energy cost and food production depicted in ten and nine maps, respectively.

Table 1 summarises calculated matrix indices for the individual maps (average values) and of the group map. The average number of connections was 25, leading to an average density index of 0.03. The group map gathered all of the variables and had a higher number of connections (C = 209) and density index (D = 0.17). The larger number of connections led to a lower number of transmitter and receiver variables in the group map, where most variables were ordinary. Therefore, the complexity index is higher in the individual maps than in the group map.

Table 1. Individual and group map indices (mean and standard deviation).

Index	Individual Maps	Group Map
Number of maps	11	1
Number of variables (N)	14 ± 2.72	35
Number of connections (C)	25 ± 9.40	209
C/N	1.81 ± 0.59	5.97
Density	0.03 ± 0.02	0.17
Number of transmitter variables (T)	2.55 ± 2.21	2
Number of receiver variables (R)	2.64 ± 2.01	1
Number of ordinary variables	8.82 ± 2.93	32
Complexity (R/T)	1.27 ± 1.57	0.5

Source: Own elaboration.

Figure 3 visualises the group map, where circles symbolise concepts of the system and lines reflect causal relationships between concepts (the group map was represented using the Pajek software (http://mrvar.fdv.uni-lj.si/pajek/)). The colour of the circle refers to the different sectors, while the size reflects the centrality of the variable within the system. The most central variables were water availability, climate change, food production, irrigated agriculture, energy cost, socio-economic factors (socio-economic factors aggregates economic development, job generation, GDP growth, equitable distribution of wealth, change in the structure of land ownership, population settlement), irrigation water use, environmental conservation, and farm performance indicators (farm performance indicators aggregate farm income, profitability, competitiveness and sustainability, insurances costs, infrastructure modernisation).

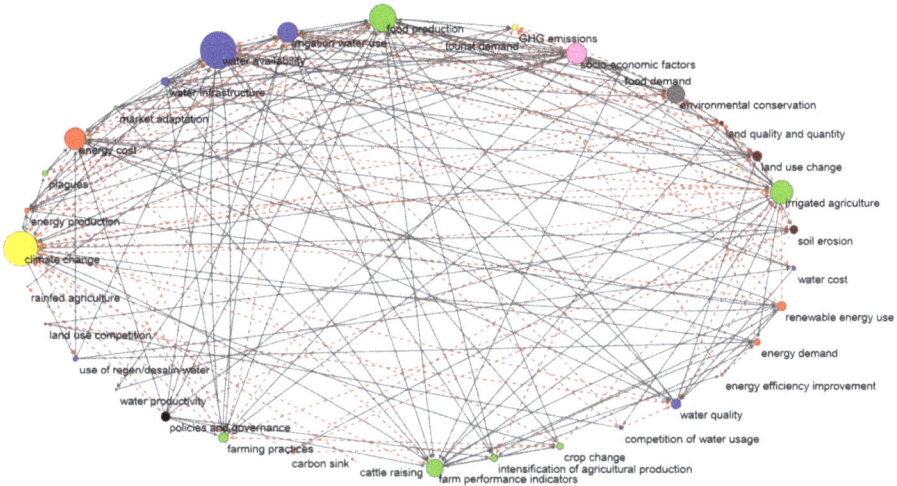

Figure 3. Visual representation of the group map. Circles represent the variables of the system, colours refer to the nexus sector (blue for water, red for energy, green for agriculture, yellow for climate, pink for socioeconomic aspects, brown for land, grey for environment and black for policies and governance) and size depicts centrality. Solid grey lines depict positive casual relationships while dotted red lines illustrate negative causal relationships. Source: Own elaboration.

As highlighted in Figure 4, most of the variables were ordinary; only policies and governance and water infrastructures were exclusively transmitter variables, while cattle raising was an entirely receiver variable. Among the most important variables in the map, climate change, water availability and energy cost had the highest outdegree, meaning that those variables were highly influencing other variables. In contrast, variables with the highest indegree, and therefore significantly affected by other variables, were food production, socio-economic factors, irrigated agriculture, environmental conservation and farm performance indicators.

Figure 5 visualizes the interconnections among the most central variables in the group map. As observed, the variables climate change, water availability and food production are highly connected to other variables, while farm performance indicators and environmental conservation have fewer connections but bear greater weights. Food production can be distinguished by the height number of incoming connections, which determines its high indegree. Looking at the type of interrelationship, the importance of energy cost and climate change are highlighted by the number of negative outgoing connections.

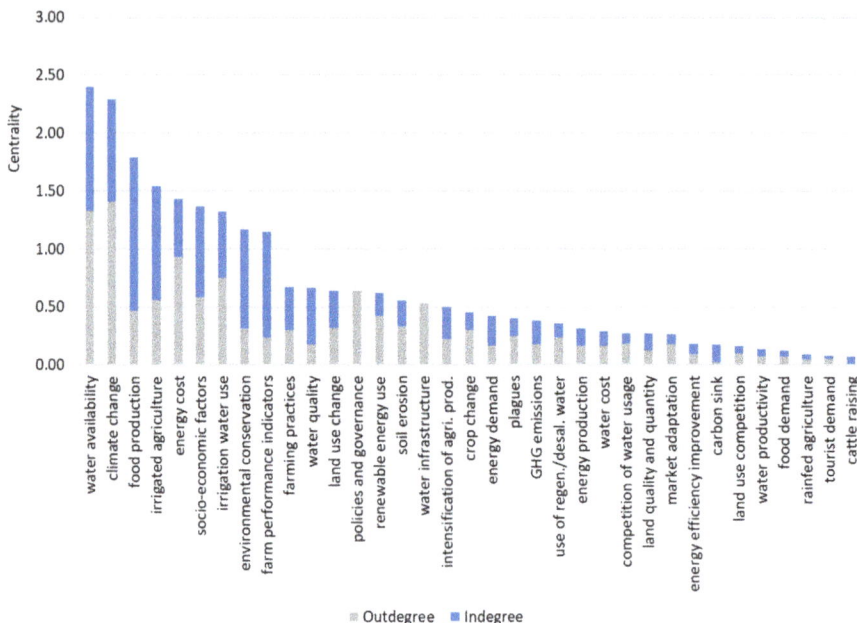

Figure 4. Importance of variables in the group fuzzy cognitive mapping (FCM) according to their centrality, represented as the sum of the outdegree and indegree. Source: Own elaboration.

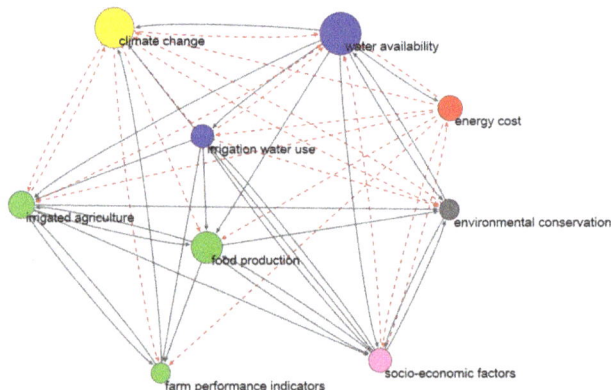

Figure 5. Visual representation of a subset of the group map. The nine variables with centrality higher than 1 are represented. Grey solid lines for positive and red dotted lines for negative connections. Source: Own elaboration.

3.2. Scenario Simulations

Figure 6 represents the results from the steady state calculation (obtained after 20 iterations) for all of the variables derived from the group FCM. The variables with the highest initial value were farm performance indicators, food production, irrigation water use and energy cost which suggest that these variables significantly influence the WEF nexus in the current situation. In contrast, the variable with the lowest initial value was water availability, which reflects the high pressure on water resources in Andalusia.

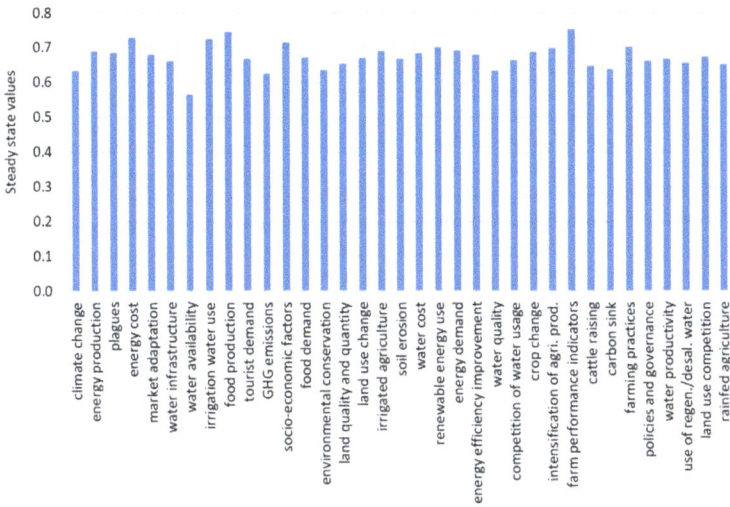

Figure 6. Steady state. Source: Own elaboration.

As shown in Figure 7, results from the climate change scenario (Scenario 1) highlight that an improvement in farming practices and the promotion of renewable energy use trigger a reduction in greenhouse gas (GHG) emissions and contribute to tackling climate change. Significant effects are also observed in the variables soil erosion and environmental conservation. Food production and farm performance indicators are also positively affected as a result of better farming practices. A negative relative change in one variable does not always imply a negative impact on the system, as can be observed for the variables climate change, GHG emissions, and soil erosion.

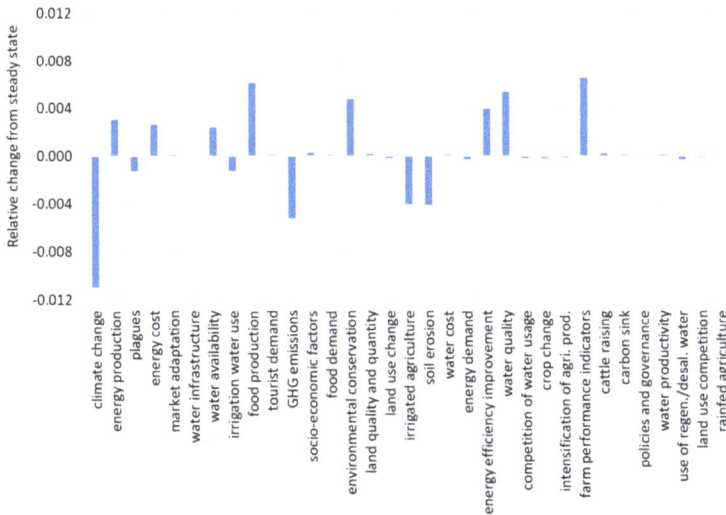

Figure 7. Climate change mitigation scenario simulation (absolute change from steady state). Source: Own elaboration.

With regards to the sustainable water management scenario (Scenario 2), as observed in Figure 8, the improvement in water infrastructures and water quality would have a positive effect on water availability and, thus, on irrigated agriculture and food production. Consequently, farm performance indicators would be positively affected. Nevertheless, these measures would imply a rise in energy costs as well as further environmental deterioration.

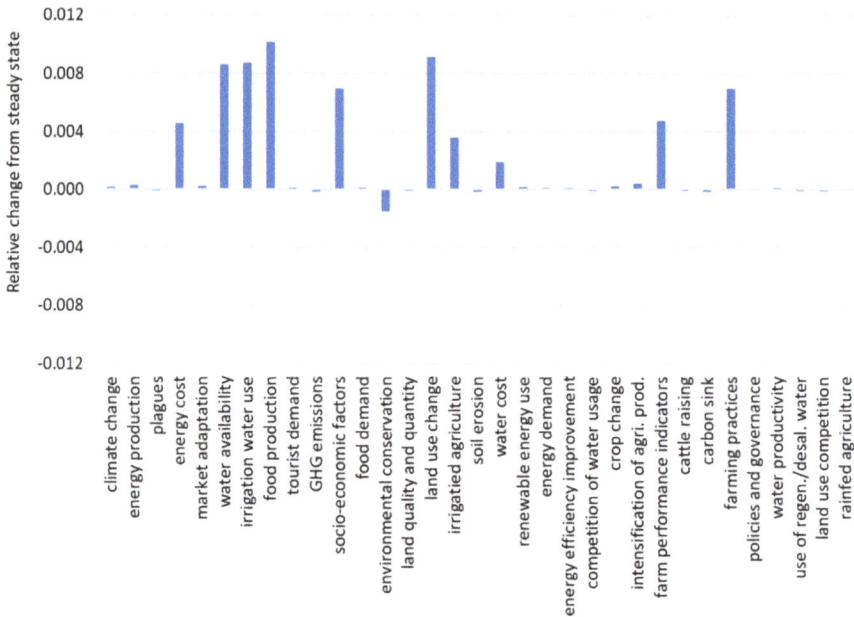

Figure 8. Sustainable water management scenario simulation (absolute change from steady state). Source: Own elaboration.

4. Discussion

Matrix indices from individual maps highlighted that stakeholders perceived the nexus as a highly complex system, having more receiver than transmitter variables. These findings contrast with other FMC studies that, overall, show more transmitter variables than receivers [47]. Notwithstanding this, in the group map, the number of these types of variables was reduced because of the increase in the number of connections. The group map presented a higher density index but lower complexity [32]. Therefore, while individual maps provided valuable information on the diversity of opinions among stakeholders, the group or social map offers a more comprehensive vision of the nexus [39].

Stakeholders agreed that climate change and water availability are key factors in the WEF nexus, as all of the maps considered them. These variables had also the highest centrality in the aggregated map, influencing significantly the behaviour of the system as drivers (highest outdegree), but also as receiver variables (highest indegree). This implies that any change in these variables (e.g., policy measure) will have a significant impact throughout the system. Food production, irrigated agriculture, environmental conservation, and farm performance indicators are also central in the system due mainly to their high indegree, suggesting that these variables are very sensitive to any intervention in the system. Energy cost highlighted by its outdegree and high number of negative connections, which means that changes in energy policy would have relevant impacts on the water and food sectors. Other studies in Mediterranean areas find similar nexus interlinkages [16].

The simulation scenarios analysed in this study addressed the policy objectives proposed by stakeholders, which are climate change and water availability. The stakeholder priorities were consistent with the results derived from the group model, where these variables were highlighted as being the most central variables and the most important drivers in the WEF nexus.

The climate change scenario illustrates that the improvement of farming practices and the promotion of renewable energies effectively lead to a reduction of GHG emissions. Better farming practices also contribute to reducing soil erosion and improving water quality, which in turn positively influences environmental conservation. Boosting renewable energies increases energy production and energy efficiency but negatively affects irrigated agriculture. This is due to the competition for water use between irrigated agriculture and energy production, mainly hydropower. Despite the negative effect on irrigated agriculture, food production is positively affected owing to the improvement of agricultural practices and, correspondingly, farm performance indicators increase.

The sustainable water management scenario demonstrates that the improvement of water infrastructure management and the reduction of water pollution leads to an increase in water availability, which in turn causes an increase in irrigation water use [15]. A negative impact on environmental conservation is also observed in this scenario. This implies that the measures to improve sustainable water management have an opposite effect than the one intended and, therefore, do not represent an appropriate policy option to be implemented.

The combination of both scenarios reinforces the impacts derived from each one. This is because, in this case, the measures analysed are complementary. Farming practices from Scenario 1 also have a positive effect on water management, as stakeholders consider that good farming practices reduce water use in agriculture.

The application of the FCM approach presents numerous advantages but also a number of disadvantages. FCMs enable systems modelling with limited data, including both quantitative and qualitative variables, whilst considering feedback relationships between concepts [37]. Furthermore, this methodology encourages stakeholder discussion, facilitates their involvement in the modelling process and increases reliance on the model [30]. Nevertheless, the outcome from the FCM model needs to be interpreted in relative terms (no absolute changes in variables can be provided). Another important drawback is the absence of temporal dimension in the model [37].

Notwithstanding this, outcomes from the FCMs provide the foundations for developing a SDM at a later stage allowing quantitative impact assessment of detailed policy scenarios in the medium- and long term. In particular, the variables and interlinkages identified within the WEF nexus will serve to develop the structure of the SDM. Scenario simulations contribute to improve knowledge on the system behaviour and facilitate the definition of policy scenarios.

5. Conclusions

In this research, we performed a participatory approach to gain insights of the WEF nexus in Andalusia. Through the application of the FCM methodology, we involved decision-makers in the identification of the main nexus interlinkages, while promoting discussion and consensus building among stakeholders from different sectors. Furthermore, the exploitation of FCM model dynamics enabled the analysis of different scenarios to improve our knowledge on the system behaviour.

Stakeholders agreed that climate change and water availability are the main drivers in the WEF nexus. Other variables with significant roles in building interlinkages in the nexus were energy cost, food production, irrigated agriculture, socio-economic factors, irrigation water use, environmental conservation and farm performance indicators.

The scenario analysis demonstrates the interdependencies among nexus sectors and the existence of unanticipated effects when changing variables in the system. Although results from scenarios only show relative variable changes, they allow improved analysis of the stakeholders' perceptions on the WEF nexus and contribute to improving understanding of the system's dynamics.

Water **2018**, *10*, 664

The participatory approach performed in this research enabled the development of a conceptual model of the WEF nexus in Andalusia which is suitable for the stakeholders that will support the nexus assessment. Outcomes from this research are being used in the development of an SDM to quantitatively assess the impact of nexus-compliant policies in Andalusia in the medium and long term.

Author Contributions: P.M. and M.B. conceived and designed the research; P.M., M.B. and B.C.-C. performed the data collection; P.M. and B.C.-C. processed the data; P.M. analysed the data and wrote the paper; M.B. supervised the work; M.B. and B.C.-C. reviewed the final manuscript.

Funding: The work described in this paper has been conducted within the project SIM4NEXUS. This project has received funding from the European Union's Horizon 2020 research and innovation programme under Grant Agreement NO 689150 SIM4NEXUS.

Acknowledgments: We would like to thank to all the stakeholders who participate in the interviews and the workshop for their invaluable contribution to the development of this research.

Conflicts of Interest: The authors declare no conflict of interest.

References

1. Hoff, H. *Understanding the Nexus: Background Paper for the Bonn 2011 Conference: The Water, Energy and Food Security Nexus*; Stockholm Environment Institute: Stockholm, Sweden, 2011; Available online: https://www.sei-international.org/publications?pid=1977 (accessed on 5 March 2018).
2. Food and Agriculture Organization (FAO). *The Water-Energy-Food Nexus: A New Approach in Support of Food Security and Sustainable Agriculture*; Food and Agriculture Organization of the United Nations: Rome, Italy, 2014; Available online: http://www.fao.org/3/a-bl496e.pdf (accessed on 5 March 2018).
3. Organisation for Economic Co-operation and Development (OECD). *The Land-Water-Energy Nexus: Biophysical and Economic Consequences*; OECD: Paris, France, 2017. [CrossRef]
4. Bhaduri, A.; Ringler, C.; Dombrowski, I.; Mohtar, R.; Scheumann, W. Sustainability in the water-energy-food nexus. *Water Int.* **2015**, *40*, 723–732. [CrossRef]
5. Lindberg, C.; Leflaive, X. The water-energy-food-nexus: The imperative of policy coherence for sustainable development. In *Coherence for Development-Better Policies for Better Lives-Organization for Economic Co-Operation and Development*; OECD: Paris, France, 2015; Volume 6, p. 12.
6. Rodriguez, D.J.; Delgado, A.; DeLaquil, P.; Sohns, A. *Thirsty Energy*; World Bank: Washington, DC, USA, 2013; Available online: http://documents.worldbank.org/curated/en/835051468168842442/Thirsty-energy (accessed on 6 March 2018).
7. Smajgl, A.; Ward, J. *The Water-Food-Energy Nexus in the Mekong Region*; Springer: New York, NY, USA, 2013.
8. Mayor, B.; Lopez-Gunn, E.; Villaroya, F.I.; Montero, E. Application of a water-energy-food nexus framework for the Duero river basin in Spain. *Water Int.* **2015**, *40*, 791–808. [CrossRef]
9. De Strasser, L.; Lipponen, A.; Howells, M.; Stec, S.; Bréthaut, C. A methodology to assess the water energy food ecosystems nexus in transboundary river basins. *Water* **2016**, *8*, 59. [CrossRef]
10. Sušnik, J.; Chew, C.; Domingo, X.; Mereu, S.; Trabucco, A.; Evans, B.; Vamkeridou-Lyroudia, L.; Savic, D.A.; Laspidou, C.; Brouwer, F. Multi-stakeholder development of a serious game to explore the water-energy-food-land-climate nexus: The SIM4NEXUS approach. *Water* **2018**, *10*, 139. [CrossRef]
11. Lopez-Gunn, E.; Zorrilla, P.; Prieto, F.; Llamas, M.R. Lost in translation? Water efficiency in Spanish agriculture. *Agric. Water Manag.* **2012**, *108*, 83–95. [CrossRef]
12. Sampedro, D.; Del Moral, L. Tres décadas de política de aguas en Andalucía. *Cuad. Geogr.* **2014**, *53*, 36–67.
13. Salmoral, G.; Garrido, A. The Common Agricultural Policy as a driver of water quality changes: The case of the Guadalquivir River Basin (southern Spain). *Bio-Based Appl. Econ.* **2015**, *4*, 103. [CrossRef]
14. Berbel, J.; Pedraza, V.; Giannoccaro, G. The trajectory towards basin closure of a European river: Guadalquivir. *Int. J. River Basin Manag.* **2013**, *11*, 111–119. [CrossRef]
15. Corominas, J.; Cuevas, R. Análisis crítico de la modernización de regadíos. Pensando el futuro: ¿cómo será el nuevo paradigma? In *Efectos de la Modernización de Regadío en España*; Berbel, J., Gutiérrez-Martín, C., Eds.; Cajamar Caja Rural: Almería, Spain, 2017; pp. 273–308. ISBN 978-84-95531-83-4. (In Spanish)
16. Hardy, L.; Garrido, A.; Juana, L. Evaluation of Spain's water-energy nexus. *Int. J. Water Resour. Dev.* **2012**, *28*, 151–170. [CrossRef]

17. De Stefano, L.; Llamas, M.R. (Eds.) *Water, Agriculture and the Environment in Spain: Can We Square the Circle?* CRC Press: Delft, The Netherlands, 2012; ISBN 978-0-415-63152-5.

18. De Marchi, B.; Ravetz, J.R. *Participatory Approaches to Environmental Policy*; EVE-Concerted Action, Policy Research Brief Number 10; Cambridge Research for the Environment: Cambridge, UK, 2001.

19. Voinov, A.; Bousquet, F. Modelling with stakeholders. *Environ. Model. Softw.* **2010**, *25*, 1268–1281. [CrossRef]

20. Barton, D.N.; Kuikka, S.; Varis, O.; Uusitalo, L.; Henriksen, H.J.; Borsuk, M.; Linnell, J.D. Bayesian networks in environmental and resource management. *Integr. Environ. Assess. Manag.* **2012**, *8*, 418–429. [CrossRef] [PubMed]

21. Bousquet, F.; Barreteau, O.; Le Page, C.; Mullon, C.; Weber, J. An environmental modelling approach: The use of multi-agent simulations. *Adv. Environ. Ecol. Model.* **1999**, *113*, 122.

22. Kosko, K. Fuzzy cognitive maps. *Int. J. Man-Mach. Stud.* **1986**, *24*, 65–75. [CrossRef]

23. Voinov, A.; Kolagani, N.; McCall, M.K.; Glynn, P.D.; Kragt, M.E.; Ostermann, F.O.; Pierce, S.A.; Ramu, P. Modelling with stakeholders–next generation. *Environ. Model. Softw.* **2016**, *77*, 196–220. [CrossRef]

24. Argent, R.M.; Sojda, R.S.; Giupponi, C.; McIntosh, B.; Voinov, A.A.; Maier, H.R. Best practices for conceptual modelling in environmental planning and management. *Environ. Model. Softw.* **2016**, *80*, 113–121. [CrossRef]

25. Basco-Carrera, L.; Warren, A.; van Beek, E.; Jonoski, A.; Giardino, A. Collaborative modelling or participatory modelling? A framework for water resources management. *Environ. Model. Softw.* **2017**, *91*, 95–110. [CrossRef]

26. Gray, S.; Voinov, A.; Paolisso, M.; Jordan, R.; BenDor, T.; Bommel, P.; Glynn, P.; Hedelin, B.; Hubacek, K.; Introne, J.; et al. Purpose, processes, partnerships, and products: Four Ps to advance participatory socio-environmental modeling. *Ecol. Appl.* **2018**, *28*, 46–61. [CrossRef] [PubMed]

27. Halbe, J.; Pahl-Wostl, C.; Adamowski, J. A methodological framework to support the initiation, design and institutionalization of participatory modeling processes in water resources management. *J. Hydrol.* **2018**, *556*, 701–716. [CrossRef]

28. Gray, S.A.; Gray, S.; De Kok, J.L.; Helfgott, A.E.; O'Dwyer, B.; Jordan, R.; Nyaki, A. Using fuzzy cognitive mapping as a participatory approach to analyze change, preferred states, and perceived resilience of social-ecological systems. *Ecol. Soc.* **2015**, *20*. [CrossRef]

29. Papageorgiou, E.; Salmeron, J.L. A review of fuzzy cognitive maps research during the last decade. *IEEE Trans. Fuzzy Syst.* **2013**, *21*, 66–79. [CrossRef]

30. Hobbs, B.F.; Ludsin, S.A.; Knight, R.L.; Ryan, P.A.; Biberhofer, J.; Ciborowski, J.J. Fuzzy cognitive mapping as a tool to define management objectives for complex ecosystems. *Ecol. Appl.* **2002**, *12*, 1548–1565. [CrossRef]

31. Kafetzis, A.; McRoberts, N.; Mouratiadou, I. Using fuzzy cognitive maps to support the analysis of stakeholders' views of water resource use and water quality policy. In *Fuzzy Cognitive Maps*; Springer: Berlin/Heidelberg, Germany, 2010; pp. 383–402.

32. Vasslides, J.M.; Jensen, O.P. Fuzzy cognitive mapping in support of integrated ecosystem assessments: Developing a shared conceptual model among stakeholders. *J. Environ. Manag.* **2016**, *166*, 348–356. [CrossRef] [PubMed]

33. Solana-Gutiérrez, J.; Rincón, G.; Alonso, C.; García-de-Jalón, D. Using fuzzy cognitive maps for predicting river management responses: A case study of the Esla River basin, Spain. *Ecol. Model.* **2013**, *360*, 260–269. [CrossRef]

34. Bosma, C.; Glenk, K.; Novo, P. How do individuals and groups perceive wetland functioning? Fuzzy cognitive mapping of wetland perceptions in Uganda. *Land Use Policy* **2017**, *60*, 181–196. [CrossRef]

35. Wildenberg, M.; Bachhofer, M.; Adamescu, M.; De Blust, G.; Diaz-Delgadod, R.; Isak, K.G.Q.; Riku, V. Linking thoughts to flows-Fuzzy cognitive mapping as tool for integrated landscape modelling. In Proceedings of the 2010 International Conference on Integrative Landscape Modelling, Montpellier, France, 3–5 February 2010.

36. Van der Sluis, T.; Arts, B.; Kok, K.; Bogers, M.; Busck, A.G.; Sepp, K.; Crouzat, E. Drivers of European landscape change: Stakeholders' perspectives through Fuzzy Cognitive Mapping. *Landsc. Res.* **2018**, 1–19. [CrossRef]

37. Kok, K. The potential of Fuzzy Cognitive Maps for semi-quantitative scenario development, with an example from Brazil. *Glob. Environ. Chang.* **2009**, *19*, 122–133. [CrossRef]

38. Vanwindekens, F.M.; Stilmant, D.; Baret, P.V. Development of a broadened cognitive mapping approach for analysing systems of practices in social–ecological systems. *Ecol. Model.* **2013**, *250*, 352–362. [CrossRef]

39. Gray, S.; Chan, A.; Clark, D.; Jordan, R. Modeling the integration of stakeholder knowledge in social–ecological decision-making: Benefits and limitations to knowledge diversity. *Ecol. Model.* **2012**, *229*, 88–96. [CrossRef]

40. Konti, A.; Damigos, D. Exploring strengths and weaknesses of bioethanol production from bio-waste in Greece using Fuzzy Cognitive Maps. *Energy Policy* **2018**, *112*, 4–11. [CrossRef]

41. Misthos, L.M.; Messaris, G.; Damigos, D.; Menegaki, M. Exploring the perceived intrusion of mining into the landscape using the fuzzy cognitive mapping approach. *Ecol. Eng.* **2017**, *101*, 60–74. [CrossRef]

42. Reckien, D. Weather extremes and street life in India—Implications of Fuzzy Cognitive Mapping as a new tool for semi-quantitative impact assessment and ranking of adaptation measures. *Glob. Environ. Chang.* **2014**, *26*, 1–13. [CrossRef]

43. Olazabal, M.; Chiabai, A.; Foudi, S.; Neumann, M.B. Emergence of new knowledge for climate change adaptation. *Environ. Sci. Policy* **2018**, *83*, 46–53. [CrossRef]

44. Ziv, G.; Watson, E.; Young, D.; Howard, D.C.; Larcom, S.T.; Tanentzap, A.J. The potential impact of Brexit on the energy, water and food nexus in the UK: A fuzzy cognitive mapping approach. *Appl. Energy* **2018**, *210*, 487–498. [CrossRef]

45. Axelrod, R. *Structure of Decision: The Cognitive Maps of Political Elites*; Princeton University Press: Princeton, NJ, USA, 1976; ISBN 9780691616988.

46. Papageorgiou, E.; Kontogianni, A. Using fuzzy cognitive mapping in environmental decision-making and management: A methodological primer and an application. In *International Perspectives on Global Environmental Change*; InTech: New York, NY, USA, 2012; ISBN 978-953-307-815-1.

47. Özesmi, U.; Özesmi, S.L. Ecological models based on people's knowledge: A multi-step fuzzy cognitive mapping approach. *Ecol. Model.* **2004**, *176*, 43–64. [CrossRef]

48. Gray, S.A.; Zanre, E.; Gray, S.R.J. Fuzzy cognitive maps as representations of mental models and group beliefs. In *Fuzzy Cognitive Maps for Applied Sciences and Engineering*; Springer: Berlin/Heidelberg, Germany, 2014; pp. 29–48.

49. Harary, F.; Norman, R.Z.; Cartwright, D. *Structural Models: An Introduction to the Theory of Directed Graphs*; John Wiley & Sons: New York, NY, USA, 1965.

50. Junta de Andalucía. *Informe Económico 2016*; Consejería de Economía y Conocimiento, Junta de Andalucía: Seville, Spain, 2017; Available online: http://www.juntadeandalucia.es/export/drupaljda/publicacion/17/08/Informe_Economico_2016_0.pdf (accessed on 9 March 2018). (In Spanish)

51. MAPAMA. *Encuesta de Superficies y Rendimientos de Cultivo*; Ministerio de Agricultura, Alimentación y Medio Ambiente, Gobierno de España: Madrid, Spain, 2015. Available online: http://www.mapama.gob.es/es/estadistica/temas/estadisticas-agrarias/boletin2015_tcm30-122275.pdf (accessed on 12 February 2018). (In Spanish)

52. European Parliament. *Research for Agri-Comitee—Agriculture in Andalusia*; Directorate General for Internal Policies: Brussels, Belgium, 2016; Available online: http://www.europarl.europa.eu/RegData/etudes/STUD/2016/573431/IPOL_STU(2016)573431_EN.pdf (accessed on 20 December 2017).

53. MAPA. *Plan Nacional de Regadíos Horizonte 2008*; Ministerio de Agricultura, Pesca y Alimentación, Gobierno de España: Madrid, Spain, 2001. Available online: http://www.mapama.gob.es/es/desarrollo-rural/temas/gestion-sostenible-regadios/plan-nacional-regadios/texto-completo/ (accessed on 12 February 2018). (In Spanish)

54. Junta de Andalucía. *Agenda del Regadío Andaluz H-2015*; Consejería de Agricultura, Pesca y Alimentación, Junta de Andalucía: Seville, Spain, 2011; Available online: http://www.juntadeandalucia.es/export/drupaljda/AGENDA_DEL_REGADIO_CONSEJO_DE_GOBIERNO_x7-4-11x.pdf (accessed on 9 March 2018). (In Spanish)

55. Fernández García, I.; Rodrigeuz Díaz, J.; Camacho Poyato, E.; Montesinos, P.; Berbel, J. Effects of modernization and medium term perspectives on water and energy use in irrigation districts. *Agric. Syst.* **2014**, *131*, 56–63. [CrossRef]

56. Corominas, J. Agua y energía en el riego en la época de la sostenibilidad. *Ing. Agua* **2010**, *17*, 219–233. (In Spanish) [CrossRef]

57. Rodríguez-Díaz, J.A.; Pérez-Urrestarazu, L.; Camacho-Poyato, E.; Montesinos, P. The paradox of irrigation scheme modernization: More efficient water use linked to higher energy demand. *Span. J. Agric. Res.* **2011**, *9*, 1000–1008. [CrossRef]

58. Rodríguez-Díaz, J.A.; Camacho-Poyato, E.; Blanco-Pérez, M. Evaluation of water and energy use in pressurized irrigation networks in Southern Spain. *J. Irrig. Drain. Eng.* **2011**, *137*, 644–650. [CrossRef]
59. Noy, C. Sampling knowledge: The hermeneutics of snowball sampling in qualitative research. *Int. J. Soc. Res. Methodol.* **2008**, *11*, 327–344. [CrossRef]
60. Bachofer, M.; Wildenberg, M. FCmapper Software. Available online: http://www.fcmappers.net/joomla/index.php (accessed on 10 January 2018).

![water logo] *water*

MDPI

Review

A Simplified Nitrogen Assessment in Tagus River Basin: A Management Focused Review

Cláudia M. d. S. Cordovil [1],*, Soraia Cruz [1], António G. Brito [1], Maria do Rosário Cameira [1], Jane R. Poulsen [2], Hans Thodsen [2] and Brian Kronvang [2]

[1] LEAF—Linking Landscape, Environment, Agriculture and Food, School of Agriculture, University of Lisbon, Tapada da Ajuda, 1349-017 Lisbon, Portugal; soraiafelix12@hotmail.com (S.C.); agbrito@isa.ulisboa.pt (A.G.B.); roscameira@isa.ulisboa.pt (M.d.R.C.)
[2] Department of Biosciences, University of Aarhus, Vejlsovej 25, 8600 Silkeborg, Denmark; jpo@bios.au.dk (J.R.P.); hath@bios.au.dk (H.T.); bkr@bios.au.dk (B.K.)
* Correspondence: cms@isa.ulisboa.pt; Tel.: +351-213-653-424

Received: 10 October 2017; Accepted: 27 March 2018; Published: 30 March 2018

Abstract: Interactions among nitrogen (N) management and water resources quality are complex and enhanced in transboundary river basins. This is the case of Tagus River, which is an important river flowing from Spain to Portugal in the Iberian Peninsula. The aim was to provide a N assessment review along the Tagus River Basin regarding mostly agriculture, livestock, and urban activities. To estimate reactive nitrogen (N_r) load into surface waters, emission factor approaches were applied. N_r pressures are much higher in Spain than in Portugal (~13 times), which is mostly because of livestock intensification. Some policy and technical measures have been defined aiming at solving this problem. Main policy responses were the designation of Nitrate Vulnerable and Sensitive Zones, according to European Union (EU) directives. Nitrate Vulnerable Zone comprise approximately one third of both territories. On the contrary, Sensitive Zones are more extended in Spain, attaining 60% of the watershed, against only 30% in Portugal. Technical measures comprised advanced urban and industrial wastewater treatment that was designed to remove N compounds before discharge in the water bodies. Given this assessment, Tagus River Basin sustainability can only be guaranteed through load inputs reductions and effective transnational management processes of water flows.

Keywords: agriculture; impact; measures; nitrogen; Sensitive Zones; Tagus River Basin; Vulnerable Zones

1. Introduction

The European Nitrogen Assessment [1] estimated that around 80% of European freshwaters exceed the nitrogen threshold for high risk to biodiversity and human health. Nitrogen can reach surface waters from point sources, which include municipal wastewater treatment plants and industrial discharges, and diffuse sources, including agricultural runoff of fertilizers and animal wastes, as well as atmospheric deposition. While point sources are relatively simple to regulate since they tend to be a continuous input over time, entering the water body at a specific location, diffuse sources are difficult to quantify and regulate because transport is via hydrologic flow paths and by wet and dry deposition [2].

Agriculture is recognized as a major diffuse source of water pollution [3]. While N fixation, atmospheric deposition and the application of treated sewage sludge can all be important; typically, the major nutrient inputs to agricultural land are from mineral fertilizers and organic manure from livestock [4]. European Union (EU) is a large food supplier and fertilizers are essential to sustain production [5]. High amounts of N are applied in natural and mineral fertilizers as compared to plant requirements, thus creating N surpluses [3,6]. The magnitude of these surpluses reflects the potential for detrimental impacts on the environment, including upon water quality. N leaching from arable fields to groundwater and surface water contributes to stream N loadings [7–9]. N leaching depends on several factors, primarily fertilization level, type, and timing of fertilizer application; the method of their application to the soil; properties of soils, types of crops and their fertilizer requirements; method of cultivation and agronomic practices; and, the level of animal production [6,10].

While farms represent the smallest operational units at which agricultural management decisions are taken and determining the fluxes of N that are associated with agriculture [11], regional watersheds or river basins, composed of a mosaic of interacting natural, semi-natural, agricultural, and urban landscapes, are the most convenient units at which to describe, but also to manage, the anthropogenic alteration of the N cycle [12,13]. Apart from land use, climatic conditions also have a crucial impact on the intensity and quantity of N leaching at his level [14]. Although certain catchment processes can attenuate much of the nutrient surplus, a significant proportion can still be transported to freshwater, and hence termed an emission. Apart from such reactive nitrogen losses, livestock production and cities metabolism are other sources of N emissions into water bodies [15].

The climatic characteristics play an essential role in the flow regime and influence land uses and water management practices within the basins, both directly and indirectly, being crucial for nutrient transformations and transport [16]. There are several studies regarding water quality in river basins using different methodologies including monitoring, simple mass balance methods, and mathematical modelling (e.g., [17–22]). Nevertheless, river basins in the Mediterranean regions show some particularities that are associated with the semi-arid climatic conditions. In this region, N dynamics has been shown to be substantially affected by flow regulation infrastructures and changes in flow-paths that are related to irrigation facilities [23]. Lassaletta et al. presented an overall N budget in the basin and detailed N calculations in different sub-catchments, within the Ebro River catchment (NE Spain), and hypothesized that agricultural and water management practices had a major influence on N retention [24].

An additional challenge that increases the complexity of N_r control is the transboundary nature of water pollution [25,26]. This is a critical issue in the western part of Iberian Peninsula, where Portugal and Spain share five international watercourses. Indeed, roughly 60% of Portuguese surface waters income from Spain and Tagus River is crucial in that regard, being the longest one. It connects the regions of both countries capitals, Madrid (Spain) and Lisbon (Portugal), being a driving force of social welfare and biodiversity. However, Tagus River Basin is also an outstanding example of water management constraints. An increasing water stress and flow diversion to southern basins are among the most significant ones [27,28]. Besides, although a more consistent delivery of minimum flows has been prescribed in the 2008 amendment of Albufeira Convention, it is doubtful if it is in accordance with ecological requirements of Tagus River. Moreover, both Tagus River Basin Management Plans that were issued in 2015 did not directly address N fate at transboundary level [29]. All of those evidences demand an urgent N assessment at Tagus River Basin scale.

Water resources contamination by intensive agriculture practices and urban wastewaters fostered EU legislative initiatives, respectively Nitrates Directive (91/676/EEC), Groundwater Directive (2006/118/EC), and Urban Wastewater Treatment Directive (91/271/EEC). The first two directives Directive intends to guarantee that nitrates concentration in groundwater remain below a threshold of 50 mg L^{-1}, supporting a low N agriculture footprint. In turn, urban wastewater treatment directive is aiming at avoiding eutrophication processes and imposing stringent limits regarding nitrogen and phosphorus content in urban discharges, thus requiring the use of advanced processes for their removal. Despite these top-down initiatives that are led by governmental agencies, both point and diffuse N pressures remain very significant in European watercourses, according to most recent Tagus River Basin Management Plans [29].

Mathematical modeling contributes more in-depth knowledge regarding complex processes in water management and related N fluxes [30–32]. The quality of primary data and fit for purpose are the main aspects in model selection and a compromise may be necessary when considering the available time and data. In some cases, experience based analysis and elementary models with few parameters may provide useful information for decision-making purposes in river basin management [33]. A similar approach is used by IPPC (Intergovernmental Panel on Climate Change), 2006 (tier 1), namely for N leaching quantification from manure and soils.

Although there are some articles that are focusing on the environmental issues of individual Mediterranean River basins, and publications summarizing hydrogeochemical and nutrient conditions, a transboundary approach was missing. Thus, aiming at understanding main N transboundary pressures and gap identification, this study provides the first assessment of N_r pressures and responses comprising Portuguese and Spanish Tagus River Basins. First, a general characterization of the Tagus River Basin is presented. Then, the main socio-economic drivers, as well as the anthropogenic pressures in the Tagus River Basin, are identified. The water quality status and the environmental consequences resulting from these conditions are described too. Key policy measures are discussed and sectorial loads are addressed using an emission factor approach. Finally, gaps are identified and recommendations for a better integration of N and water resources management are indicated.

2. The Case of the Tagus River Basin

2.1. General Caracterization

Tagus river spring is located in the Albarracín hills in Spain, at a height of 1593 m above sea level. It then flows 1102 km down through Spain and Portugal connecting Madrid and Lisbon on its way to the Atlantic Ocean where it forms one the world's largest estuaries with 320 km^2 (Figure 1) [34]. This river flows between the latitudes 40°19′28″ N and 38°55′17″ N, and the longitudes 1°41′26″ O and 9°00′38″ O. In Spain, the Tagus Basin holds six main tributaries: Jarama, Alberche, Tietar, Alagon, Guadiela, and Almonte rivers. In Portugal, it receives six tributaries too: Erges, Pônsul, Ocreza, Sever, Zêzere, and Sorraia rivers, being the last two the most important ones due to the size of their river basins (4980 and 7520 km^2, respectively), totaling around 50% of the Portuguese basin area [34].

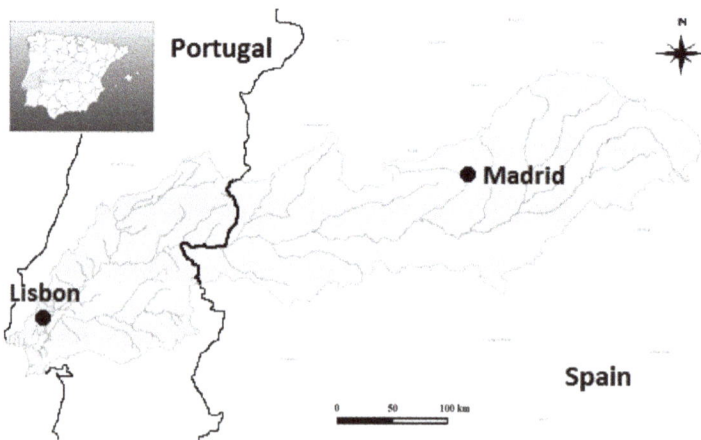

Figure 1. Tagus River Basin in the Iberian Peninsula (built with QGIS) [34,35].

The Tagus River Basin presents an area of 80,500 km^2. Roughly two-thirds of the catchment are Spanish territory (69%), the remaining one-third is Portuguese (31%) [34]. The climate is temperate Mediterranean, with a dry period of two months for July and August. The average annual temperature varies between 7.4 °C (in areas further north and higher altitude) and 16.9 °C (in estuary area), and the annual rainfall is between 2744 mm (in the northern part of the region and at an altitude of more than

1300 m) and the 524 mm (in the south, near the coast). In wet years, the annual rainfall is about 130% of the precipitation at normal year, while in dry year, this only reaches about 70% of normal precipitation. The annual potential evapotranspiration varies between around 500 mm in areas further north and the higher altitude, while the higher values, at around 800 mm, are found on the South side of the river basin district. The Thornthwaite climate classification shows great variability, ranging from climate super-wet to sub-humid dry (C1). In natural regime, the average discharge is 600 m^3 s^{-1} and the total volume per year is on average 19 km^3, with 66% being generated in the Spanish basin and 34% in the Portuguese basin. At present, approximately nine million people live in the basin, which contains the capital cities of both countries. The river is highly regulated with a large number of dams, creating a total storage capacity of nearly 14 km^3, of which 80% in Spain. Installed hydropower potential amounts to 3300 MW and the mean annual power production is approximately 5000 GWH. In the Portuguese territory, there are almost 2150 dams, providing a theoretical useful storage capacity of about 2.5 km^3. Most were built for hydropower production, but given the characteristics of hydrological variability, the volumes of water stored in reservoirs add some resilience under scarcity periods [34–36].

The Spanish Tagus Basin attains almost twice the Portuguese Tagus Basin area and number of inhabitants (Table 1). Cultivated area of temporary crops in the Spanish Tagus River Basin is four times larger than that of Portugal [37]. In terms of water demand, 80% of the water use is related to agricultural needs and the 20% remaining for drinking-water production, since both Madrid and Lisbon use the Tagus River Basin as the source of their water supply. Water abstractions for agriculture purposes attain 1929 hm^3 y^{-1} and urban water supply reaches 741 hm^3 y^{-1} in the Spanish Tagus River Basin [34]. In Portugal, Tagus water is strongly demanded for the irrigation (~1173 hm^3 y^{-1}) of intensive rice paddies, orchards and arable crops area (1482 km^2), and for drinking water supply (~392 hm^3 y^{-1}). It should be pointed that Lisbon drinking water is abstracted from a reservoir that is located in a Portuguese Tagus tributary, Zêzere River. Table 1 shows the Tagus River Basin main dimensions and land use characteristics [34,35,37,38].

Table 1. Tagus River and land use characteristics.

Countries	Drainage Area (km²)	River Length (km)	Total Population (10⁶)	Field Crops (10⁶ ha)	Trees and Vines (10⁶ ha)	Forests (10⁶ ha)	Pastures (10⁶ ha)
Portugal	25.026	275	3.5	0.4	0.2	0.7	0.008
Spain	55.781	827	7	1.6	0.2	1.25	0.05
Total	80.807	1102	10.5	2	0.4	1.95	0.06

Note: Weighted average according to River Basin area information [34,35,37,38].

2.2. Socio Economic Drivers and Pressures

Economic activities and most of the population in the Tagus River Basin are concentrated in the Madrid and Lisbon areas. In clear contrast to the urbanized areas, the remaining part of the Tagus River Basin is dominated by the use of water for extensive agriculture. Yet, the availability of these resources is disputed by the more productive irrigated agriculture in the Segura River basin that is close to the Mediterranean coastline, and which is connected to the Tagus by a water transfer channel. Indeed, part of the Tagus flow in upper Spain is diverted to the Segura basin for irrigation and food production. This is a main threat on river Tagus sustainability and a very complex pressure to manage, namely during drought events [39].

2.2.1. Portuguese Tagus Basin

Looking to all of the Portuguese hydrographic regions distribution, a stronger concentration of economic activities point out Tagus Basin as hosting more than 55% of the product, activity, and investment, and almost 50% of the existing establishments and employment in Portugal. This is the largest Portuguese basin representing 31% of the mainland area, 39% of the population, 48% of the employment, and 57% of the production [40].

The urban sector in Tagus Basin is responsible for an annual water consumption of about 297 million m^3 (46% of the mainland's total), with an average per capita around 205 L d^{-1}, when considering both resident and a strong influx of non-permanent population. Inside the Tagus Basin, Lisbon is the region with the highest household income (40,090 € y^{-1}). Despite the highest household income in the context of the eight hydrographic regions (1st place in 8) and representing about 39% of the mainland's population, it includes regions of great social fragility [34,40].

In 2012, the agricultural sector represented only 1.1% of Gross Value Added (GVA) (a productivity metric that measures the contribution to an economy of an individual producer, industry, sector, or region) and 2.3% of the employment in Tagus Basin, placing the region in the last position in terms of economic relevance. However, about 40% of the region's total area is dedicated to agriculture, corresponding to 11.221 km^2 of utilized agricultural area (UAA). The irrigated area counts 148.148 ha (13% of the utilized agricultural area), with around 25,000 farms, corresponding to smaller but more productive farms (for 46% of irrigated GVA only 20% of annual work unit-AWU) than average in the mainland. Economic indicators show that irrigation activity in this region is of much less intensive labor: 0.19 AWU ha^{-1} when compared to 0.30 AWU ha^{-1} in mainland, and much more productive: per unit area (144% mainland's mean), per unit work (233% of the mainland's mean), and per unit of m^3 of used water (134% of the mainland's mean). The 25,000 mentioned above consume close to 1.173 million m^3 y^{-1} of water (about 34% of total irrigation consumption in the mainland), making the intensity of water use in the agricultural sector higher than in mainland average: mainly per farms (201% of the mainland's values), but per irrigated area unit too (108% of the mainland's consumption per ha), although as seen before water productivity is also higher than mainland's average (134%). Investments in irrigation infrastructure have contributed to improved water storage and distribution capacity, as well as to promote and use precise irrigation technologies, while playing an increasing role to reduce environmental pressures and to adapt to climate change [34,40].

In 2012, industry represented 14% of GVA and 11.5% of employment in Tagus Basin, placing it in the second position in terms of economic relevance. Tagus is the most important industrial basin of the country, with a weight that remained at 36% along the analysed years (2007–2012), with corresponding pressures into the river basin. The annual "product" of energy sector represents about 2% of the Gross Domestic Product (GDP) (a monetary measure of all the finished goods and services produced in a specific time period) of the country. During 2007–2012, this sector registered a strong activity expansion, in contrast with the dominant depressive trajectory, which was reflected in a GVA increase of 19% (annual average of 3.8%) supported by the increase in both turnover (more 38%) and number of establishments (plus 20% between 2008 and 2012). The importance of Tagus Basin in the "energy" sector is distinctly marked in economic terms: it has an overwhelming weight representing between 80% and 90% of GVA and employment volume, placing third in the ranking of the hydrographic regions in terms of hydroelectric power generation, with an average of 12% of the total in 2010–2014. It should be noted that the productivity levels in the use of water and the intensity of its use are qualitatively below the national average [34,40].

In 2012, the tourism sector represented 3.5% of GVA and 8% of employment in Tagus Basin, placing it in the fifth position in terms of economic relevance. Tourism and recreation is associated to various activities related to nature, landscape, and rural areas, linked to historical heritage or cultural and sporting events. Lisbon and surrounding areas plays an increasing role as a tourist attraction and the whole coastline is conducive to new activities.

Aquaculture production that is directly attributable to Tagus Basin has been increasing (274 tonnes in 2011 as compared to 509 tonnes in 2014). About 50% of aquaculture production is an extensive regime, corresponding to less significant pollutant loads as compared to the intensive and semi-intensive regimens [40].

2.2.2. Spanish Tagus Basin

In the Spanish side of Tagus Basin, the densely populated Madrid metropolitan area (6,271,000 inhabitants) has the greatest impact in the region. Here, we find most of the urban water uses, and also the two irrigated areas that are involved in the assessed water transfers [35,39,41]. Madrid registered a population grown by almost 1.5 million people within the last 15 years, from 5,022,000 inhabitants in 1996 to 6,458,000 inhabitants in 2010, at an average annual rate of 2.04%. Population density has also risen from 625 to 805 inhabitants km^{-2}. The attracting power of the Madrid area is explained by the rapid economic growth at an average rate of 3.3% until the end of 2007 [39,41]. Even accounting for three years of economic decline after 2007, GDP per capita had a positive growth rate and increased from EUR 19 755 in 1996 to EUR 23 636 in 2010. In fact, GDP have a mean increased rate around 3% every year resulting in a total GDP growth rate around 60% between 1995 and 2010 (Figure 2). The GVA of Madrid community represents 89% of the total basin Gross Value Added [41]. The main engine of growth up to 2007 was the construction sector but the service sector also grew, and is actually still growing more than average, with a current share of circa four fifths of the regional GDP. Agriculture has never been an important source of growth and its contribution to the overall added value is declining, representing less than 0.6% of GDP (Figure 2). The other potential water user, the manufacturing industry, has being shrinking for more than a decade and its output is nowadays 10% lower than in 2000. The share of the overall regional production has been consequently declining (from nearly 15% in 1995) to less than 9% in 2010 [39].

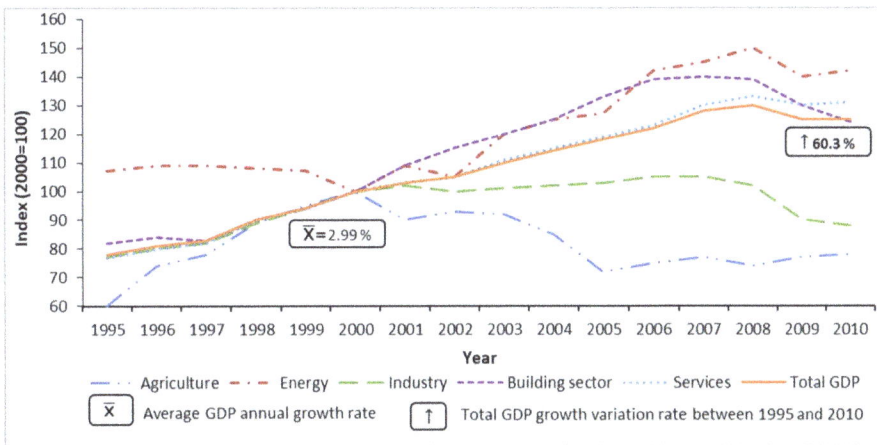

Figure 2. Total Gross Domestic Product (GDP) growth and sectorial evolution, total GDP growth variation rate between the sixteen analysed years and average GDP annual growth rate in Spanish Tagus Basin (1995–2010), chained indexes (2000 = 100) [39].

Economic growth in Madrid was led by the services sector, which explains more than 60% of the total economic growth and 93% of employment generation in the area for the period 1995–2010. The provision of the necessary inputs (such as water) to this sector has laid the foundations to generate most of the economic and employment growth. Agriculture, which is the sector from which water rights are exchanged from, has a minor importance for employment (0.4% of total employment in the region in 2010 and even decreasing during the period 1995–2010), and output (0.1% of total GDP in 2010, also decreasing). Given the low quality of the soils in the Madrid area, agriculture is a receding activity, and, in some areas, water allowances are higher than the effective demand for irrigation water [39]. Allowing for transfers of water rights from agricultural to urban uses may allow for using use water resources that are not being effectively used in the present.

3. Simplified Nitrogen Assessment

3.1. Datasets

Nitrogen data from the ecological potential and status of surface waters in Tagus Basin was obtained mostly from the last River Basin Management Plans reports. In Portugal, these plans were reported by the Portuguese Environment Agency ("Agência Portuguesa do Ambiente", APA) [34], while in Spain, they were reported by the Spanish Hydraulic Administration ("Confederación Hidráulica del Tajo", CHT) [35]. Datasets about monitoring stations and reservoirs in Tagus Basin, regarding total N and NO_3^- concentrations were obtained from CHT [42] for Spain and from SNIRH, a National Water Resources Information System [43], for Portugal. Portuguese geographic files used in QGis were taken mostly from SNIAmb data (National Environmental Information System) of Portuguese authorities [44]. Spanish maps were adapted through the online GeoPortal (access portal to geographic information of Spain), which uses IDEE data (Infraestrutura de Datos Espaciales de España) of Spanish authorities [45].

Potential N loads in Tagus River Basin were estimate using available emissions factors, datasets, and statistics that were published in Decree-laws. Agriculture diffuse N loads to surface waters were estimated according to land use information from the Corine Land Cover [37] and emission factors of agriculture practices [46] (Table 2).

Table 2. Nitrogen export coefficients, according to agriculture practices.

Sources	Sub-Categories	Emission Factors (kg N_r ha^{-1} y^{-1})	Ref.
	Temporary rainfed crops, irrigated crops and rice paddies	5.0	
Agriculture practices	Vineyards, orchards, olive grooves	2.7	[46]
	Permanent grassland	1.5	

N loads originated by livestock production were based on the Portuguese agricultural census that was carried out by the Portuguese national statistics institution (INE) [47] and on the Spanish CHT data, based on the Impress II data of Corine Land Cover and livestock facilities inventories [35]. Excretion factors and correspondence normal head were taken from the Decree-Law 214/2008 of 10 November (republic diary number 218) and Ordinance 259/2012 of 28 August (republic diary number 166), both in force in Portugal.

For the urban sector, *per capita* N discharge factors used were the ones reported by Johnes [33] regarding urban Wastewater Treatment Plants with biological secondary processes and on-site anaerobic septic tanks: 2.14 and 2.49 kg N y^{-1}, respectively. An additional removal efficiency of 86% was considered if the advanced biological processes are implemented [48].

3.2. State and Impacts: Water quality status in Tagus River Basin

Nitrates are considered in Water Framework Directive Annex VIII [29], and they belong to physico-chemical parameters that are required for an integrative ecological status assessment. Figure 3 shows surface waters ecological potential along Tagus River Basin and the two N monitoring points used.

In the Spanish part of Tagus watershed, 63% of rivers and 71% of lakes present Good ecological status, 31% of rivers and 29% of lakes present Moderate status, and 6% of rivers display Poor status. Concerning the ecological potential, 59% of the artificial/heavily modified water bodies display Good or Higher status, 35% Moderate status, and 2% Poor status (4% remain unknown). For groundwater bodies, 75% are classified with Good Global status and 25% have Less than Good [35]. Concerning the Portuguese part (including the west side), 53% of surface water bodies have Good ecological status, 26% Moderate status, and 21% Poor status. As for the ecological potential of artificial/heavily modified water bodies, 14% present Good status, 46% Moderate status, and 27% Poor status (13% remain unknown). For groundwater bodies, 90% are classified with Good Global status and 10% have Poor classification [34].

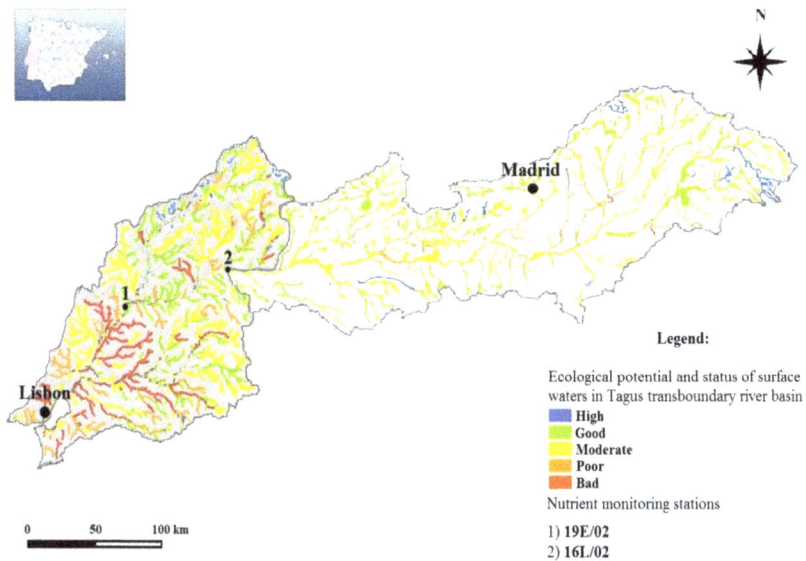

Figure 3. Surface waters ecological potential and status along Tagus watershed reported in 2015. N monitoring stations: (1) "Valada do Tejo" (code: 19E/02); and (2) "Monte Fidalgo – Cedillo" (code: 16L/02) (built with QGIS) [34,35].

3.3. Nitrogen Loads in Tagus River Basin

3.3.1. Livestock

N loads derived from livestock activity were estimated from the average amount of nutrients that were excreted annually by "normal head" for each animal species and after having the normal head numbers per specie and per parish (only parishes with 50% or more of land within the Tagus Basin were considered), total N load was estimated based on the average amount of total N that was excreted annually by normal head. Calculations taking account that the N loads emitted by manure vary according to the animal category, species, nutritional habits, and weight of the animal, along with manure management characteristics [49]. An average concentration of 25 g N kg^{-1} in manure was considered [50]. Finally, to determine N leaching into water bodies, after manure deposition in the soil, an approach consisting in export taxes was used. These taxes range from a mean for N around 10–17% considered by different authors [33,51,52]. It was assumed that 17% of the N load reaches the water bodies of the river basin where the livestock farm is located, and already take into account the percentage of manure that is applied on land after storage losses [33]. In Portugal, around 9% of the total holdings with livestock have manure storage facilities, while in Spain, this percentage is almost four times higher (~44%) [53].

3.3.2. Agriculture

The methodology that was used to estimate N loads from agriculture was based on emission factors for each land use classes, corresponding to the diffuse N load that will be transported through the surface runoff from the drainage area for each water body. Identification and spatial distribution of land use classes in the Tagus Basin were determined using the land use map [37], which allowed for the use of a geographic information system (QGIS) to define the exact percentage of each land use classes. Then, N export coefficients [46] were applied to these percentages of land.

3.3.3. Urban

The approach used to determine the rejected N loads was based on removal coefficients, according to Wastewater Treatment Plants performance. Most of these urban treatments existing in the Tagus Basin have secondary treatment only (80%). The more advanced treatment has also disinfection or a N removal step. Tagus Basin has about 14% of this treatment degree. Urban wastewater discharges without treatment were considered to be 8% in Portugal and 4% in Spain [54]. Annual N load at the international border between Spain and Portugal was estimated based on the "Monte Fidalgo–Cedillo" reservoir (Figure 3) monitoring between 2010 and 2015 and considering an annual average flow of 210 m^3 s^{-1} at "Monte da Vinha", which is a hydrometric station located too on the border between Portugal and Spain [42,43].

3.3.4. Intersectorial Assessment

As mentioned previously, N pressure to surface waters derives mostly from agriculture and urban settlements in the Portuguese Tagus River Basin. On the contrary, in the Spanish Tagus River Basin, the main challenge is potential livestock emissions: estimated N load into Spanish Tagus river is ~242 kt y^{-1}, which is much higher (~13 times) than Portuguese discharge, only ~18 kt y^{-1}. The main contributor to N losses in Portugal is agriculture (Figure 4), while in Spain, it is livestock, which is more than 50 times higher than in Portugal. N pressures in urban and agriculture sectors in Spain are higher than in Portugal too. Riverine export from Spain to Portugal (14 kt y^{-1}) is slightly smaller than the total N load discharged from the three main sectors in Portugal (18 kt y^{-1}). However, only 3 kt y^{-1} is discharged into the estuary (Figure 4). The estimates of potential N loads from different sectors contribute to set up a simplified N mass balance along Tagus River Basin, which is depicted in Figure 4.

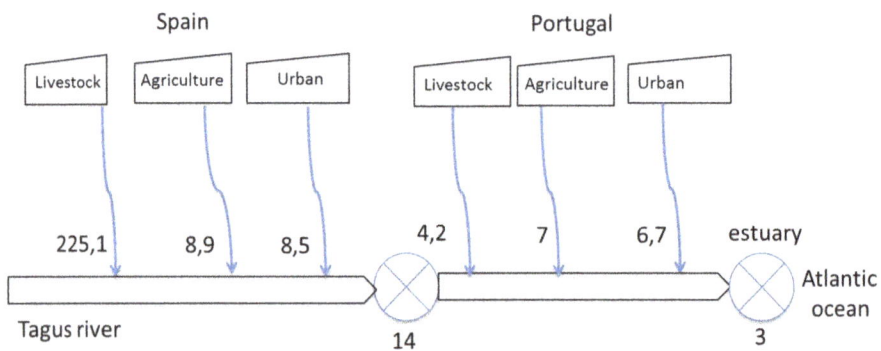

Figure 4. Estimated potential N loads (kt y^{-1}) into Tagus River Basin surface waters per sector (livestock, agriculture, and urban), pressure at the Spain/Portugal border, and the discharge at the Tagus River estuary, in Atlantic Ocean for 2009–2015 [34,35,37,42,43,54].

This integrated N assessment of Tagus River Basin confirms the presence of significant challenges in both countries, Portugal and Spain. However, their magnitude is different. Agriculture, livestock and urban sectors discharge higher N loads in Spanish Tagus than in the Portuguese Tagus. Livestock is the most important sector, but, while Portuguese reports a livestock density of 1.1 ha^{-1} per hectare, Spain reports values that are ten times higher, 11.4 units ha^{-1} [35] that is spread along a higher area, leading to a global livestock load that is ~120 times higher than in Portugal [35,47]. Thus, when considering the same N exports coefficients in both countries, one could expect water body's quality to be the worst in Spain. However, this is not the case. One reason may be the type of livestock holdings and the number of manure storage facilities that are much higher (~4 times) in Spain than in Portugal [53]. Good manure management practices and effective wastewater treatment may support Spain's good ecological quality status (Figure 3). In terms of agricultural N loads, the difference in crop area in both of the countries explains the higher magnitude in Spain, e.g., for temporary crops. Finally, regarding the urban sector, it could be expected that Spain displayed a higher N load than Portugal because of a higher population (Table 2). This effect is attenuated by an increased N removal that is imposed by an extensive Sensitive Zone designation, which aims at a significant pressure reduction from Wastewater Treatment Plants loads.

Regarding N loads pattern in Portugal in 2014 [55] and when compared with the values that were estimated in the present work, it is possible to identify an important livestock load increase (~2.5 to 4.2 kt y^{-1}), while agriculture had a slight rise only (~6.5 to 7 kt y^{-1}) and the urban sector a significant shrinkage (~12.9 to 6.7 kt y^{-1}). In fact, according to Portuguese National Statistics [47,56], animal production in 2015 increased in all sectors (meat, eggs, milk, and processed dairy products). Production of livestock meat increased by 5.0% in 2015, a general trend extended to all livestock species. Not so pronounced, but with a similar trend too, is the N net balance in the soil with 100,000 tonnes of N in 2015, which is equivalent to 27 kg N ha^{-1} of agricultural area (26 kg N ha^{-1} in 2014) [56]. When compared to 2014, N net balance increased by 4.6%. This evolution was justified by the increase of 2.3% N incorporation in soil when compared to 2014 (more than 7.2 thousand tonnes of N), in particular due to the incorporation of manure in the soil (more 2.6%). On the other hand, in 2015, the N removal from the soil by agricultural crops, fodder, and pasture increased by 1.6% (more 2.8 thousand tonnes of N) [47,56]. Between the two periods, total N_r pressure in Tagus River Basin decreased substantially in Portugal (22 to 18 kt y^{-1}). Several reasons can contribute to such a decrease, but data reliability is not very conclusive. Nevertheless, a main reason could be the gradual improvement of wastewater treatment services [34].

3.4. Total Nitrogen, Nitrate Fate and Transboundary Pressures

A downward trend of the Total N concentration was observed in Portuguese-Spanish border at "Monte Fidalgo" monitoring station from 2013, although there was an increase in NO_3^- N concentration in 2015, probably from agricultural activities intensification. Between 2010 and 2015, the total N concentration was always higher in the monitoring station "Valada do Tejo" than in the upstream station (Figure 5B). In 2015, total N concentration dropped significantly below the detection limit that was reported by CHT, which is 2 mg L^{-1} for total N and 2.5 mg L^{-1} for nitrates [35]. Until 2012, nitrate concentration in "Valada do Tejo" was smaller than "Monte Fidalgo" (Figure 5A). In 2013, it remained the same for both stations, and, since then, "Monte Fidalgo" recorded values below the other station with significance differences in 2015. In order to assess transboundary impacts, average total N and NO_3^- concentrations in the two N monitoring points of Tagus River Basin are displayed in Figure 5.

Water quality monitoring in Tagus mainstream shows a biased pattern. For the study years, "Valada do Tejo" has higher N concentration than upstream "Monte Fidalgo", but the same did not happen with NO_3^- concentration, being more expressive in 2011 and 2012, possibly because the intensive agriculture that occurs at Valada Zone (Figure 5). As a rule, N concentration in the Portuguese-Spanish border of Tagus is around 1 mg L^{-1}, which is probably due to the contribution of Tagus tributaries near the border causing a dilution effect (e.g., Alburrel, Aurela and Salor streams) and to the N consumption in Alcántara and Cedillo [57]. Annual riverine export monitored in Spain-Portugal border is still high (14 kt y^{-1}) (average 2010–2015), an amount similar to the N discharged downstream by Portuguese agriculture, livestock, and urban sectors (~18 kt y^{-1}). The difference in source discharges and transport at border and final input at coast attains 94% in Spain and 91% in Portugal and in total for the entire Tagus Basin is 99%. In Denmark, the national average N retention in groundwater and surface water accounts to 72% from source to coast. Biological reaction may be the reason, as several reservoirs exist along Tagus River: higher hydraulic residence times and warmer climate with excess carbon could fuel denitrification in the surface waters. These results show that the monitoring gaps in Tagus Basin are quite important regarding N fluxes, thus hindering a comprehensive calibration of any phenomenological model. Better information on water quality coupled with surface and groundwater fluxes quantification, besides biological activity and atmospheric emissions, are necessary. In addition, N fate in Tejo estuary could be also meaningful to induce anoxia in coastal areas, but the issue has not been addressed yet.

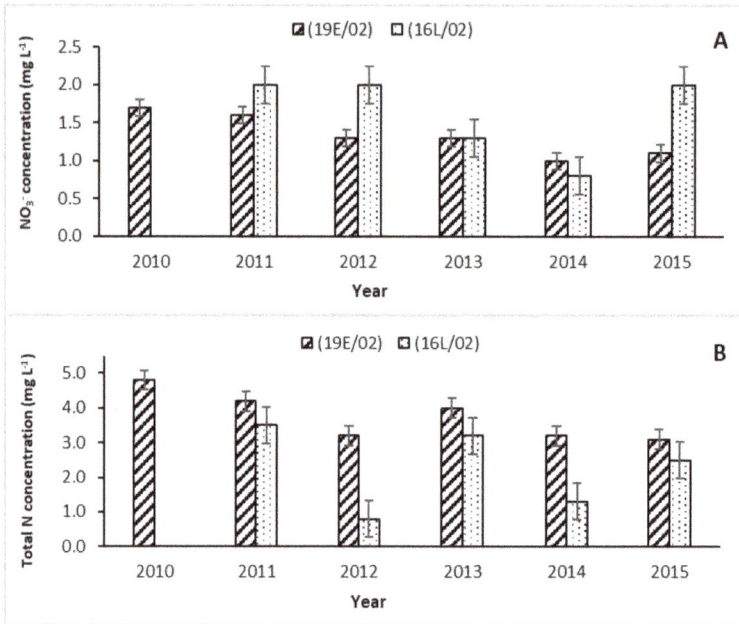

Figure 5. Average NO_3^- (**A**) and total N (**B**) concentration (mg L^{-1}) in Tagus River Basin between 2010 and 2015. (Monitoring stations: code: (19E/02)—"Valada do Tejo"; code: (16L/02)—"Monte Fidalgo") [42,43].

4. Policy Measures

Despite the fact that Tagus is an international basin, each country issued its own Management Plans in 2015 and not an international one, as foreseen by Water Framework Directive in its article 13(2). Even so, the Programme of Measures foreseen in both Tagus River Basin Management Plans comprises a set of initiatives that are related with N issues [34,35]. Among them are agriculture best practices and further wastewater treatment upgrade for nutrient removal. The 2016–2021 planning cycle is expected to spend 229 M€ for Programme of Measures implementation in Portuguese Tagus River Basin Management Plans, including 174 M€ specifically for sanitation and pollutants removal [58]. To achieve a decrease of diffuse and point sources in Spanish Tagus River Basin for the same planning cycle will require a higher investment in N control than in Portugal, 1.452 M€ [35].

Besides the initiatives that were encompassed in Tagus River Basin Management Plans, an effective cooperation between Portugal and Spain is necessary in accordance to Albufeira Convention goals. Although a joint co-management is difficult, some steps could be implemented, such as, for example, joint best agricultural management practices for the entire Peninsula, including the appropriate measures for the edafo-climatic common regions. In the Programme of Measures definition, a cost-effectiveness analysis will be advisable in order to find the best upstream-downstream measures [59]. Joint monitoring systems would be a step ahead for achieving water quality compromises [60], as well as a common interpretation of the EU Directives. Above all, transparency and a far-reaching coherence regarding the policies in Tagus transboundary river basin is required in order to prevent negative trade-offs [61].

4.1. Nitrogen and Water Resources Protection Policies: Protected Areas

Nitrates Directive (91/676/EEC) and Urban Wastewater Treatment Directive (91/271/EEC) are central EU legal tools embedding anticipative water management policies. In Portugal, the first legal initiative on groundwater protection was issued in 1997, with the release of Decree-Law 235/1997 of 3 September (republic diary number 203). Ground waters with more than 50 mg L^{-1} of nitrate concentration are considered to be Vulnerable Zones [62]. Tagus Nitrate Vulnerable Zone was implemented only in 2004 through the publication of Ordinance 1100/2004 of September 3 (republic diary number 208). This Nitrate Vulnerable Zone covers ~36% of river watershed and is the largest one in the country, with a total area of 2416.86 km^2. The other Nitrate Vulnerable Zone was Estremoz-Cano issued in 2010 through the publication of Ordinance 164/2010 of March 16 (republic diary number 52) and attains 207.1 km^2. In both cases, an intensive agricultural activity is present, namely maize, tomato, and rice [62]. Despite such measures, from 2012 to 2015, more than 50% of monitoring data stations recorded a nitrate concentration average value of 36 mg L^{-1} and less than 25% recorded an average value of 44 mg L^{-1} in aquifer (0–5 m) of Tagus Vulnerable Zone. In the karst aquifer of the Estremoz-Cano Vulnerable Zone, more than 50% stations registered an average of 37.5 mg L^{-1} during the same period and less than 25% registered an average of 12.5 mg L^{-1} [34,62]. In Spain, 7 Vulnerable Zones that spread 31% of Tagus watershed area (17,064 km^2) are in force [35,63]. Vulnerable Zone effectiveness is controlled by 46 network stations. From 2008 to 2014, five groundwater bodies presented at least a representative point of analysis above 50 mg L^{-1}. This accounts for 50% of total groundwater bodies [35,63].

Regarding Sensitive Zones in terms of nutrients and eutrophication risks, only in 2008 the first zones were designated in Portuguese Tagus: Pracana and Maranhão reservoirs and associated watersheds, covering 13% of the Tagus River Basin (Figure 6) [34]. On the contrary, a stronger policy was adopted in Spain, namely downstream Madrid region: 53 Sensitive Zones were identified that influence areas to a sum total of 32 815 km^2 (60% of the river watershed) require biological N removal in wastewater treatment plants [63]. Vulnerable and Sensitive Zones along the Tagus River watershed are shown in Figure 6.

Portuguese water authorities are concerned with nitrate concentration in the Tagus Vulnerable Zone [62]. Regarding the Estremoz-Cano Vulnerable Zone, the Action Program, which was issued by Portugal in Ordinance 259/2012 of 28 August, may induce a water quality improvement in the next 10 years [62]. On the other hand, Vulnerable Zone in Spain Tagus does not go through this problem despite some records of nitrate contamination [35]. The effectiveness of the designation of large Sensitive area in Spain (53 zones) seems evident by the water ecological status in the Spanish Tagus River Basin [63]. This is a good result that deserves further considerations because other European countries apply environmental protection policies with similar results. For instance, the whole Denmark territory is designated as a Nitrate Vulnerable Zone, and due to the application of stringent wastewater treatment removal (N and P) over all of its territory, Denmark was not required to identify Sensitive Zones for the purposes of the Directive [29,64]. The good water status goal has been achieved in Spain, despite a highly regulated water flow regime in Tagus headwaters, namely the aforementioned water abstractions towards Segura and Jucar river basins (Figure 3) [65]. Such water transfers to adjacent river basins from upstream Tagus River Basin supported strong debates over pros and cons, confronting the agricultural needs in Spanish southeast zones with nature conservancy goals in Tagus River Basin, including in Portugal [66].

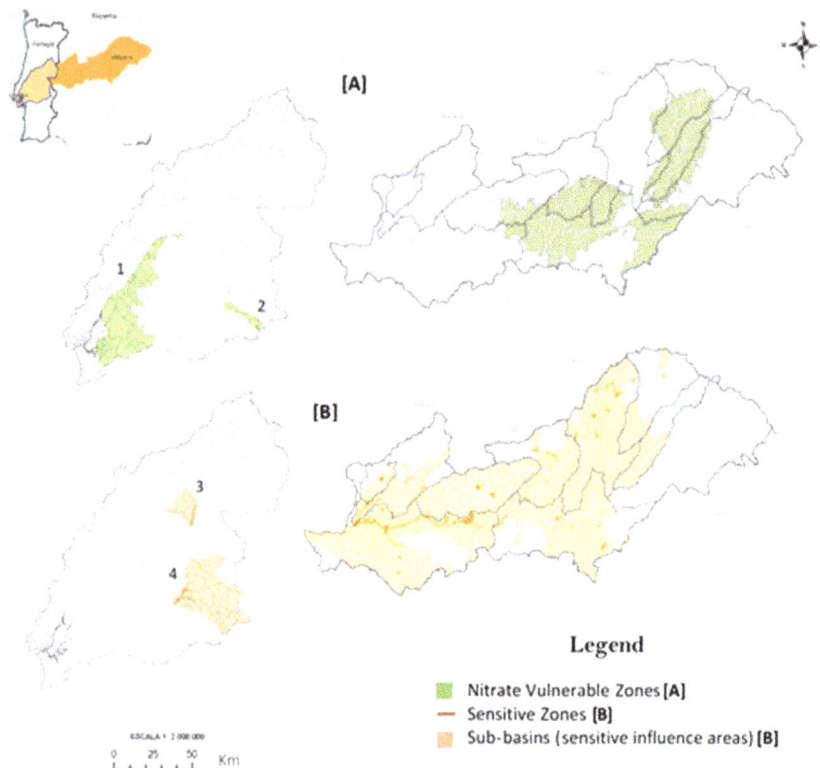

Figure 6. Nitrogen Vulnerable [A] and Sensitive [B] Zones in the Portuguese and Spanish sides of the Tagus River Basin (built with QGIS) [34,35,44,45]. Numbers correspond to Portuguese Tagus sections of Vulnerable (1-Tagus; 2-Estremoz-Cano) and Sensitive (3-Pracana reservoir; 4-Maranhão reservoir) Zones.

4.2. Best Management Practices

The Water Framework Directive [29] and its daughter Directives recognize the urgent need to adopt specific measures against the contamination of water by individual pollutants or a group of pollutants that present a significant risk to the quality of water. There are some useful tools and techniques that can be used by farmers, technicians or government entities to improve their knowledge and mitigate this problem on Tagus Basin.

One of them is the use of mathematical models to assess the fate of N in the soil–crop environment from the field to the watershed scales. The Root Zone Water Quality Model (RZWQM) [67–69] was already successfully used at the field scale in some areas of the Tagus basin. The model simulates N transformations, uptake, and transport in the soil–crop systems, making it possible to evaluate field measured soil hydraulic properties and to predict N related variables for different boundary conditions (irrigation and fertilization) with a good accuracy. Thus, through scenario analysis, the best management practices that were protecting surface and groundwater quality can be selected. As for every process based mathematical model, each time that RZWQM is used in a new soil–crop system, it has to be calibrated. Despite this, it is essential to first have a good estimation of the water balance components (evapotranspiration, leaching, and soil water storage) before applying the model [68].

Researchers of the Marine Research Institute, an ecological modelling center in Portugal, have demonstrated that salt marshes vegetation could be a great attenuator of the effects of N enrichment in several coastal systems, like in the Tagus estuary, and used as a measure of the system susceptibility to nutrient enrichment [70]. Their model results show that after phytoplankton, salt marsh plants have the highest productivity in the estuary. The model also suggests that the Tagus estuarine sediments act as a source of ammonium to the water column, but that its diffusion flux is minimized by the growth of salt marsh plants. So, the ammonium diffused to the water is reduced by about 15% in the Tagus salt marsh sediments.

Not only models can be a useful tool for N control. Probability maps, showing that the nitrate concentrations exceed a legal threshold value in any location of the aquifer, can be used to assess risk of groundwater quality degradation from intensive agricultural activity in aquifer. The DK (Disjunctive Kriging) technique is an example of that mapping type. These kinds of tools are very useful for the decision-makers because they can reinforce the implementation of agri-environmental measures in Vulnerable Zones, so as to ensure good compliance with the Nitrate and Groundwater Directives in the EU zone [71].

5. Conclusions

Nitrogen pollution from agriculture, livestock, and urban discharges is a key environmental pressure on the Tagus River Basin. Nitrogen load in Spain is $\approx$13 times higher than in Portugal, being the livestock sector the major input. Nitrogen concentration in the Spanish Tagus River is low, but the input to Portugal is significant (14 kt y^{-1}), accounting for approximately the same amount as the total sectorial load in the Portuguese region (18 kt y^{-1}). However, reported water quality status was slightly better in Spain than in Portugal. Around 67% of natural surface waters (63% of rivers and 71% of lakes) showed good ecological status, while for Portugal, this status accounted for only 53%. Gaps in monitoring hinder the construction of a comprehensive mass balance at basin level and foster data uncertainties that hinder an integrated water resources management.

The main policy responses to N pressures are related to the designation of water resources protection areas in both countries. In Spain, Sensitive Zones attained 60% of Tagus River Basin area, while in Portugal, these are limited to 13% of the basin. Vulnerable Zones attain near 30% of the territory in both countries, but the enlargement of these zones could be beneficial in terms of water quality in both counties. Iberian countries administration of their own Tagus water resources is driven most by water security reasons and not by benefits sharing in water and agriculture related activities. Stronger collaboration was though by Water Framework Directive aiming at a shared effort to attain environmental objectives for transboundary waters. The adoption of Best Management Practices built under a common strategy, would help to control and mitigate the nitrate problems in the Peninsula. A joint systematic information exchange and interaction will lead to a better transboundary N management and water resources protection in the Iberian Peninsula.

The simplified methodology presented in this paper allowed for analyzing and characterizing the current state of the Tagus Basin and understand the main driving forces and responses. Nevertheless, for a more comprehensive analysis of the N processes and dynamics at the basin level, a process-based model is being developed and will be presented in a near future.

Acknowledgments: The authors acknowledge NitroPortugal, H2020-TWINN-2015, EU coordination and support action n. 692331 for funding. Authors would like to thank APA Portuguese Environmental Agency support, in particular Felisbina Quadrado, and Sofia Batista. Guilherme Gonçalves is acknowledged for the participation in the initial data collection and first figures elaboration.

Author Contributions: Cláudia M. d. S. Cordovil, António G. Brito, Maria do Rosário Cameira and Soraia Cruz have written, coordinated and reviewed the paper and finalized the data collection. Soraia Cruz reviewed and corrected all the data for all the necessary calculations within the N assessment in the river basin along both countries, to ensure the accuracy of the numbers presented in this paper, and redraw paper figures. Jane R. Poulsen, Hanns Thoden and Brian Kronvang contributed to refine the paper structure and to improve the scientific aspects, as per their twinning role in the NitroPortugal project.

Conflicts of Interest: The authors declare no conflict of interest.

References

1. Sutton, M.A.; Howard, C.; Erisman, J.W.; Billen, G.; Bleeker, A.; Grennfelt, P.; Grizzetti, B. Assessing our Nitrogen Inheritance. Chapter 1. In *The European Nitrogen Assessment Perspectives*; Sutton, M., Howard, C., Erisman, J.W., Billen, G., Bleeker, A., Grennfelt, P., van Grinsven, H., Grizzetti, B., Eds.; Cambridge University Press: Cambridge, UK, 2011; pp. 1–8.
2. Allan, J.D. Landscapes and Riverscapes: The Influence of Land Use on Stream Ecosystems. *Ann. Rev. Ecol. Evol. Syst.* **2004**, *35*, 257–284. [CrossRef]
3. Billen, G.; Garnier, J.; Lassaletta, L. The nitrogen cascade from agricultural soils to the sea: Modelling nitrogen transfers at regional watershed and global scales. *Phil. Trans. R. Soc. B* **2013**, *368*, 20130123. [CrossRef] [PubMed]
4. Butterbach-Bahl, K.; Gundersen, P.; Ambus, P.; Augustin, J.; Beier, C.; Boeckx, P.; Kitzler, B. Nitrogen Processes in Terrestrial Ecosystems. In *The European Nitrogen Assessment: Sources, Effects and Policy Perspectives Perspectives*; Sutton, M., Howard, C., Erisman, J.W., Billen, G., Bleeker, A., Grennfelt, P., van Grinsven, H., Grizzetti, B., Eds.; Cambridge University Press: Cambridge, UK, 2011; pp. 99–125.
5. Erisman, J.W.; Galloway, J.N.; Seitzinger, S.; Bleeker, A.; Dise, N.B.; Roxana Petrescu, A.M.; Leach, A.M.; Vries, W. Consequences of human modification of the global nitrogen cycle. *Phil. Trans. R. Soc. B* **2013**, *368*. [CrossRef] [PubMed]
6. Kyllmar, K.; Stjernman Forsberg, L.; Andersson, S.; Mårtensson, K. Small agricultural monitoring catchments in Sweden representing environmental impact. *Agric. Ecosyst. Environ.* **2014**, *198*, 25–35. [CrossRef]
7. Hatano, R.; Nagumo, T.; Kuramochi, K. Impact of nitrogen cycling on stream water quality in a basin associated with forest, grassland, and animal husbandry, Hokkaido, Japan. *Ecol. Eng.* **2005**, *24*, 509–515. [CrossRef]
8. Garnier, M.; Recanatesi, F.; Ripa, M.N.; Leone, A. Agricultural nitrate monitoring in a lake basin in central Italy: A further step ahead towards an integrated nutrient management aimed at controlling water pollution. *Environ. Monit. Assess.* **2010**, *170*, 273–286. [CrossRef] [PubMed]
9. Bryan, B.A.; Kandulu, J.M. Designing a Policy Mix and Sequence for Mitigating Agricultural Non-Point Source Pollution in a Water Sup Catchment. *Water Resour. Manag.* **2011**, *25*, 875–892. [CrossRef]
10. Bechmann, M. Nitrogen losses from agriculture in the Baltic Sea region. *Agric. Ecosyst. Environ.* **2014**, *198*, 13–24. [CrossRef]
11. Jarvis, S.; Hutchings, N.; Brentrup, F.; Olesen, J.; van der Hoek, K. Nitrogen Flows in Farming Systems across Europe. In *The European Nitrogen Assessment: Sources, Effects and Policy Perspectives*; Sutton, M., Howard, C., Erisman, J.W., Billen, G., Bleeker, A., Grennfelt, P., van Grinsven, H., Grizzetti, B., Eds.; Cambridge University Press: Cambridge, UK, 2011; Chapter 10; pp. 211–228.
12. Billen, G.; Garnier, J.; Mouchel, J.-M.; Silvestre, M. The Seine system: Introduction to a multidisciplinary approach of the functioning of a regional river system. *Sci. Total Environ.* **2007**, *375*, 1–2. [CrossRef] [PubMed]
13. Billen, G.; Silvestre, M.; Grizzetti, B.; Leip, A.; Garnier, J.; Voss, M.; Howarth, R.; Bouraoui, F.; Lepisto, A.; Kortelainen, P.; et al. Nitrogen Flows from European Regional Watersheds to Coastal Marine Waters. In *The European Nitrogen Assessment: Sources, Effects and Policy Perspectives*; Sutton, M., Howard, C., Erisman, J.W., Billen, G., Bleeker, A., Grennfelt, P., van Grinsven, H., Grizzetti, B., Eds.; Cambridge University Press: Cambridge, UK, 2011; Chapter 13; pp. 271–297.
14. Jiang, R.; Hatano, R.; Zhao, Y.; Woli, K.P.; Kuramochi, K.; Shimizu, M.; Hayakawa, A. Factors controlling nitrogen and dissolved organic carbon exports across timescales in two watersheds with different land uses. *Hydrol. Process.* **2014**, *28*, 5105–5121. [CrossRef]
15. Oenema, O.; Oudendag, D.; Velthof, G.L. Nutrient losses from manure management in the European Union. *Livest. Sci.* **2007**, *112*, 261–272. [CrossRef]
16. Aguilera, R.; Marcé, R.; Sabater, S. Detection and attribution of global change effects on river nutrient dynamics in a large Mediterranean basin. *Biogeosciences* **2015**, *12*, 4085–4098. [CrossRef]
17. Hirt, U.; Kreins, P.; Kuhn, U.; Mahnkopf, J.; Venohr, M.; Wendland, F. Management options to reduce future nitrogen emissions into rivers: A case study of the Weser river basin, Germany. *Agric. Water Manag.* **2012**, *115*, 118–131. [CrossRef]

18. Deelstra, J.; Iital, A.; Povilaitis, A.; Kyllmar, K.; Greipsland, I.; Blicher-Mathiesen, G.; Lagzdins, A. Reprint of "Hydrological pathways and nitrogen runoff in agricultural dominated catchments in Nordic and Baltic countries". *Agric. Ecosyst. Environ.* **2014**, *198*, 65–73. [CrossRef]

19. Lawniczak, A.E.; Zbierska, J.; Nowak, B.; Achtenberg, K.; Grześkowiak, A.; Kanas, K. Impact of agriculture and land use on nitrate contamination in groundwater and running waters in central-west Poland. *Environ. Monit. Assess.* **2016**, *188*, 172. [CrossRef] [PubMed]

20. Ding, J.; Jiang, Y.; Fu, L.; Liu, Q.; Peng, Q.; Kang, M. Impacts of land use on surface water quality in a subtropical River Basin: A case study of the Dongjiang River Basin, Southeastern China. *Water* **2015**, *7*, 4427–4445. [CrossRef]

21. Giri, S.; Qiu, Z.; Prato, T.; Luo, B. An integrated approach for targeting critical source areas to control nonpoint source pollution in watersheds. *Water Resour. Manag.* **2016**, *30*, 5087–5100. [CrossRef]

22. Blaas, H.; Kroeze, C. Excessive nitrogen and phosphorus in European rivers: 2000–2050. *Ecol. Indic.* **2016**, *67*, 328–337. [CrossRef]

23. Törnqvist, R.; Jarsjö, J.; Thorslund, J.; Rao, P.S.C.; Basu, N.B.; Destouni, G. Mechanisms of basin-scale nitrogen load reductions under intensified irrigated agriculture. *PLoS ONE* **2015**, *10*, e0120015. [CrossRef] [PubMed]

24. Lassaletta, L.; Romero, E.; Billen, G.; Garnier, J.; Garcia-Gomez, H.; Rovira, J.V. Spatialized N budgets in a large agricultural Mediterranean watershed: High loading and low transfer. *Biogeosciences* **2012**, *9*, 57–70. [CrossRef]

25. Van Rijswick, M.; Gilissen, H.K.; van Kempen, J. The need for international and regional transboundary cooperation in European river basin management as a result of new approaches in EC water law. *ERA Forum* **2010**, *11*, 129–157. [CrossRef]

26. Varol, M. Temporal and spatial dynamics of nitrogen and phosphorus in surface water and sediments of a transboundary river located in the semi-arid region of Turkey. *Catena* **2012**, *100*, 1–9. [CrossRef]

27. Nevado, J.J.B.; Martín-Doimeadios, R.C.R.; Guzmán Bernardo, F.J.; Jiménez Moreno, M.; Ortega Tardío, S.; Sánchez-Herrera Fornieles, M.; Martín-Nieto Rios, S.; Doncel Pérez, A. Integrated pollution evaluation of the Tagus River in Central Spain. *Environ. Monit. Assess.* **2011**, *156*, 461–477. [CrossRef] [PubMed]

28. Garrote, L.; Granados, A.; Iglesias, A. Strategies to reduce water stress in Euro-Mediterranean river basins. *Sci. Total Environ.* **2016**, *543*, 997–1009. [CrossRef] [PubMed]

29. European Commission. Report from the Commission to the European Parliament and the Council on the Implementation of the Water Framework Directive (WFD) (2000/60/EC)—River Basin Management Plans. 2012. Available online: http://eur-lex.europa.eu/LexUriServ/LexUriServ.do?uri=COM:2012:0670:FIN:EN: PDF (accessed on 21 December 2016).

30. Martins, G.; Ribeiro, D.; Pacheco, D.; Cruz, J.V.; Cunha, R.; Gonçalves, V.; Nogueira, R.; Brito, A.G. Prospective scenarios for water quality and ecological status in Lake Sete Cidades (Portugal): The integration of mathematical modelling in decision processes. *Appl. Geochem.* **2008**, *23*, 2171–2181. [CrossRef]

31. Kronvang, B.; Hoffmann, C.C.; Droge, R. Sediment deposition and net phosphorus retention in a hydraulically restored lowland river floodplain in Denmark: Combining field and laboratory experiments. *Mar. Freshw. Res.* **2009**, *60*, 638–646. [CrossRef]

32. Malagó, A.; Bouraoui, F.; Vigiak, O.; Grizzetti, B.; Pastori, M. Modelling water and nutrient fluxes in the Danube River Basin with SWAT. *Sci. Total Environ.* **2017**, *603–604*, 196–218. [CrossRef] [PubMed]

33. Johnes, P.J. Evaluation and management of the impact of land use change on the nitrogen and phosphorus load delivered to surface waters: The export coefficient modelling approach. *J. Hydrol.* **1996**, *183*, 323–349. [CrossRef]

34. Agencia Portuguesa do Ambiente (APA). *Plano de Gestão da Região Hidrográfica do Tejo e Ribeiras do Oeste (RH5), Parte 2—Caracterização e Diagnóstico*; Autoridades Portuguesas: Lisboa, Portugal, 2015; p. 176. (In Portuguese)

35. Confederación Hidrográfica del Tajo (CHT). *Plan Hidrológico de la Parte Española de la Demarcación Hidrográfica del Tajo—Memoria. Ministério de Agricultura, Alimentación y Medio Ambiente*; Gobierno de España: Madrid, Spain, 2015; p. 128. (In Spanish)

36. Kilsby, C.G.; Tellier, S.S.; Fowler, H.J.; Howels, T.R. Hydrological impacts of climate change on the Tejo and Guadiana Rivers. *Hydrol. Earth Syst. Sci. Discuss.* **2007**, *11*, 1175–1189. [CrossRef]

37. CLC, Corine Land Cover. Version 18.5.1. 2006. Available online: http://land.copernicus.eu/pan-european/ corine-land-cover/clc-2006/view (accessed on 14 October 2016).

38. López-Moreno, J.I.; Vicente-Serrano, S.M.; Begueria, S.; García-Ruíz, J.M.; Portela, M.M.; Almeida, A.B. Dam effects on droughts magnitude and duration in a transboundary basin: The lower river Tagus, Spain and Portugal. *Water Resour. Res.* **2009**, *45*, W02405. [CrossRef]

39. Gómez, C.M.; Delacámara, G.; Pérez, C.D.; Ibáñez, E.; Solanes, M. *EPI WATER: Evaluating Economic Policy Instruments for Sustainable Water Management in Europe*; WP3 EX-POST Case Studies: Water Transfers in the Tagus River Basin (Spain), Deliverable no.: D3.1—Review Reports; EPI-WATER: Milan, Italy, 2011.

40. Agencia Portuguesa do Ambiente (APA). *Plano de Gestão da Região Hidrográfica do Tejo e Ribeiras do Oeste (RH5), Parte 3—Análise Económica das Utilizações da Água*; Autoridades Portuguesas: Lisboa, Portugal, 2015; p. 179. (In Portuguese)

41. Confederación Hidrográfica del Tajo (CHT). *Plan Hidrológico de la Parte Española de la Demarcación Hidrográfica del Tajo-Anejo 3 de la Memoria: Usos y Demandas de Agua. Ministério de Agricultura, Alimentación y Medio Ambiente*; Gobierno de España: Madrid, Spain, 2015; p. 73. (In Spanish)

42. Confederación Hidrográfica del Tajo (CHT). Ministério de Agricultura, Alimentación y Medio Ambiente, Gobierno de España. 2016. Available online: www.chtajo.es/Informacion%20Ciudadano/Calidad_Vertidos/Resultados_Informes/Documents/AguasSuperficiales/Red%20ICA/Informes_ICA.htm (accessed on 9 January 2017). (In Spanish)

43. SNIRH, Sistema Nacional de Informação de Recursos Hídricos. 2016. Available online: http://snirh.apambiente.pt/index.php?idMain= (accessed on 14 December 2016). (In Portuguese)

44. SNIAmb, Sistema Nacional de Informação de Ambiente. 2017. Available online: https://sniamb.apambiente.pt/content/cat%C3%A1logo?language=pt-pt (accessed on 21 July 2017). (In Portuguese)

45. GeoPortal, Geographic Portal Information of Spain. 2017. Available online: http://sig.mapama.es/geoportal/ (accessed on 21 July 2017).

46. Novotny, V.; Olem, H. *Water Quality: Prevention, Identification and Management of Diffuse Pollution*; Van Nostrand Reinhold: New York, NY, USA, 1994; Volume 9, pp. 507–572. ISBN 9780442005597.

47. Instituto Nacional de Estatística (INE). Efectivo Animal (nº) da Exploração Agrícola por Localização Geográfica (NUTS–2002) e Espécie animal. Recenseamento Agrícola—Séries Históricas. 2009. Available online: https://www.ine.pt/xportal/xmain?xpid=INE&xpgid=ine_indicadores&indOcorrCod=0004460&contexto=bd&selTab=tab2 (accessed on 3 July 2017). (In Portuguese)

48. Lau, P.S.; Tam, N.F.Y.; Wong, Y.S. Wastewater nutrients removal by *Chlorella vulgaris*: Optimization through acclimation. *Environ. Tecnol.* **1996**, *17*, 183–189. [CrossRef]

49. Tanik, A.; Ozalp, D.; Seker, Z. Practical estimation and distribution of diffuse pollutants arising from watershed in Turkey. *J. Environ. Sci. Technol.* **2013**, *10*, 221–230. [CrossRef]

50. Cordovil, C.M.; de Varennes, A.; Pinto, R.; Fernandes, R.C. Changes in mineral nitrogen, soil organic matter fractions and microbial community level physiological profiles after application of digested pig slurry and compost from municipal organic wastes to burned soils. *Soil Biol. Biochem.* **2011**, *43*, 845–852. [CrossRef]

51. Haygarth, P.; Johnes, P.; Butterfield, D. *Land Use for Achieving "Good Ecological Status" of Water Bodies in England and Wales: A Theoretical Exploration for Nitrogen and Phosphorus*; HMSO: London, UK, 2003; 30p.

52. Erturk, A.; Gurel, M.; Ekdal, A.; Tavsan, C.; Seker, D.Z.; Çokgor, E.U.; Insel, G.; Mantas, E.P.; Aydin, E.; Ozgun, H.; et al. Estimating the Impact of Nutrients Emissions via Water Quality Modelling in the Melen Watershed. In Proceedings of the IWA 11th Diffuse Pollution Conference, Chiang, Thailand, 19–22 November 2007; Volume 167, pp. 26–31.

53. Eurostat Statistics Explained. Share of Holdings with Livestock Which Have Manure Storage Facilities in Total Holdings with Livestock by Size of the Holding in Livestock Units, EU-28, IS, NO, CH and ME. 2010. Available online: http://ec.europa.eu/eurostat/statistics-explained/index.php/File:Share_of_holdings_with_livestock_which_have_manure_storage_facilities_in_total_holdings_with_livestock_by_size_of_the_holding_in_livestock_units,_EU-28,_IS,_NO,_CH_and_ME,_2010_.png (accessed on 4 July 2017).

54. European Environment Agency (EEA). Changes in Wastewater Treatment in Southern European Countries between 1980s and 2009. 2013. Available online: http://www.eea.europa.eu/data-and-maps/figures/changes-in-wastewater-treatment-in-countries-of-europe-between-1980s-and-2005-southern-3 (accessed on 12 December 2016).

55. Agência Portuguesa do Ambiente (APA). *Questões Significativas da Gestão da Água (QSiGA)*; Participação Pública: Lisboa, Portugal, 2014; 138p. (In Portuguese)

56. Instituto Nacional de Estatística (INE). *Estatísticas Agrícolas 2015*; Instituto Nacional de Estatística: Lisboa, Portugal, 2016; pp. 7–60. ISBN 978-989-25-0360-78.

57. Confederación Hidrográfica del Tajo (CHT). *Proposta do Plano Hidrológico do Lado Espanhol da Região Hidrográfica do Tejo, Ciclo de Planificação 2015-2021: Efeitos Ambientais Transfronteiriços Espanha-Portugal*; Ministerio de Agricultura, Alimentación y Medio Ambiente, Gobierno de España: Madrid, España, 2015; p. 26, (In Spanish/Portuguese).

58. Agência Portuguesa do Ambiente (APA). *Plano de Gestão da Região Hidrográfica do Tejo e Ribeiras do Oeste (RH5), Parte 6—Programa de Medidas*; Autoridades Portuguesas: Lisboa, Portugal, 2016; p. 190. (In Portuguese)

59. Christianson, L.; Tyndall, J.; Helmers, M. Financial comparison of seven nitrate reduction strategies for Midwestern agricultural drainage. *Water Resour. Econ.* **2013**, *2–3*, 30–56. [CrossRef]

60. Brito, A.G.; Maia, R.; Silva, C.; Fernandes, T.; Lacerda, M. The Portuguese-Spanish Cooperation on Transboundary Water Governance: The Way Forward. In Proceedings of the 8th International Conference of the European Water Resources Association (EWRA), Porto, Portugal, 26–29 June 2013.

61. Roebeling, P.; Alves, H.; Rocha, J.; Brito, A.G.; Mamede, J. Gains from trans-boundary water quality management in linked catchment and coastal socio-ecological systems: A case study for the Minho region. *Water Resour. Econ.* **2014**, *8*, 32–42. [CrossRef]

62. Agência Portuguesa do Ambiente (APA); Direcção-Geral de Agricultura e Desenvolvimento Rural (DGADR). *Poluição Provocada por Nitratos de Origem Agrícola–Directiva 91/676/CEE, de 12 de dezembro. Relatório 2012–2015*; Autoridades Portuguesas: Lisboa, Portugal, 2016; p. 235. (In Portuguese)

63. Confederación Hidrográfica del Tajo (CHT). *Plan Hidrológico de la Parte Española de la Demarcación Hidrográfica del Tajo-Anejo 4 de la Memoria: Registro de Zonas protegidas. Ministério de Agricultura, Alimentación y Medio Ambiente*; Gobierno de España: Madrid, Spain, 2015; p. 563. (In Spanish)

64. European Environment Agency (EAA). Nitrate Vulnerable Zones. 2009. Available online: http://eea.europa.eu/data-and-maps/figures/nitrate-vullnerable-zones-eu (accessed on 15 February 2017).

65. Gil, A.M.; Amorós, A.M.R.; Hernández, M.H. El trasvase Tajo-Segura. *Obs. Medioambient.* **2005**, *8*, 73–110.

66. Lorenzo-Lacruz, J.; Vicente-Serrano, S.M.; López-Moreno, J.I.; Beguería, S.; García-Ruíz, J.M.; Cuadrat, J.M. The impact of droughts and water management on various hydrological systems in the headwater of the Tagus River (central Spain). *J. Hydrol.* **2010**, *386*, 13–26. [CrossRef]

67. Cameira, M.R.; Fernando, R.M.; Ahuja, L.; Pereira, L. Simulating the fate of water in field soil–crop environment. *J. Hydrol.* **2005**, *315*, 1–24. [CrossRef]

68. Cameira, M.R.; Fernando, R.M.; Ahuja, L.R.; Ma, L. Using RZWQM to simulate the fate of nitrogen in field soil–crop environment in the Mediterranean region. *Agric. Water Manag.* **2007**, *90*, 121–136. [CrossRef]

69. Cameira, M.R.; Pereira, A.; Ahuja, L.R.; Ma, L. Sustainability and environmental assessment of fertigation in an intensive olive grove under Mediterranean conditions. *Agric. Water Manag.* **2014**, *146*, 346–360. [CrossRef]

70. Simas, T.C.; Ferreira, J.G. Nutrient enrichment and the role of salt marshes in the Tagus estuary (Portugal). *Estuar. Coast. Shelf Sci.* **2007**, *75*, 393–407. [CrossRef]

71. Mendes, M.P.; Ribeiro, L. Nitrate probability mapping in the northern aquifer alluvial system of the river Tagus (Portugal) using Disjunctive Kriging. *Sci. Total Environ.* **2010**, *408*, 1021–1034. [CrossRef] [PubMed]

water

MDPI

Article

Monthly Prediction of Drought Classes Using Log-Linear Models under the Influence of NAO for Early-Warning of Drought and Water Management

Elsa Moreira [1],*, Ana Russo [2] and Ricardo M. Trigo [2]

[1] Center of Mathematics and Applications (CMA), Faculty of Sciences and Technology, New University of Lisbon, 2829-516 Caparica, Portugal
[2] Instituto Dom Luiz (IDL), Faculty of Sciences, University of Lisbon, 1749-016 Lisboa, Portugal; acrusso@fc.ul.pt (A.R.); rmtrigo@fc.ul.pt (R.M.T.)
* Correspondence: efnm@fct.unl.pt

Received: 3 November 2017; Accepted: 10 January 2018; Published: 12 January 2018

Abstract: Drought class transitions over a sector of Eastern Europe were modeled using log-linear models. These drought class transitions were computed from time series of two widely used multiscale drought indices, the Standardized Preipitation Evapotranspiration Index (SPEI) and the Standardized Precipitation Index (SPI), with temporal scales of 6 and 12 months for 15 points selected from a grid over the Prut basin in Romania over a period of 112 years (1902–2014). The modeling also took into account the impact of North Atlantic Oscillation (NAO), exploring the potential influence of this large-scale atmospheric driver on the climate of the Prut region. To assess the probability of transition among different drought classes we computed their odds and the corresponding confidence intervals. To evaluate the predictive capabilities of the modeling, skill scores were computed and used for comparison against benchmark models, namely using persistence forecasts or modeling without the influence of the NAO index. The main results indicate that the log-linear modeling performs consistently better than the persistence forecast, and the highest improvements obtained in the skill scores with the introduction of the NAO predictor in the modeling are obtained when modeling the extended winter months of the SPEI6 and SPI12. The improvements are however not impressive, ranging between 4.7 and 6.8 for the SPEI6 and between 4.1 and 10.1 for the SPI12, in terms of the Heidke skill score.

Keywords: drought classes; Standardized Precipitation and Evapotranspiration Index (SPEI); Standardized Precipitation Index (SPI); North Atlantic Oscillation (NAO); log-linear modeling; persistence

1. Introduction

The social, environmental and economic impacts of drought, as well as its features, have been studied in numerous studies [1–7]. Since 1950, a positive tendency of drought patterns has been observed over some regions of the globe [8–10], in part due to global warming [11]. Within the European context, the positive drought trends seem to be particularly acute in the Mediterranean area [10,12,13], where increased dryness [14–16] and temperature is more notorious [6,17]. As shown in a recent study by Dumitrescu et al. [18] covering all of Romania over the period 1961–2013, the most striking results concern the number of summer days, which has increased by 95% in of the considered stations, while an increase in the duration of warm spells of 83% has been observed in the same stations. In a paper by Spinoni et al. [19], the drought events in the last five decades in the Carpathian region (ranging from the Czech Republic to Serbia, encompassing Slovakia, Poland, Hungary, Ukraine, and Romania) were analyzed. The most intense droughts took place in 1990, 2000, and 2003, followed by other 10 notable events. In general, for that region, the drought frequency is slightly increasing, however, the rise has not been confirmed by significance tests [19].

Under the current global warming process [17], in the next decades an increase in the frequency, duration, and affected area of drought events in the drought-prone areas like the Mediterranean is expected [11,16,20–22]. The Carpathian region shows a dual behavior, as in the past it was reported as being generally affected by increasingly frequent and severe droughts [19] and, according to future drought projections, more extreme changes are expected in the Carpathian Mountains than in the surrounding regions. Besides the regional differences, seasonal differences are also expected, with more frequent droughts in spring and summer and a stronger decrease of winter droughts [23]. Changes in the agriculture forest sector in Romania have been shown to be directly affected by the weather and climatic conditions [24,25], as recognized within the Romanian National Climate Change Strategy (2013–2020), in its adaptation component [26]. In particular, drought is a major cause of unexpected crop failure [27–29]. Future changes of climatic conditions in Romania will drive additional changes in crop production in the region [24,25,30] and the sector is expected to continue to be affected in the future [30,31]. Therefore, for countries and stakeholders considering the implementation of appropriate preparedness and mitigation measures, it becomes increasingly important to develop robust early warning systems [3,32]. However, under the current state of the art, it is extremely challenging to forecast the early stages of a drought, as well as its expected conclusion, particularly in the middle latitudes [33]. Nevertheless, considering that droughts evolve slowly, it is increasingly feasible to release a timely prediction allowing for some measures to be adopted to mitigate the effects of the drought [32,34,35]. In this context, drought predictions at a monthly scale can be beneficial so that resource managers and water users can prepare for the occurrence of various stages of a drought, including onset, duration, or conclusion.

The assessment of drought severity can be achieved with a number of indices developed for that purpose, such as the Palmer Drought Severity Index (PDSI) [36], and the MedPDSI [37,38], etc. In recent years, it has been shown that the use of multi-scalar drought indices is quite useful, including the Standardized Precipitation Index (SPI) [39] and the Standardized Precipitation Evapotranspiration Index (SPEI) [40]. In this study, the SPI and SPEI were the chosen indices because of their multi-scalar nature that allows assessing drought impacts, for different time scales, on soil moisture, snowpack, reservoir storage, groundwater, and stream-flow [39,40]. Although droughts are mostly controlled by the temporal variability of precipitation, in a context of global warming, the temperature variable may play an increasing role in drought assessments, and therefore the SPEI was also considered. Thus, while the SPI is computed only based on precipitation, the SPEI is sensitive to changes in the evaporation demand (caused by temperature fluctuations and trends) because it considers the effect of potential evapotranspiration (PET) on drought severity [10,40].

On the other hand, atmospheric circulation plays a major role in terms of precipitation occurrence but also on its frequent inhibition of precipitation leading to drought on different time and spatial scales (e.g., [41,42]). The main features of the atmospheric circulation for the European region are captured by a set of large-scale patterns (and corresponding indices) that are usually computed at the monthly/seasonal temporal scales [43]: the North Atlantic Oscillation (NAO), the East-Atlantic (EA) pattern, the Scandinavian (SCAND) pattern, and the East-Atlantic Western Russia (EAWR) pattern. The temporal indices associated to these patterns are freely available at the National Centers for Environmental Prediction (NCEP) website. It is now well established that the main pattern controlling wet and dry rainfall regimes over Western Europe is the NAO mode [44–46]. However, this effect extends to the eastern part of the Mediterranean. In fact, when the NAO is in its positive phase (i.e., NAO+), winter storms coming from the Atlantic are deflected northward, resulting in wet winters over the northern part of Europe, dry winters over the southern and eastern part of Europe [47,48], and also lower-than-average temperatures [45]. Thus, the NAO+ regime is associated to below-average precipitation over most southern and central European regions and higher-than average values in Northern Europe [45–48]. In contrast, when the NAO is in its negative phase (i.e., NAO−) storm tracks are shifted southwards, promoting wet and mild winters over the southern and eastern parts of Europe and dry winters over the northern part of Europe [45–48].

The region focused on this study is the Prut River basin in Eastern Europe (Figure 1). The Prut River source is located in the Carpathian Mountains, Ukraine, and runs into the Danube River. It is circa 1000 km long, and its catchment area is 27,540 km^2. The basin of the Prut River, being a transboundary basin, is located in the territory of three countries: 28% in Moldova; 33% in Ukraine; and 39% in Romania (Figure 1). The Romanian territory is divided by the Carpathian Mountains into two groups of regions: the intra-Carpathian regions and extra-Carpathian regions. The extra-Carpathian regions cover the southern, eastern, and southeastern parts of Romania, while areas located inside the mountain chain and those located westward from the mountains are considered the intra-Carpathian regions (Figure 1). The Carpathian Mountains form an orographic border which imposes a separation between the regions influenced by mild oceanic (south and west) and continental (north and east) climates. In the last 20 years, a shift from oceanic to continental climate can be seen, especially in the Romanian part of the Carpathians [19]. The extra-Carpathian regions (delimited area in Figure 1), which include most of the Prut basin, are more often influenced by southern tropical or eastern continental air masses [49]. The maximum flow of the Prut River is recorded during the beginning of the warm season, often in early April due to the significant quantities of precipitation overlapping with snowmelt, and manifests itself as spring–summer floods [50]. Although the Prut watershed has a relatively low water stress under present climate, a more detailed assessment is needed of this basin since its watershed has been exposed to landscape changes that may increase the runoff and reduce reservoir capacity [51]. Corduneanu et al. [50] studied the Prut river flow recordings during the period 1978–2012 and noticed that two extremely severe droughts occurred, one spanning from 1982 to 1988 and more recently another that began in 2011 and continued until the end of 2012. In addition, Croitoru et al. [49] studied the spatiotemporal distribution of aridity indices based on temperature and precipitation in the extra-Carpathian region of Romania and found that the southeastern region, including the Prut basin, are most vulnerable to semi-aridity, especially during the warm period of the year.

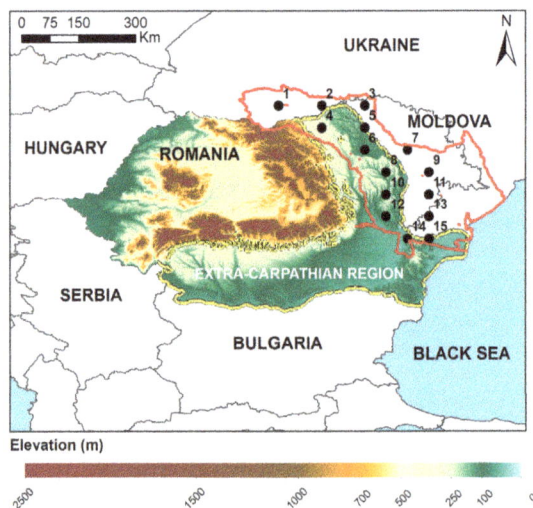

Figure 1. Map of Romania indicating the 15 grid points (from the Climatic Research Unit (CRU) database) covering the study area (delimited by a red line) and showing the Carpathians Mountains as well as the extra-Carpathian region [45] delimited here by a yellow line.

In terms of the influence of climate variability, Tomozeiu et al. [52] found a significant association between Romania's precipitation variability in the winter months and large-scale circulation patterns, such as the NAO and the blocking phenomenon in the Atlantic-European sector. Bojariu and Giorgi [5] observed fine-scale features of the NAO precipitation signal over the Iberian, Scandinavian, Alpine, Balkan, and Carpathian regions, demonstrating the important role played by the topographic forcing in modulating the large-scale NAO variability over Europe. Additionally, a decreasing trend in winter precipitation was discovered, especially in the extra-Carpathian region, during the last two decades of the 20th century, likely related with an intensification of the NAO+ and a decrease in the frequency of blocking especially after 1980. The winter NAO signal was found by other authors to be stronger in the extra-Carpathian regions, as result of the orographic force on the atmospheric flow in the Carpathian Mountains [53]. Moreover, regions in Central and Eastern Europe were identified as having a source of NAO-related predictability [53], which results from the positive correlation present in the month of November between the thermal anomalies and the NAO index. This source of predictability might be used to predict the onset of the NAO in the following winter [53].

Several approaches to the have been proposed for the forecasting of droughts on a monthly basis, namely statistical, physical-based techniques, or hybrid techniques. Examples of those hybrid approaches which conjugate statistics and physical-based techniques are the works by Mishra et al. [54] and Kim and Valdes [55], which use respectively a hybrid stochastic and neural network model and the later conjugates wavelet transforms and neural networks for a nonlinear model. However, other examples are the model combination of the wavelet and fuzzy logic models used in Ozger et al. [56] and the adaptive neuro-fuzzy inference used in Bacanli et al. [57]. Among the techniques used for drought forecasting, statistical models are chosen many times, since they are simple to implement, do not have a high computational burden, and produce useful predictions [58]. There are a variety of statistical methodologies available which can be applied for the intended purpose, namely autoregressive integrated moving average (ARIMA)-type approaches [59,60], artificial neural network (ANN) models [61,62] or even other types of stochastic and probability models, such as Markov chains [63], log-linear models [64,65], and others [66,67]. A thorough discussion on various methodologies used for drought modeling and prediction showing the limitations and advantages of each modeling/technique was done by Mishra and Singh [58].

One approach often used to include useful information related to climate dynamics in drought forecasting models considers (as covariates) atmospheric–oceanic monthly or seasonal anomaly indices. Some examples of this type of approach include the use of ANN models and time series of drought indices, including the NAO index as a covariate [68], or the use of probabilistic models resulting from evaluating conditional probabilities of future SPI classes based on current SPI and NAO classes [67].

The combined effects of the NAO index and SPI time series' information have been well explored in the context of Portuguese drought predictability, specifically to predict drought class transitions some months in advance based on the SPI of the previous months [57,58,62]. These models, which are three-dimensional log-linear, rely exclusively on probabilities of transition between states and are fitted to drought class transitions counts separately for the NAO+ and NAO− phases. Taking into account the NAO state in the current month, the most probable drought class transition for the next month and their confidence intervals are computed. This approach has proved its usefulness in predicting the SPI drought class one or two months in advance [64,65,69].

Considering the challenges in climate change, and the reported NAO influence on precipitation and, consequently, on drought occurrence in Romania and the Prut basin, the aim of this work is to assess the predictability of drought at a monthly scale for the Prut basin, with the objective of providing drought warnings for agricultural management. For this purpose, the same log-linear models are used fitted to drought indicators, the SPI, and for the first time also to SPEI class transitions, under the influence of the NAO index.

Additionally, we aim to go a step further from previous studies using log-linear models [64,65,69], identifying the predictors that can better predict changes of drought classes one or two months in advance, namely with the use of a four-month NAO average as a predictor instead of the single lagged NAO value as in a previous study [65]. An analysis of the extended winter months (November–May) and the complete annual period was performed, as well a study of the use of different SPI (SPEI) drought class intervals, aiming at a better assessment of the Prut basin climate. Finally, with the aim of improving the study, the frequency of severe and extreme droughts and also wet events in three periods with similar numbers of years were statistically compared using an analysis of variance technique.

2. Data and Methods

2.1. Standardized Drought Indicators and the NAO

The climatic data used in this study were retrieved from the 0.5 × 0.5 gridded resolution CRU (Climatic Research Unit) time-series (TS) 3.23 produced by the Climatic Research Unit (CRU) at the University of East Anglia. The CRU TS 3.23 variables used for SPI and SPEI computation were monthly precipitation and mean temperature, for the period January 1901–December 2014, located over the Prut basin (Romania, Ukraine and Moldavia) (Figure 1). Details regarding this dataset are available at the CRU high-resolution gridded database (https://crudata.uea.ac.uk/cru/data/hrg/).

The choice of the baseline period relative to which standardized drought indicators are calculated is a key factor [23]. If a shorter period (e.g., 30 years) characterized by frequent and severe droughts is used as a baseline, this may impinge influence on the indicator's trend over the entire period [23]. This study focuses on class transitions over a long-term period, so we decided to use the entire period (1901–2014) as the baseline, thus ensuring more robust comparisons. Moreover, drought is different from other natural disasters and takes place when a significant deviation from the normal hydrologic conditions of an area occurs [36]. Thus, using as a baseline the past decades for estimating droughts over the years 1901–2014 might be problematic, as the last decades have shown an intensification of drought events relative to the beginning of the century, which might be unrealistic.

Moreover, the long-term period range, from 1901 to 2014, is considered an important pre-requisite since the log-linear modeling approach benefits from using long and recent data to better parameterize the models. Six- and twelve-month time-scales of both SPI and SPEI (SPI6, SPEI6, SPI12 and SPEI12) were computed from precipitation and temperature time series correspondent to the 15 grid points shown in Figure 1. The use of these two time-scales relies on the fact that each one reflects the accumulated drought conditions of longer or shorter periods. Specifically, the 12-month time scale is able to identify long-term dry and wet periods, being more representative of the impacts of drought on the hydrologic regimes [70]. In contrast, shorter time scales are useful to detect agricultural droughts [25], reflecting a better change of class instead of its persistence [70]. We would, however, like to highlight that the terms "short-term accumulation periods" and "long-term accumulation periods" in SPI/SPEI drought indicators here refer to meteorological droughts, as they use meteorological inputs and not, for example, soil moisture (agricultural drought), or low flow data (hydrological droughts).

For the SPI computation, the gamma distribution was used to model precipitation and for the SPEI, Penman–Monteith's PET was used, as well as the log-logistic distribution to model the water deficit (D) [40].

The categories usually used to classify drought, and wetness classes are presented in Table 1. However, to avoid problems in the model fitting [64] caused by too many class transitions with value zero, several classes were grouped. We have considered a total of four classes in our modeling analysis as in recent studies [65] albeit not necessarily exactly defined in the same way. Thus, a new unique wetness class grouping all the wetness categories was included, since our main focus in this study was drought prediction. The normal and the near-normal drought classes are grouped together, since, being near-normal, the first category of drought has low impacts, and therefore we do not find necessary to distinguish a single class for it. Lastly, both the severe and extreme drought classes are grouped due to the number of transitions attributable to the extremely severe drought classes, since they occur considerably less frequently than the other classes. It must be noted that the number of classes and their arrangement in different works reflect, to a certain extent, the different frequency of these classes. Therefore, we are confident that the classification presented herein is more balanced for modeling purposes and may respond better to the climate features of the Prut basin.

Table 1. Drought and wetness classes of the Standardized Precipitation Index/Standardized Precipitation Evapotranspiration Index (Adopted from Mckee et al. [39]).

Code	Classes	SPI/SPEI Interval
ew	Extreme wetness	$[2, +\infty[$
sw	Severe wetness	$[1.5, 2[$
mw	Moderate wetness	$[1, 1.5[$
n	Normal	$[-1, 1[$
md	Moderate drought	$[-1.5, -1[$
sd	Severe drought	$[-2, -1.5[$
ed	Extreme drought	$]-\infty, -2[$

The monthly drought classes' time series were categorized considering the limits defined in Table 2 reflecting the modifications described before, relative to both SPI and SPEI time series with 6- and 12-month time scales.

Table 2. Drought and wetness classes of SPI/SPEI for modeling purposes (modified from Mckee et al. [39]).

Code	Classes	SPI/SPEI Interval
1	Wet	$[1, +\infty[$
2	Normal/Near-Normal	$[-1, 1[$
3	Moderate	$[-1.5, -1[$
4	Severe/Extreme	$]-\infty, -1.5[$

To cover the full period of available precipitation data (1901–2014), an extended historical record (starting in 1864) of a station-based NAO was computed for this study. This long NAO dataset relies upon the difference of normalized Sea Level Pressure (SLP) between Lisbon (Portugal) and Reykjavik (Iceland) and overcomes the problem of the available National Centers for Environmental Prediction (NCEP) Climate Prediction Center monthly NAO, which only dates back to 1950.

2.2. Correlations between SPEI/SPI and NAO

As a preliminary step, a simple correlation analysis was performed to evaluate the association between the SPEI (SPI) and the NAO index monthly time series for each time scale and grid point (Table 3). The correlations with other large-scale indices like the EAWR and EA patterns influencing Central and Eastern Europe were also tested, and the highest correlations were always obtained with the NAO. Besides that, different NAO averages and lags with the SPI/SPEI were tested. For SPEI6/SPI6, the maximum inverse correlation was obtained with the NAO index average of the last four months, i.e., the NAO value at month t considered for the modeling is the simple four-month average $((NAO\_(t-3) + NAO\_(t-2) + NAO\_(t-1) + NAO\_t)/4)$. If that NAO > 0, then the NAO state is assigned positive (NAO+), otherwise it is considered negative (NAO−). For the SPEI12/SPI12, the maximum inverse correlation is obtained also with the last four-month average but with a lag of 1 month, i.e., with $(NAO\_(t-4) + NAO\_(t-3) + NAO\_(t-2) + NAO\_(t-1))/4$. This correlation between the SPEI/SPI and the NAO was computed twice considering initially all the months of the year and afterward only the winter and spring months from November to May, commonly called the extended winter months. The analysis for the extended winter months was performed taking into account that the impact of the NAO pattern in Southern Europe is usually better defined in this time of the year. The inverse correlation between the SPEI (SPI) and the NAO considering all the months of the year is, on average, lower by approximately 0.08.

Table 3. Correlation between the SPEI (SPI) with 6- and 12-month time scales and the North Atlantic Oscillation (NAO) average of the last 4 months, considering only the extended winter months (November–May). The highest inverse correlation values are highlighted in bold.

Grid Point	Latitude	Longitude	Correlation			
			SPEI6	SPEI12	SPI6	SPI12
1	48.25	25.25	−0.30	−0.27	−0.31	−0.29
2	48.25	26.25	−0.28	−0.25	−0.29	−0.27
3	48.25	27.25	−0.30	−0.28	−0.30	−0.29
4	47.75	26.25	−0.28	−0.25	−0.28	−0.26
5	47.75	27.25	−0.30	−0.28	−0.31	−0.30
6	47.25	27.25	−0.29	−0.27	−0.29	−0.28
7	47.25	28.25	−0.34	−0.34	−0.32	−0.33
8	46.75	27.75	−0.32	−0.32	−0.30	−0.30
9	46.75	28.75	−0.37	−0.36	−0.34	−0.34
10	46.25	27.75	−0.35	−0.34	−0.33	−0.32
11	46.25	28.75	−0.38	−0.37	−0.35	−0.33
12	45.75	27.75	−0.36	−0.34	−0.33	−0.32
13	45.75	28.75	−0.38	−0.35	−0.34	−0.32
14	45.25	28.25	−0.38	−0.35	−0.35	−0.33
15	45.25	28.75	−0.39	−0.35	−0.35	−0.32
Average			−0.34	−0.31	−0.32	−0.31
Minimum			−0.39	−0.37	−0.35	−0.34
Maximum			−0.28	−0.25	−0.28	−0.26

As can be observed in Table 3, the correlations values between NAO and SPEI (or SPI) in the Prut basin range in the interval [−0.39, −0.25], and the use of different drought indices does not provide added information, as no significant differences (for either time scale 6 or 12) can be observed between results attained with SPEI and SPI. Moreover, all the correlation values presented in Table 3 were found to be statistically different from zero at a 5% level.

2.3. Modeling

The number of two-step monthly transitions between drought classes was counted separately for NAO+ and NAO−, evaluated at the current month (t), to form two three-dimensional (4 × 4 × 4) contingency tables [65] with *N* = 64 cells each. The referred contingency NAO− and NAO+ tables present three categories and four levels each, respectively the drought class at months t − 1, t, and t + 1, and drought classes 1, 2, 3, and 4, previously defined in Table 2. Contingency tables for the SPEI6 and SPEI12 are presented in Tables 4 and 5. On the other hand, SPI contingency tables, which are similar, are not shown for the sake of simplicity.

Log-linear modeling inputs correspond to the number of times that in each month the drought class i was followed by the drought class j in the following month and then by the drought class k in the month after that (two-step transitions) called the observed frequencies. The model computes the expected frequencies.

Modeling more than two step transitions with log-linear models is very difficult since we will get too many observed frequencies equal to zero, which brings fitting problems [71]. On the other hand, separately modeling the negative and positive phase of the NAO leads to simpler models, with fewer parameters and therefore easier adjustment.

Table 4. Counts of two consecutive transitions between drought classes of SPEI6 (t − 1 → t → t + 1) for NAO− and NAO+.

NAO−	Drought Class Month t + 1				Drought Class Month t + 1				Drought Class Month t + 1				Drought Class Month t + 1			
	1				2				3				4			
Drought Class	Drought Class Month t				Drought Class Month t				Drought Class Month t				Drought Class Month t			
Month t − 1	1	2	3	4	1	2	3	4	1	2	3	4	1	2	3	4
1	52	3	0	0	39	43	0	0	0	3	0	0	0	1	0	0
2	41	49	0	0	21	246	7	1	1	25	9	1	0	7	7	3
3	0	2	0	0	0	19	5	10	0	1	7	2	0	0	5	2
4	0	1	0	0	0	6	5	1	0	2	4	4	0	0	0	4

NAO+	Drought Class Month t + 1				Drought Class Month t + 1				Drought Class Month t + 1				Drought Class Month t + 1			
	1				2				3				4			
Drought Class	Drought Class Month t				Drought Class Month t				Drought Class Month t				Drought Class Month t			
Month t − 1	1	2	3	4	1	2	3	4	1	2	3	4	1	2	3	4
1	27	1	0	0	19	42	0	0	0	1	1	0	0	0	0	0
2	17	37	0	0	15	293	19	1	0	35	16	3	0	7	12	6
3	0	1	0	0	0	33	16	6	0	2	7	6	0	0	10	9
4	0	1	0	0	0	17	6	9	0	1	6	6	0	0	1	13

Table 5. Counts of two consecutive transitions between drought classes of SPEI12 (t − 1 → t → t + 1) for NAO− and NAO+.

NAO−	Drought Class Month t + 1				Drought Class Month t + 1				Drought Class Month t + 1				Drought Class Month t + 1			
	1				2				3				4			
Drought Class	Drought Class Month t				Drought Class Month t				Drought Class Month t				Drought Class Month t			
Month t − 1	1	2	3	4	1	2	3	4	1	2	3	4	1	2	3	4
1	71	7	0	0	31	30	0	0	0	0	0	0	0	0	0	0
2	28	26	0	0	10	297	8	0	0	32	7	0	0	2	6	2
3	0	1	0	0	0	15	8	1	0	6	6	2	0	0	6	5
4	0	0	0	0	0	4	8	0	0	1	3	8	0	0	0	20

NAO+	Drought class Month t + 1				Drought Class Month t + 1				Drought Class Month t + 1				Drought Class Month t + 1			
	1				2				3				4			
Drought Class	Drought Class Month t				Drought Class Month t				Drought Class Month t				Drought Class Month t			
Month t − 1	1	2	3	4	1	2	3	4	1	2	3	4	1	2	3	4
1	44	4	0	0	23	33	0	0	0	0	0	0	0	0	0	0
2	26	34	0	0	10	323	12	0	0	19	19	0	0	2	7	2
3	0	2	0	0	0	28	13	1	0	1	10	4	0	0	5	13
4	0	0	0	0	0	4	5	7	0	0	3	7	0	0	2	29

Contingency Tables 4 and 5 show that the most frequent drought transitions correspond to a persistent condition which implies the maintenance of the previous drought classes, whereas the change of class is much less frequent. When comparing the SPEI6 with the SPEI12, we can see that the persistence increases and the transitions that imply the change of class decrease a bit for the SPEI12, which is related with the larger accumulation period implicit for 12 months that highlights the tendency for drought class maintenance. In addition, when comparing the negative and positive NAO tables, we can verify that the transitions involving drought class 1 (wet) are in general larger in number for the NAO− than for the NAO+. The opposite occurs with respect to drought classes 3 and 4 (moderate and severe/extreme), which is justified by the fact that NAO+ is associated with below-average precipitation over Southern and Central Europe and in Mediterranean regions, and the NAO− with above-average precipitation also in the same regions [45].

Quasi-association (QA) log-linear models [71] have been shown that they are the ones that better fit to similar three-dimensional contingency tables [64,65]. Thus, QA log-linear models were also herein used to fit the observed frequencies for NAO− and NAO+ and for each grid point. After the model fitting, ratios between expected transition frequencies (odds) are computed. Odds specify the proportion between the probabilities of transition to one state over another, and its values range between 0 and +∞ [71]. Here, odds represent the number of times that it is more, less, or equally probable that the occurrence of a drought class transition takes place over another. For these odds, confidence intervals associated with a probability $1 - \alpha = 0.95$ are computed. Odds confidence intervals, besides reflecting the sampling variability of the observed drought transitions internal to each time series, also indicate if given odds are significantly different from 1. For obtaining the most probable class transition for the following month (t + 1), the odds for the three closest class transitions are computed, as well as their confidence intervals. The most probable transition is chosen. Technical details on QA log-linear model fitting, odds, and its confidence intervals can be found in Moreira et al. [65].

The modeling described above was done considering initially all the months of the year and then only the extended winter months (November–May). This distinction was performed with the aim of evaluating if there is any improvement in the predictions when considering the extended winter months since the magnitude of their correlation coefficient is higher.

2.4. Model Performance

The model performance was assessed using two statistical measures which account for the total number of agreements within the total number of months, namely the proportion correct (PC) and the Heidke skill score (HSS) [72,73]. The HSS measures the percentage of improvement of the forecast over a random prediction and ranges in [−∞, 1]. Negative or null values indicate that the model prediction is at the most as good as a random forecast, while a perfect forecast obtains a HSS of 1.

The PC and HSS of the persistence were also calculated, i.e., considering that the predicted drought class for the following month (t + 1) equals the drought class that was in the preceding month (t). Then, the fractional improvement of the log-linear modeling forecast over the persistence forecast, (the latter considered as a reference), was computed. The persistence forecast is often used as a reference in measuring the degree of skill of forecasts produced by other methods, especially for short-term predictions, and reflects the tendency for the pre-existing weather conditions to be maintained, i.e., the occurrence of a specific event at a given time is more probable if that same event has occurred in the immediately preceding time period.

Moreover, the performance of the modeling divided in NAO− and NAO+ was compared to the modeling without the NAO influence.

3. Results and Discussion

3.1. Drought Class Analysis

The temporal evolution of the original drought and wetness classes for the study period is shown in Figure 2 (SPEI6) and Figure 3 (SPEI12), considering the classification proposed by McKee et al. [39] with seven categories (Table 1). In these graphs, the statistical mode is considered since we are working with categorical variables (the SPEI classes) and the mode retrieves the most frequent drought or wetness class in that month within the 15 grid points that constitute the Prut basin (Figure 1). Consequently, the retrieval of the statistical mode results in a single class for each month and, as expected, most of the months were categorized as 'normal' as can be seen by the dense green line in Figures 2 and 3.

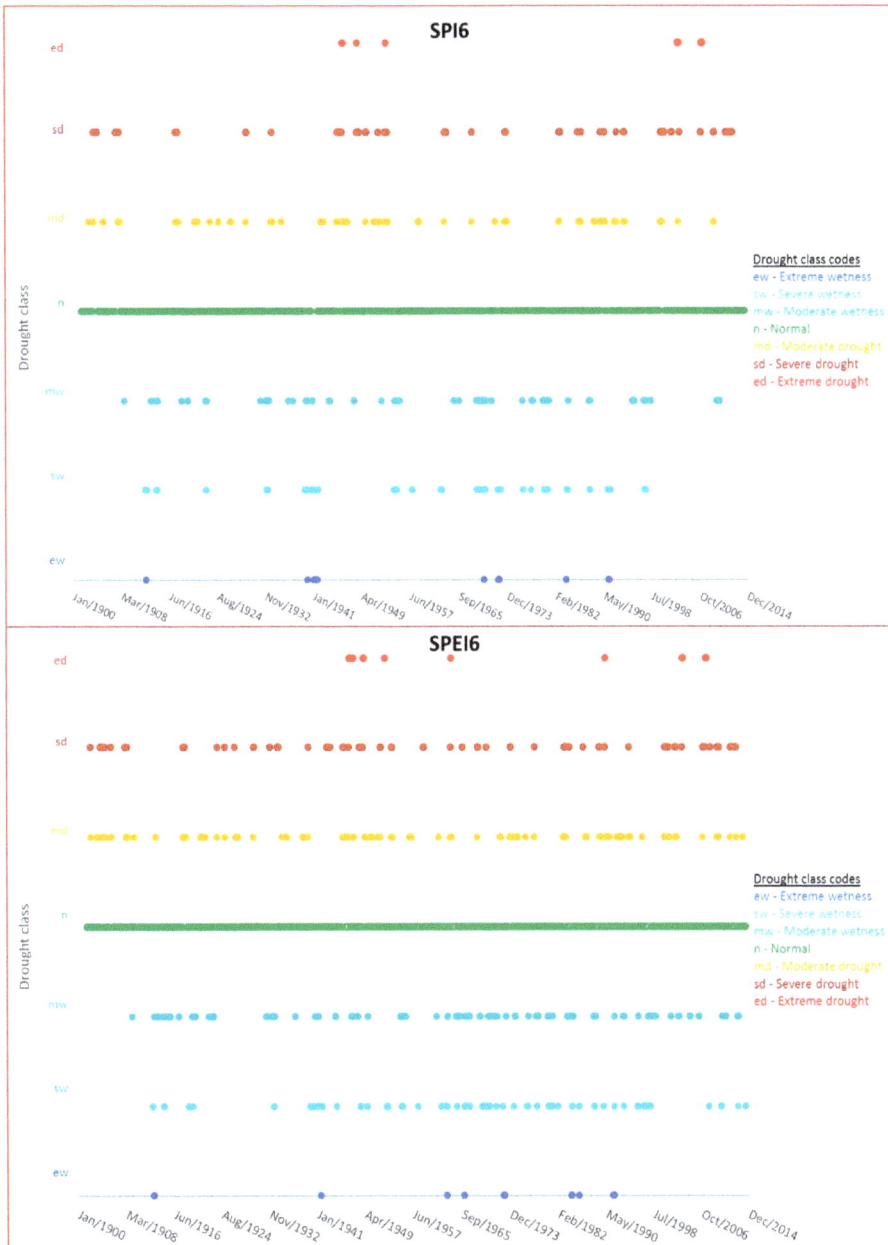

Figure 2. Most frequent classes by month for SPI6 and SPEI6.

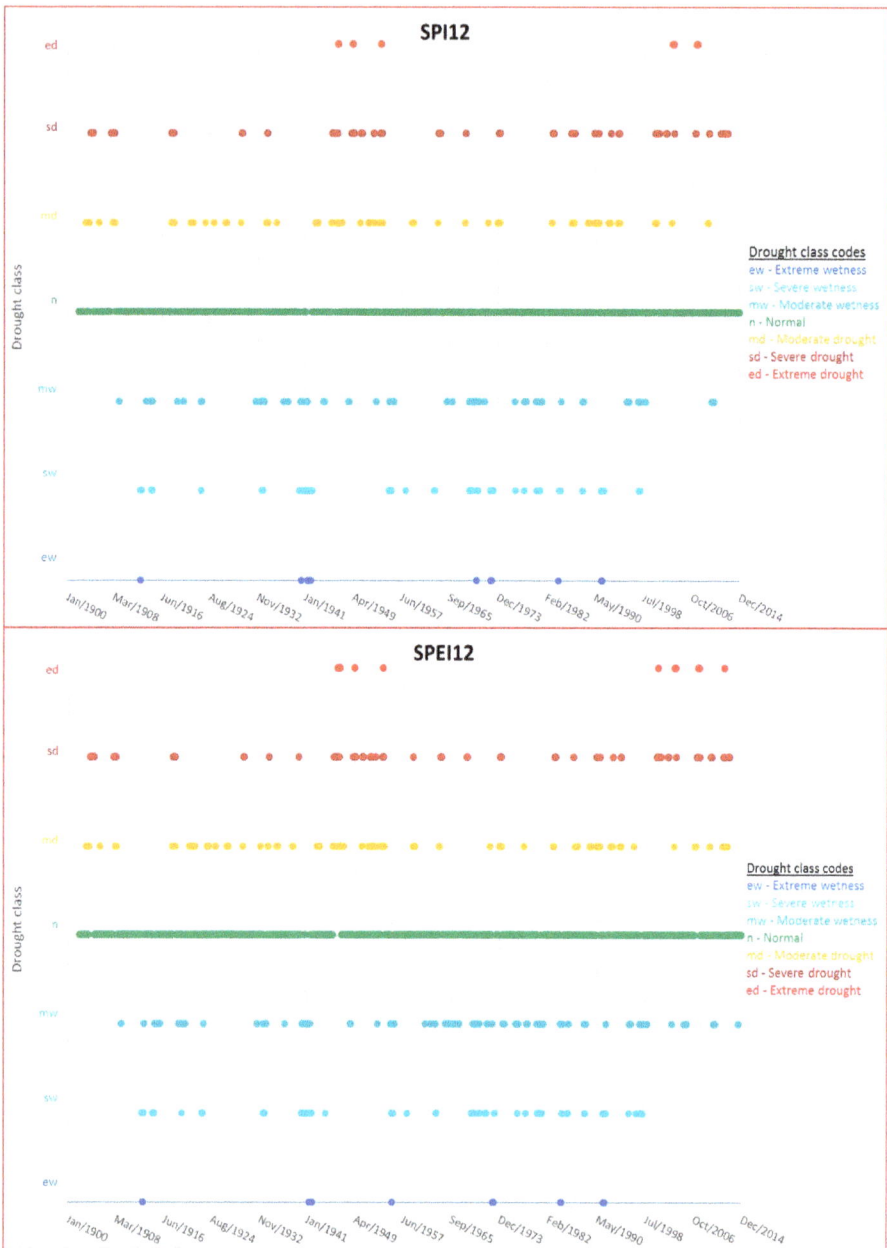

Figure 3. Most frequent classes by month for the SPI12 and SPEI12.

When comparing the SPI6 with the SPEI6 (Figure 2), it is evident that the SPEI6 identifies slightly more severe and extreme droughts than the SPI6. The same seems to happen for severe and extreme wetness conditions. Regarding the temporal 12-month time scale (Figure 3), the same is true but the difference between both is slighter.

The full period range 1901–2014 (114 years) of available data was divided into four similar length sub-periods of 28–29 years each: the first sub-period was from 1901 to 1930; the second sub-period was from 1931 to 1958; the third sub-period was from 1959 to 1986, and the fourth sub-period was from 1987 to 2014. For each of the 15 grid points and 4 sub-periods, the number of times each class occurred was computed and the mean values compared between sub-periods. This division into four equal/similar sub-periods is done in order to obtain a reliable comparison of the drought class occurrences between sub-periods and because four is the minimum number of points necessary to find a curvature and its counter-curve and so to find an eventual cyclicity in drought class frequency. The analysis focused only on the sum of severe and extreme drought (classes sd + ed) and the sum of severe and extreme wetness classes (classes sw + ew) as these extreme classes are the ones with higher socio-economic impacts. The choice of joining classes sw and ew (and also classes sd and ed) results from the fact that the most extreme classes have only a few cases for the analyzed period and joining both classes will allow for a more robust analysis. Figure 4 presents the mean values of the occurrences for these sums of classes by the sub-period obtained for the SPEI6 (left) and the SPEI12 (right).

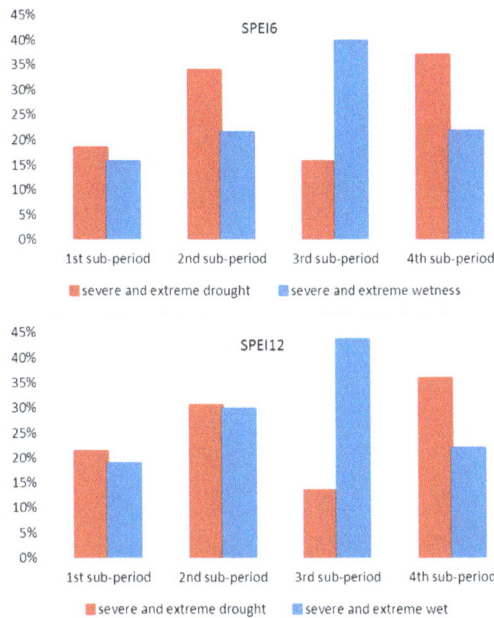

Figure 4. Mean frequencies of the 15 grid points for the occurrence of severe and extreme drought (red column) and severe and extreme wetness (blue column) in each sub-period for the SPEI6 and SPEI12.

Regarding the evolution of severe and extreme droughts (classes sd + ed), Figure 4 reveals an increase from the first to the second sub-period, followed by a decrease in the third sub-period and again an intense increase in the fourth sub-period. Concerning equivalent performance of severe and extreme wetness events (classes sw + ew) the behavior is different, showing an increase tendency from the first to the third sub-periods followed by a decrease in the fourth sub-period. Moreover, these distinct behaviors are similar for both the SPEI6 and SPEI12. To elucidate if the differences between the four sub-periods are statistically significant, a simple one-way analysis of variance (ANOVA) followed by a multiple comparison Turkey test [74] with a confidence level $\alpha = 0.05$, was performed for each case (classes sd + ed and classes sw + ew, for both SPEI6 and SPE12).

Results for classes sd + ed of SPEI6 show that the increases in the number of severe and extreme drought occurrences from the first to second, first to fourth, and third to fourth sub-periods are statistically significant, as well as the decrease in number from the second to the third sub-periods. Between the first and the third and between the second and the fourth sub-periods, the differences are nonsignificant. With respect to the SPEI12, all changes between sub-periods are statistically significant, i.e., all the increases and decreases in the number of severe and extreme drought occurrences are statistically significant. These results point to an overall increase from the first to the fourth sub-periods but also to a strong multi-decadal evolution with possible cyclicity of the severe and extreme drought frequency in the Prut basin. In particular, if we restrict ourselves to the last 50–60 years (corresponding to the third and fourth sub-periods) the notorious rise from the third to the fourth sub-period is corroborated by several studies that reveal significant positive trends in drought frequency, severity, and duration for the Mediterranean and Carpathian region [6,7,19]. However, when using a longer span period, a study regarding Portugal also agrees with the hypothesis that severe and extreme droughts may present a cyclic component with the duration ranging from 26 to 30 years [75].

Concerning the number of severe and extreme wetness events (classes sw + ew) for SPEI6, ANOVA results reveal a statistical significance between all the sub-periods except between the second and the fourth. For the SPEI12, the exceptions (nonsignificant differences) occur between the first and the fourth and the second and the fourth sub-periods. Therefore, these finds point to a significant increase in severe and extreme wetness events from the first period to the third and then a significant decrease in the last sub-period. Apart from some minor differences, the results obtained for the SPI are similar to those attained for SPEI and are not presented for the sake of simplicity.

3.2. Prediction Analysis

A comparison between the observed the SPEI6 (or SPEI12) drought class at month t + 1 (from now on denoted as OBS) and the prediction using the modeling with and without the NAO influence (from now on denoted respectively as PRED and PRED_NAO) was performed for each grid point (Figure 1, Table 3). In order illustrate the prediction made for each grid point, an example is presented in Tables 6 and 7 for two distinct temporal periods. In those tables the comparison between the observed SPEI6 (or SPEI12) drought class at month t + 1 (OBS), and the prediction using the modeling with and without the NAO influence (PRED and PRED_NAO) for a chosen grid point is shown for two periods, respectively from January 2012 to June 2013 (Table 6) and the extended winter months between 2011 and 2013 (Table 7). The observed SPEI6 (SPEI12) drought class at months t − 1 and t, as well as the average NAO value for the last four months considered in the modeling are also shown. It should be noted that the period selected for the comparison refers to a period with several class changes involving all four drought classes. When two drought classes are equally probable, then the predicted drought class is identified for instance as "1 or 2", indicating that the transition probabilities into either class 1 or class 2 are similar. The cells in Tables 6 and 7 are highlighted in grey when the predictions do not match the observations.

Table 6. The SPEI6 and SPEI12: Comparison between observed (OBS) and predicted drought class transitions (with and without the NAO influence; "PRED" and "PRED_NAO", respectively) for grid point 13 (Table 3) during the period from January 2012 to June 2013.

SPEI6 Date	NAO 4-Month Average	Drought Class at Month t − 1	Drought Class at Month t	Drought Class at Month t + 1 OBS	Drought Class at Month t + 1 PRED	Drought Class at Month t + 1 PRED_NAO
Jan-12	2.09	4	4	2	4	3 or 4
Feb-12	1.86	2	2	2	2	2
Mar-12	1.82	2	2	2	2	2
Apr-12	2.08	2	2	2	2	2
May-12	0.69	2	2	4	2	2
Jun-12	−0.03	2	4	3	2	2
Jul-12	−1.00	2	4	4	4	3 or 4
Aug-12	−1.77	4	3	4	2	3 or 4
Sep-12	−1.29	3	4	4	3 or 4	3 or 4
Oct-12	−1.44	4	4	3	2	3 or 4
Nov-12	−1.60	4	4	2	4	3 or 4
Dec-12	−1.55	3	3	1	2	2
Jan-13	−1.29	2	2	1	1	1
Feb-13	−0.66	1	1	1	1 or 2	1 or 2
Mar-13	0.08	1	1	2	1 or 2	1 or 2
Abr-13	−0.58	1	2	2	2	2
May-13	−0.73	2	2	2	2	2
Jun-13	−0.69	2	2	2	2	2

SPEI12 Date	NAO 4-Month Average	Drought Class at Month t − 1	Drought Class at Month t	Drought Class at Month t + 1 OBS	Drought Class at Month t + 1 PRED	Drought Class at Month t + 1 PRED_NAO
Jan-12	1.25	4	4	4	4	4
Feb-12	2.09	4	4	2	2	4
Mar-12	1.86	2	2	2	2	2
Apr-12	1.82	2	2	3	2 or 3	2
May-12	2.08	2	3	4	2	2 or 3
Jun-12	0.69	3	2	4	2	2
Jul-12	−0.03	2	4	3	3 or 4	3 or 4
Aug-12	−1.00	4	3	4	2 or 3	3 or 4
Sep-12	−1.77	3	4	4	3 or 4	3 or 4
Oct-12	−1.29	4	4	3	4	4
Nov-12	−1.44	4	4	3	4	4
Dec-12	−1.60	4	3	2	2 or 3	3 or 4
Jan-13	−1.55	3	2	2	2	2
Feb-13	−1.29	2	2	2	2	2
Mar-13	−0.66	2	2	1	1	1 or 2
Apr-13	0.08	2	1	2	2	2
May-13	−0.58	1	2	2	2	2
Jun-13	−0.73	2	2	2	2	2

Table 7. The SPEI6 and SPEI12: Comparison between observed (OBS) and predicted drought class transitions ("PRED" and "PRED_NAO") for grid point 13 (Table 3) during the periods November 2011 to May 2012, November 2012 to May 2013, and November 2013 to December 2013.

SPEI6 Date	NAO 4-Month Average	Drought Class at Month t − 1	Drought Class at Month t	Drought Class at Month t + 1 OBS	Drought Class at Month t + 1 PRED	Drought Class at Month t + 1 PRED_NAO
Nov-11	0.21	4	4	4	4	3 or 4
Dec-11	1.25	4	4	4	4	3 or 4
Jan-12	2.09	4	4	2	2	3 or 4
Feb-12	1.86	2	2	2	2	2
Mar-12	1.82	2	2	2	2	2
Apr-12	2.08	2	2	2	2	2
May-12	0.69	2	3	2	2	2 or 3
Nov-12	−1.60	4	4	3	4	4
Dec-12	−1.55	3	3	2	3 or 4	4
Jan-13	−1.29	2	2	1	2	2

SPEI12 Date	NAO 4-Month Average	Drought Class at Month t − 1	Drought Class at Month t	Drought Class at Month t + 1 OBS	Drought Class at Month t + 1 PRED	Drought Class at Month t + 1 PRED_NAO
Nov-11	−0.55	3	3	4	3	3 or 4
Dec-11	0.21	3	4	4	3 or 4	3 or 4
Jan-12	1.25	4	4	2	4	4
Feb-12	2.09	4	4	2	4	4
Mar-12	1.86	2	2	2	2	2
Apr-12	1.82	2	2	3	2	2
May-12	2.08	2	3	2	3	2 or 3
Nov-12	−1.44	4	4	3	4	4
Dec-12	−1.60	4	3	2	3 or 4	2 or 3
Jan-13	−1.55	3	2	2	2	2

Table 7. Cont.

SPEI6

SPEI6 Date	NAO 4-Month Average	Drought Class at Month t − 1	Month t	Drought Class at Month t + 1 OBS	PRED	PRED_NAO
Feb-13	−0.66	1	1	1	1 or 2	1 or 2
Mar-13	0.08	1	1	2	2	2
Apr-13	−0.58	1	2	2	2	2
May-13	−0.73	2	2	2	2	2
Nov-13	0.94	2	2	2	2	2
Dec-13	0.32	2	2	2	2	2

SPEI12

SPEI12 Date	NAO 4-Month Average	Drought Class at Month t − 1	Month t	Drought Class at Month t + 1 OBS	PRED	PRED_NAO
Feb-13	−1.29	2	2	2	2	2
Mar-13	−0.66	2	2	1	2	2
Apr-13	0.08	2	1	2	3 or 4	3 or 4
May-13	−0.58	1	2	2	2	2
Nov-13	1.38	2	2	2	2	2
Dec-13	0.94	2	2	2	2	2

Table 8. Proportion correct (PC) and Heidke skill score (HSS) (%) for the SPEI6 and SPEI12 modeling in the extended winter months for PRED (1) and its fractional improvement over the persistence forecast; for PRED_NAO (2) and its fractional improvement over the persistence forecast; and for the difference between both modeling approaches.

SPEI6

Grid Point	PRED (1) PC	HSS	Persistence PC	HSS	PRED_NAO (2) PC	HSS	Persistence PC	HSS	Difference: (2)−(1) PC	HSS
1	75.4	45.6	31.9	20.8	78.0	51.5	39.2	29.5	2.6	6.0
2	78.5	54.8	40.2	32.3	78.8	56.7	40.9	35.1	0.3	1.9
3	76.4	48.2	30.0	19.3	79.0	54.0	37.9	28.4	2.6	5.8
4	76.8	50.5	31.7	21.5	78.6	54.1	37.5	27.5	1.8	3.6
5	76.5	52.4	23.4	18.2	80.2	59.2	35.1	29.9	3.7	6.8
6	79.2	57.1	32.2	26.2	81.7	62.5	40.2	35.4	2.5	5.4
7	78.0	54.3	33.3	25.5	79.7	58.3	38.8	32.3	1.8	4.0
8	80.6	58.5	40.3	33.0	80.8	59.8	37.9	31.9	0.2	1.3
9	78.3	56.6	28.1	22.8	81.7	62.9	39.4	34.1	3.4	6.3
10	76.7	50.7	32.8	23.9	78.3	55.8	37.5	31.8	1.6	5.1
11	78.2	53.5	30.0	22.6	79.5	57.0	34.2	28.3	1.3	3.5
12	80.6	60.6	41.9	35.0	82.6	64.8	47.9	41.9	2.0	4.2
13	76.9	52.4	25.0	19.8	79.4	57.8	33.2	28.8	2.5	5.4
14	77.4	52.1	35.8	26.7	80.2	56.4	43.6	33.4	2.8	4.3
15	76.7	52.2	25.4	19.5	79.8	58.8	35.2	30.8	3.1	6.6
Mean	77.7	53.3	32.1	24.5	79.9	58.0	38.6	31.9	2.1	4.7
Max	80.6	60.6	41.9	35.0	82.6	64.8	47.9	41.9	3.7	6.8
Min	75.4	45.6	23.4	18.2	78.0	51.5	33.2	27.5	0.2	1.3

SPEI12

Grid Point	PRED (1) PC	HSS	Persistence PC	HSS	PRED_NAO (2) PC	HSS	Persistence PC	HSS	Difference: (2)−(1) PC	HSS
1	77.7	58.1	12.1	9.8	83.0	68.0	32.2	31.1	5.3	9.9
2	79.2	60.6	13.3	11.5	82.0	66.1	25.0	23.7	2.8	5.5
3	78.5	59.8	13.8	11.4	79.9	61.9	19.5	15.9	1.4	2.1
4	82.2	66.1	20.6	18.1	83.7	69.2	26.9	25.7	1.5	3.1
5	83.5	68.2	21.8	19.6	83.8	68.8	23.2	21.1	0.3	0.6
6	82.8	68.0	11.8	11.7	87.4	75.6	35.3	32.6	4.6	7.6
7	82.6	67.2	20.5	18.7	82.2	66.2	18.7	16.4	−0.4	−1.0
8	82.5	65.9	20.3	18.5	85.1	71.1	32.0	30.8	2.6	5.2
9	84.8	68.9	23.2	19.8	87.5	73.9	36.8	32.9	2.7	5.0
10	84.3	69.4	30.9	26.8	83.0	68.4	25.3	24.6	−1.3	−1.0
11	81.7	64.5	17.8	15.3	82.6	66.5	21.8	20.1	0.9	2.0
12	86.8	72.6	36.8	31.8	85.6	70.8	30.7	27.4	−1.2	−1.8
13	81.6	62.8	19.1	15.7	84.4	68.9	31.5	29.4	2.8	6.1
14	82.4	64.5	20.2	16.5	83.1	66.9	23.7	22.2	0.7	2.4
15	80.7	61.4	25.6	19.8	82.6	66.5	33.0	30.4	1.9	5.1
Mean	82.1	65.2	20.5	17.7	83.7	68.6	27.7	25.6	1.6	3.4
Max	86.8	72.6	36.8	31.8	87.5	75.6	36.8	32.9	5.3	9.9
Min	80.7	61.4	11.8	9.8	79.9	61.9	18.7	15.9	−1.3	−1.8

From the analysis of these tables, it may be observed that the modeling performs very well in predicting the maintenance of the drought classes, but not particularly well when a change of class occurs (decrease or increase of the drought class category) breaking with the drought class established in the preceding two months. Furthermore, for the example shown in Tables 6 and 7, we can see that it seems more difficult to predict a class decrease or increase when using the SPEI12 than the SPEI6, likely due to the SPEI6 having quicker response to the variability of precipitation in periods with frequent changes of drought classes. Besides that, when comparing Tables 6 and 7, some decrease in the number of disagreements is perceived for the case that considers the modeling restricted to the extended winter months: this probably due to the slightly higher magnitude of the correlations between NAO and SPEI for the wetter months. Finally, the agreements between the predictions and observations are a little higher when the "PRED_NAO" modeling is used instead of just "PRED", for both the SPEI6 and SPEI12, indicating that some cases of class change could be better predicted with the "PRED_NAO", namely those with negative NAO. Results for the other grid points can present some variability, but the overall behavior will not show significant differences. This analysis was repeated for all the grid points in Figure 1 and the results were aggregated.

To assess the performance of the log-linear modeling influenced by the NAO comparative to modeling without that influence and the percentage of improvement of both modeling over the persistence forecast, the proportion correct (PC) and Heidke skill score (HSS) were computed for the entire period of the time series as explained in Section 2.3. The results corresponding to the SPEI6 and the SPEI12 modeling of the extended winter months are shown in Table 8, respectively. For the SPI6 and SPI12, as well as for the modeling of all the months of the year, only the mean, maximum, and minimum values are presented instead of the results per grid point (Tables 9–11) for the sake of simplicity.

To better visualize the performance of the log-linear modeling without and with the NAO influence, graphical representations of the PC and HSS skill score results are presented in Figures 5–8. In those figures, it is shown for the SPEI6 (SPI6) (above) and SPEI12 (SPI12) (bellow) the PC (left) and HSS (right) the improvement of the "PRED_NAO" over the "PRED" forecast, corresponding to the difference between both. In some cases, there is no improvement in certain grid points, indeed there is a decrease.

The information given in Tables 8–11 allows us to conclude that the log-linear modeling performs consistently better than the persistence forecast, indicating that the model can predict more accurately than the simple maintenance of the drought class of the preceding month, that is, a considerable number of class changes are well predicted.

Table 9. Mean, maximum and minimum of the PC and HSS skill scores (%) for the SPI6 (above) and SPI12 (below) for the extended winter months.

SPI6	PRED (1)		Persistence		PRED_NAO (2)		Persistence		Difference: (2)–(1)	
	PC	HSS	PC	HSS	PC	HSS	PC	HSS	PC	HSS
Mean	80.2	55.1	36.7	27.9	80.8	57.2	38.7	31.3	0.6	2.1
Max	82.8	58.4	41.0	33.1	82.0	59.5	44.1	34.7	2.4	5.2
Min	77.8	49.6	26.7	19.6	78.6	54.8	32.5	26.6	−1.3	−0.6

SPI12	PRED (1)		Persistence		PRED_NAO (2)		Persistence		Difference: (2)–(1)	
	PC	HSS	PC	HSS	PC	HSS	PC	HSS	PC	HSS
Mean	82.4	63.8	16.0	13.3	84.4	67.9	25.9	23.2	2.1	4.1
Max	85.3	69.0	27.1	22.2	86.1	70.4	35.1	31.7	5.4	10.1
Min	79.8	58.9	5.5	1.3	81.0	61.7	18.6	17.3	0.3	0.7

Table 10. As in Table 9 but relative to all months and SPEI6 (above) and SPEI12 (below).

SPEI6	PRED (1)		Persistence		PRED_NAO (2)		Persistence		Difference: (2)–(1)	
	PC	HSS	PC	HSS	PC	HSS	PC	HSS	PC	HSS
Mean	78.7	54.8	33.5	25.7	79.1	56.0	34.2	27.2	0.3	1.2
Max	82.5	62.7	41.3	33.9	81.1	60.6	41.9	34.5	2.5	6.4
Min	75.4	44.8	24.9	17.1	76.4	51.2	24.7	18.9	−2.5	−4.9

SPEI12	PRED (1)		Persistence		PRED_NAO (2)		Persistence		Difference: (2)–(1)	
	PC	HSS	PC	HSS	PC	HSS	PC	HSS	PC	HSS
Mean	81.6	64.0	18.2	15.5	82.6	65.9	22.5	20.0	1.0	1.9
Max	83.8	68.2	29.2	25.6	84.3	69.6	26.5	23.0	2.5	5.2
Min	79.5	60.6	7.6	6.1	80.9	62.8	16.9	15.7	−0.6	−1.2

Table 11. As in Table 9 but relative to all months and SPI6 (above) and SPI12 (below).

SPI6	PRED (1)		Persistence		PRED_NAO (2)		Persistence		Difference: (2)–(1)	
	PC	HSS	PC	HSS	PC	HSS	PC	HSS	PC	HSS
Mean	79.0	52.0	34.0	24.3	80.2	54.5	37.8	28.3	1.2	2.5
Max	81.6	57.1	41.2	32.2	82.8	60.0	45.0	37.0	5.0	9.2
Min	75.3	44.0	23.4	13.6	78.5	48.5	32.0	20.8	−0.6	−2.0

SPI12	PRED (1)		Persistence		PRED_NAO (2)		Persistence		Difference: (2)–(1)	
	PC	HSS	PC	HSS	PC	HSS	PC	HSS	PC	HSS
Mean	81.8	62.3	19.0	15.6	82.4	63.4	21.6	18.8	0.6	1.1
Max	84.6	66.8	25.8	21.2	85.2	69.1	34.1	29.4	4.0	6.1
Min	79.4	57.7	7.3	6.2	80.4	59.6	15.7	14.7	−1.6	−1.9

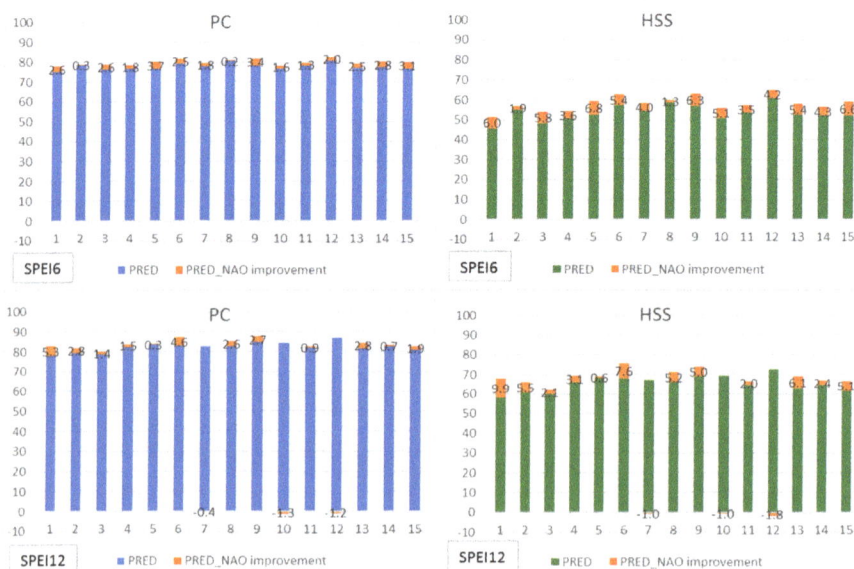

Figure 5. PC and HSS improvement (%) considering the influence of NAO on the SPEI6 and SPEI12 modeling for the extended winter months for each of the 15 grid points.

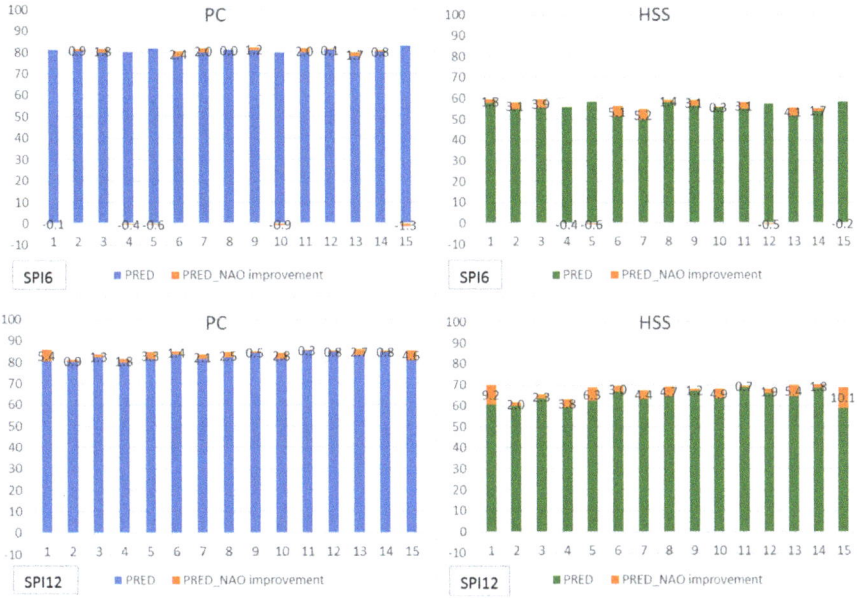

Figure 6. PC and HSS improvement (%) considering the influence of NAO on the SPI6 and SPI12 modeling for the extended winter months for each of the 15 grid points.

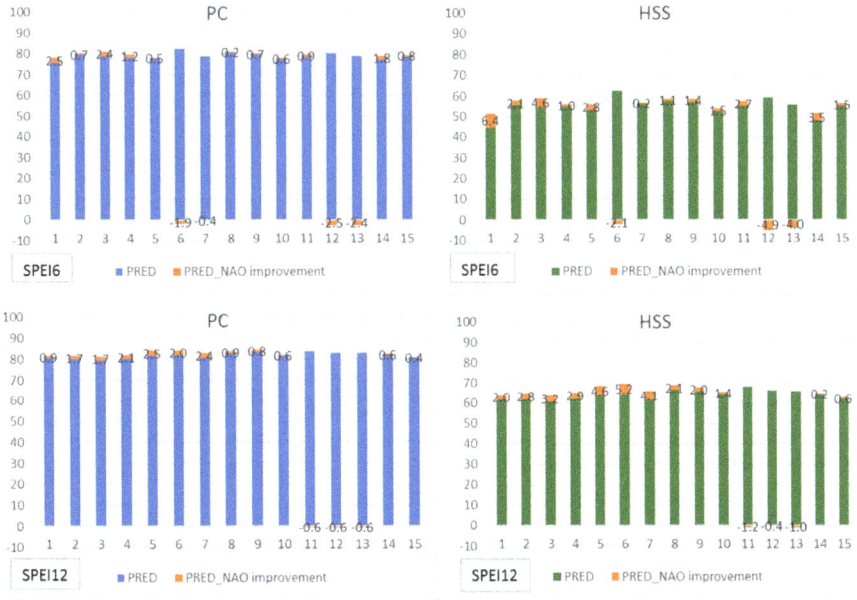

Figure 7. PC and HSS improvement (%) considering the influence of NAO on the SPEI6 and SPEI12 modeling for all months for each of the 15 grid points.

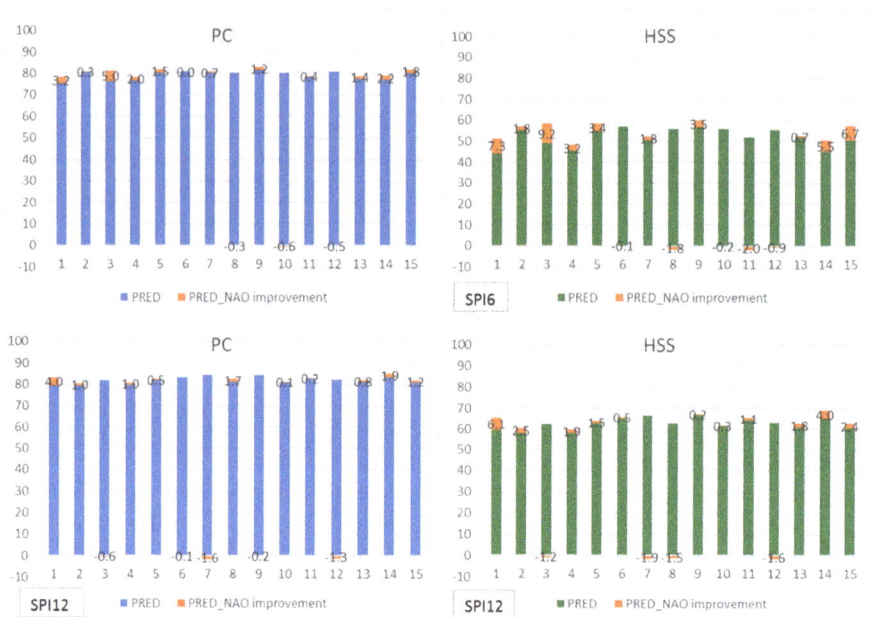

Figure 8. PC and HSS improvement (%) considering the influence of NAO on the SPI6 and SPI12 modeling for all months for each of the 15 grid points.

Furthermore, better overall performances are obtained when the modeling is applied to the 12-month time scale (average range PC: 82–84; HSS: 62–69), in contrast with the 6-month time scale (average range PC: 77–80; HSS: 52–58) (Tables 8–11, Figures 5–8). This is likely to be related to the less frequent change of classes within the 12-month time scale and favors the ability of easily identifying capturing the behavior of changes in drought classes in the preceding months.

When the modeling is restricted only to the extended winter months instead of all months of the year, higher skill scores are obtained in most of the cases (Figure 5 vs. Figure 7, Figure 6 vs. Figure 8), the increase being however not very expressive (Table 8 vs. Table 10, Table 9 vs. Table 11). The relatively small increase of the correlation coefficient between the SPEI (SPI) and the NAO when considering only the extended winter months over all the months of the year may justify these small gains in the skill scores.

Concerning the introduction of the NAO predictor in the modeling, the highest improvements obtained in the skill scores with the "PRED_NAO" are obtained when modeling the extended winter months of the SPEI6 and SPI12 (Figures 5 and 6). Those improvements are however not impressive, which may be due to the low correlations between the SPEI (SPEI) and the NAO index (Table 3). They range between 0.2 and 3.7 (PC) and between 4.7 and 6.8 (HSS) for the SPEI6, and between 0.3 and 5.4 (PC) and between 4.1 and 10.1 (HSS) for the SPI12 (Tables 8 and 9, Figures 5 and 6). For the other cases, there are several grid points for which there is no improvement, i.e., the inclusion of the NAO influence does not bring any benefit. This situation may be justified by the fact that when modeling the drought classes transitions separately (NAO+ and NAO−), the accuracy of the prediction is somewhat diminished in comparison with modeling without the NAO because the data is divided into two groups with about half of the observations each. This accuracy reduction is not strongly compensated by the relatively low correlations, resulting in minor improvements of the predictions with the model "PRED_NAO" and occurs mainly for the transitions that imply a jump of 2 or 3 in the drought category, for which there are fewer observations.

Regarding the SPEI versus SPI comparison, no significant differences in terms of performance are found, since sometimes better performances are obtained for the SPEI, but the opposite also occurs (Figure 5 vs. Figure 6, Figure 7 vs. Figure 8).

The same modeling was applied to data from Portugal [63], however, there are several differences between both studies that do not allow a true comparison. First, just the SPI6 and SPI12 for all months of the year were modeled. Then, the NAO value considered in the PRED_NAO was not the mean of the four months before, but the value at the month $t - l$, where l is the lag that maximizes the correlation between the SPI and NAO. The skill scores obtained in that study are a little higher than the ones obtained herein, and the improvements in the predictions brought by the NAO introduction are also a bit larger, which is in agreement with the higher correlations between the SPI and the NAO for Portugal.

4. Conclusions

Monthly predictability of the SPEI and SPI drought classes under the influence of NAO was studied for the Prut basin using three-dimensional log-linear models, which are probabilistic models that learn from the preceding two months with the aim of predicting the drought class in the following month. These models have the ability to predict the persistence of the drought classes and some class changes that may be captured in the preceding two months. However, when a sudden change of class occurs (decrease or increase), it becomes particularly difficult to predict it, even when the NAO phase influencing the transition is known. Some reasons for that may be related to the not sufficiently high correlations between the SPEI (SPEI) and the NAO, for the Prut basin. However, some improvements in the predictions are obtained by the NAO introduction in the modeling, mainly when applied to the SPEI6 and the SPI12 and only to the extended winter months. Nevertheless, the overall performances of the log-linear modeling are good. Consequently, it can be concluded that the log-linear modeling has good performance skills when predicting the drought class one month ahead based on the information of drought classes of the two previous months. Nevertheless, this method presents the fragility of failing to predict many class transitions, which are different than the classes in the previous two months.

With respect to the drought class analysis the full study period was divided into four sub-periods of similar lengths. The findings, corroborated by other studies for the Mediterranean region, point to an overall increase in drought frequency from the first to the fourth sub-period but also to a strong multi-decadal evolution, with possible cyclicity of severe and extreme drought frequency in the Prut basin. In particular, for the last 50–60 years, the increment is very pronounced. Concerning the number of severe and extreme wetness events, the findings point to a significant increase from the first period to the third, and then a significant decrease in the last sub-period.

We believe that a methodology which can predict drought classes a few months in advance will beneficiate stakeholders, farmers, and also insurance companies so they are able to react accordingly and take proper action plans. In this sense, even with the mentioned limitations, we think the presented methodology can be part of an early warning system aiming to inform water users (especially farmers) in their water management decisions.

Acknowledgments: This work was partially supported by the projects IMDROFLOOD—Improving Drought and Flood Early Warning, Forecasting and Mitigation using real-time hydroclimatic indicators (WaterJPI/0004/2014) and project UID/MAT/00297/2013 (Centro de Matemática e Aplicações), both funded by funded by Fundação para a Ciência e a Tecnologia, Portugal (FCT). Ana Russo thanks also FCT by the Post-Doc research grant SFRH/BPD/99757/2014.

Author Contributions: E.M. conceived the work, performed the drought class calculations, analyzed the data, and wrote the paper; A.R. calculated the drought indicators, contributed to the materials and wrote the paper; and R.M.T. conceived the work, contributed to the materials, analyzed the data and wrote the paper.

Conflicts of Interest: The authors declare no conflict of interest. The founding sponsors had no role in the design of the study; in the collection, analyses, or interpretation of data; in the writing of the manuscript, and in the decision to publish the results.

References

1. Lloyd-Hughes, B.; Saunders, M.A. Drought climatology for Europe. *Int. J. Climatol.* **2002**, *22*, 1571–1592. [CrossRef]
2. Bordi, I.; Fraedrich, K.; Sutera, A. Observed drought and wetness trends in Europe: An update. *Hydrol. Earth Syst. Sci.* **2009**, *13*, 1519–1530. [CrossRef]
3. Pereira, L.S.; Cordery, I.; Iacovides, I. *Coping with Water Scarcity, Addressing the Challenges*; Springer: Berlin, Germany, 2009.
4. Mishra, A.; Singh, V. A review of drought concepts. *J. Hydrol.* **2010**, *391*, 202–216. [CrossRef]
5. Vogt, J.V.; Barbosa, P.; Hofer, B.; Magni, D.; De Jager, A.; Singleton, A.; Horion, S.; Sepulcre, G.; Micale, F.; Sokolova, E.; et al. Developing a European drought observatory for monitoring, assessing and forecasting droughts across the European continent. In *AGU Fall Meeting Abstracts 1*; NH24A-07; American Geophysical Union: Washington, DC, USA, 2011.
6. Spinoni, J.; Naumann, G.; Vogt, J.; Barbosa, P. European drought climatologies and trends based on a multi-indicator approach. *Glob. Planet. Chang.* **2015**, *127*, 50–57. [CrossRef]
7. Spinoni, J.; Naumann, G.; Vogt, J.; Barbosa, P. The biggest drought events in Europe from 1950 to 2012. *J. Hydrol. Reg. Stud.* **2015**, *3*, 509–524. [CrossRef]
8. Dai, A. Drought under global warming: A review. *WIREs Clim. Chang. Adv. Rev.* **2011**, *2*, 45–65. [CrossRef]
9. Spinoni, J.; Naumann, G.; Carrao, H.; Barbosa, P.; Vogt, J. World drought frequency, duration, and severity for 1951–2010. *Int. J. Climatol.* **2014**, *34*, 2792–2804. [CrossRef]
10. Vicente-Serrano, S.; Lopez-Moreno, J.; Beguería, S.; Lorenzo-Lacruz, J.; Sanchez-Lorenzo, A.; García-Ruiz, J.; Azorin-Molina, C.; Morán-Tejeda, E.; Revuelto, J.; Trigo, R.; et al. Evidence of increasing drought severity caused by temperature rise in southern Europe. *Environ. Res. Lett.* **2014**, *9*, 044001. [CrossRef]
11. Trenberth, K.E.; Dai, A.; Van der Schrier, G.; Jones, P.D.; Barichivich, J.; Briffa, K.R.; Sheffield, J. Global warming and changes in drought. *Nat. Clim. Chang.* **2014**, *4*, 17–22. [CrossRef]
12. Briffa, K.R.; Van der Schrier, G.; Jonesa, P.D. Wet and dry summers in Europe since 1750: Evidence of increasing drought. *Int. J. Climatol.* **2009**, *29*, 1894–1905. [CrossRef]
13. Paltineanu, C.; Mihailescu, I.F.; Prefac, Z.; Dragota, C.; Vasenciuc, F.; Claudia, N. Combining the standardized precipitation index and climatic water deficit in characterizing droughts: A case study in Romania. *Theor. Appl. Climatol.* **2009**, *97*, 219–233. [CrossRef]
14. Sousa, P.M.; Trigo, R.M.; Aizpurua, P.; Nieto, R.; Gimeno, L.; Garcia Herrera, R. Trends and extremes of drought indices throughout the 20th century in the Mediterranean. *Nat. Hazards Earth Syst. Sci.* **2011**, *11*, 33–51. [CrossRef]
15. Hoerling, M.; Eischeid, J.; Perlwitz, J.; Quan, X.; Zhang, T.; Pegion, P. On the Increased Frequency of Mediterranean Drought. *J. Clim.* **2012**, *25*, 2146–2161. [CrossRef]
16. Spinoni, J.; Naumann, G.; Vogt, J. Pan-European seasonal trends and recent changes of drought frequency and severity. *Glob. Planet. Chang.* **2017**, *148*, 113–130. [CrossRef]
17. Intergovernmental Panel on Climate Change (IPCC). Climate Change 2014, Synthesis Report, WMO and UNEP, Geneva. Available online: https://www.ipcc.ch/pdf/assessment-report/ar5/syr/SYR_AR5_FINAL_full_wcover.pdf (accessed on 10 January 2018).
18. Dumitrescu, A.; Bojariu, R.; Birsan, M.; Marin, L.; Manea, A. Recent climatic changes in Romania from observational data (1961–2013). *Theor. Appl. Climatol.* **2015**, *122*, 111–119. [CrossRef]
19. Spinoni, J.; Antofie, T.; Barbosa, P.; Bihari, Z.; Lakatos, M.; Szalai, S.; Szentimrey, T.; Vogt, J. An overview of drought events in the Carpathian Region in 1961–2010. *Adv. Sci. Res.* **2013**, *10*, 21–32. [CrossRef]
20. Wang, Q.; Wu, J.; Lei, T.; He, B.; Wu, Z.; Liu, M.; Mo, X.; Geng, G.; Li, X.; Zhou, H.; et al. Temporal-spatial characteristics of severe drought events and their impact on agriculture on a global scale. *Quat. Int.* **2014**, *349*, 10–21. [CrossRef]
21. Trenberth, K.E. Changes in precipitation with climate change. *Clim. Res.* **2011**, *47*, 123–138. [CrossRef]
22. Trenberth, K.E.; Fasullo, J.T.; Shepherd, T.G. Attribution of climate extreme events. *Nat. Clim. Chang.* **2015**, *5*, 725–730. [CrossRef]
23. Spinoni, J.; Vogt, J.V.; Naumann, G.; Barbosa, P.; Dosio, A. Will drought events become more frequent and severe in Europe? *Int. J. Climatol.* **2017**. [CrossRef]

24. Cuculeanu, V.; Marica, A.; Simota, C. Climate change impact on agricultural crops and adaptation options in Romania. *Clim. Res.* **1999**, *12*, 153–160. [CrossRef]
25. Marica, A.C.; Busuioc, A. The potential of climate change on the main components of water balance relating to maize crop. *Romanian J. Meteorol.* **2004**, *6*, 50–57.
26. Romanian National Climate Change Strategy (2013–2020), Adaptation Component, Report. Available online: http://climate-adapt.eea.europa.eu/countries-regions/countries/romania (accessed on 10 January 2018).
27. Falco, S.D.; Adinolfi, F.; Bozzola, M.; Capitanio, F. Crop Insurance as a Strategy for Adapting to Climate Change. *J. Agric. Econ.* **2014**, *65*, 485–504. [CrossRef]
28. Lesk, C.; Rowhani, P.; Ramankutty, N. Influence of extreme weather disasters on global crop production. *Nature* **2016**, *529*, 84–87. [CrossRef] [PubMed]
29. Ontel, I.; Vladut, A. Impact of drought on the productivity of agricultural crops within the Oltenia Plain, Romania. *Geogr. Pannonica* **2015**, *19*, 9–19. [CrossRef]
30. Kang, Y.; Khan, S.; Ma, X. Climate change impacts on crop yield, crop water productivity and food security—A review. *Prog. Nat. Sci.* **2009**, *19*, 1665–1674. [CrossRef]
31. Nguyen, T.; Mula, L.; Cortignani, R.; Seddaiu, G.; Dono, G.; Virdis, S.G.P.; Pasqui, M.; Roggero, P.P. Perceptions of Present and Future Climate Change Impacts on Water Availability for Agricultural Systems in the Western Mediterranean Region. *Water* **2016**, *8*, 523. [CrossRef]
32. Pozzi, W.; Sheffield, J.; Stefanski, R.; Cripe, D.; Pulwarty, R.; Vogt, J.V.; Heim, R., Jr.; Brewer, M.J.; Svoboda, M.; Westerhoff, R. Toward global drought early warning capability: Expanding international cooperation for the development of a framework for monitoring and forecasting. *Am. Meteorol. Soc.* **2013**, *94*, 776–785. [CrossRef]
33. National Drought Mitigation Center (NDMC): What is Drought? Available online: http://drought.unl.edu/DroughtBasics/WhatisDrought.aspx (accessed on 10 January 2018).
34. Pulwarty, R.S.; Sivakumar, M. Information systems in a changing climate: Early warnings and drought risk management. *Weather Clim. Extremes* **2014**, *3*, 14–21. [CrossRef]
35. Turco, M.; Ceglar, A.; Prodhomme, C.; Soret, A.; Toreti, A.; Doblas-Reyes, F. Summer drought predictability over Europe: Empirical versus dynamical forecasts. *Environ. Res. Lett.* **2017**, *12*, 084006. [CrossRef]
36. Palmer, W.C. *Meteorological Drought*; US Department of Commerce: Washington, DC, USA, 1965.
37. Pereira, L.S.; Rosa, R.D.; Paulo, A.A. Testing a modification of the Palmer drought severity index for Mediterranean environments. In *Methods and Tools for Drought Analysis and Management*; Rossi, G., Vega, T., Bonaccorso, B., Eds.; Springer: Dordrecht, The Netherlands, 2007; pp. 149–167.
38. Martins, D.S.; Raziei, T.; Paulo, A.A.; Pereira, L.S. Spatial and temporal variability of precipitation and drought in Portugal. *Nat. Hazards Earth Syst. Sci.* **2012**, *12*, 1493–1501. [CrossRef]
39. McKee, T.B.; Doesken, N.J.; Kleist, J. Drought monitoring with multiple time scales. In Proceedings of the 9th Conference on Applied Climatology, Dallas, TX, USA, 15–20 January 1995; pp. 233–236.
40. Vicente-Serrano, S.M.; Beguería, S.; López-Moreno, J.I. A multi-scalar drought index sensitive to global warming: The Standardized Precipitation Evapotranspiration Index—SPEI. *J. Clim.* **2010**, *23*, 1696–1718. [CrossRef]
41. Vautard, R. Multiple weather regimes over the North Atlantic: Analysis of precursors and successors. *Mon. Weather Rev.* **1990**, *118*, 2056–2081. [CrossRef]
42. Trigo, R.; DaCamara, C. Circulation weather types and their influence on the precipitation regime in Portugal. *Int. J. Climatol.* **2000**, *20*, 1559–1581. [CrossRef]
43. Trigo, R.M.; Valente, M.A.; Trigo, I.F.; Miranda, P.M.; Ramos, A.M.; Paredes, D.; García-Herrera, R. North Atlantic wind and cyclone trends and their impact in the European precipitation and Atlantic significant wave height. *Ann. N. Y. Acad. Sci.* **2008**, *1146*, 212–234. [CrossRef] [PubMed]
44. Hurrell, J.W. Decadal trends in the North Atlantic Oscillation: Regional temperatures and precipitation. *Science* **1995**, *269*, 676–679. [CrossRef] [PubMed]
45. Trigo, R.; Osborn, T.; Corte-Real, J. The North Atlantic Oscillation influence on Europe: Climate impacts and associated physical mechanisms. *Clim. Res.* **2002**, *20*, 9–17. [CrossRef]
46. Cleary, D.M.; Wynn, J.G.; Ionita, M.; Forray, F.L.; Onac, B.P. Evidence of long-term NAO influence on East-Central Europe winter precipitation from a guano-derived δ15N record. *Sci. Rep.* **2017**, *7*, 14095. [CrossRef] [PubMed]

47. Busuioc, A.; Von Storch, H. Changes in the winter precipitation in Romania and its relation to the large-scale circulation. *Tellus* **1996**, *48*, 538–552. [CrossRef]

48. Croitoru, A.; Piticar, A.; Imbroane, A.; Burada, D. Spatiotemporal distribution of aridity indices based on temperature and precipitation in the extra-Carpathian regions of Romania. *Theor. Appl. Climatol.* **2013**, *112*, 597–607. [CrossRef]

49. Corduneanu, F.; Bucur, D.; Cimpeanu, S.M.; Apostol, I.C.; Strugariu, A. Hazards resulting from hydrological extremes in the upstream catchment of the Prut River. *Water Resour.* **2016**, *43*, 42–47. [CrossRef]

50. Bojariu, R. Prut River Basin. Available online: http://imdroflood.csic.es/wp-content/uploads/2016/09/Bojariu.pdf (accessed on 31 July 2017).

51. Tomozeiu, R.; Stefan, S.; Busuioc, A. Winter precipitation variability and large-scale circulation patterns in Romania. *Theor. Appl. Climatol.* **2005**, *81*, 193–201. [CrossRef]

52. Bojariu, R.; Giorgi, F. The North Atlantic Oscillation signal in a regional climate simulation for the European region. *Tellus A* **2005**, *57*, 641–653. [CrossRef]

53. Bojariu, B.; Paliu, D. North Atlantic Oscillation Projection on Romanian Climate Fluctuations in the Cold Season. In *Detecting and Modelling Regional Climate Change*; Springer: Berlin, Germany, 2001; pp. 345–356.

54. Mishra, A.; Desai, V.; Singh, V. Drought forecasting using a hybrid stochastic and neural network model. *J. Hydrol. Eng.* **2007**, *12*, 626–638. [CrossRef]

55. Kim, T.; Valdes, J.B. Nonlinear model for drought forecasting based on a conjunction of wavelet transforms and neural networks. *J. Hydrol. Eng.* **2003**, *8*, 319–328. [CrossRef]

56. Ozger, M.; Mishra, A.; Singh, V. Long lead time drought forecasting using a wavelet and fuzzy logic combination model: A case study in Texas. *J. Hydrometeorol.* **2012**, *13*, 284–297. [CrossRef]

57. Bacanli, U.; Firat, M.; Dikbas, F. Adaptive neuro-fuzzy inference system for drought forecasting. *Stoch. Environ. Res. Risk Assess.* **2009**, *23*, 1143–1154. [CrossRef]

58. Mishra, A.; Singh, V. Drought modeling—A review. *J. Hydrol.* **2011**, *403*, 157–175. [CrossRef]

59. Mishra, A.; Desai, V. Drought forecasting using stochastic models. *Stoch. Environ. Res. Risk Assess.* **2005**, *19*, 326–339. [CrossRef]

60. Han, P.; Wang, P.; Zhang, S.; Zhu, D. Drought forecasting based on the remote sensing data using ARIMA models. *Math. Comput. Model.* **2010**, *51*, 1398–1403. [CrossRef]

61. Mishra, A.; Desai, V. Drought forecasting using feed-forward recursive neural network. *Ecol. Model.* **2006**, *198*, 127–138. [CrossRef]

62. Bierkens, M.; Van Beek, L. Seasonal predictability of European discharge: NAO and hydrological response time. *J. Hydrometeorol.* **2009**, *10*, 953–968. [CrossRef]

63. Paulo, A.A.; Pereira, L.S. Prediction of SPI drought class transitions using Markov chains. *Water Resour. Manag.* **2007**, *21*, 1813–1827. [CrossRef]

64. Moreira, E.E.; Coelho, C.A.; Paulo, A.A.; Pereira, L.S.; Mexia, J.T. SPI-based drought category prediction using log-linear models. *J. Hydrol.* **2008**, *354*, 116–130. [CrossRef]

65. Moreira, E.E.; Pires, C.L.; Pereira, L.S. SPI Drought Class Predictions Driven by the North Atlantic Oscillation Index Using Log-Linear Modeling. *Water* **2016**, *8*, 43. [CrossRef]

66. Cancelliere, A.; Di Mauro, G.; Bonaccorso, B.; Rossi, G. Drought forecasting using the Standardized Precipitation Index. *Water Resour. Manag.* **2007**, *21*, 801–819. [CrossRef]

67. Bonaccorso, B.; Cancelliere, A.; Rossi, G. Probabilistic forecasting of drought class transitions in Sicily (Italy) using Standardized Precipitation Index and North Atlantic Oscillation. *J. Hydrol.* **2015**, *526*, 136–150. [CrossRef]

68. Morid, S.; Smakhtin, V.; Bagherzadeh, K. Drought forecasting using artificial neural networks and time series of drought indices. *Int. J. Climatol.* **2007**, *27*, 2103–2111. [CrossRef]

69. Moreira, E.E. SPI drought class prediction using log-linear models applied to wet and dry seasons. *Phys. Chem. Earth* **2016**, *94*, 136–145. [CrossRef]

70. Paulo, A.A.; Pereira, L.S. Drought concepts and characterization: Comparing drought indices applied at local and regional Scales. *Water Int.* **2006**, *31*, 37–49. [CrossRef]

71. Agresti, A. *Categorical Data Analysis*; John Willey & Sons: New York, NY, USA, 1990.

72. Jolliffe, I.; Stephenson, D. *Forecast Verification: A Practitioner's Guide in Atmospheric Science*, 2nd ed.; Willey: Indianapolis, IN, USA, 2011.

73. Wilks, D.S. *Statistical Methods in the Atmospheric Sciences*, 3rd ed.; Academic Press: Cambridge, MA, USA, 2011.

74. Montgomery, D.C. *Design and Analysis of Experiments*, 5th ed.; John Willey & Sons: New York, NY, USA, 1997.

75. Moreira, E.E.; Mexia, J.T.; Pereira, L.S. Are drought occurrence and severity aggravating? A study on SPI drought class transitions using log-linear models and ANOVA-like inference. *Hydrol. Earth Syst. Sci.* **2012**, *16*, 3011–3028. [CrossRef]

Review

Nitrogen in Water-Portugal and Denmark: Two Contrasting Realities

Soraia Cruz [1],*, Cláudia M.d.S. Cordovil [1], Renata Pinto [1], António G. Brito [1], Maria R. Cameira [1], Guilherme Gonçalves [1], Jane R. Poulsen [2], Hans Thodsen [2], Brian Kronvang [2] and Linda May [3]

[1] School of Agriculture, University of Lisbon, LEAF—Linking Landscape, Environment, Agriculture and Food, Tapada da Ajuda, 1349-017 Lisbon, Portugal; cms@isa.ulisboa.pt (C.M.d.S.C.); renatamspinto@sapo.pt (R.P.); agbrito@isa.ulisboa.pt (A.G.B.); roscameira@isa.ulisboa.pt (M.R.C.); guilherme964@gmail.com (G.G.)
[2] Department of Biosciences, University of Aarhus, Vejlsovej 25, 8600 Silkeborg, Denmark; jpo@bios.au.dk (J.R.P.); hath@bios.au.dk (H.T.); bkr@bios.au.dk (B.K.)
[3] Centre for Ecology & Hydrology, Edinburgh, Bush Estate, Penicuik, Midlothian EH26 0QB, UK; lmay@ceh.ac.uk
* Correspondence: scruz@isa.ulisboa.pt; Tel.: +351-913-507-286

Received: 11 March 2019; Accepted: 4 May 2019; Published: 28 May 2019

Abstract: Agricultural activities are responsible for most of the nitrogen (N) inputs that degrade water quality. To elucidate the drivers leading to N pressures on water, we examined the resulting state of surface waters in terms of N concentrations, the impact of this on water quality status and policy responses to these constraints across different climatic and management conditions. Portugal and Denmark were chosen as contrasting case studies for the Driver-Pressure-State-Impact-Response (DPSIR) analysis. Our results showed reductions of 39% and 25% in the use of mineral fertilizer in Portugal and Denmark, respectively, between 2000 and 2010. The N surplus in Portugal varied between 15 and 30 kg N ha^{-1} between 1995 and 2015. In Denmark, in 2015, this amount was 70 kg N ha^{-1}, representing a 53% decrease from the 1990 value. The average amount of total N discharged to surface waters was 7 kg ha^{-1} for mainland Portugal in 2015 and 14.6 kg ha^{-1} for Denmark in 2014. These reductions in the N surplus were attributed to historical policies aimed at N pressure abatement. In Denmark, N losses are expected to decline further through the continuation or improvement of existing national action plans. In Portugal, they are expected to decline further due to the expansion of Nitrate Vulnerable Zones and the introduction of targeted policies aimed at improving N use efficiency and reducing losses to water.

Keywords: agricultural intensification; DPSIR; nitrogen; pressures; policies; surface water pollution

---

1. Introduction

On many occasions, agriculture has been highlighted as a major contributor to nitrogen (N) pollution to water, although the livestock and urban sectors are important contributors as well. The low use efficiency of N applied as chemical fertilizers and manures leads to a N surplus that can be lost to ground and surface waters, reducing their quality and putting pressure onto receiving waterbodies [1]. Indeed, the European Environmental Agency [2] reported that pollution pressures from diffuse sources (especially agriculture) were affecting 30% to 50% of surface waters in the European Union (EU). However, the negative impacts of N inputs to water from agricultural sources vary across the EU due to significant spatial differences in climate, soil types, geology, topography, and agricultural management activities. In many cases, natural conditions determine land use, N attenuation and, hence, potential N losses [3]. European crop and livestock farms have increased their productivity in recent decades, contributing to the level water contamination by nitrates [4,5]. This is the case in the two countries under study, which have contrasting agricultural systems [6,7]. Agriculture accounts for approximately 47% of the land use in mainland Portugal [8] and about 60% in Denmark.

Denmark, one of the most intensively farmed countries in Europe [9], became aware at an early stage that nutrient emissions from diffuse sources, especially in agricultural areas, were causing algal blooms and promoting hypoxia in coastal waters [10,11]. Since 1985, the negative impacts of nutrient loads to water in Denmark have triggered a series of "nutrient action plans" such as the designation of the whole country as a Nitrate Vulnerable Zone (NVZ). These have significantly improved agricultural N use efficiency and reduced N leaching, even though animal production has increased [12].

In contrast, Portugal has extensive areas of its land (~86%) under agricultural use. However, the implementation of the Nitrates Directive (ND) was not translated into NVZ designation until 1998, with only 4% of the territory included by 2012. Local factors controlling nitrate leaching are the high temperatures during half of the year that promote the volatilization of N and reduce losses to water [13]. On the other hand, alternating dry and wet soil conditions stimulate soil organic matter mineralization and promote nitrate leaching from mineralized nitrogen compounds [14].

The main objective of this paper is to compare how the current N pressures may affect the environmental state and water quality in two contrasting countries, Portugal and Denmark. This paper also hypothesizes that different policies, and their implementation, have influenced agricultural N pressures and, consequently, the environmental condition of water and the effectiveness of various mitigation measures that have been implemented to reduce N losses from agriculture in both countries. Although knowledge exists already about water related environmental issues in Portugal and Denmark, there has been no focused comparison between these contrasting countries in terms of impacts on water quality and policy responses.

2. Materials and Methods

A brief description of the main differences between Denmark and Portugal, in terms of agriculture and livestock production, urban pressures, water monitoring systems, climate, policy responses, etc., is essential to understand better the causes of water quality degradation in each country. Figure 1 shows where the case studies, Portugal and Denmark, are in Europe, and the main differences between these countries are presented below. The data used for the analysis in each country was obtained from the respective National Statistics Institutions.

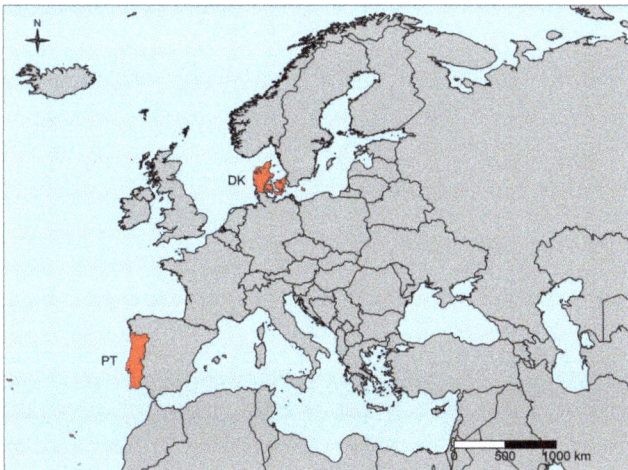

Figure 1. Location of Portugal (PT) and Denmark (DK) in Europe.

2.1. Site Study—Brief Characterization of the Countries

The mainland of Portugal (89,060 km^2) is in the south-west of the Iberian Peninsula (Figure 1); it also has two major archipelagos, the Azores (2333 km^2) and Madeira (801 km^2). Only the mainland, which borders Spain (north and east) and the Atlantic Ocean (south and west), was considered in this study. A Mediterranean climate, characterized by hot, dry summers and cool, wet winters, is the dominant climatic type in this area. Land use is dominated by agriculture and agroforestry (44%), with the remaining area being covered by forests (34%), urban areas (4.7%), surface waters and wetlands (1.4%), pastures and other land uses (15.9%) [15]. The north and south of Portugal differ in terms of the complexity of their river networks, which are denser in the northern region due to the wetter climate and geology.

Denmark is in a coastal, temperate and humid part of northern Europe and covers an area of 43,100 km^2 (Figure 1). Denmark consists, almost entirely, of sedimentary deposits, although the island of Bornholm, in the Baltic Sea, consists mainly of bedrock. It comprises 58 major catchments, each draining to a specific marine area. Runoff varies between 600 mm y^{-1} in the wet and sandy south-western part of the country and 150 mm y^{-1} in the drier and more clayey south-east [16]. According to the Danish Area Information System, land use is predominantly agricultural (67%, including pastures), while the remaining area includes a mixture of forests (12%), urban areas (7%), surface waters and wetlands (4%), and other land uses (10%) [17].

In both countries studied, agricultural activities include intensive and extensive systems, livestock production, and forestry. However, the importance of these activities varies between the countries. Although the two countries possess different geology, topography, and hydrology, they face similar problems in relation to declining water quality caused by N inputs from similar sources.

2.2. Nitrogen in Waterbodies

The pressures associated with excess N from different sources on the waterbodies within each country (PT and DK) were analyzed to understand surface water status. The impact of each N pressure on the waterbodies in each country was highlighted to identify the main drivers of the environmental state of water quality. For that, a review of the water monitoring network of each country was performed and an environmental assessment framework was adapted to analyze all the data collected from each country. Furthermore, the policy response to N water pollution was examined to understand better the success or failure of the mitigation measures implemented by each country. Both European and national scale measures were considered in this analysis.

2.2.1. Water Monitoring Network

Water quality was evaluated across a network of monitoring stations covering the entire EU28 territory, including Portugal and Denmark, in accordance with the related EU Directives. These data are used by the National Environmental Agencies of both countries to classify and document changes in the status of waterbodies under the rules of the Water Framework Directive (WFD). Monitoring of surface waters comprises common European rules of ecological, chemical, physical and hydro morphological variables to determine whether the quality objectives for the River Basin Management Plans (RBMPs) are being fulfilled and to document the effects of pressures, including those caused by N losses to water. Nitrogen contamination enters waterbodies from both diffuse and point sources.

The WFD mandates a "one out all out" approach; this means that if a waterbody fails on one parameter, it fails the overall water quality assessment. In addition to the mandatory monitoring, some countries, such as Denmark, have also implemented several country-specific assessment measures.

To understand better the condition of water bodies in relation to N pressures, it is crucial to know the size of the total N (TN) load that flows from the different parts of the catchment (urban, livestock and agriculture) to those water bodies. With this purpose, the water monitoring network related to N of each country was characterized and analyzed (Table 1).

Table 1. Summary of features and water sampling criteria for Portuguese (RBD1–RBD8) and Danish (DK1–DK4) River Basin Districts (RBD) for the 2010–2015 WFD planning cycle [6,18,19].

| RBD and DK | RBD Area (km²) | Quantity and Length of Waterbodies Monitored | | | N° of TN$_C$** Samples Station^{-1} y^{-1} (Average 2010–2015) | N° of Stations for TN$_L$* Calculations |
		Number of Transitional/Coastal Zones	N° of Lakes/Reservoirs	Length of Streams/Rivers (km)		
RBD1	2465	7	2	628	1.8	0
RBD2	3584	5	7	850	1.7	0
RBD3	19,218	4	15	5678	1.7	0
RBD4	12,144	14	6	4070	1.9	0
RBD5	30,502	10	20	7503	1.7	0
RBD6	12,149	12	10	2580	0.8	0
RBD7	11,611	5	18	3606	1	0
RBD8	5511	10	5	977	2.1	0
DK1	32,000	84	595	14,157	15	186
DK2	9310	33	217	2443	20	71
DK3	588	2	11	368	20	1
DK4	1100	0	33	411	40	4

*TN$_L$—total nitrogen load/discharged from the different N pressures on land to the waterbodies; **TN$_C$—total nitrogen concentration in waterbodies.

In contrast to Denmark, in Portugal, TN$_C$ monitoring stations are not co-located with the hydrometric stations that conduct daily streamflow measurements. No specific TN determination is available for any monitoring station in Portugal (Table 1). The sampling frequency for each River Basin District (RBD) follows the guidelines established under the WFD: each water quality variable can range from continuous sampling to once every 6 years. Therefore, as Portugal does not have specialized stations for TN calculations, the loads discharged into receiving waters need to be estimated annually by the Portuguese Environment Agency (APA), to evaluate the real TN$_L$ and source of the N pressure into the waterbodies.

Denmark has organized monitoring stations (Table 1) where the real annual TN transport to the waterbodies is calculated using daily rates of discharge [19]. The daily discharge values from the hydrometric monitoring stations cover the whole range of agricultural land uses, soil types (sandy to loamy) and levels of precipitation (dry to wet) [20]. Point source loads are reported annually from individual sources, which are grouped into urban wastewater treatment plants (WWTP), industrial sources, sewer storm overflows, freshwater fish farms, and marine fish farms. Stream water samples are gathered at biweekly to monthly intervals at each monitoring station. Each water sample is analyzed, among other parameters, for TN$_C$, NO$_3$$^-$, NO$_2$-N and NH$_4$-N concentrations, in accordance with Danish standards. The ecological condition of Danish waterbodies is evaluated annually by the competent Official Body of Denmark, using the monitoring data on stream biological status and the calculated indices for macroinvertebrates, fish and macrophytes [21,22].

2.2.2. Driver-Pressure-State-Impact-Response (DPSIR) Framework

An adapted Driver-Pressure-State-Impact-Response (DPSIR) environmental framework [23] was used to perform a general analysis of the data collected from the two countries, regarding the effect of N to water quality (Table 2). This novel comparison helps us to understand how different pressures, under different conditions, respond to similar policy application measures. In this analysis, the main driving forces and pressures responsible for the quality of waterbodies and its impacts (number of eutrophic areas/areas at risk of becoming eutrophic), and any current mitigation measures (e.g., NVZs and EU-27), were identified. A comprehensive set of indicators was defined to facilitate the discussion. For some of these indicators, the data collected were used to build maps and graphics so that the differences between Portugal and Denmark could be compared.

Data on Driver and Pressure indicators (Table 2) were extracted from Eurostat [24–31] for both countries and the EU. The gross N balance was usually estimated from N inputs (mineral and organic) and atmospheric deposition (in harvested crops and grazing). These values provided an indication of the potential N surplus on agricultural land and, consequently, the potential N losses responsible for the contamination and quality degradation of water.

In Portugal, data on TN$_L$ from agriculture, livestock production and urban sectors to waterbodies were collected from the RBMPs submitted to the EU Commission by APA in 2015 [6]. APA calculated these loads using N export coefficients and national scale statistics on crops, livestock, and urban population. The TN$_C$ (mg N L^{-1}) in the Portuguese waterbodies were obtained from the National Water Resources Information System database [18]. Data from the five most downstream water quality monitoring stations in each RBD were selected to provide TN$_C$ for the period 2010–2015.

In Denmark, data on N leaching and N discharges from point sources, and loads to surface waters, were obtained from the national monitoring program of the aquatic environment [7,19]. Data on the TN$_L$ from Danish catchments to coastal waters were collected from Danish statistics, Denmark's Miljøportal [32]. The Danish Environmental Agency determines the TN$_L$ by source, mobilization, attenuation, and transport pathways between the source and the receiving water. The impact on waterbodies translates into a status classification with which water quality was assessed with official data from the RBMPs in the case of Portugal [6] and from water monitoring for Denmark [7,19].

Response to the impacts is translated into policies and legislation. One of the most important policies relating to N is the ND [33], which gave birth to the creation of NVZ [34] in EU countries, using different national approaches. Information about NVZs was collected from Portuguese [6,35] and Danish legislation [12], and from Eurostat and the European Environment Agency (EEA) for EU data [34,36].

For wastewater treatment, data was collected from APA for Portugal [6], from Denmark's Miljøportal [32], Wiberg-Larsen et al. [19] and Danish Nature Agency [37] for Denmark, and from EEA [2,38] for EU.

Management practices implemented by each country, regarding N water pollution, were compared and highlighted. To analyze the success or failure of the mitigation measures implemented by each country, historical N surplus levels and TN$_C$ were analyzed.

N surplus data were collected from APA [6] and Cameira et al. [39] for Portugal and from Blicher-Mathiesen et al. [40] for Denmark.

2.3. Uncertainties and Shortcoming of the Study

Although we believe this study to be useful in comparing how two different water management systems operate and produce results for use in both countries, there are some uncertainties. Data were not collected in similar ways in both countries, despite the mandatory regulations and procedures from the European Union with which Portugal and Denmark must comply. Besides the mandatory measures imposed at European level, each country has different local characteristics that benefit from different approaches to water quality monitoring. Denmark is particularly prolific in what concerns environmental legislation to protect water quality, contrary to Portugal that relies almost exclusively on adapted European legislation. There was no available data for the same years in both countries so, the data considered for analysis belongs to the closest dates possible. Nevertheless, the procedures in each country regarding the adoption of the mandatory EU regulations, accompanied by national measures will allow evaluating the efficacy of such policies.

3. Results and Discussion

Table 2 shows the outputs of the DPSIR framework analysis [23], with a set of indicators focused on agricultural and N management in Portugal and Denmark. This information provides the basis for the discussion in the subsections below regarding the steps in the DPSIR chain (Driving Forces, Pressures, States, Impacts, and Responses).

Table 2. Trends of driving forces, pressures, state, impacts, and responses of Portuguese and Danish nitrogen (N) losses from land to water [23].

	Indicator	Portugal (PT)		Denmark (DK)		EU	
Driving Forces *	Fertilizer consumption	Mineral fertilizer used by agriculture (2000–2010)	170×10^6 kg (2000) 103×10^6 kg (2010) [24]	39% ↓	252×10^6 kg (2000) 190×10^6 kg (2010) [24]	25% ↓	10.03×10^9 kg (2000) 7.85×10^9 kg (2010) [24]	22% ↓
	Land use	Agricultural area (2005–2010)	3.68×10^6 ha (2005) 3.67×10^6 ha (2010) [25]	0.3% ↓	2.7×10^6 ha (2005) 2.6×10^6 ha (2010) [25]	2.3% ↓	1.76×10^8 ha (2005) 1.74×10^8 ha (2010) [25]	1% ↓
	Arable land	Arable area (2005–2010)	1.24×10^6 ha (2005) 1.17×10^6 ha (2010) [26,30]	6% ↓	2.50×10^6 ha (2005) 2.42×10^6 ha (2010) [26,30]	3% ↓	1.04×10^8 ha (2005) 1.03×10^8 ha (2010) [26,30]	1.4% ↓
	Livestock patterns	Livestock Unit (LSU) (2005–2010)	2.07×10^6 LSU (2005) 2.21×10^6 LSU (2010) [27]	7% ↑	4.57×10^6 LSU (2005) 4.92×10^6 LSU (2010) [27]	8% ↑	1.37×10^8 LSU (2005) 1.34×10^8 LSU (2010) [27]	1.9% ↓
	Population	Total population (2005–2017)	$10{,}562 \times 10^3$ persons (2005) $10{,}320 \times 10^3$ persons (2017) [31]	2.3% ↓	5419×10^3 persons (2005) $5\ 764 \times 10^3$ persons (2017) [31]	6.4% ↑	$495{,}517 \times 10^3$ persons (2005) $511{,}876 \times 10^3$ persons (2017) [31]	3.3% ↑
Pressures *	Nitrogen loss	Gross N balance (2005–2014)	45 kg N ha^{-1} (2005) 43 kg N ha^{-1} (2014) [28]	4.4% ↓	87 kg N ha^{-1} (2005) 79 kg N ha^{-1} (2014) [28]	9% ↓	54 kg N ha^{-1} (2005) 51 kg N ha^{-1} (2013) [28]	5.6% ↓
	Ammonia (NH$_3$) emissions	NH$_3$ emissions from agriculture (1990–2010)	51×10^6 kg NH$_3$ (1990) 43×10^6 kg NH$_3$ (2010) [29]	16.7% ↓	113×10^6 kg NH$_3$ (1990) 72×10^6 kg NH$_3$ (2010) [29]	36.7 ↓	4.79×10^9 kg NH$_3$ (1990) 3.36×10^9 kg NH$_3$ (2010) [29]	29.8 ↓
State	Water quality	Average TN$_L$ from land to water (PT: 2015; DK: 2014)	7 kg N ha^{-1} to surface waters [6]	-	14.6 kg N ha^{-1} to marine waters [7,19]	-	-	-
		Total TN$_L$ WWTP to water (2015)	21.5 (t N y^{-1}) [6]	-	3.48 (t N y^{-1}) [7,19]	-	-	-
Impacts	Eutrophication	N in eutrophic waterbodies and areas at risk of becoming eutrophic	10 reservoirs 2 transitional waters [6]	-	459 lakes, 82 transitional and coastal waters [7,19]	-	-	-
Responses	Nitrates Directive Nitrate Vulnerable Zones (NVZ)	Total NVZ area	4011 km^2 [6,35]	-	26,086 km^2 [12]	-	1.7×10^6 km^2 [34,36]	-
		Percentage NVZs of total country/continent area	4% [6,35]	-	100% [12]	-	40.9% [34,36]	-
	Wastewater treatment (WWTP)	% without treatment	1% [6]	-	0% [19,32,37]	-	N-0.9%, C-0.2%	-
		% with secondary treatment	74% [6]	-	2.1% [19,32,37]	-	S-2%, E-13.1% (recovered) [2,39] N-2.3%, C-19.54%	-
		% with tertiary treatment	11% [6]	-	97.9% [19,32,37]	-	S-32.8%, E-10.6% (recovered) [2,39] N-76.6%, C-77.36% S-60.3%, E-51.5% (recovered) [2,39]	-

*Arrows in Driving forces and Pressures indicate the tendency for a specific indicator to increase ↑ or decrease ↓ between the years into brackets.

3.1. Nitrogen Drivers, Pressures, and State of Receiving Waters

The N pressures on waterbodies in Portugal and Denmark result from three main sources: (i) agriculture, (ii) livestock and (iii) urban drivers. These pressures result in different water states depending on their intensity and on the nature of the associated river basins. The latter, presenting characteristics that favor TN drainage from leaching and runoff into the waterbodies were analyzed and the results are presented in Figure 2 for Portugal and Figures 3 and 4 for Denmark.

In Portugal, the designation of an NVZ is restricted to river basins with geographical and soil characteristics that increase drainage potential and favor an increased load of nitrate to water; so, NVZ action plans [35] apply to these areas, only. In contrast, in Denmark, 100% of the land was considered to be vulnerable and there is a national scale action plan for the whole country [12]. As shown in Table 1, the hydrological coverage in Denmark is far more extensive than in Portugal, which favors situations of potential nitrate losses into water basins, and increased denitrification potential. In Portugal, the alternation of dry and wet periods promotes the mineralization of soil organic matter leading to nitrate formation and, potentially, greater losses to water.

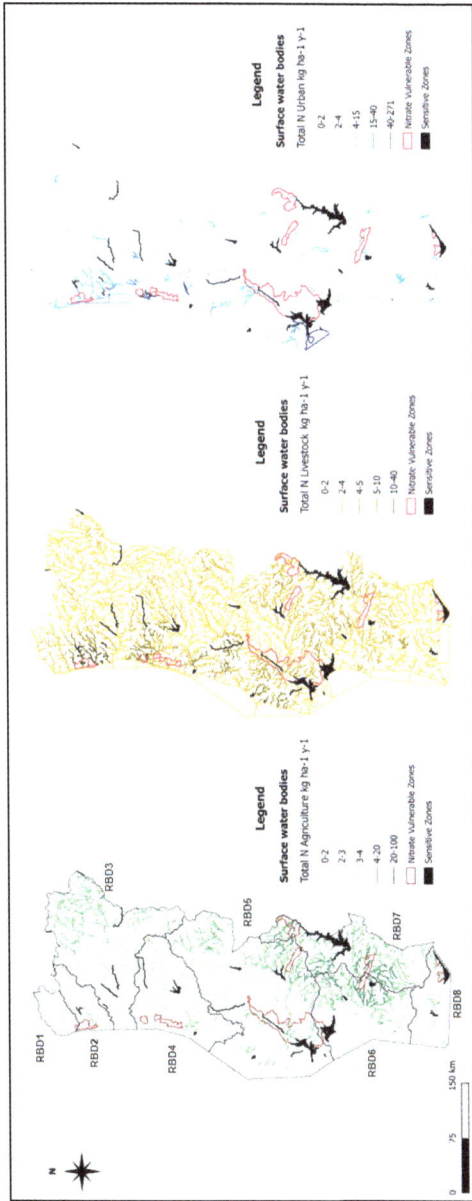

Figure 2. Total N loads from agriculture (**left**), livestock (**center**) and urban (**right**) sources to surface waters in 2015 (kg ha^{-1} y^{-1}) in mainland Portugal in Datum 73/UTM 29N. Designated Nitrate Vulnerable Zones (NVZ) (red closed lines) and Sensitive Zones (SZ) (shaded in black) are also shown. River Basin Districts (RBD) (1 to 8) are display on the agricultural pressures map (**left**) limited by grey lines: RBD1—Minho e Lima rivers (NVZ1); RBD2—Cávado, Ave e Leça rivers; RBD3—Douro river; RBD4—Vouga, Mondego and Lis rivers (NVZ2); RBD5—Tagus river (NVZ5 and NVZ7); RBD6—Sado e Mira (NVZ4); RBD7—Guadiana (NVZ6); RBD8—Algarve rivers (NVZ3 and NVZ8) [6,18,35].

Figure 2 shows the loads of TN discharged to surface waters from agriculture, livestock, and urban sources in mainland Portugal. The currently designated NVZs and Sensitive Zones (SZs) are also shown. Despite the enhanced emissions of TN_L from urban sources, particularly in RBD5 (271 kg ha^{-1} y^{-1}), agriculture is still the major contributor of TN_L to surface waters on a national scale (~40% from agriculture and ~35% from livestock compared to ~22% from the urban sector). The primary sector is quite important in Portugal as it contributes to a great percentage of the Gross Domestic Product (GDP), but it also poses a source for potential pollution. A more detailed allocation source of TN_L to water from all the different sectors is of great importance at the regional and local scales, especially when focusing on better control of N emissions in NVZs. NVZs are always associated with agricultural regions, but include some urban areas too.

Surface waters in RBD1, located in the north of Portugal, are mostly affected by TN_L from livestock (2.5 kg N ha^{-1} y^{-1}) and agriculture (2.0 kg N ha^{-1} y^{-1}). In the case of surface waters in RBD2 (associated with NVZ1) livestock (6.6 kg N ha^{-1} y^{-1}) and urban (6.1 kg N ha^{-1} y^{-1}) drivers produce the predominant pressures, whereas in RBD4 which is located in the Aveiro estuary (associated with NVZ2) all drivers exert similar, but low, pressures (0.25–0.31 kg N ha^{-1} y^{-1}). In this region in particular, the hydrology and topography determine the possibility of denitrification occurring. The surface water catchments in RBD3, RBD5, RBD6 and RBD7 are associated with the highest agricultural areas (16%, 31%, 18% and 22%, respectively) and relatively high total losses of N from agriculture (2.6, 2.3, 1.9 and 3.1 kg N ha^{-1} y^{-1}, respectively) (Table 2). Despite the vast agricultural area, RBD3 has the heaviest urban pressures due to increasing urbanization and tourist activities in the Douro Valley. Including the NVZ3 and NVZ8, RBD8 receives mostly urban discharges, with a N load of 2.3 kg N ha^{-1} y^{-1}. The agricultural sector contributes an N load of 1.8 kg N ha^{-1} y^{-1}.

In Portugal, agricultural areas are occupied by non-irrigated cereal crops (e.g., wheat, barley, oats, rye), irrigated crops (vegetables, maize, paddy rice), olive groves, vineyards, and orchards. Livestock production is also common throughout the country. More specifically, dairy cow production dominates in the north of Portugal (in small and large intensive production units), and sheep and goats dominate in the southern provinces (on extensive grassland farms with less nitrate leaching potential). In RBDs 5, 6 and 7 especially in "Ribatejo" (Tagus vulnerable zone 5; TVZ) and "Alentejo" (NVZ7) provinces, intensive agricultural activities pose an important nitrate pollution risk because maize, which requires large quantities of mineral fertilizer and frequent irrigation, is the most common crop in the area [41]. In "Ribatejo", processing of tomato and other vegetable crops is also responsible for a large share of N losses to water, thus contributing to the degradation of surface water state [42].

Fertilization practices vary significantly from one region to another due to different cropping systems. However, as shown by Cameira et al. 2018 [39] for the TVZ, the ND (91/676/EEC) [33] related measures that have been put in place are relatively effective in areas with intensively irrigated crops, while they are less efficient in areas where livestock predominates due to intensive nature of the production units. Besides the identification of a N surplus in parts of the TVZ related to agricultural activity and land-use change in recent years, additional concern was raised about the recent increase in intensively managed olive grove areas in Alentejo (associated with the NVZ6), which may constitute an additional driver of TN discharge and losses to the water bodies [43].

Figure 3 shows point source discharges of TN (kg ha^{-1}) from 2010 to 2015 (left) and N leaching from land to water in Denmark in 2010. In general, the N leaching pattern was higher in Jutland, in the south-western part of the peninsula (60–75 kg N ha^{-1} and >75 kg N ha^{-1}). This was due to a combination of high livestock densities (dairy and pig production), high mean annual rainfall (900–1100 mm y^{-1}) leading to high percolation of water and runoff into streams (600 mm y^{-1}) and a dominance of sandy soils. These conditions associated with intensive agricultural practices led to higher N leaching in this province than in the drier and loamy eastern islands of Funen and Zealand (Figure 3; right). Point source discharges of N to surface waters are very low because tertiary treatment of sewage effluent is implemented in 97.9% of all the WWTP in Denmark [37]. Therefore, high levels of point source discharges are found only near major cities, such as Copenhagen, Aarhus, Odense and Aalborg (Figure 3). This does not correspond to the situation in Portugal, where only secondary treatment is implemented in nearly all the WWTP [6], transforming urban pressures into a high concern in relation to N and other losses.

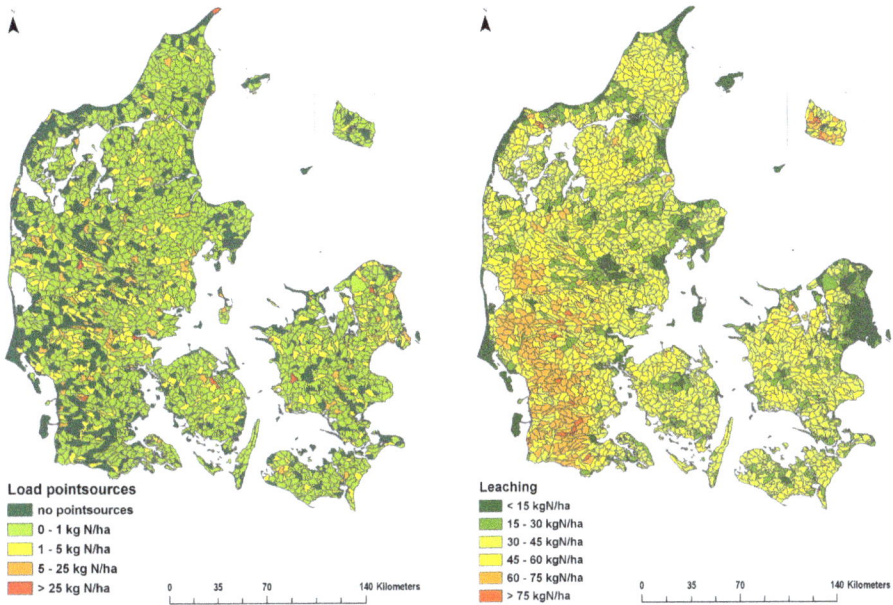

Figure 3. Average annual point source discharges of TN (kg ha^{-1}) in 2010 to 2015 (**left**) and level of N leaching from land to water in 2010 (**right**), for Denmark [7,19,32,37].

Danish territory is highly dependent on, and linked to, the sea and corresponding coastal interaction. This contrasts with Portugal, where only the west and south coast connect to the sea, with some small saltwater intrusions. Three of the NVZs, however, are in these areas (Figure 2).

The amount of TN lost from land to water is an indicator of the likely level of N contamination by receiving waters. In Portugal, the value of N losses is estimated to be an average of 7 kg N ha^{-1} to surface waters, whereas the corresponding value for discharges to coastal waters around Denmark is estimated to be about 14.6 kg N ha^{-1} (Figure 4) [32]. Although the discharge to coastal waters in Portugal is not presented due to lack of statistical information, there is a clear difference in N discharge in both countries. While in Denmark the hydrological conditions and flat topography favor denitrification and could reduce N losses, in reality leaching losses were found to be higher than in Portugal. Water quality policies appear not to be reducing the loss of nitrate to water significantly, despite the heavier legislation requirements in Denmark.

The TN$_L$ from Danish catchments to coastal waters in 2015 is shown in Figure 4. Nitrogen loads to surface waters are particularly high (>15 kg N ha^{-1}) in livestock production areas with high levels of precipitation (i.e., southern and northern Jutland and Funen), whereas the N load is lower in the predominantly plant producing areas of eastern Funen and western Zealand. High N loadings (>12.5 kg N ha^{-1}) from point sources occur within the Copenhagen region (eastern Zealand) (Figure 4).

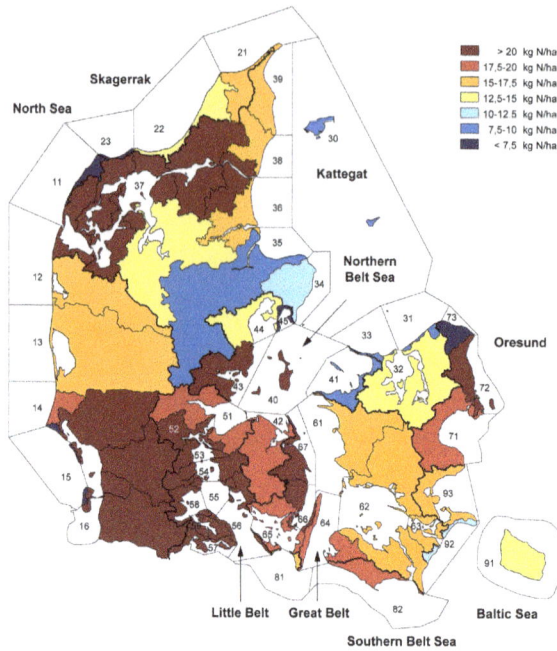

Figure 4. Total Nitrogen loads (kg ha^{-1}) from Danish catchments to coastal waters in 2015. [7,19,32].

3.2. Impacts on Water Quality and Ecology

The pressure of N discharges from the different drivers, as shown in Figures 2 and 3 respectively for Portugal and Denmark, results in impacts on water quality that can be seen in Figure 5. To understand better the impact of N on waterbodies, TN$_C$ (mg N L^{-1}) in the surface waters of Portugal and Denmark were analyzed between the years of 2010 and 2015 (Figure 5). Then, the ecological status of rivers, lakes/reservoirs and coastal/transitional waters were analyzed and compared for both countries (Figure 6).

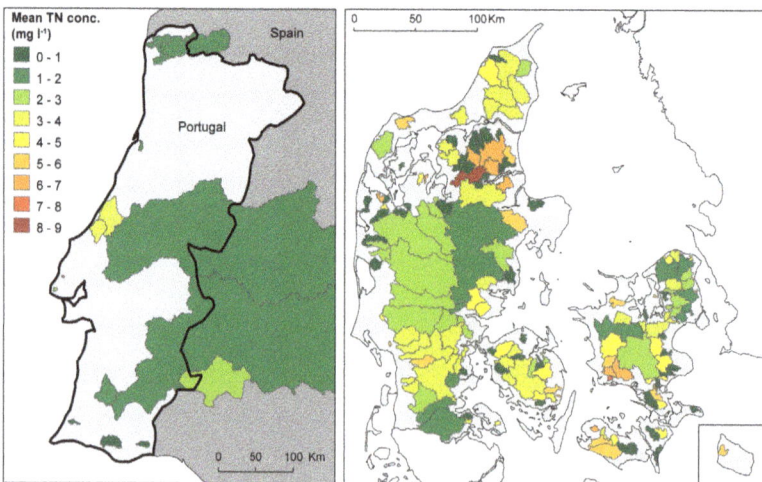

Figure 5. Average concentration of TN$_C$ (mg N L^{-1}) in surface water in Portugal (**left**) and Denmark (**right**), 2010–2015 [18,32].

Total (N) and (NO$_3$-N) concentrations in ground and surface waters are important because they determine the impact on chemical water quality of these waterbodies as defined by the ND and WFD. Monitoring of TN$_C$ and daily flows at surveillance stations is important for providing reliable data on TN loadings in surface waterbodies.

In Portugal, TN$_C$ are only available for 30% of the mainland due to the limited sampling and inconsistency of sampling dates compared to monitoring sites. This does not provide reliable information on water (N) concentrations (Table 1 and Figure 5). Nitrogen transport from Spain to Portugal plays an important role in detremining the level of total N loading and (N) concentrations in some Portuguese rivers, such as the Tagus [44]. Total nitrogen concentrations in Portuguese rivers exhibit a pattern of generally low (N) concentrations (Figure 5), which does not necessarily reflect reality, up to where the extent of samplings permitted estimation (i.e., ≥8 months). Hotspots for nitrate pollution will only be identified through modelling at higher resolution, while at the broader scale policy implementation seems to be effective in controlling nitrate [39], despite the loads in the discharges coming from Spain.

In Denmark measured TN$_C$ in surface waters were generally higher in the loamy and drier eastern part of the country (Funen and Zealand) than in the northern part of Jutland which, with its chalk groundwater aquifers, is characterized by low N attenuation (Figure 5) [45].

In Portugal similar assessments to those performed in Denmark cannot be made due to the almost complete absence of specific monitoring stations to assess N loading (Table 1).

The patterns observed in the two countries seem to vary markedly from each other—with the Danish surface waters displaying higher flow-weighted total N concentrations (~4 mg N L^{-1}) than the Portuguese rivers (~2.5 mg N L^{-1}) (Figure 5). A possible explanation for the much higher TN$_C$ in Danish surface waters compared to the Portuguese rivers may be the 2.6 times higher N surplus in Danish agriculture compared to Portuguese agriculture (Table 2). Due to the lack of a comprehensive monitoring network in Portugal, the identification of N surplus is only possible through the downscaled analysis of specific river basins [39]. However, while WWTP with tertiary treatment in Portugal are non-existant [6], almost 100% of the WWTP in Denmark have high nutrient recovery (Table 2). This leads to the assumption that the N loads to Danish waters should be lower, which is not the case. In fact, in Denmark, the N load from the WWTP is measured as a pressure on water quality whereas, in Portugal, it is simply unknown.

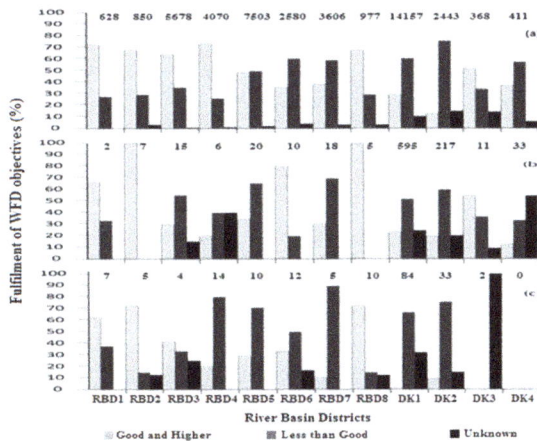

Figure 6. Overall status of (**a**) rivers, (**b**) lakes/reservoirs and (**c**) coastal/transitional waters in the 8 Portuguese and 4 Danish River Basin Districts (RBD/DK). Values above columns refer to the length of rivers (km), number of lakes/reservoirs and coastal/transitional waters for (**a**), (**b**) and (**c**), respectively [6,32].

For rivers, three out of the four Danish RBDs have an overall WFD status of less than good, while the same status only occurs in three of the eight RBDs in Portugal (Figure 6a). The differences between the two countries is probably influenced by the fact that all Danish rivers are considered irrespective of their size, so a significant number of small streams (<2 m width) are included in the determination of status. These small streams are more vulnerable to pollution than larger streams that have a higher dilution rate and greater potential for recolonisation by biota. In contrast, in Portugal, only streams that are monitored and have WFD targets for ecological and physical conditions are considered. The three Portuguese RBDs that do not meet good ecological status—RBDs 5, 6 and 7 (Figure 6a)—are located in areas dominated by irrigated and intensive cropping systems (e.g., rice and tomato) with high fertilizer application rates (RBD 5) or non-irrigated production intensive cropping systems (e.g., maize) and vegetable production (e.g., horticultural crops) (RBDs 6 and 7). In Portugal, agricultural intensification is often associated with livestock production, which also contributes to water quality degradation [6].

In relation to lakes and reservoirs, three out of the eight Portuguese RBDs have a less than good status; the comparable figure for Denmark is two out of four (Figure 6b). However, phopshorus (P) loading rather than N loading is the dominant nutrient pressure on lakes and reservoirs, and in most inland waterbodies P loading from the lake sediments is also significant due to chemical and biological factors [46]. In Portugal, few studies have classified the trophic state of lakes and reservoirs. However, Diogo (2008) [47] found that 64% of the eutrophicated reservoirs in Portugal are also affected by excessive inputs of P. In Denmark, considerable efforts have been made to reduce the P load to lakes and reservoirs [48–50]. However, in Portugal, P is just starting to be a concern in terms of water quality, since the attention of policy has focused so far on N as a major driver for water contamination.

In terms coastal/transitional waters in Denmark, in two of the four DKs, 70–80% of the waters have less than good status according to their WFD classification (Figure 6c). In Portugal, in four out of the eight RBDs, 50–90% of coastal/transitional waters have less than good status (Figure 6c). In Denmark, excessive N loading is the main reason for the poor state of its coastal and transitional waters, with the severe eutrophication of these systems acting as one of the main drivers for the implementation of seven action plans that have been implemented over the last 30 years to improve the aquatic environment [12,45]. In Portugal, although several local studies have investigated the impact of nutrient input on water quality status in transitional and coastal waters [51–53], uncertainty still exists about which nutrient is causing excessive algal growth and, thereby, influencing trophic state and ecological quality status in these areas.

Although knowledge of the quality status of Portuguese waterbodies is increasing, several gaps in knowledge remain. These are mainly due to the low number of water quality monitoring stations within each RBD and the non-existent sites that are suitable for TN_C calculation (Table 1). Moreover, the importance of coastal waters within the Danish territory is far more significant than in Portugal, due to the inherent characteristics of the territories themselves.

3.3. Responses to Nitrogen Impacts

The ND [33] was the first European response to water pollution associated with nitrate from agricultural and livestock rearing activities. To mitigate and abate N losses from agricultural sources, all of the Danish territory has been designated as a NVZ [12] while in Portugal, the NVZ areas occupy only 4% of the land (Figure 2) [35]. This means that in Denmark, the implemented action plan on waterbodies covers the entire country, and strict regulations have been imposed on all farmers since the mid-1980s. This has resulted in large reductions in N leaching [12]. In Portugal, the first three NVZs were mapped in 1997 and these increased in number and areal coverage until 2012. New areas are now likely to become NVZs and, therefore, require further regulation. For instance, some may be designated as SZs with strict regulation associated with discharges.

Despite the heavier regulation, nitrate losses in Denmark are higher per unit area of land due to the more intensive type of agriculture compared to the more extensive agriculture found in Portugal. As well as the land characteristics being important, the type of farming affects the water quality, too.

Also, Denmark has a much higher number of lakes and transitional coastal waters that are likely to be impacted by N losses from land compared to Portugal, which is mainly formed of continuous areas of land with a defined and substantially less extensive river basin network. This results in a less vulnerable water network with correspondingly lower N loads to surface waters. Agricultural intensity is also a main driver of the higher N excess in Danish waters, due to higher agricultural and husbandry pressures (Table 2). In theory, Danish conditions could lead to additional denitrification, which would reduce the level of nitrate leaching into water. However, since this is not the case, agricultural systems are probably the main factors that are determining N losses.

Where the soil-water system is defined by fast-flow processes in an oxic groundwater zone, there is a risk that a large part of the N surplus may be discharged to surface waters [45]. Nitrogen loads in streams are also influenced by manure and fertilizer use, runoff and soil type [3,54,55]. These may be enhanced by the discharge of poorly treated wastewaters and/or effluents from different sources, and the level of sophistication of the process in the WWTP may affect water quality, too.

3.4. Nitrogen Management and Policies Implemented

To analyze and compare the success or failure of the mitigation and abatement measures implemented by each country, the historical levels of N surplus (Figure 7) and TN_C (Figure 8) were analyzed. Policy responses for each country were identified (Table 3).

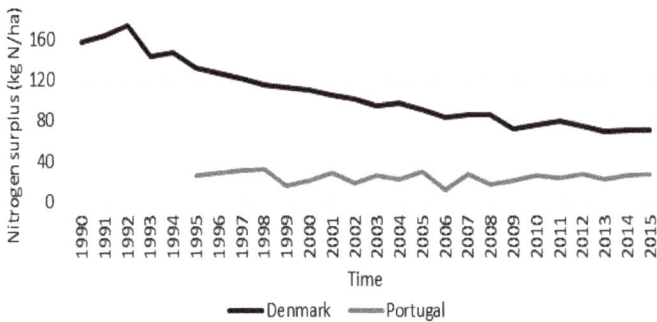

Figure 7. Comparison of historical levels of nitrogen (N) surplus in Denmark (1990–2015) and Portugal (1990–2015) [6,39,40].

The N surplus in Portugal and Denmark between 1995 and 2015 (where available) are shown in Figure 7. According to the Portuguese Administration [6], the N surplus varied between 15 and 30 kg N ha^{-1} over this period, but showed no obvious trend. The more or less constant, but low, N surplus in Portugal was apparently unaffected by the incorporation of European Directives into Portuguese law—i.e., the Nitrates Directive [56], the Water Framework Directive [57] and the Wastewater Treatment Directive [58] in 1997; the designation of NVZs in 1998, 2005 and 2010; and, the introduction of a common agricultural policy in 2003. It is likely that this occurred because the effect of these new regulations was masked by a higher level of incorporation of manure into the soil and an increase in the area of land used for vegetable cultivation and pasture over this period [59]. Despite the mandatory policy regulations in Portugal, best practices remain at the discretion of the farmers and are so dependent on the farmer's ability to protect any streams that have an impact on the RBDs.

A decreasing trend is evident in Denmark, where a reduction from 157 kg N ha^{-1} in 1990 to 70 kg N ha^{-1} in 2015 has been achieved. This significant downward trend can be ascribed to the seven aquatic action plans that have been implemented in Denmark since 1985 (Table 3) [45].

Data from the Valada Tejo (19E/02) monitoring station [18], located on the downstream part of the Tagus River in Portugal, was selected to analyze the trend in TN_C over the period 2002–2015. Total N concentrations decreased from 1.75 mg N L^{-1} to 1.1 mg N L^{-1}, which is equivalent to a reduction of about 38% (Figure 8). A much steeper downward trend in average TN$_C$ over time was detected for monitored streams in Denmark (Figure 8) [19], with flow-weighted TN$_C$ decreasing from 7.5 mg N L^{-1} in 1990 to 4.3 mg N L^{-1} in 2015, i.e., a reduction of about 43% (Figure 8). It should be noted, however, that the starting point for TN$_C$ was higher in Danish waters than in Portuguese waters.

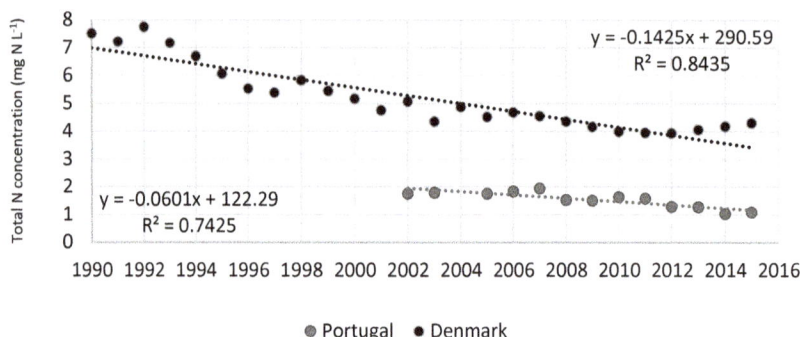

Figure 8. Historical concentrations of TN$_C$ based on data from the Valada Tejo (19E/02) monitoring station in Portugal (2002–2015) and on Danish flow-weighted total N concentrations between (1990–2015) [18,19].

In Portugal and Denmark, TN$_C$ in water were affected differently by the implementation of N abatement policies that are shown in chronological order in Table 3. Higher precipitation occurring in Denmark, compared to Portugal, could have also contributed to a dilution effect, thus enhancing the reduction of TN$_C$ in Danish waters. However, it should be noted that the two time series are not exactly comparable because the Danish time series represents the whole country (120 monitoring stations), whereas the Valada Tejo station shows the trend at a single station, only. Also, 80% of the catchment that drains into this monitoring site is in Spain [44]. Nitrogen monitoring is clearly insufficient in Portugal, although the reason for this is unknown. As mentioned before, officially recorded N concentrations in streams do not necessarily show the presence or absence of a N surplus. Further downscaling of the data is needed to identify N hotspots [39].

Table 3. Historical nitrogen abatement policies in Portugal and Denmark.

Country	Year	Policies	Main Recommendations
DK	1986	NPO Program	Ban on point source pollution from slurry tanks; Control at farm level; Limit application to 265/230 kg organic N for cattle and pig farms.
	1987	Action Plan for the Aquatic Environment	49% reduction in NO$_3$ leaching from agriculture by 1993; Code of good agriculture practice.
	1989	National Monitoring Program for the Aquatic Environment	Advisory services to the EPA; Environmental research; Monitoring.
	1991	Action Plan for Sustainable Agriculture	Mandatory N standards for crops; Controlled use of N in organic manure; Ban on slurry application from harvest until 1st February.
	1998	Action Plan for the Aquatic Environment II	Catch crops on 6% of land area; N standards set at least 10% below economic optimum; 170 kg N ha^{-1} for cattle and pig farms.
	2004	Action Plan for the Aquatic Environment III	10 m buffer strips along rivers and lakes; 13% reduction of N leaching from agriculture by 2015.
	2008	Green Growth	19 k tons reduction in emission of N to marine waters; 10 m buffer strips along watercourses; 140 k ha target for catch crops.
PT	1997	Transposition of Nitrates Directive into Portuguese law (Decree-Law 235/97)	Designation of 3 NVZs; Code of good agricultural practice; Control of NO$_3$ concentrations in surface and ground waters; Evaluation of eutrophication status of waters.
	1997	Transposition of Waste Water Treatment Directive into Portuguese law (Decree-Law 152/97)	Designation of SZ and less sensitive zones; Regulation of N and P levels in waste waters discharge.
	2002	National Water Plan (Decree-Law 112/2002)	Strategy for national water policy; Creation of river basin management plans.
	2003	Common Agriculture Policy	
	2005	Water Law 58/2005	Institutional framework for sustainable water management.
	2007	Decree 214/2007 and Order 8277/2007	National strategy for livestock and agri-industrial effluents.
	2009	Ordinance 631/2009	Rules for the management of livestock effluents.
	2010	Ordinance 164/2010	Designation of 8 NVZs.
	2012	Ordinance 259/2012	Action plan for the 8 NVZs in mainland.
	2013	Decree 81/2013	New strategy for livestock activities; Management.

3.5. Evaluation and Recommendations for Policy Making

Due to the maintenance and improvement of the Danish action plans, N losses are expected to decline even further. In Portugal, the introduction of several additional measures has been suggested to control N losses to water. These include expansion of the NVZs, a more extensive surveillance monitoring network that includes high-frequency measurements of water flow and chemical parameters at co-located gauging stations, and policies targeted at improving N use efficiency. Due to the very diverse nature of the agricultural systems throughout Portugal, a revision of ND measures is suggested for the existing NVZs, with the aim of producing targeted measures. The results from this study also highlight a need to ensure that farmers are complying with the ND related programs of measures.

Water quality matters for agricultural use, to some extent, but is more important where water is used for drinking purposes as it affects human health. Research on this topic has not had a sufficiently high influenced enough on policy making or implementation, due to the differences in countries or regions characteristic in terms of natural aspects but also regarding the sectors that drive N losses into water. Barriers that prevent science-based information to further influence EU policy making are mostly national and regional and relate to the lack of political will, of incentives to adoption and of implementation of command and control legislation [60]. Moreover, the lack of current understanding of the effects of the measures on the real water quality improvement, further complicates science contribution to policy making [61].

4. Conclusions and Future Perspectives

The carrying capacity of watercourses is often determined by site specific factors that cause variations in water quality responses to N inputs. Therefore, to protect water resources effectively, environmental policies need to be introduced that take into account many complex issues and address several economic and societal challenges. Moreover, the preventive approach is more cost-effective than remediation. For example, nitrate removal from contaminated drinking water is very costly, which is of great concern to water utility companies.

The DPSIR framework provides a strong framework for presenting water quality indicators to policy makers and raising awareness of the importance of considering environmental quality in the political choices made, now and in the future. A comparison of drivers, pressures, state, impacts, and responses in relation to N contamination of waterbodies showed that the implementation of policy measures promoted a significant reduction of N losses from land to water in Denmark, which resulted in a decrease of 25% in fertilizer use and a reduction of 53% in N surplus. In Portugal, the corresponding reductions were 39% for fertilizer use and 15 to 30 kg N ha^{-1} for the N surplus, with no significant trend. Average total N discharge to surface waters was calculated to be 7 kg ha^{-1} for mainland Portugal in 2015 and 14.6 kg ha^{-1} for Denmark in 2014. In both countries, the agricultural sector was found to be the main source of the N pressure on waterbodies.

Following the implementation of the Water Framework Directive in 2010, "less than good" water quality status was recorded in three out of eight RBDs in Portugal and in two out of four RBDs in Denmark. However, data suggest that the main differences in water quality status between the two countries is probably due to lower frequency of sampling for the determination of TN_C, the less extensive monitoring network and the higher agriculture production intensity that is found in Denmark compared to Portugal.

Reduction in the levels of N in waterbodies was attributed to the historical implementation of policies aimed at reducing the N losses from land to water in both countries, but climatic factors such as precipitation can also have a role in reducing the N concentration in waters through a dilution effect. In Portugal, policies were the transposition of the ND and Waste Water Treatment Directives into law in 1997, the Water Framework Directive into the Water Law in 2005 (with successive modifications until 2012), and an action plan for NVZ designation, which now covers 4% of the Portuguese mainland. In Denmark, the observed reduction can be ascribed to the various "National Action Plans for the Aquatic

Environment" that have been introduced since 1985. These have led to improved N use efficiency and reduced N pollution, culminating in the designation of the entire country as an NVZ.

Author Contributions: Conceptualization—C.M.d.S.C., A.G.B.; investigation—S.C.; writing—original draft preparation—S.C., R.P., J.R.P., H.T., C.M.d.S.C., A.G.B., M.R.C.; data collection and figures—G.G.; data correction and figures—S.C.; writing—review and editing—C.M.d.S.C., A.G.B., B.K., L.M.

Funding: This research was funded by NitroPortugal, H2020-TWINN-2015, EU coordination and support action n. 692331 and the FCT, UID/AGR/04129/2013.

Acknowledgments: The authors acknowledge the NitroPortugal, H2020-TWINN-2015, EU coordination and support action n. 692331 and the FCT, UID/AGR/04129/2013 for funding and the Portuguese Environmental Agency (APA) for data supply.

Conflicts of Interest: The authors declare no conflict of interest.

References

1. Sutton, M.A.; Oenema, O.; Erisman, J.W.; Leip, A.; van Grinsven, H.; Winiwarter, W. Too much of a good thing. *Nature* **2011**, *472*, 159–161. [CrossRef]
2. European Environment Agency (EEA). *European Waters—Assessment of Status and Pressures*; Report N°8/2012; European Environment Agency: Copenhagen, Denmark, 2012; ISSN 1725-9177.
3. Vagstad, N.; Stålnacke, P.; Andersen, H.E.; Deelstra, J.; Jansons, V.; Kyllmar, K.; Loigu, E.; Rekolainen, S.; Tumas, R. Regional variations in diffuse nitrogen losses from agriculture in the Nordic and Baltic regions. *Hydrol. Earth Syst. Sci.* **2004**, *8*, 651–662. [CrossRef]
4. Di, H.J.; Cameron, K.C. Nitrate leaching in temperate agrosystems: Sources, factors and mitigation strategies. *Nutr. Cycl. Agroecosyst.* **2002**, *46*, 237–256. [CrossRef]
5. Oenema, O.; Oudendag, D.; Velthof, G.L. Nutrient losses from manure management in the European Union. *Livest. Sci.* **2007**, *112*, 261–272. [CrossRef]
6. Agência Portuguesa do Ambiente (APA). *Planos de Gestão das Bacias Hidrográficas Portuguesas (RH 1-8)*; Agência Portuguesa do Ambiente: Lisbon, Portugal, 2015. (In Portuguese)
7. Jensen, P.N.; Boutrup, S.; Fredshavn, J.R.; Svendsen, L.M.; Blicher-Mathiesen, G.; Wiberg-Larsen, P.; Johansson, L.S.; Hansen, J.W.; Nygaard, B.; Søgaard, B. *NOVANA. Tilstand og Udvikling—Faglig Sammenfatning*; Videnskabelig Rapport fra DCE—Nationalt Center for Miljø og Energi nr. 170. Scientific report no. 170; Aarhus University, DCE—National Center for Environment and Energy: Aarhus, Denmark, 2015; p. 96. (In Danish)
8. Corine Land Cover (CLC). Version 18.5.1. 2006. Available online: http://land.copernicus.eu/pan-european/corine-land-cover/clc-2006/view (accessed on 15 July 2016).
9. Bos, J.; Smit, A.; Schröder, J. Is agricultural intensification in The Netherlands running up to its limits? *NJAS Wagening. J. Life Sci.* **2013**, *66*, 65–73. [CrossRef]
10. Kronvang, B.; Ærtebjerg, G.; Grant, R.; Kristensen, P.; Hovmand, M.; Kirkegaard, J. Nationwide monitoring of nutrients and their ecological effects. State of the Danish Aquatic Environment. *Ambio* **1993**, *22*, 176–187.
11. Martins, G.; Ribeiro, D.; Pacheco, D.; Cruz, J.V.; Cunha, R.; Gonçalves, V.; Nogueira, R.; Brito, A.G. Prospective scenarios for water quality and ecological status in Lake Sete Cidades (Portugal): The integration of mathematical modelling in decision processes. *Appl. Geochem.* **2008**, *23*, 2171–2181. [CrossRef]
12. Dalgaard, T.; Hansen, B.; Hasler, B.; Hertel, O.; Hutchings, N.L.; Jacobsen, B.H.; Jensen, L.S.; Kronvang, B.; Olesen, J.E.; Schjorring, J.K.; et al. Policies for agricultural nitrogen management—Trends, challenges and prospects for improved efficiency in Denmark. *Environ. Res. Lett.* **2014**, *9*, 115002. [CrossRef]
13. Pagans, E.; Barrena, R.; Font, X.; Sánchez, A. Ammonia emissions from the composting of different organic wastes. Dependency on process temperature. *Chemosphere* **2006**, *62*, 1534–1542. [CrossRef] [PubMed]
14. Burgos, P.; Madejón, E.; Cabrera, F. Nitrogen mineralization and nitrate leaching of a sandy soil amended with different organic wastes. *Waste Manag. Res.* **2006**, *24*, 175–182. [CrossRef]
15. Direcção Geral do Teritoório (DGT). *Uso e Ocupação do Solo em Portugal Continental: Avaliação e Cenários Futuros*; Projecto LANDYN; Direcção Geral do Território: Lisboa, Portugal, 2006; ISBN 978-989-98477-9-8EC.
16. Kronvang, B.; Windolf, J.; Grant, R.; Andersen, H.E.; Thodsen, H.; Ovesen, N.B.; Larsen, S.E. Linking monitoring and modelling for river basin management: Danish experience with combating nutrient loadings to the aquatic environment from point and non-point sources. *Sci. China Ser. E Technol. Sci.* **2009**, *52*, 3335–3347. [CrossRef]

17. Nielsen, K.; Stjernholm, M.; Olsen, B.Ø.; Muller-Wohlfeil, D.I.; Madsen, I.L.; Kjeldgaard, A.; Groom, G.; Hansen, H.S.; Rolev, A.M.; Hermansen, B.; et al. *Areal Informations Systemet*; Aarhus Universitet: Aarhus, Denmark, 2000; Volume 200. (In Danish)

18. Sistema Nacional de Informação de Recursos Hídricos (SNIRH). 2016. Available online: http://snirh.apambiente.pt/index.php?idMain= (accessed on 22 December 2016).

19. Wiberg-Larsen, P.; Windolf, J.; Bøgestrand, J.; Larsen, S.E.; Tornbjerg, H.; Ovesen, N.B.; Niel-sen, A.; Kronvang, B.; Kjeldgaard, A. *Vandløb. NOVANA*; Scientific report no. 165; Aarhus University, DCE—National Center for Environment and Energy: Aarhus, Denmark, 2015; p. 54.

20. Windolf, J.; Thodsen, H.; Troldborg, L.; Larsen, S.E.; Bogestrand, J.; Ovesen, N.B.; Kronvang, B. A distributed modelling system for simulation of monthly runoff and nitrogen sources, loads and sinks for ungauged catchments in Denmark. *J. Environ. Monit.* **2011**, *13*, 2645–2658. [CrossRef]

21. Kristensen, E.A.; Jepsen, N.; Nielsen, J.; Pedersen, S.; Koed, A. *Dansk Fiskeindeks for Vandløb (DFFV)*; Scientific report no. 95; Aarhus University, DCE—National Center for Environment and Energy: Aarhus, Denmark, 2014; p. 58. (In Danish)

22. Baattrup-Pedersen, A.; Larsen, S.E. *Udvikling af Planteindeks til Brug i Danske Vandløb. Vurdering af Økologisk Tilstand (fase I)*; Scientific report no. 60; Aarhus University, DCE—National Center for Environment and Energy: Aarhus, Denmark, 2013; p. 32. (In Danish)

23. European Environment Agency (EEA). DPSIR Framework (Driving Forces, Pressure, State, Impact, Response). 2018. Available online: https://www.eea.europa.eu/publications/TEC25 (accessed on 6 April 2016).

24. Eurostat Statistics Explained. Agri-Environmental Indicator—Mineral Fertilizer Consumption. 2012. Available online: http://ec.europa.eu/eurostat/statistics-explained/index.php/Agri-environmental_indicator_-_mineral_fertiliser_consumption (accessed on 9 July 2016).

25. Eurostat Statistics Explained. Agri-Environmental Indicator—Consumption of Pesticides. 2012. Available online: http://ec.europa.eu/eurostat/statistics-explained/index.php/Agri-environmental_indicator_-_consumption_of_pesticides (accessed on 9 July 2016).

26. Eurostat Statistics Explained. Agri-Environmental Indicator—Cropping Patterns. 2012. Available online: http://ec.europa.eu/eurostat/statistics-explained/index.php/Agri-environmental_indicator_-_cropping_patterns (accessed on 9 July 2016).

27. Eurostat Statistics Explained. Agri-Environmental Indicator—Livestock Patterns. 2012. Available online: http://ec.europa.eu/eurostat/statistics-explained/index.php/Agri-environmental_indicator_-_livestock_patterns (accessed on 9 July 2016).

28. Eurostat Statistics Explained. Agri-Environmental Indicator—Gross Nitrogen Balance. 2012. Available online: http://appsso.eurostat.ec.europa.eu/nui/submitViewTableAction.do (accessed on 9 July 2016).

29. Eurostat Statistics Explained. Agri-Environmental Indicator—Ammonia Emissions. 2012. Available online: http://ec.europa.eu/eurostat/statistics-explained/index.php/Agri-environmental_indicator_-_ammonia_emissions (accessed on 9 July 2016).

30. Eurostat Statistics Explained. Agri-Environmental Indicator—Irrigation. 2016. Available online: http://ec.europa.eu/eurostat/statistics-explained/index.php/Agri-environmental_indicator_-_irrigation (accessed on 9 July 2016).

31. Base de dados de Portugal Continental (PORDATA). População Residente Total. 2016. Available online: https://www.pordata.pt/DB/Europa/Ambiente+de+Consulta/Tabela (accessed on 9 July 2016). (In Portuguese)

32. Danish Statistics. Danmarks Miljøportal. 2016. Available online: https://www.dst.dk/da/Statistik/emner/landbrug-gartneri-og-skovbrug/afgroeder;http://www.miljoeportal.dk/borger/Sider/Borger.aspx (accessed on 20 December 2016).

33. European Environment Agency (EEA). Nitrates Directive. 2010. Available online: http://ec.europa.eu/environment/pubs/pdf/factsheets/nitrates.pdf (accessed on 9 July 2016).

34. European Environment Agency (EEA). Nitrate Vulnerable Zones. 2009. Available online: http://eea.europa.eu/data-and-maps/figures/nitrate-vullnerable-zones-eu (accessed on 9 August 2016).

35. *Ordinance n° 259/2012. Portaria n° 259/218 de 28 de Agosto. Diário da República n° 166—I Série*; Ministério da Agricultura, do Mar, do Ambiente e do Ordenamento do Território: Lisboa, Portugal, 2012. (In Portuguese)

36. Eurostat Statistics Explained. Nitrate Vulnerable ZonesEu-27. 2010. Available online: http://ec.europa.eu/eurostat/statisticsexplained/index.php?title=File:Nitrate_Vulnerable_Zones_(NVZ),_EU-27,_2009.png&oldid=274885 (accessed on 2 December 2016).

37. Nature Agency. *Point Sources 2014*; Danish Ministry of Environment and Food: Copenhagen, Denmark, 2015; Volume 135. (In Danish)

38. European Environment Agency (EEA). Changes in Wastewater Treatment in Regions of Europe between 1990 and 2012. 2013. Available online: http://www.eea.europa.eu/data-and-maps/figures/changes-in-wastewater-treatment-in-regions-of-europe-between-1990-and-2 (accessed on 2 December 2016).

39. Cameira, M.R.; Rolim, J.; Valente, F.; Faro, A.; Dragosits, U.; Cordovil, C.M.d.S. Spatial distribution and uncertainties of nitrogen budgets for agriculture in the Tagus river basin in Portugal—Implications for effectiveness of mitigation measures. *Land Use Policy* **2019**, *84*, 278–293. [CrossRef]

40. Blicher-Mathiesen, G.; Rasmussen, A.; Rolighed, J.; Andersen, H.E.; Jensen, P.G.; Wienke, J.; Hansen, B.; Thorling, L. *Landovervågningsoplande*. *NOVANA*; Scientific report no. 164; Aarhus University, DCE—National Center for Environment and Energy: Aarhus, Denmark, 2015; p. 150.

41. Associação Nacional dos Produtores de Milho e Sorgo (Anpromis). 2014. Available online: http://www.anpromis.pt/images/eventos/FeiraMilho/anpromis2014.pdf (accessed on 9 July 2016). (In Portuguese).

42. INE (Instituto Nacional de Estatística). *Estatísticas Agrícolas 2015*; INE: Lisboa, Portugal, 2016; pp. 7–60. ISBN 978-989-25-0360-78. (In Portuguese)

43. Cameira, M.R.; Pereira, A.; Ahuja, L.; Ma, L. Sustainability and environmental assessment of fertigation in an intensive olive grove under Maditerranean conditions. *Agric. Water Manag.* **2014**, *146*, 346–360. [CrossRef]

44. Cordovil, C.M.d.S.; Cruz, S.; Brito, A.G.; Cameira, M.R.; Poulsen, J.R.; Thodsen, H.; Kronvang, B. A simplified nitrogen assessment in Tagus river basin: A management focused review. *Water* **2018**, *10*, 406. [CrossRef]

45. Windolf, J.; Blicher-Mathiesen, G.; Carstensen, J.; Kronvang, B. Changes in nitrogen loads to estuaries following implementation of governmental action plans in Denmark: A paired catchment and estuary approach for analysing regional responses. *Environ. Sci. Policy* **2012**, *24*, 24–33. [CrossRef]

46. Ribeiro, D.; Martins, G.; Nogueira, R.; Brito, A.G. Mineral cycling and pH gradient related with biological activity under transient anoxic-oxic conditions: Effect on P mobility in volcanic lake sediments. *Environ. Sci. Tech.* **2014**, *48*, 9205–9210. [CrossRef]

47. Diogo, P. *Fontes de Fósforo Total e o Estado Trófico de Albufeiras em Portugal Continental*; Dissertação apresentada para a obtenção do grau de Mestre em Engenharia do Ambiente, perfil de Gestão de Sistemas Ambientais; Faculdade Ciências e Tecnologia da Universidade Nova de Lisboa: Almada, Portugal, 2008. (In Portuguese)

48. Jeppesen, E.; Søndergaard, M.; Jensen, J.P.; Havens, K.E.; Anneville, O.; Carvalho, L.; Coveney, M.F.; Deneke, R.; Dokulil, M.T.; Foy, B.; et al. Lake responses to reduced nutrient loading—An analysis of contemporary long-term data from 35 case studies. *Freshw. Biol.* **2005**, *50*, 1747–1771. [CrossRef]

49. Kronvang, B.; Jeppesen, E.; Conley, D.; Søndergaard, M.; Larsen, S.E.; Ovesen, N.B.; Carstensen, J. An analysis of pressure, state and ecological impacts of nutrients in Danish streams, lakes and coastal waters and ecosystem responses to nutrient pollution reductions. *J. Hydrol.* **2005**, *304*, 274–288. [CrossRef]

50. Kronvang, B.; Audet, J.; Baattrup-Pedersen, A.; Jensen, H.S.; Larsen, S.E. Phosphorus Load to Surface Water from Bank Erosion in a Danish Lowland River Basin. *J. Environ. Qual.* **2012**, *41*, 304–313. [CrossRef]

51. European Environment Agency (EEA). *Water quality and pollution by nutrients*; European Environment Agency: Copenhagen, Denmark, 2018.

52. Caetano, M.; Raimundo, J.; Nogueira, M.; Santos, M.; Mil-Homens, M.; Prego, R.; Vale, C. Defining benchmark values for nutrients under the Water Framework Directive: Appli-cation in twelve Portuguese estuaries. *Mar. Chem.* **2016**, *185*, 27–37. [CrossRef]

53. Vasconcelos, R.P.; Reis-Santos, P.; Fonseca, V.; Maia, A.; Ruano, M.; França, S.; Vinagre, C.; Costa, M.J.; Cabral, H. Assessing anthropogenic pressures on estuarine fish nurseries along the Portuguese coast: A multi-metric index and conceptual approach. *Sci. Total Environ.* **2007**, *374*, 199–215. [CrossRef]

54. Whitall, D.; Bricker, S.; Ferreira, J.; Nobre, A.M.; Simas, T.; Silva, M. Assessment of eutrophication in estuaries: Pressure-State-Response and nitrogen source apportionment. *Environ. Manag.* **2007**, *40*, 678–690. [CrossRef] [PubMed]

55. Andersen, H.E.; Kronvang, B.; Larsen, S.E.; Hoffman, C.C.; Jensen, T.S.; Rasmussen, E.K. Climate-change impacts on hydrology and nutrients in a Danish lowland river basin. *Sci. Total Environ.* **2006**, *365*, 223–237. [CrossRef] [PubMed]

56. Decree-Law 235/97. *Decreto-Lei n° 235/97 de 3 de Setembro*; Diário da República, I Série-A—N°203—3 de Setembro; Ministério do Ambiente: Lisbon, Portugal, 1997. (In Portuguese)

57. European Commission (EC). Report from the Commission to the European Parliament and the Council on the Implementation of the Water Framework Directive (WFD); (2000/60/EC)—River Basin Management Plans. 2012. Available online: http://eur-lex.europa.eu/LexUriServ/LexUriServ.do?uri=COM:2012:0670:FIN:EN:PDF (accessed on 19 June 2016).
58. Decree-Law 152/97. *Decreto-Lei nº 152/97 de 19 de Junho*; Diário da República, I Série A—Nº139—19 de Junho; Ministério do Ambiente: Lisbon, Portugal, 1997. (In Portuguese)
59. Sistema Nacional de Informação de Ambiente (SNIAmb). Relatório de Estado do Ambiente. 2016. Available online: http://sniamb.apambiente.pt/infos/geoportaldocs/REA/REA2016/REA2016.pdf (accessed on 19 June 2016). (In Portuguese).
60. Glavan, M.; Zeleznikar, S.; Velthof, G.; Boekhold, S.; Langaas, S.; Pintar, M. How to Enhance the Role of Science in European Union Policy Making and Implementation: The Case of Agricultural Impacts on Drinking Water Quality. *Water* **2019**, *11*, 492. [CrossRef]
61. Wuijts, S.; Driessen, P.P.J.; Van Rijswick, H.F.M.W. Towards More Effective Water Quality Governance: A review of Social-Economic, Legal and Ecological Perspectives and Their Interactions. *Sustainability* **2018**, *10*, 914. [CrossRef]

MDPI

St. Alban-Anlage 66

4052 Basel

Switzerland

Tel. +41 61 683 77 34

Fax +41 61 302 89 18

www.mdpi.com

Water Editorial Office

E-mail: water@mdpi.com

www.mdpi.com/journal/water

www.ingramcontent.com/pod-product-compliance
Lightning Source LLC
Chambersburg PA
CBHW051707210326

41597CB00032B/5395